Studies
in the History of Mathematics and Physical Sciences

12

B.A. Rosenfeld

A History of
Non-Euclidean Geometry

Evolution of the Concept of
a Geometric Space

Translated by Abe Shenitzer
With the Editorial Assistance of Hardy Grant

With 114 Illustrations

Springer-Verlag
New York Berlin Heidelberg
London Paris Tokyo

B. A. Rosenfeld
Institute for the History of Science and Technology
103012 Moscow
U.S.S.R.

Abe Shenitzer (*Translator*)
Department of Mathematics
York University
North York, Ontario M3J 1P3
Canada

Mathematics Subject Classifications (1980): 01A20, 01A30, 01A35, 01A40, 01A45, 01A50, 01A55, 01A60, 51-03, 51-MXX

Library of Congress Cataloging-in-Publication Data
Rozenfel'd, [Rosenfeld] B. A. (Boris Abramovich)
 A history of non-euclidean geometry.
 (Studies in the history of mathematics and physical
sciences; 12)
 Translation of: Istoriĩa neevklidovoĭ geometrii.
 Bibliography: p.
 Includes index.
 1. Geometry, Non-Euclidean—History. I. Title.
II. Title: History of non-euclidean geometry.
III. Series.
QA685.R6713   1988   516'.9'09   87-9455

61, 323

Original Russian edition: *Istoriya Neevklidovoĭ Geometrii: Razvitie ponyatiya o geometričeskom prostranstve*. Nauka, Moscow, 1976.

Typeset by Asco Trade Typesetting, Ltd., Hong Kong.
Printed and bound by R. R. Donnelley and Sons, Harrisonburg, Virginia.
Printed in the United States of America.

9 8 7 6 5 4 3 2 1

ISBN 0-387-96458-4 Springer-Verlag New York Berlin Heidelberg
ISBN 3-540-96458-4 Springer-Verlag Berlin Heidelberg New York

To the memory of
my friend

Isaac Yaglom
(1921–1988)

one of the creators of
modern non-Euclidean geometry

Foreword

The Russian edition of this book appeared in 1976 on the hundred-and-fiftieth anniversary of the historic day of February 23, 1826, when Lobačevskiĭ delivered his famous lecture on his discovery of non-Euclidean geometry.

The importance of the discovery of non-Euclidean geometry goes far beyond the limits of geometry itself. It is safe to say that it was a turning point in the history of all mathematics.

The scientific revolution of the seventeenth century marked the transition from "mathematics of constant magnitudes" to "mathematics of variable magnitudes." During the seventies of the last century there occurred another scientific revolution. By that time mathematicians had become familiar with the ideas of non-Euclidean geometry and the algebraic ideas of group and field (all of which appeared at about the same time), and the (later) ideas of set theory. This gave rise to many geometries in addition to the Euclidean geometry previously regarded as the only conceivable possibility, to the arithmetics and algebras of many groups and fields in addition to the arithmetic and algebra of real and complex numbers, and, finally, to new mathematical systems, i.e., sets furnished with various structures having no classical analogues. Thus in the 1870's there began a new mathematical era usually called, until the middle of the twentieth century, the *era of modern mathematics*. Now, however, in the wake of the modern scientific and technological revolution, and the related appearance of computers and the significant development of finite mathematics, it is necessary to rename that earlier age. In view of the tremendous importance of the discovery of non-Euclidean geometry we might call it the *era of non-Euclidean mathematics*.

In the present volume we investigate the mathematical and philosophical factors underlying the discovery of non-Euclidean geometry and the extension of the concept of space, the history of these discoveries, and their subsequent development.

Of the ten chapters in the book the sixth, "Lobačevskian Geometry," plays a central role. It deals with the history of the discovery of hyperbolic geometry, the struggle for its acceptance, and the history of its more important interpretations. Chapters 1–5 are devoted to the prehistory of hyperbolic geometry and of the ideas associated with the evolution of the concept of space. Specifically, in chapter 1 ("Spherical Geometry") we study the first geometry different from Euclidean geometry and the one which later served as the basis for elliptic geometry. In chapter 2 ("The Theory of Parallels") we give a detailed account of the attempts to prove Euclid's fifth postulate, the so-called parallel postulate; these attempts led directly to Lobačevskiĭ's discovery. Chapter 3 ("Geometric Transformations") deals with the history of geometric transformations, an evolution that subsequently led to the theory of groups of transformations—the latter providing the basis for the contemporary non-Euclidean geometries and others related to them. In chapter 4 ("Geometric Algebra and the Prehistory of Multidimensional Geometry") we study the history of geometric calculi, which later served as the basis for multidimensional geometry, and the theory of algebras, also closely related to non-Euclidean geometry. In chapter 5 ("Philosophy of Space") we consider the history of the philosophical views on space that were important to Lobačevskiĭ's discovery as well as to later generalizations of non-Euclidean geometry. Chapters 7 and 8 ("Multidimensional Spaces" and "The Curvature of Space") are devoted to the subsequent development of the concept of space—the appearance of the concepts of multidimensional flat and curved spaces and their later generalizations. In chapter 9 ("Groups of Transformations") we study the history of groups of transformations whose importance for geometry was disclosed by Klein's famous "Erlangen Program," and the history of Lie groups—in particular, the theory of simple Lie groups—that has given rise to the most natural generalizations of the classical non-Euclidean geometries. Chapter 10 ("Application of Algebras") deals with the history of associative and nonassociative algebras and their different geometric applications—in particular, the history of the emergence of spaces over various algebras.

The English edition is a corrected and supplemented edition. In chapter 1, I added a section on spherical trigonometry in medieval India and in the treatises of al-Khwārizmī; this material provides the necessary links between ancient and later Arabic mathematics. In chapter 2, I added information on new publications devoted to the history of the theory of parallel lines. In chapter 3, I added a section on Apollonius' geometric transformations. In chapter 4, I added a section on three-dimensional geometric algebra. In chapter 5 of the Russian edition, I considered only pre-nineteenth century philosophy. In the English version of chapter 5, I brought the discussion up to our own time and added relevant material on medieval India. In chapters 9 and 10, I added sections on parabolic spaces (flag manifolds) and their connections with the theories of groups and algebras, on finite geometries, and on the applications of geometries to physics. The present bibliography includes references to books and papers that have appeared in the last ten years.

The author is very thankful to the translator, Abe Shenitzer, for performing a very difficult task, to J. P. Hogendijk for his transcription of Arabic names, titles, and terms, and to Adolf Pavlovič Yushkevič, Piama Pavlovna Gaidenko, and the late Walter Kaufmann-Bühler for valuable advice.

Contents

Chapter 1
Spherical Geometry

The Rise of Spherical Geometry

The first geometry other than Euclidean geometry was spherical geometry, or, as the ancients called it, *Sphaerica*. This geometry appeared after plane and solid Euclidean geometry. The main stimulus for the rise of plane and solid geometry was the need to measure the areas of fields and other plane figures and the capacities of vessels and storehouses of various shapes, that is, the volumes of different solids. The main stimulus for the rise of spherical geometry was the study of the starry heavens.

Observation of heavenly bodies was carried out already in ancient Egypt and Babylon, largely with a view to making a calendar. The Egyptians divided the day into 24 hours. (The original "temporal" hours were equal to 1/12 of daylight or darkness and were thus of unequal duration. They were later replaced by the generally accepted equal hours.) *[384, pp. 83–86]*. The Babylonians made a more significant contribution to the development of astronomy: in his *Almagest*, Ptolemy cites not the observations of his fellow Egyptians but rather the Babylonian observations of eclipses and stars dating back to the first centuries of the *era of Nabonassar* that began in the eighth century B.C. *[441, p. 166]*. It was the Babylonian astronomers who introduced the division of the ecliptic into the 12 signs of the Zodiac, the division of each sign into 30°, and the sexagesimal division of the degree into minutes and seconds. They described the motion of planets along the ecliptic by means of step and zigzag functions. They were also the founders of astrology, which played an important role in the solution of many problems of astronomy. The ancient Greeks became familiar with Babylonian astronomy not later than the fourth century B.C. At that time, following Babylonian usage, they replaced the earlier names of the planets with names of gods. The modern names of the planets are Latin translations of the latter.[1] The astronomy expounded

---
[1] The new names of the planets appeared in Plato's *Epinomis* [427, vol. 6, p. 400].

by Ptolemy in the *Almagest* was the result of centuries of development which absorbed the traditions of Babylonian astronomers and Greek geometers.

The apparent celestial sphere is a sphere whose center is the center of the earth. The most important circles of this sphere are the *horizon*, the fixed great circle parallel to the tangent plane of the earth at the observer's position; the *celestial equator*, the great circle that goes over into itself during the apparent daily rotation of the celestial sphere, and the *ecliptic*, the great circle of the apparent motion of the Sun whose position varies during the daily apparent rotation of the celestial sphere. The poles of the horizon are called the *zenith* (in the upper hemisphere) and the *nadir* (in the lower hemisphere). The line joining the celestial poles is the *celestial axis*. The Sun crosses the celestial equator at the spring and autumn equinoxes and is farthest from it at the summer and winter solstices; the points of the ecliptic corresponding to these days trace the *tropics of Cancer and Capricorn*[2] during the daily rotation of the celestial sphere. The ecliptic and celestial equator form a fixed angle of approximately 23°. The horizon and celestial equator form an angle that complements the geographic latitude (in particular, they are perpendicular at the earth's equator and coincide at a pole).

The *Sphaerica* of Autolycus

The earliest mathematical work of antiquity to come down to our time is *On the rotating sphere* (Peri kinoumenēs sphairas) by Autolycus *[31]*, who lived at the end of the fourth century B.C. The subject of this work is, essentially, the celestial sphere, considered (in conformity with the traditions of Plato's school) in very abstract form. Autolycus' work consists of definitions and 12 propositions. The definitions refer to uniform motion. Proposition 1 is that when a sphere rotates uniformly on its axis, then all of its points not on the axis describe parallel circles with the same poles as the sphere, and the planes of these circles are perpendicular to the axis of the sphere. Here the term *circle* denotes a circular disk, and the expression "a point describes a circle" means that the point traverses the boundary of a disk. Parallel circles are circles on the sphere that determine parallel planes. Proposition 2 is that the points of a sphere in uniform motion describe in equal times similar arcs. In proposition 4 the "bounding circle" is introduced as that which separates the "visible points" of the sphere from the invisible ones—in other words, it is the horizon (the Greek *horizōn* means "bounding"). The proposition deals with the so-called parallel sphere, which arises when the horizon coincides with the celestial equator (which is what happens at a geographic pole). It is shown that, in that case, no point of the celestial sphere "rises or sets," that is, in-

---

[2] During the spring and autumn equinoxes the Sun appears in the constellations Aries and Libra, and during the summer and winter solstices in Cancer and Capricorn.

tersects the "bounding circle." Proposition 5 deals with the "upright sphere" (*sphaera recta*), which arises when the "bounding circle" passes through the poles. Then the horizon is perpendicular to the celestial equator (which is what happens at the geographic equator). It is shown that, in that case, all the points of the celestial sphere "rise and set." Proposition 6 deals with the "inclined sphere" (*sphaera obliqua*), which arises when the "bounding circle" is inclined to the axis. It is shown that, in that case, some points of the sphere are "always visible," some are "always invisible," and some "rise and set." The proofs of most of the propositions use motions: the proposition is negated, and then the sphere is rotated and it is noted that the negation contradicts the state of affairs resulting from the rotation of the sphere.

Autolycus' second treatise, *On the rising and setting of stars*, deals with the rising and setting of the points of a certain inclined circle that can be easily identified with the ecliptic.

The *Sphaerica* of Theodosius

The first systematic account of spherical geometry to come down to us is the *Sphaerica* (*Sphairika*) of Theodosius *[578]*, a native of Bithynia in Asia Minor, who lived in the second and first centuries B.C. in Tripoli, in Phoenicia. In addition to the *Sphaerica*, Theodosius wrote two astronomical treatises— *On inhabited places* and *On days and nights*—devoted to risings and settings of the Sun and to the duration of days and nights in various inhabited places of the earth.

The *Sphaerica* of Theodosius consists of three books. The first book contains 6 definitions and 23 propositions, the second, 1 definition and 23 propositions, and the third, 14 propositions.

The first definition is

A sphere is a solid figure contained in a single surface such that all lines falling on it from a single point inside the figure are equal *[578, p. 1]*.

The *center* of the sphere is defined as the point mentioned in the first definition. A *diameter* of the sphere is a line through its center ending at its surface. Alternatively, it is a fixed line about which the sphere can rotate. There follow definitions of *poles of the sphere* (ends of its axis), a *pole of a circle on the sphere* (a point on the surface of the sphere such that all the lines from it to the circle are equal), *equal inclination of two planes* to two other planes, and (at the beginning of book II) *tangent circles* on the sphere.

The first three definitions of Theodosius are direct generalizations of definitions 15–17 of book I of Euclid's *Elements* (the definitions of a circle, its center, and a diameter) *[173, vol. 1, p. 153]*. On the other hand, Euclid defined a *sphere* (definition 14, book XI, *[173, vol. 3, p. 261]*) as a solid obtained by revolving a semicircle about its diameter.

Just as in Euclid's *Elements*, the propositions of Theodosius are theorems and construction problems. For example, proposition 1 is the theorem that the curve of intersection of a sphere and a plane is a circle, and proposition 2 deals with the construction of the center of a sphere. To construct it, one finds the intersection of the sphere with a plane, erects a perpendicular to the plane at the center of the circle of intersection, extends the perpendicular in both directions to the surface of the sphere, and bisects the resulting diameter.

Most of the propositions of the *Sphaerica* are stereometric theorems and constructions. When Theodosius speaks of circles on a sphere as meeting at a certain angle or being parallel he thinks of their planes' meeting at that angle or being parallel. When he speaks of the mutual bisection of two circles on a sphere he has in mind the mutual bisection of plane figures. Some of these propositions, however, can be easily reformulated in terms of geometry on the surface of a sphere. The relevant propositions are proposition 6 of book I (that the largest circles on a sphere are those passing through its center (the "great circles")), and that of the remaining ("small") circles (those equidistant from the center are equal and the more distant ones are smaller); propositions 11–12 (that two intersecting great circles bisect each other, and conversely); propositions 13–15 (that if a great circle on a sphere intersects a circle on that sphere at right angles, then it bisects it and passes through its poles, and conversely (two converses)); propositions 1–2 of book II (that "parallel circles" on a sphere have common poles, and conversely); and propositions 3–5 (that if two circles on a sphere meet at a point of a great circle that contains their poles, then they are tangent, and conversely (two converses)).

In addition to stereometric propositions, the *Sphaerica* of Theodosius contains propositions formulated in terms of geometry on the surface of the sphere. Relevant examples are propositions 20–21 of book I (construction of a great circle passing through two points on a sphere, and the construction of a pole of a circle on a sphere; the latter construction enables one to draw great circles perpendicular to a given circle, or, what amounts to the same thing, great circles passing through the poles of a given circle); proposition 10 of book II (that if two great circles pass through the poles of two parallel circles, then the arcs of the small circles cut off by the great circles are similar, and the arcs of the great circles cut off by the small circles are equal); proposition 16 (the converse of proposition 10); propositions 14–15 (construction of a great circle tangent at a given point to a small circle and passing through a given point of its exterior); and proposition 5 of book III (that if a number of parallel small circles cut off equal arcs on an inclined great circle, then they cut off unequal arcs on a great circle perpendicular to them).

The *Sphaerica* of Menelaus

Menelaus lived in Alexandria at the end of first century A.D. In the year 98 he carried out astronomical measurements in Rome. The spherical geometry contained in his treatise *On the sphere* (*Peri sphairas*) is more advanced than its predecessors. There exist only edited Arabic versions of this work. Of these, the best are those of Abū Naṣr ibn ʿIrāq *[361]* and Naṣīr al-Dīn al-Ṭūsī *[595, vol. 2, part 10]*. Menelaus' *Sphaerica* consists of three books containing (in the Ibn ʿIrāq version) 39, 21, and 25 propositions, respectively. In the introduction to book I Menelaus defines a *spherical triangle* ("a three-sided figure")—that is, part of a spherical surface bounded by three arcs of great circles each less than a semicircle—and the angles of such a triangle. Whereas most of the propositions in Theodosius' *Sphaerica* are stereometric, Menelaus' book expounds geometry on the surface of a sphere in a way analogous to Euclid's exposition of plane geometry. Proposition 1 of book I is the construction of an arc of a great circle at a given angle to a given arc of a great circle. Propositions 2 and 3 are the theorem stating the equality of the base angles of an isosceles spherical triangle and its converse (they are the analogues of propositions 5 and 6 of book I of Euclid's *Elements*). In proposition 4 of book I it is proved that if the sides of two spherical triangles are pairwise equal, then so are their angles (analogue of proposition 8 of book I of the *Elements*). In proposition 5 of book I it is proved that the sum of two sides of a spherical triangle is always greater than the third (analogue of proposition 20 of book I of the *Elements*). In propositions 7 and 9 it is proved that opposite a greater angle of a spherical triangle lies a greater side, and conversely (analogues of propositions 18 and 19 of book I of the *Elements*). Of the propositions that are not analogues of propositions of plane geometry we mention propositions 10 and 11, which imply that the sum of angles of a spherical triangle exceeds two right angles.

The tenth proposition. *If two sides of a three-sided figure are together less than a semicircle, then the exterior angle adjacent to one of these sides is greater than the opposite interior angle adjacent to the remaining side; if two sides are together greater than a semicircle, then the exterior angle is smaller than the opposite interior angle; and if the two sides are together equal to a semicircle, then the exterior angle is equal to the opposite interior angle.*

Let the sides *AB* and *BC* of a three-sided figure *ABC* be together less than a semicircle (Figure 1). Then I claim that angle *BCD* is greater than angle *BAC*.

We extend the arcs *AB* and *AC* to *D*; each of them is a semicircle. Since the sides *AB* and *BC* are together less than a semicircle, they are less than the arc *ABD*. But then the remaining arc *BD* is greater than the arc *BC*. But then, as shown in the ninth proposition, the angle *BDC*, which is equal to the angle *BAC*, is smaller than the angle *BCD*. Similarly, if the arcs *AB* and *AC* are together greater than a semicircle, then the arc *BC* will be greater than the remaining arc *BD*. But then the angle *BDC*, which is equal to the angle *BAC*, is greater than the mentioned angle *BCD*. Similarly, it is clear that if the arcs *AB* and *BC* are together

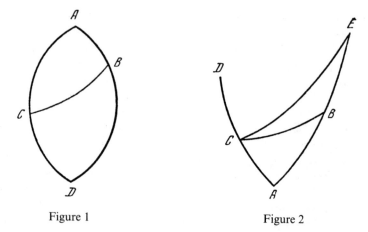

Figure 1   Figure 2

equal to a semicircle, then the arc *BC* will be equal to the arc *BD*, and, as was shown in the second proposition, the angle *BAC* will be equal to the angle *BCD*.

In just the same way one proves the converse of what we stated, namely: if the angle *BCD* is equal to the angle *BAC*, then what was set forth proves that *BA* and *BC* are together equal to a semicircle; if the exterior angle *BCD* is smaller than the opposite interior angle, then what was set forth proves that *AB* and *BC* are together greater than a semicircle, for the angle *BDC* is greater than the angle *BCD*, and hence the arc *BD* is smaller than the arc *BC*. Similarly, if the angle *BCD* is greater than the angle *BAC*, then it is greater than *BDC*, so that *BD* is greater than *BC*, and *BA* and *BC* are together less than a semicircle. And this is what we wished to prove.

The eleventh proposition. *An exterior angle of any three-sided figure is smaller than both interior angles opposite to it.*

Let the figure *ABC* be a triangle. We extend *AC* to *D* (Figure 2). Then I claim that the exterior angle *BCD* is smaller than both angles at the points *A* and *B* opposite to it.

At the point *C* of the arc *CD* we construct an angle equal to the angle *BAC*—that is, the angle *DCE*. We extend *AB* from the point *B* to the point *E* at which it meets *CE*. Since the angle *DCE* (an exterior angle of the triangle *ACE*) is equal to the angle *BAC*, the sum of *AE* and *EC* is a semicircle—as shown in the tenth proposition. But then *BAC* and *ABC* together are greater than the exterior angle *BCD* of the triangle.

From this [it is clear that] the angles *A*, *B*, *C* [exceed two right angles], for the two angles on both sides of *BC*, that is, the angles *BCA* and *BCD*, are equal to two right angles, and the angles *A*, *B*, *C* are greater than these two angles. That is what we wished to prove *[361, pp. 8–9]*.

The difference between the sum of the angles of a spherical triangle and two right angles is called the *angular excess* of the spherical triangle. Thus if the radian measures of the angles are α, β, γ, then the angular excess of the spherical triangle is $\alpha + \beta + \gamma - \pi$. It is easy to check that if a spherical triangle is divided into two triangles, then the angular excess of the large triangle is equal to the sum of the angular excesses of the two small triangles. Since, as just indicated, the angular excess is additive, and also has the obvious properties of invariance under rotation of the sphere and of being positive for every spherical triangle, it follows that *the angular excess of a spherical triangle is proportional to its area.*

Menelaus' Theorems

Proposition 1 in book III of Menelaus' work has played a special role in the history of spherical geometry and trigonometry. It proves the plane and spherical cases of a theorem now called *Menelaus' theorem* or *the theorem of the complete quadrilateral*. A *complete quadrilateral* is a quadrilateral whose sides have been extended to intersection. Since in the Arabic versions of Menelaus' work this theorem was subjected to extensive modernization, we give here the version from chapter 12, book I, of Ptolemy's *Almagest*. First we set down the proofs of two variants of the plane theorem of Menelaus.

To two straight lines AB and AC are drawn two straight lines BE and CD that intersect at a point G (Figure 3). I claim that *the ratio of CA to AE is composed of the ratios of CD to DG and GB to BE.* Through E is drawn EH parallel to CD (see Figure 3a). Since CD and EH are parallel, the ratio of CA to EA is the same as the ratio of DC to EH. Let us bring in GD. Then the ratio of CD to HE will be composed of the ratios of CD to DG and DG to HE. Thus the ratio of CA to AE is composed of the ratios of CD to DG and DG to HE. But the ratio of

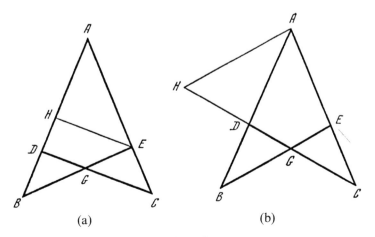

(a)   (b)

Figure 3

DG to *HE* is the same as the ratio of *GB* to *BE* owing to the parallelism of *EH* and *GD*. Therefore the ratio of *CA* to *AE* is composed of the ratios of *CD* to *DG* and *GB* to *BE*, which was to be proved.

In much the same way we shall prove that the ratio of *CE* to *EA* is composed of the ratios of *CG* to *DG* and *DB* to *BA*. We draw through *A* a parallel to *EB* and extend to that parallel the straight line *CDH* (see Figure 3b). Since *AH* is again parallel to *EG*, *CE* is to *EA* as *CG* is to *GH*. If we bring in *GD*, then the ratio of *CG* to *GH* will be composed of the ratios of *CG* to *GD* and *DG* to *GH*. But the ratio of *DG* to *GH* is the same as the ratio of *DB* to *BA*, for *BA* and *GH* are drawn between the parallels *AH* and *GB*. Therefore the ratio of *CG* to *GH* is composed of the ratios of *CG* to *DG* and *DB* to *BA*. But the ratio of *CG* to *GH* is the same as the ratio of *CE* to *EA*, so that the ratio of *CE* to *EA* is composed of the ratios of *CG* to *DG* and *DB* to *BA*, which was to be proved *[442, vol. 1, pp. 50–51]*.

The ancient mathematicians used the term *composite ratio* for our product of ratios; and to prove that the ratio *A* : *B* is composed of the ratios *C* : *D* and *E* : *F* they would find magnitudes *L*, *M*, *N* such that $A/B = L/N$, $C/D = L/M$, $E/F = M/N$ (here, the role of the magnitudes *L*, *M*, *N* is played in the first case by *CD*, *GD*, and *EH*, and in the second case by *CG*, *CD*, and *GH*). Hence the theorems established previously can be written as

$$\frac{CA}{EA} = \frac{CD}{DG}\frac{GB}{BE},\tag{1.1}$$

and

$$\frac{CE}{EA} = \frac{CG}{DG}\frac{DB}{BA}.\tag{1.2}$$

Then Menelaus proves the lemma that if *A*, *B*, *C* are three points of a circle with center *D* and each of the arcs *AB*, *BC* is less than a semicircle, and if *E* is the point of intersection of the chord *AC* with the diameter *DB*, then

$$\frac{\text{chord } 2AB}{\text{chord } 2BC} = \frac{AE}{EC}, \qquad \frac{\text{chord } 2CA}{\text{chord } 2AB} = \frac{CE}{BE}$$

(Menelaus and Ptolemy refer to the chord of the arc *AC* as "the straight line under the arc *AC*.")

What follows is Ptolemy's exposition of the spherical variant of Menelaus' theorem.

We describe on the surface of the sphere arcs of great circles such that two arcs *BE* and *CD*, drawn to two arcs *AB* and *AC*, intersect at a point *G*. Let each of these arcs be smaller than a semicircle; this we shall assume for all such constructions. I claim that *the ratio of the straight line under twice the arc* CE *to the straight line under twice* EA *is composed of the ratio of the straight line under twice* CG *to the straight line under twice* GD *and the ratio of the straight line under twice* DB *to the straight line under twice* BA.

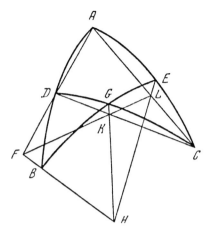

Figure 4

In fact, we take the center of the sphere, and let it be H (Figure 4). We draw straight lines HB, HG, and HE from H to the points B, G, E of intersection of the circles. Then we extend the connecting straight line AD and suppose that it intersects the extension of HB at the point F. Similarly, let the connecting straight lines DC and AC intersect HG and HE at the points K and L. The points F, L, and K lie on the same straight line, for they are simultaneously in two planes—that of the triangle ACD and the circle BGE. The straight line joining them, together with the straight lines FA and CA, gives the two straight lines FL and CD, drawn to them, that intersect at the point K. Therefore the ratio CL to LA is composed of the ratios of CK to KD and DF to FA. But the ratio of CL to LA is the same as that of the straight line under twice CE to the straight line under twice EA, the ratio of CK to KD is the same as that of the straight line under twice CG to the straight line under twice GD, and the ratio of FD to FA is the same as that of the straight line under twice DB to the straight line under twice BA. It follows that the ratio of the straight line under twice CE to the straight line under twice EA is composed of the ratio of the straight line under twice CG to the straight line under twice GD and the ratio of the straight line under twice DB to the straight line under twice BA [442, vol. 1, pp. 54–55].

This theorem can be written as

$$\frac{\text{chord } 2CE}{\text{chord } 2EA} = \frac{\text{chord } 2CG}{\text{chord } 2DG} \frac{\text{chord } 2DB}{\text{chord } 2BA}$$

Starting with theorem (1.1), one can prove in the same way that

$$\frac{\text{chord } 2CA}{\text{chord } 2AE} = \frac{\text{chord } 2CD}{\text{chord } 2DG} \frac{\text{chord } 2GB}{\text{chord } 2BE}.$$

Since the chord of an arc 2α is twice the line of the sine of the arc α, these theorems can be written as

$$\frac{\sin CE}{\sin AE} = \frac{\sin CG}{\sin DG}\frac{\sin DB}{\sin BA}, \tag{1.3}$$

and

$$\frac{\sin CA}{\sin AE} = \frac{\sin CD}{\sin DG}\frac{\sin GB}{\sin BE}. \tag{1.4}$$

This shows that, essentially, the spherical theorem of Menelaus is a theorem of spherical trigonometry. We note that the equations (1.3) and (1.4) are obtained from the equations (1.2) and (1.1) by replacing segments by the sine lines of the corresponding arcs.

Ptolemy's Spherical Trigonometry

On a number of occasions we have mentioned the name of Ptolemy, the famous astronomer of antiquity, and his *Almagest*. Ptolemy worked in Alexandria in the second century A.D. At the present time, *Almagest* is the generally accepted name of his fundamental astronomical work *The Mathematical Composition* (*Mathēmatikē Syntaxis*). *Almagest* is the Latin distortion of the book's Arabic name *Al-Majisṭī*, derived from one of its Greek names *Megistē Syntaxis—The Greatest Composition*. The *Almagest* is a summary of all ancient astronomy which, as mentioned earlier, grew out of the union of the Babylonian and Greek traditions.

The *Almagest* is important because it is the first preserved work containing an exposition of spherical trigonometry. Chapters 10, 11, and 13 of book I of the *Almagest* are devoted to mathematics. Chapter 10 contains the geometric theorems needed for the computation of tables of chords, including the well-known "theorem of Ptolemy" for a quadrilateral inscribed in a circle. Chapter 11 contains a table of chords, that is, a table of values of the function $s = r\sin\frac{\alpha}{2}$, where the arc α varies by $1/2°$ and the chords s are computed to three sexagesimals. Chapter 13 contains the proofs of the plane and spherical versions of Menelaus' theorem quoted previously ((1.1)–(1.4)). Ptolemy applies the last two theorems to solve concrete problems of spherical astronomy. In chapter 14 of book I Ptolemy computes the length of the spherical perpendicular from a point G of the ecliptic BGE to the celestial equator BDA (see Figure 4). To do this, he completes the spherical triangle BDG, formed by the arc BG of the ecliptic, the arc BD of the celestial equator, and the arc of the spherical perpendicular GD, to the full quadrilateral $AECGBD$ and considers the special case of theorem (1.4) when $CA = EB = CD = 90°$. Then theorem (1.4) takes the form

$$\frac{\sin 90°}{\sin B} = \frac{\sin GB}{\sin DG}.$$

This is the special case of the *spherical sine theorem*, usually written in the form

$$\frac{\sin a}{\sin A} = \frac{\sin b}{\sin B} = \frac{\sin c}{\sin C}, \tag{1.5}$$

for the right spherical triangle BDG with the right angle D (and known angle B between the ecliptic and the celestial equator). In chapter 16 of the same book Ptolemy computes the right ascension of the point G of the ecliptic, that is, the side BD of the same triangle BDG. He again completes this triangle to the quadrilateral $AECGBD$ and considers the special case of theorem (1.3) when $CE + EA = CD + DG = AB = 90°$. In that case theorem (1.3) takes the form

$$\frac{\cos EA}{\sin EA} = \frac{\cos GD}{\sin GD} \frac{\sin DB}{\sin 90°},$$

that is, becomes the *spherical tangent theorem*, usually written as

$$\tan a = \tan A \cdot \sin c, \tag{1.6}$$

for the right spherical triangle with right angle B.

Spherical Trigonometry in India

It was in the medieval Near and Middle East that spherical trigonometry came close to its modern form. The trigonometry of Ptolemy and of other Alexandrian astronomers reached the Muslim East through India. In the fourth century A.D. there turned up in India the astronomer Paulos, a refugee from Alexandria, who became known in India under the name of Pauliśa. We can trace back to him the Indian astronomical work known as the *Pauliśa siddhānta*. Many other Indian astronomical *siddhāntas* are also of Hellenistic origin. Whereas the Greeks called chords simply "straight lines under arcs," the Indians used more colorful terminology. They called an arc "bow," a chord "bowstring" (*jīva*), and the perpendicular from the center of the arc to its chord "arrow." In the eighth century there appeared the Arabic work *Sindhind*, an adaptation of the Indian *siddhāntas*. The Arabic terms *qaws*, *watar*, and *sahm* were the Arabic words for "bow," "bowstring," and "arrow," respectively. In turn, these words were rendered by 12th-century Latin translators as *arcus*, *chorda*, and *sagitta*. Hence our "chord." The Indians introduced in their *siddhāntas* the half-chord, which they called at first *ardha jīva* and then simply *jīva* or *jyā*. When *jīva* referred to a half-chord, the Arabs transliterated it as *jīb* or *jaib* (instead of translating it as *watar* (chord)). Since *jaib* means "cavity, pocket," the Latin translators rendered it as *sinus*.

In the Indian *siddhāntas* there are many astronomical rules equivalent to formulas of spherical trigonomentry such as the spherical sine theorem (1.5) and the *spherical cosine theorem*

$$\cos a = \cos b \cos c + \sin b \sin c \cos A. \tag{1.7}$$

For example, verse 13 of chapter 3 of the astronomical treatise *Khaṇḍakhādyaka* (The sweetest teaching) of Brahmagupta (sixth century A.D.) reads:

> Multiply the *Antyā* by the hypotenuse at noon and divide by the hypotenuse at the given shadow. Substract the result from the *Antyā*. Consider the remainder as the *Utkramajyā* and find the corresponding arc. The result is the number of *Prāṇas* in the *Natakala* measured from midday *[77a, p. 61]*.

Here *Antyā* is the versed sine (sinus versus) of the arc of half the day and "shadow" is the shadow of the vertical gnomon in the horizontal plane. If the length of the gnomon is l, then the length of the shadow is $l \cot h$, where h is the altitude of the Sun (the Sun's spherical distance from the horizon). "Hypotenuse" is the hypotenuse of the right triangle in the plane whose other sides are the gnomon and the shadow, hence its length is $l \csc h$. *Natakala* is the hour angle t and *prāṇa* (literally: sigh) is a measure of time. If we denote the arc of half the day by t_0, then we can express Brahmagupta's rule by the formula

$$r \sin \text{vers} \, t_0 - \frac{r \sin \text{vers} \, t_0 \cdot l \csc h_{\max}}{l \csc h} = r \sin \text{vers} \, t,$$

which is equivalent to

$$\sin \text{vers} \, t_0 \left(1 - \frac{\sin h}{\sin h_{\max}} \right) = \sin \text{vers} \, t,$$

or

$$\frac{(\sin \text{vers} \, t_0 - \sin \text{vers} \, t) \sin h_{\max}}{\sin \text{vers} \, t_0} = \sin h. \tag{1.8}$$

This formula connects the Sun's altitude h, its maximal altitude $h_{\max}$, the arc of half the day t_0, and the hour angle t, that is, the Sun's distance from the meridian measured along its diurnal parallel circle. Since the hour angle is proportional to time, it can be measured in hours and fractions of hours. Figure 5 shows the celestial equator EMW and the celestial pole P, the horizon $ESWN$ and the zenith Z, and the celestial meridian ZLS. K denotes the Sun, and KL its diurnal parallel circle. The altitude h of the Sun is given by the arc KH, and the side ZK of the spherical triangle PZK is $90° - h$. The declination δ is given by the arc KM, and the side PK of PZK is $90° - \delta$. The arc ZP is the complement (to $90°$) of the geographic latitude φ. The hour angle t equals the angle ZPK. The Sun's noon altitude, that is, its meridian altitude, equals

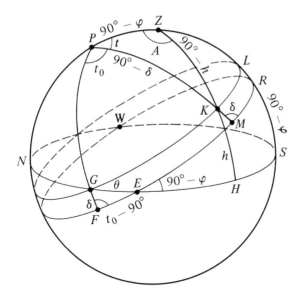

Figure 5

the arc SL. Since $PL = PK$ and $SZ = 90°$, it follows that $SL = SZ + ZP - PL = 90° - \varphi + \delta$. The *versed sine* (*sinus versus*) is defined by the relation $\sin \text{vers } t = 1 - \cos t$ if $t < 90°$ and $= 1 + \cos t$ if $t > 90°$.

The arc of half the day is the arc FR and its surplus over $90°$ is the arc FE, a side of the right spherical triangle EFG whose side FG is the declination δ of the Sun and whose hypotenuse GE is the ortive amplitude θ (the spherical distance between the point G of the sunrise and the point E of the east). Since the angle E of this triangle is equal to the difference $90° - \varphi$ between $90°$ and the latitude φ, the spherical sine theorem (1.5) for this triangle gives

$$\sin \theta = \frac{\sin \delta}{\cos \varphi},\tag{1.9}$$

and the spherical cosine theorem (1.7) for the same triangle ("spherical Pythagoras theorem" $\cos c = \cos a \cos b$) gives

$$\cos \theta = \cos \delta \sin t_0.\tag{1.10}$$

Eliminating θ from (1.9) and (1.10) we obtain

$$\cos t_0 = \tan \delta \tan \varphi.$$

Therefore Brahmagupta's rule is equivalent to the formula

$$\sin h = \frac{\tan \delta \tan \varphi + \cos t}{1 + \tan \delta \tan \varphi}(\cos \delta \cos \varphi + \sin \delta \sin \varphi)$$

or

$$\sin h = \sin \delta \sin \varphi + \cos t \cos \delta \cos \varphi,$$

which is the spherical cosine theorem (1.7), with $a = 90° - h$, $b = 90° - \delta$, $c = 90° - \varphi$, $A = t$, for the triangle PZK.

In *Pañcasiddhāntika* ("Five *siddhāntas*") of Varāhamihira (fifth century A.D.), we find the rule (1.10) in verse 34 of chapter 4 *[599a, vol. 1, p. 63]*, the rule (1.9) in verse 39 of the same chapter *[599a, vol. 1, p. 65]*, the rule equivalent to the rule (1.8) of Brahmagupta in verses 41–43 of the same chapter *[599a, vol. 1, p. 65]* (in this edition we find the term "difference of sines" instead of "difference of versed sines." The same error occurs in the commentaries to these verses; see *[599a, vol. 2, p. 43]*). The correct interpretation of this rule is found in the book of Braunmühl *[77b, vol. 1, p. 41]*. In verses 53–54 of the same chapter Varāhamihira gives the rule for the determination of the azimuth of the Sun, that is, the angle between the two great circles of the celestial sphere joining the Sun and the zenith and the zenith and the celestial pole; this angle is equal to the angle PZK in Figure 5. This rule can be written as

$$\cos A = \frac{\sin \theta - \dfrac{\sin \varphi}{\cos \varphi} \sin h}{\cos h}, \tag{1.11}$$

or, by expressing θ in terms of δ and φ, by (1.8):

$$\cos A = \frac{\sin \delta - \sin \varphi \sin h}{\cos \varphi \cos h}$$

This coincides with the spherical cosine theorem (1.7), with $a = 90° - \delta$, $b = 90° - \varphi$, $c = 90° - h$, for the triangle ZPK and its angle Z.

The Spherical Trigonometry of al-Khwārizmī and Thābit ibn Qurra

In the ninth century we encounter the Arabic rules for the solution of astronomical problems equivalent to the spherical sine theorem (1.5) and the spherical cosine theorem (1.7). We find them in the astronomical treatises of the great Baghdad mathematician and astronomer Muḥammad ibn Mūsā al-Khwārizmī (ab.780–ab.850), known in medieval Europe as *Algorithmus* or *Algorismus* (the Latin form of his name indicating that he originated from Khwārizm), who introduced in his arithmetical treatise "Indian numerals," the ancestors of our *Arabic numerals*, and founded *algebra* as a branch of mathematics in his book *al-Jabr wa'l-muqābala*. He was also the author of one of the first Arabic *zījes* (astronomical tables). These treatises are in an Istanbul manuscript (library Aya Sofya, no. 4830, folios 183^r–199^v, 288^v–235^r). They have recently been published (see *[10a]*). In the treatise *Determination of the Ortive Amplitude in Each Town* (Maʿrifa siʿa al-mashriq fī kull balad) al-Khwārizmī formulates the rule (1.9) and in the *Geometric Construction of the Ortive Amplitude for Every Sign in Every Latitude* ('Amal si'a ayy mashriq shi'ta min al-burūj fī ayy ard shi'ta bi'l-handasa) he gives a graphical method, based on rule (1.9), for constructing an arc equal to the ortive amplitude. In the treatise *Determination of the Azimuth According to Altitude* (Ma'rifa samt min qabl al-irtifā') he formulates the rule (1.11) equivalent to the spherical cosine theorem. In the treatise *Determination of the Action with Azimuth, Shadow, and Altitude* (Ma'rifa 'amal bi'l-samt bi'l-zill wa bi'l-irtifā') he formulates Brahmagupta's rule (1.8), another equivalent of the spherical cosine theorem, for finding the altitude of the Sun given t, t_0, and h_{max}.

In the same century, Muhammad al-Māhānī (c. 825–888), in his *Book on the Determination of the Azimuth at Any Time and in Any Place* (Maqāla fī ma'rifat al-samt li-ayya sā'atin aradta wa-fī ayyi mawdi'in aradta) *[341]*, gave a graphical solution of the problem of determination of the azimuth of the Sun according to al-Khwārizmī's rule that was analogous to his graphical method.

We also find the same rule for the determination of the Sun's altitude, and a simpler rule for the Sun's azimuth equivalent to the spherical sine theorem (1.5), in the *Book on time instruments called sundials* (Kitāb fī ālāt al-sā'āt allatī tusammā rukhāmāt) *[241; 492, pp. 252–266]* of the famous Baghdad mathematician and astronomer Thābit ibn Qurra (836–901). Ibn Qurra, a native of Harrān in northern Syria (now in Turkey), belonged to the heathen sect of star worshippers, Sābians, descendants of the ancient Babylonians. He translated many Greek and Syrian works into Arabic, including a number of treatises of Archimedes and books V–VII of Apollonius' *Conics*. The latter are available only in Ibn Qurra's translation. Ibn Qurra is the author of specialized works on Menelaus' theorem, the theory of composite ratios, the theory of parallel lines, and methods of integration.

In the treatise on sundials Ibn Qurra solves the problem of finding the altitude of the Sun and its azimuth. Ibn Qurra's solutions follow.

> Let us take the distance from the Sun to the middle of heaven [that is, the celestial meridian] along a small circle, at a convenient time, in hours or their fractions. Take its versed sine, multiply it by the sine of the complement of the declination of the Sun in degrees and divide the product by the largest sine, multiply the quotient by the sine of the complement of the latitude of the locality, divide the product by the largest sine and keep the quotient in mind. Then subtract that from the sine of the altitude of the Sun and take the arc of the remainder, and that is the altitude. . . .
>
> If you want to know the azimuth, take the sine of the distance of the Sun from the middle of the heaven along a small circle, multiply it by the sine of the complement of the declination of the Sun in degrees, and divide the product by the sine of the complement of the altitude. Take the arc of the quotient, and that is the azimuth in the southern or northern direction *[241, pp. 16–17]*.

Ibn Qurra's second rule can be written as

$$\frac{\sin A}{\cos \delta} = \frac{\sin t}{\cos h},$$

or, putting $90° - \delta = a$, $90° - h = b$, $t = B$, as

$$\frac{\sin A}{\sin a} = \frac{\sin B}{\sin b},$$

which is one of the three equalities comprising the spherical sine theorem (1.5).

Neither the Indian astronomers nor al-Khwārizmī, al-Māhānī and Ibn Qurra give proofs of their rules. We also find rules (1.8) and (1.11) in the *Ṣābian Astronomical Tables* (al-Zīj al-Ṣābī) of al-Battānī *[381, vol. 1, pp. 23, 31; vol. 3, pp. 33–34, 46]* and in later Arabic zījes.

The Spherical Trigonometry of Ibn ʿIrāq and al-Bīrūnī

For Ptolemy and the astronomers of the early Middle Ages the spherical sine theorem (1.5) and the spherical tangent theorem (1.6) were rules for the solution of concrete problems of spherical astronomy. In the course of the 10th century, these theorems (in the case of the sine theorem, first the sine theorem for a right spherical triangle and then for a general spherical triangle) became independent theorems of spherical trigonometry. At the end of the 10th century, two scholars of Khwārizm (the already mentioned) Abū Naṣr Manṣūr ibn ʿIrāq (d. 1036) and his pupil, the great scholar-encyclopedist Abū'l-Rayḥān Aḥmad al-Bīrūnī (973–1048), played a major role in this development. Ibn ʿIrāq belonged to the dynasty of the Afrigids, who ruled in Khwārizm from the 5th to the 10th century and in the 10th century referred to themselves as Banī ʿIrāq ("the sons of ʿIrāq"). The capital of the Afrigids was Kāth, now the city of Beruni (named after al-Bīrūnī) in the Qaraqalpaq Republic (in Soviet Uzbekistan). Ibn ʿIrāq took part in the education of al-Bīrūnī, a native of Kāth. Himself eminent in mathematics, astronomy, and the construction of astronomical instruments, Ibn ʿIrāq guided al-Bīrūnī in these areas. After the conquest of Khwārizm by Maʾmūn, ruler of Gurganj (the present Kunya-Urgenč in Soviet Turkmenia), which became the capital of Khwārizm, al-Bīrūnī emigrated to Gurgān in northern Iran, but returned to Gurganj after Maʾmūn's death. He was employed at the court of Maʾmūn's son, Maʾmūn ibn Maʾmūn, where he collaborated with Ibn ʿIrāq and the great Ibn Sīnā (Avicenna) from Bukhārā. After the conquest of Khwārizm by the sultan Maḥmūd Ghaznawī, al-Bīrūnī and Ibn ʿIrāq were summoned by Maḥmūd to his capital Ghazna (now in Afghanistan), where they stayed for the rest of their lives. The most important works of Ibn ʿIrāq are *The book of azimuths* (Kitāb al-sumūt), written in Kāth and the (previously mentioned) edition of Menelaus' *Sphaerica* completed in Gurganj. While in Kāth, al-Bīrūnī wrote *An exhaustive account of all possible ways of constructing astrolabes*. In Gurganj he wrote *A chronology of ancient nations [58; 60, vol. 1]*, devoted to chronology and astronomy, and a treatise on spherical geometry (discussed later). In Ghazna, during Maḥmūd's lifetime, he wrote *The book of instruction in the elements of the art of astrology [59; 53, vol. 6]*; a large treatise on India and Indian science, treatises on geometry, plane trigonometry and geodesy; and, under Maḥmūd's successor Masʿūd, he completed his main astronomical work *The Canon of Masʿūd on astronomy and the stars [57; 53, vol. 5]*. In the last years of his life he wrote works on mineralogy and pharmacognosy.

The trigonometric work of al-Bīrūnī is called *The book of the keys to astronomy—what occurs upon the surface of a sphere* (Kitāb maqālīd ʿilm al-hayʾa mā yahduthu fī basīt al-kura) *[138]*. The introduction to this treatise contains a dedication to Marzubān ibn Rustam, prince of Gīlān and Ṭabaristān, who at that time lived in Gurgān. Then there is a reference to the astronomical treatise of al-Bīrūnī's older contemporary Abū Saʿid al-Sijzī, which contained unproved rules, including the spherical sine and tangent theorems, used to solve problems of spherical astronomy. Following this reference, al-Bīrūnī writes:

> My lord Abū Naṣr Manṣūr ibn ʿAlī ibn ʿIrāq exerted his efforts to finding proofs of such calculations.... He asked me to check their validity and to discover the reasons for obtaining proofs and the reasons for choosing among them. I did that. Abū Naṣr wrote a book about this matter which he called *[The book] of azimuths [138, p. 104]*.

Al-Bīrūnī goes on to say that he sent a copy of *The book of azimuths* to Abū-l-Wafāʾ al-Būzjānī in Baghdad, who, in a special treatise and in a subsequently written adaptation of the *Almagest*[3] gave a simpler proof of this theorem and communicated the theorem to Abū Maḥmūd al-Khujandī and Kūshyār ibn Labbān al-Jīlī, then working in Iran, who also presented it in their astronomical books. Al-Jīlī called the theorem "a proposition that frees one from the figure of secants" (that is, from Menelaus' theorem), and al-Khujandī called it "the rule of astronomy." Al-Bīrūnī formulates ibn ʿIrāq's theorem as follows:

> Abū Naṣr said: Let there be given arcs *AB* and *AD*, each a quarter of a circle, and arcs *CB* and *CHG* each less than, greater than, or equal to a quarter [of a circle]. Then I say that the sine of *DH* is in the same ratio to the sine of *GB* as the sine of *CH* is to the sine of *CG [138, p. 140]*.

Al-Bīrūnī reproduces Ibn ʿIrāq's drawing (Figure 6), which shows the familiar complete spherical quadrilateral. But ibn ʿIrāq proves his theorem without relying on Menelaus' theorem. Since $AB = AD = 90°$ and B and D are right angles, Ibn ʿIrāq's theorem is "the rule of four magnitudes," which is equiva-

---

[3] Partly translated into French by Carra de Vaux *[93]*.

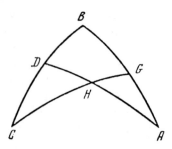

Figure 6

lent to the sine theorem for a right triangle. This rule is usually formulated as follows: If ABC and $AB'C'$ are two right spherical triangles with common angle A and right angles B and B' at the arc ABB', then

$$\frac{\sin a}{\sin b} = \frac{\sin a'}{\sin b'} \qquad (1.12)$$

(the sine theorem implies that the common value of these two ratios is $\dfrac{\sin A}{\sin 90°}$. If $b' = c'$, then the arc $B'C'$ equals A, and our theorem yields the sine theorem).

Shortly thereafter, Ibn ʿIrāq *[239, part 8; 341a]* gave a proof of the spherical sine theorem (1.5) for an arbitrary triangle. This appeared in a small treatise entitled *Treatise on the determination of arcs of the celestial spheres, one in terms of the other, by a method different from the method of composite ratios* (Risāla fī maʿrifat al-qisī al-falakiyya baʿdihā min baʿdin bi-tarīq ghayr tarīq al-nisbat al-muʾallafa). Like *The book of azimuths*, this treatise, written under the name of al-Bīrūnī, actually resulted from the cooperation of the two scholars. In the treatise of al-Bīrūnī cited previously the spherical sine theorem is formulated as follows:

> I shall begin by recalling the method found by Abū Naṣr in the treatise he sent me; I have in mind the freeing proposition. ... It consists in the following: ... the sines of the sides in a triangle made up of arcs of great circles on the surface of a sphere are in the same ratio, one to another, as are, one to another, the respective sines of the opposite angles *[138, p. 118]*.

We see that al-Bīrūnī extends the label "proposition that frees one from the figure of secants" to the general law of sines. We present a short proof of this theorem, found in chapter 9, book III, of al-Bīrūnī's *The Canon of Masʿūd on astronomy and the stars* (Al-Qānūn al-Masʿūdī fī ʾl-hayʾa waʾ-l-nujūm) *[57; 53, vol. 5]* and similar to the proof of the plane sine theorem in chapter 8 of the same book:

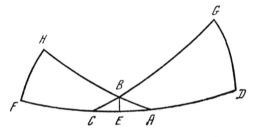

Figure 7

We say that what we put forward as a premise for right triangles holds precisely for triangles made up of arcs of great circles. Namely: the sines of the arc sides are proportional to the sines of the angles that are opposite those sides [if] we take each of them [in the ratio] to its corresponding side, for example, in the triangle ABC, whose sides are [arcs of] great circles, the sine of AB is in the same ratio to the sine of BC as the sine of the angle C is to the sine of the angle A.

The proof. We extend each of the [arcs] AH, AF, CD and CG to a quadrant (Figure 7). From the poles A and C we draw arcs HF and GD at a distance equal to the side of [the inscribed] square [that is, arcs of great circles]. They measure the two mentioned angles. We draw the arc BE of a great circle perpendicular to AC. In view of what was said above [that is, in view of Ibn 'Irāq's theorem proved previously], the ratio of the sine of AB to the sine of BE is the same as the ratio of the sine of the quadrant AH to the sine of HF. Similarly, the ratio of the sine of BE to the sine of BC is the same as the ratio of the sine of DG to the sine of the quadrant GC. By the rule of mixed proportion [that is, by the rule of transition from the proportions $A:B = L:M$ and $B:C = K:L$ to the proportion $A:C = K:M$], the ratio of the sine of AB to the sine of BC is the same as the ratio of the sine of [the arc] DG, the magnitude of the angle C, to the sine of [the arc] HF, the magnitude of the angle A [57, pp. 355–356; 53, vol. 5, part 1, p. 306].

In *The book of keys* al-Bīrūnī describes a priority argument among Ibn 'Irāq, al-Būzjāni, and al-Khujandī concerning the discovery of the spherical sine theorem and says that the priority belongs to Ibn 'Irāq. Actually, as we saw, the theorem in question was simultaneously discovered by Ibn 'Irāq and al-Bīrūnī. In the same work al-Bīrūnī proves the spherical tangent theorem (1.6) and points out that this theorem was formulated as a separate theorem of spherical trigonometry by al-Būzjāni. This theorem was also frequently referred to as "the proposition that frees one from the figure of secants."

In *The book of keys* al-Bīrūnī does not give a general proof of the spherical cosine theorem (1.7). In chapters 20 and 22 of book IV of *The Canon of Mas'ūd*, however, he solves problems of determining an arc of the parallel circle traversed by the Sun or a star from rising to a definite moment (that is, the problem of determining time by the position of the Sun or a star) and gives the solution of such problems in the form of rules equivalent to theorem (1.7) *[57, pp. 477, 486; 53, vol. 5, part 1, pp. 386–387, 394–395]*. But unlike al-Khwārizmī, al-Māhāni, Ibn Qurra and the authors of zījes, al-Bīrūnī gives complete proofs of these rules. At the same time, when dealing in chapter 2 of book V of *The Canon of Mas'ūd* with the problem of finding the difference of the longitudes of two towns from their latitudes and distance, a problem that also reduces to theorem (1.7), al-Bīrūnī solves it not by relying on a rule equivalent to (1.7) but by reducing it to the solution of two right spherical triangles.

The Spherical Trigonometry of Naṣīr al-Dīn al-Ṭūsī

Between the 11th and 13th centuries there appeared a number of works whose authors used Menelaus' theorem and "propositions that free one from the figure of secants," that is, the spherical sine and tangent theorems, to give systematic presentations of all six cases of solving spherical triangles given three elements. The first of these treatises is the anonymous *Collection of rules of the science of astronomy* (Jami' qawānīn 'ilm al-hay'a) *[545]* (see also *[273]*) written in Iṣfahān and dedicated to a certain 'Amīd al-Mulk Abū Naṣr Manṣūr ibn Muḥammad. This is the name of Kundurī (1025–1064), the vizier of Sultan Toghrul Bēk (1056–1063), who founded the Seljuq dynasty. It is possible that the author of the treatise was the eminent mathematician Abū-l-Ḥasan 'Alī al-Nasāwī (ab. 970–ab. 1070), a native of Nasa (near the modern Ashkhabad), who worked at various times in Rayy, Iṣfahān, and Ghazna, but worked at the time in Iṣfahān. The term *rules of the science of astronomy* is undoubtedly derived from the name given by al-Khujandī to the theorem of Ibn 'Irāq. The *Collection of rules* consists of three parts: *On composite ratios, On the figure of secants* (that is, Menelaus' theorem), and *On a proposition that frees one from secants*. The last part opens with a presentation of all six cases of the solution of a spherical triangle given three of its elements. In this work there appears the first reference to the so-called polar triangle, that is, the triangle $A'B'C'$ whose vertices are the poles of the sides of a given spherical triangle ABC. But the author of the *Collection of rules* is apparently unaware of its involutory property, that is, that the polar triangle of the triangle $A'B'C'$ is the triangle ABC.

The most complete exposition of spherical geometry in the East in the Middle Ages is found in the treatise *Disclosing the secrets of the figure of secants* (Kashf al-qinā' 'an asrār al-shakl al-qaṭṭā') *[590]* by the greatest mathematician and astronomer of the 13th century, Naṣīr al-Din al-Ṭūsī (1201–1274). A native of Ṭūs in Khorasān, al-Ṭūsī worked for some time in the "state of Assassins," which operated a terrorist sect that was the bane of the Near and Middle East. When the Mongols conquered that state in 1256, al-Ṭūsī became court astrologer and adviser to the Mongol prince Hūlāgu Khān. In Marāgha, the capital of Hūlāgu (in southern Azerbaijan), al-Ṭūsī organized an excellent observatory and scientific school for all the mathematicians and astronomers of the lands conquered by the Mongols. Al-Ṭūsī's work was originally written in Persian in the state of Assassins, but in 1260 he published a shorter variant in Arabic. Each consists of five books: *On the composite ratio, On the plane figure of secants, Introduction to the spherical figure of secants, On the spherical figure of secants,* and *On propositions that free one from the figure of secants.*

The structure of the treatise *Disclosing the secrets* is very close to that of the *Collection of rules.* It is possible that an intermediate link between the two was the treatise of Ḥusām al-Dīn al-Sālār (d. 1262), al-Ṭūsī's predecessor in the post of court astrologer of Hūlāgu Khān. Al-Ṭūsī refers in his treatise to the treatise of al-Sālār and writes that the latter "gave the necessary rules for all cases, but supplied neither proofs nor examples" *[590, p. 30]*, which is why he, al-Ṭūsī, decided to provide rigorous proofs of these rules.

Al-Ṭūsī also presents all six cases of the solution of a spherical triangle given three of its elements. In the solution of a triangle given its three angles he also makes use of a polar triangle but is aware of its involutory property. We reproduce al-Ṭūsī's proof.

All angles in a triangle are known. In the triangle *ABC* we extend the sides *AB* and *AC* to arcs *AE* and *AD* each equal to a quarter of the circle. We also extend the sides *BA* and *BC* to arcs *BF* and *BH* each equal to a quarter of the circle, and the sides *CA* and *CB* to arcs *CG* and *CK* each equal to a quarter of the circle. We draw the arcs of the great circles *DE, FH,* and *GK.* The points of intersection of these circles will be *L, M,* and *N,* and we obtain the triangle *LMN* whose sides are arcs of great circles (Figure 8). Since the angles *A, B, C* are known, the arcs *DE, FH, GK* are also known. Since *K* and *M* are right angles, *L* is a pole of *KH.* Similarly, *M* is a pole of *GD* and *N* is a pole of *FE.* The

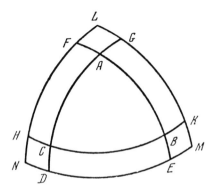

Figure 8

arcs *GL* and *KM* are also known, for they are complements of the arc *KG*. But then the arc *LM* is known. In the same way we find the arcs *LN* and *MN*. This means that we know all three sides of the triangle *LMN* and therefore, in view of the previous case, all three angles of that triangle. This implies that we know the arcs *KH*, *DG*, *EF*. Since each of the arcs *KC*, *BH* is a quarter of the circle, the complement of *HC* is *BC*, so that the side *BC* is known. The same applies to the sides *AB* and *BC*. We have thus determined all sides of the triangle *ABC* [590, pp. 197–198].

If we denote the side *BC*, *CA*. *AB* of the triangle *ABC* by *a*, *b*, *c*, and the sides *MN*, *NL*, *LM* of the triangle *LMN* by *l*, *m*, *n*, then the connection between the sides and the angles of these triangles established by al-Ṭūsī can be written in the following symmetric form:

$$l = \pi - A, \qquad m = \pi - B, \qquad n = \pi - C;$$

$$L = \pi - a, \qquad M = \pi - b, \qquad N = \pi - c.$$

Substituting these values of *l*, *m*, *n*, *N* for *a*, *b*, *c*, *C* in formula (1.7) of the spherical cosine theorem we obtain the relation

$$\cos(\pi - C) = \cos(\pi - A)\cos(\pi - B) + \sin(\pi - A)\sin(\pi - B)\cos(\pi - c),$$

that is,

$$\cos C = -\cos A \cos B + \sin A \sin B \cos c, \tag{1.13}$$

which expresses the *dual spherical cosine theorem*. This theorem permits us to find the sides of a spherical triangle given its angles. But in the mediaeval East only the spherical sine theorem was formulated as a distinct theorem of spherical trigonometry, and neither the spherical cosine theorem (1.7) nor its dual (1.13) was so formulated—this despite the fact that mathematicians and astronomers used rules equivalent to the latter theorems.

The Spherical Trigonometry of Regiomontanus

The spherical cosine theorem first appeared as a distinct theorem of spherical trigonometry in the work *Five books on all manner of triangles* (De triangulis omnimodis libri quinque. Nürnberg, 1533) [450]. Its author was the German mathematician and astronomer Regiomontanus (Johann Müller, 1436–1476), a native of Königsberg in Franconia who worked in Nürnberg. The first book of his work contains auxiliary geometric theorems and the elements of the study of plane triangles, the second deals with plane trigonometry, the third with elements of spherical geometry, the fourth and fifth with spherical trigonometry. The fourth book contains an exposition, modeled on Arabic works, of all six cases of the solution of spherical triangles. This dependence is made apparent by the fact that instead of denoting the vertices of a triangle as *A*,

B, C, Regiomontanus denotes them by A, B, G, the Latin letters commonly used to transcribe the Arabic letters, a, b, and j. In the fifth book Regiomontanus proves a number of new theorems, including the spherical cosine theorem (1.7). He formulates the latter as follows:

> In every spherical triangle made up of arcs of great circles, the ratio of the versed sine of each angle to the difference of the versed sine of the side that subtends the angle and the versed sine of the difference of the sides that bound it is the same as the ratio of the square of the complete right sine to the rectangle under the sines of the arcs bounding the indicated angle [450, pp. 270–271].

Here the "versed sines" are sin vers $C = 1 - \cos C$, sin vers $c = 1 - \cos c$, and sin vers$(a - b) = 1 - \cos(a - b) = 1 - \cos a \cos b - \sin a \sin b$; the "complete right sine" is the same as the "largest sine" of Ibn Qurra, that is, the radius of the circle; and the "rectangle under the sines of the arcs" is the product $\sin a \sin b$. Hence Regiomontanus' theorem can be written as

$$\frac{\text{sin vers } C}{\text{sin vers } c - \text{sin vers}(b - c)} = \frac{1}{\sin a \sin b}$$

or as

$$(1 - \cos C)\sin a \sin b = \cos a \cos b + \sin a \sin b - \cos c,$$

which reduces to the spherical cosine theorem (1.7). Regiomontanus' drawing (Figure 9) differs from the drawing which we used to illustrate Ibn Qurra's theorem (see Figure 5) only in that the key triangle is denoted in the former as ABC and in the latter as PKZ. This makes it clear that the rule that served Regiomontanus as a model for his theorem was analogous to Ibn Qurra's rule.

In Western European literature the spherical cosine theorem is sometimes referred to as *Albategnius' theorem*. This is due to the fact that a rule equivalent to this theorem (and very close to Ibn Qurra's rule) is found in the pre-

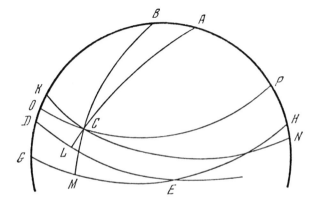

Figure 9

viously mentioned Ṣābian astronomical tables of al-Battāni, known in Western Europe as Albategnius.

Copernicus' Spherical Trigonometry of Chords

Whereas Regiomontanus used and developed the contributions to spherical trigonometry made in the medieval East, another eminent scholar of the 15th and 16th centuries, the great Polish astronomer Nicolaus Copernicus (1473–1543), started directly from Ptolemy. In his famous treatise On the revolutions of celestial spheres (De revolutionibus orbium coelestium. Nürnberg, 1543) [126], in which he replaced Ptolemy's geocentric system with the heliocentric one, Copernicus went back to Ptolemy's names for the constellations, and in the mathematical chapters (chapters 12–14 of book I) to Ptolemy's trigonometry of chords. In chapter 12 of book I Copernicus states the geometric theorems needed for the computation of tables of chords and gives tables of chords, in steps of 10', in parts of the radius, to which he assigns the value 100,000. In chapter 13 Copernicus presents theorems of plane trigonometry, and in chapter 14 theorems of spherical geometry and trigonometry. In proposition 3 of chapter 14 Copernicus proves a theorem equivalent to the sine theorem for a right spherical triangle:

> In right spherical triangles the ratio of the chord subtending twice the side opposite the right angle to the chord subtending twice either one of the sides including the right angle is equal to the ratio of the diameter of the sphere to the chord subtending twice the angle included on a great circle of the sphere between the remaining side and the hypotenuse [126, p. 43].

Chapter 14 also contains rules for solving spherical triangles given any three elements using the theorem just quoted and Euclid's stereometric propositions. Specifically, in proposition 5 Copernicus proves that

> If the angles of a [spherical] triangle are given and one of them is a right angle, the sides are given [126, p. 45]

and in proposition 15 he proves that

> if all the angles of a [spherical] triangle are given even though none of them is a right angle, all the sides are given [126, p. 50].

Viète's Spherical Trigonometry

Close-to-the-modern versions of the spherical cosine theorem (1.7) and the dual spherical cosine theorem (1.9) first appeared in The eighth book of answers to various mathematical questions (Variorum de rebus mathematicis responsorum liber VIII. Tours, 1593) *[603, pp. 347–436]* of Francois Viète (1540–1603). Viète, a French mathematician mainly known for his discoveries in algebra, published in Paris in 1579 extensive tables (*Canon mathematicus*) consisting largely of trigonometric tables in which the radius of the circle was taken to be 100,000. Already in his *Canon*, but especially in the 19th chapter of *The eighth book*, Viète formulates without proofs the whole system of plane and spherical trigonometry. Propositions 15 and 16 in that chapter are his formulations of the two spherical cosine theorems:

XV. *Given the three sides of an arbitrary spherical triangle we can find its angles.* Let the side opposite the required angle be the first side. Now apply to the complete sine two rectangular figures of which one is to be the same as [the rectangle] under the sines of the complements of the second and third sides, and the other the same as [the rectangle] under the sines of the second and third sides themselves. Then the width extracted from the second application is [in the same ratio] to the sum or difference of the width to be extracted from the first application and the sine of the complement of the first side as the complete sine is to the sine of the complement of the required angle.

The case of a sum [occurs] if the first side is less than a quarter of the circle and the properties of the second and third sides are different, or if the first side is greater than a quarter and the properties of the remaining sides are the same. In the first case the required angle is acute and in the second it is obtuse.

On the contrary, the case of a difference [occurs] if the first side is less than a quarter and the properties of the remaining [sides] are the same, call this the third case, or if the specified side is greater than a quarter and the properties of the remaining sides are different, call this the fourth case, and in the third case if the width extracted from the first application is less than the sine of the complement of the first side and the required angle is acute, or if it is greater and the angle is obtuse, and in the fourth case if, to the contrary, the width extracted from the first application is less than that same sine and the required angle is obtuse, or if it is greater and the angle is acute. If the difference is zero, then this implies that the required angle is a right angle. . . .

XVI. *Given the three angles of an arbitrary spherical triangle we can find its sides.* Let the angle opposite the required side be the first. Now apply to the complete sine two rectangular figures of which one is to be the same as [the rectangle] under the sines of the complements of the second and third angles, and the other the same as [the rectangle] under the sines of the second and third angles. Then the width extracted from the second application is [in the same ratio] to the sum or difference of the width to be extracted from the first application and the sine of the complement of the first angle as the complete sine is to the sine of the complement of the required side.

The case of a sum [occurs] if the first angle is obtuse and the properties of the remaining angles are different, or if the first angle is acute and the properties of the remaining angles are the same. In the first case the required side is greater than a quarter and in the second it is less.

On the contrary, the case of a difference [occurs] if the first angle is obtuse and the properties of the other [angles] are the same, call this the third case, or if the first angle is acute and the properties of the remaining [angles] are different, call this the fourth case, and in the third case if the width extracted from the first application yields to the sine of the complement of the first angle and the required side is greater than a quarter, or if it exceeds and [the side] is less, and in the fourth case if, to the contrary, the width extracted from the first application yields to that same sine and the required side is less than a quarter or if it exceeds it and the side is greater. If the difference is null, then this implies that the required side is equal to a quarter [of the circle] *[603, pp. 407–408]*.

The following "dictionary" will make the meaning of Viète's propositions clear.

Let α be the required angle and let the second and third angles be β and γ. Let a, b, c denote their opposite sides. Viète's "sines" are the quantities $r \sin \alpha$, $r \sin a$, and so on; his "sines of the complements" are the quantities $r \cos \alpha$, $r \cos a$, and so on; "the rectangle under the sines of the second and third sides" is the product $r \sin b \cdot r \sin c$; "the rectangle under the sines of the complements of the second and third sides" is the product $r \cos b \cdot r \cos c$; the "application" of these rectangles to the "complete sine," that is, the radius, is the construction on the radius of rectangles with correspondingly equal areas; and "width extracted from the application" is the width of the resulting rectangle, that is, in one case the segment $\dfrac{r \sin b \cdot r \sin c}{r}$ and in the other the segment $\dfrac{r \cos b \cdot r \cos c}{r}$.

With this "dictionary" in mind, we can write Viète's proposition 15 as the proportion

$$\frac{r \sin b \sin c}{r \cos a - r \cos b \cos c} = \frac{r}{r \cos \alpha}.$$

Here Viète considers the following four cases:

(1) $\cos a > 0$, $\cos b \cos c < 0$, $\cos \alpha > 0$;

(2) $\cos a < 0$, $\cos b \cos c > 0$, $\cos \alpha < 0$;

(3) $\cos a > 0$, $\cos b \cos c > 0$, $\cos b \cos c < \cos a$, $\cos \alpha > 0$
    or $\cos b \cos c > \cos a$, $\cos \alpha < 0$;

(4) $\cos a < 0$, $\cos b \cos c < 0$, $|\cos b \cos c| < |\cos a|$, $\cos \alpha < 0$
    or $|\cos b \cos c| > |\cos a|$, $\cos \alpha > 0$.

Similarly, Viète's proposition 16 can be written as the proportion

$$\frac{r \sin \beta \sin \gamma}{r \cos \alpha + r \cos \beta \cos \gamma} = \frac{r}{r \cos a}$$

Here Viète considers four cases:

(1) $\cos \alpha < 0$, $\cos \beta \cos \gamma < 0$, $\cos a < 0$;

(2) $\cos \alpha > 0$, $\cos \beta \cos \gamma > 0$, $\cos a > 0$;

(3) $\cos \alpha < 0$, $\cos \beta \cos \gamma > 0$, $\cos \beta \cos \gamma < \cos \alpha$, $\cos a < 0$
or $\cos \beta \cos \gamma > \cos \alpha$, $\cos a > 0$;

(4) $\cos \alpha > 0$, $\cos \beta \cos \gamma < 0$, $|\cos \beta \cos \gamma| < |\cos \alpha|$, $\cos a > 0$
or $|\cos \beta \cos \gamma| > |\cos \alpha|$, $\cos a < 0$.

The complete analogy between these two propositions shows that Viète was fully aware of the connection between the two cosine theorems and may possibly have known that the second could be obtained from the first by means of a polar triangle.

Area of a Spherical Triangle and Polygon in the Work of Girard

Formulas expressing the area of a spherical triangle and polygon in terms of their respective angular excesses appeared in print for the first time in the paper *On a newly discovered measure of area of spherical triangles and polygons* (De la mesure de la superfice des triangles & polygones sphericques, nouvellement inventée), published as an appendix to *A new invention in algebra* (Invention nouvelle en l'algèbre, Amsterdam, 1629 *[204]* by the Flemish mathematician Albert Girard (1595–1632). Girard begins with a "new hypothesis" according to which the surface of a sphere is subdivided into 720 "surface degrees" (degrez superficieles), each of which is subdivided into 60 minutes, and so on; that is, he considers the surface of a sphere, which is equal to 4π for a sphere of unit radius, to be equal to $4 \cdot 180° = 720°$. Next Girard defines a "buckle" (*fibulle*)—a spherical triangle two of whose sides are quadrants. The buckle is said to be acute, obtuse, or right in accordance with the size of the angle between the indicated sides. There follow three lemmas:

(1) Let A be a pole of a great or small circle on the surface of the sphere and let AB and AC be two great arcs issuing from the pole (Figure 10a). Then the part of four right angles constituted by A is equal to the part of the spherical surface bounded by the circle BC constituted by the triangle ABC; the proof is obvious *[204, f. G 2 v.]*;
(2) if α and β are two arcs of the same circle not exceeding a quadrant and $\alpha > \beta$, then

$$\frac{\tan \alpha}{\tan \beta} > \frac{\alpha}{\beta} > \frac{\sin \alpha}{\sin \beta}$$

(Girard points out that the second half of this assertion is equivalent to

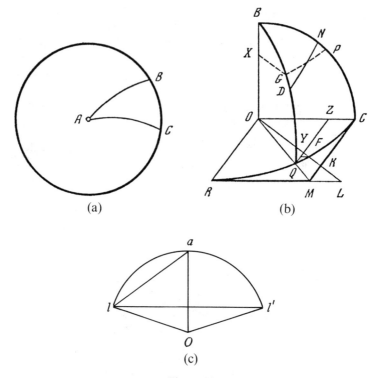

Figure 10

an assertion in book I of Ptolemy's *Almagest* and in book I of Copernicus'
On the revolutions of celestial spheres[4]);
(3) the sum of the interior angles of a rectilinear n-gon is equal to $2n - 4$
right angles.

Then Girard formulates the theorem:

> Every spherical polygon enclosed between arcs of great circles contains
> as many surface degrees as the amount by which the sum of its internal
> angles exceeds the sum of the internal angles of the rectilinear polygon
> with the same name; here the surface of the sphere is assumed to contain
> 720 surface degrees *[204, f. G 3 v.]*.

Next Girard gives some examples and proves the following proposition.

> Proposition. *A spherical triangle with three great arcs contains as
> many surface degrees as the amount by which the sum of its three angles
> exceeds 180 degrees [204, f. H 1].*

Girard gives his proof in stages: (1) For a "buckle": if the buckle is bounded

---

[4] The second of these inequalities is proved in chapter 10 of book 1 of Ptolemy's *Almagest [441,
p. 54]* and in theorem 6 of book I of Copernicus' *On the revolutions of celestial spheres [126].*

by two quadrants AB and AC, then the angles B and C are right angles, their sum is equal to $180°$, and the sum of all three angles exceeds $180°$ by the angle A. By lemma 1, the part of $360°$ constituted by A is the same as the part of 360 surface degrees constituted by the buckle. (2) For a right spherical triangle each of whose sides is less than a quadrant, Girard considers the right spherical triangle BND with right angle N and extends its sides BN and BD to quadrants BC and BQ (Figure 10b). Then the arc CQ contains as many degrees as the angle B, and the angle D contains more degrees than the arc QR, the complement of the arc CQ. Therefore if we lay off on the quadrant RC the arc RF containing as many degrees as the angle D, then the point F will end up on the arc QC, and the arc QF contains as many degrees as the excess of the angle sum of the triangle BND above $180°$.

Now Girard applies the dual spherical cosine theorem (1.13) to the right spherical triangle BND. In this case, the theorem can be written as $\tan B = \tan D/\sec BD$. In turn, this can be written as the proportion $\dfrac{OC}{CM} = \dfrac{RL}{\sec BD}$, so that $\sec BD = \dfrac{CM \cdot RL}{OC}$. But, by lemma 2, $\dfrac{MC}{CK} > \dfrac{QC}{CF}$, so that $\left.\dfrac{RL \cdot MC}{OC}\right|$ $\dfrac{RL \cdot CK}{OC} > OC \left|\dfrac{OC \cdot CF}{QC}\right.$. Since $RL \cdot CK = \tan C \cdot \cot D$ is the square of the radius, that it, OC^2, and $\dfrac{RL \cdot MC}{OC} = \sec BD$, it follows that $\sec BD/OC >$ $OC \left|\dfrac{OC \cdot CF}{QC}\right.$. But $\sec BD \cos BD = OC^2$, or, since $\cos BD = \sin DQ$, $\sec BD \sin DQ = OC^2$. Hence $\sec BD/OC = OC/\sin DQ$. But then $OC \cdot CF/QC > \sin DQ$.

Let $\dfrac{OC \cdot CF}{QC} = \sin GQ = OX$. Girard drops perpendiculars QZ to OC and FY to QZ and once more applies the dual cosine theorem to the triangle BND, this time in the form of the proportion radius$/\sin B = \sec D/\sec BN$, which he rewrites as $OC/QZ = OL/\sec BN$, that is, $\sec BN = QZ \cdot OL/OC$. By lemma 2, $QZ/YZ < QC/CF$, so that $\left.\dfrac{OL \cdot QZ}{OC}\right|\dfrac{OL \cdot YZ}{OC} < OC \left|\dfrac{OC \cdot CF}{QC}\right.$. Since $OL \cdot YZ = OC^2$ and $OL \cdot QZ/OC = \sec BN$, we see that $\sec BN/OC < OC \left|\right.$ $\dfrac{OC \cdot CF}{QC}$. Now $\sec BN \cos BN = OC^2$. Since $\cos BN = \sin NC$, it follows that $\sec BN \sin NC = OC^2$, or $\sec BN/OC = OC/\sin NC$. Hence $OC \cdot CF/QC < \sin NC$. Since $OC \cdot CF/QC = \sin GQ$, we have $DQ < GQ < NC$.

Now Girard describes from the pole B an arc through G. This arc intersects the arcs DN and NC, the latter at the point P.

Since $OC \cdot CF/QC = OX$ and $OC = OB$, we have the proportion $QC/CF = BO/OX$. Since $QF = QC - CF$ and $BX = OB - OX$, this proportion yields the proportion $CQ/QF = OB/BX$. Then Girard *[204, f. H 3]* writes that "from the works of Archimedes we can conclude that the ratio of OB to BX is equal to the ratio of the area of the buckle QBC to the area of GBP." Girard is alluding to proposition 42 in book I of Archimedes' *On the sphere and cylinder*, which asserts that

If *lal'* be a segment of a sphere less than a hemisphere (Figure 10c) and *Oa* the radius perpendicular to the base of the segment, the surface of the segment is equal to a circle whose radius is equal to *al [25, p. 52]*.

It follows that the surface area of a segment that is a spherical circle with spherical radius r on a sphere of radius R is

$$S = 2\pi R^2[1 - \cos(r/R)],$$

where $R[1 - \cos(r/R)]$ is the height of the segment.

In the case considered by Girard, the area of the buckle QBC is to the area of the triangle GBP as the area of the hemisphere is to the area of the spherical circle with spherical radius BG, that is, as the radius OB of the sphere to the height BX of the segment. It follows that CQ is to PG as the area of the buckle QBC is to the area of the triangle GBP. On the other hand, the ratio of these arcs is the same as the ratio of the buckles QBC and QBF, where BF is an arc of the great circle through B and F. It follows that the area of the buckle QBF is the same as that of the triangle GBP. Girard notes that the sum of the angles of the buckle QBF is equal to the sum of the angles of the triangle BND, for that sum exceeds two right angles by the magnitude of the arc QF. Girard claims further that

BGP is equal to the triangle *BND* for it always overlaps it, and that because *GP* always intersects *DN*. It follows that the buckle *QBF* is equal to the triangle *BDN* whose three angles exceed two right angles by the magnitude of the arc *QF* in degrees, and in view of what was said about buckles in the first lemma, the truth of the theorem is clear and plausible *[204, f. H 3]*.

Girard's assertion about the equality of the areas of the spherical sector BGP and the spherical triangle BND is entirely correct, but his argument is not a rigorous proof. What it does is to establish the approximate equality of the areas of the indicated figures.

It seems that Girard himself was not quite satisfied with his argument, for he adds that

even if *ND* if infinitely small and *BD* is almost a quadrant, so that *GD* and *NP* are very small, nevertheless *DN* always intersects *GP*, so that *BGP* will always be equal to the triangle *BND*, in accordance with the assertion of the theorem. Note that I have proved it for two different cases, when *GD* is greater than twice *NP* and when *BP* or *BG* is smaller than the harmonic mean of *DB* and *BN* *[204, ff. H 3–H 3 v]*.

This shows that Girard felt that a rigorous proof of his assertion called for infinitesimal arguments, which he did not have at his disposal.

Starting with the special case of his assertion for a right spherical triangle Girard proved it for an arbitrary spherical triangle by dividing the latter into two right triangles and by noting that the angular excess above 180° of the given triangle is the sum of the angular excesses of the component right triangles. Girard reduced the theorem on spherical polygons to the case of triangles by subdividing the n-gon by means of diagonals into $n-2$ triangles.

Girard's assertion for a spherical triangle can be written as

$$\delta/4\pi r^2 = (\alpha + \beta + \gamma - 180°)/720°.$$

A letter written by Briggs to Kepler in 1625 indicates that a similar rule was found by Thomas Harriot (1560–1621) in 1603 *[271, vol. 18, pp. 228–229]*. The same rule was published by Bonaventura Cavalieri (1598–1647) in his *General handbook of astronomical measurements* (Directorium generale uranometricum. Bologna, 1632) *[101]*.

If the measures of the angles of a triangle are expressed in radians rather than in degrees, then its area is given by

$$S = r^2(\alpha + \beta + \gamma - \pi). \tag{1.14}$$

Next we give the simplest proof of this theorem, which is due to Euler.

Euler's Spherical Geometry and Trigonometry

During the 17th and 18th centuries there appeared a great many monographs devoted to spherical trigonometry, such as *The construction of spherical trigonometry* (Trigonometriae sphericae constructio. Rome, 1737) of Roger Josip Boscovich (1711–1787) *[75]*, a native of Dubrovnik (now in Yugoslavia), who worked for the most part in Rome and Milan. Boscovich gave graphical solutions, far simpler than the earlier ones, of spherical triangles with three given elements.

The modern form of spherical trigonometry, as well as of all trigonometry, is due to the great Leonhard Euler (1707–1783), a native of Basel, who worked in Petersburg and Berlin. Whereas trigonometry before Euler was concerned with trigonometric *lines* Euler's trigonometry dealt with trigonometric *functions*, which he linked to the exponential function by means of the well-known formula bearing his name. This banished from trigonometric formulas the *sinus totus*, the complete sine, that is, the radius of the circle, and replaced it with unity.

In his *On the measure of solid angles* (De mensura angulorum solidorum. Petersburg, 1781) *[176, vol. 26, pp. 204–223]*, Euler gave a remarkably simple proof of Girard's theorem on the area of a spherical triangle. Euler considers on the unit sphere a triangle *ABC* and an equal triangle *abc* whose vertices

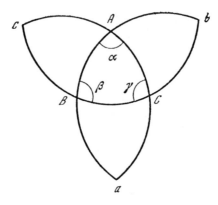

Figure 11

a, b, c are antipodal to A, B, C (Figure 11). Let α, β, γ be the angles at the vertices A, B, C, respectively. Then the areas of the 2-gons $ABaC$, $BCbA$, and $CAcB$ are, respectively, 2α, 2β, and 2γ. If S denotes the area of the triangle ABC common to all three 2-gons, then the areas of the remaining parts of the 2-gons, that is, the triangles aBC, bAC, and cAB, are $2\alpha - S$, $2\beta - S$, and $2\gamma - S$. The areas of the triangles abc, Abc, Bac, and Cab are, respectively, equal to the areas of the triangles ABC, aBC, bAC, and cAB. Since these eight triangles cover the surface of the sphere, the sum of their areas is 4π, the area of the unit sphere. Hence

$$2S + 4(\alpha + \beta + \gamma) - 6S = 4(\alpha + \beta + \gamma) - 4S = 4\pi.$$

But then

$$S = \alpha + \beta + \gamma - \pi.$$

In the case of a sphere of radius r the corresponding formula (1.14) is obtained in a similar manner.

In *Various investigations concerning the area of spherical triangles* (Varia speculationes super area triangulorum sphaericorum, Petersburg, 1797) *[176, vol. 29, pp. 253–266]*, Euler gave a number of expressions for the area of a spherical triangle on a unit sphere in terms of its sides. The simplest of these expressions is

$$\cos S/2 = \frac{1 + \cos a + \cos b + \cos c}{4 \cos a/2 \cos b/2 \cos c/2}.$$

In the same paper Euler showed that the locus of the vertices of spherical triangles with the same base and area consists of arcs of two small circles whose endpoints are antipodal to the endpoints of the base.

Many of Euler's papers are devoted to the solution of problems of spherical geometry that are analogous to problems of plane geometry. For example, in *Construction of a problem of Pappus of Alexandria* (Problematis Pappi Alexandrini constructio. Petersburg, 1783) Euler *[176, vol. 26, pp. 237–242]*, having solved by means of analytic geometry the problem of constructing a triangle inscribed in a circle whose sides, or extensions of the sides, pass through given points (a generalization of a theorem of Pappus in which the three points are collinear), solved that problem for a spherical triangle. Again, in the paper, *Something geometric and spherical* (Geometrica et sphaerica quodam. Petersburg, 1815) Euler *[176, vol. 26, pp. 344–358]*, having proved the theorem of plane geometry on the ratio of the segments into which a given point divides the cevians of a triangle passing through that point, proved the corresponding theorem for a spherical triangle.

The following two of Euler's papers are devoted to spherical trigonometry: *Principles of spherical trigonometry deduced from the method of maxima and minima* (Principes de la trigonométrie sphérique tirés de la méthode des plus grands et plus petits. Berlin, 1755) *[176, vol. 27, pp. 277–308]*, and *Universal spherical trigonometry derived from first principles in a brief and simple manner* (Trigonometria sphaerica universa ex primis principiis breviter et lucide derivata. Petersburg, 1782) *[176, vol. 26, pp. 224–236]*. In the first of these papers Euler constructs spherical trigonometry as the intrinsic geometry of the surface of the sphere. He expresses the line element *ds* of the surface of the sphere in terms of the longitude and latitude of a point, defines the great circles as curves that minimize the integral of the line element, and, in connection with the determination of the minimum of a side of a spherical triangle, derives 10 equations of spherical trigonometry. After the discovery that the shape of the earth is that of a spheroid, Euler extended his methods to spheroids in the paper *Elements of spheroidal trigonometry derived from the method of maxima and minima* (Eléments de la trigonométrie sphéroidique tirés de la méthode de plus grands et plus petits. Berlin, 1755) *[176, vol. 27, pp. 309–339]*.

In the second paper on spherical trigonometry Euler developed this subject in its entirety. He employed a solid angle that he cut by various planes and used plane trigonometry to investigate the resulting plane triangles. In this paper Euler deduced very many of the formulas of spherical trigonometry. It should be noted that Euler made use of polar triangles.

A significant number of papers on spherical geometry are due to Euler's students, the Russian academicians Andrei Ivanovič Lexell (1741–1784) and Nicolai Ivanovič Fuss (1755–1826). In particular, in the paper *Solution of a geometrical problem of spherical geometry* (Solutio problematis geometrici ex doctrina sphaericorum. Petersburg, 1784) *[321]*, Lexell, independently of Euler, proved the theorem on the locus of the vertices of spherical triangles with the same base and area (we have pointed out that Euler's paper was published in 1797). Lexell also gave an analytic solution of this problem. In the paper *On properties of circles drawn on spherical surfaces* (De proprietatibus circulorum in superficie sphaerica descriptorum. Petersburg, 1786) *[322]*, Lexell showed that in a spherical quadrilateral inscribed in a small circle of a sphere the sums of opposite angles are equal, and that in the circumscribed quadrilateral the sums of opposite sides are equal. The first of these theorems is the spherical analogue of a plane theorem, and the second theorem is its dual, that is, the result of interchanging great circles and their poles. Of the many other new results in this paper we mention the formula

$$\tan \varepsilon/4 = \sqrt{\tan\frac{s-a}{2}\tan\frac{s-b}{2}\tan\frac{s-c}{2}\tan\frac{s-d}{2}} \qquad (1.15)$$

for the angular excess of a spherical quadrilateral inscribed in a small circle. Here a, b, c, d are the sides of the quadrilateral and $s = \dfrac{a+b+c+d}{2}$.

By Girard's theorem on the area of a spherical triangle, this excess is proportional to the area of the spherical quadrilateral, and Lexell's theorem is an analogue of the well-known theorem of Brahmagupta for the area $S = \sqrt{(s-a)(s-b)(s-c)(s-d)}$ of a plane quadrilateral inscribed in a circle, as well as (for $d = 0$) of the classical Archimedes-Heron theorem for a triangle. (The special case of formula (1.15) for $d = 0$ that expresses the angular excess of a triangle in terms of its sides is known as *L'Huillier's theorem*.)

As for Fuss, we mention his paper *On certain properties of ellipses drawn on the surface of a sphere* (De proprietatibus quibusdam ellipses in superficie sphaerica descriptae. Petersburg, 1788) *[193]*, in which Fuss defines a spherical ellipse as the locus of points on the sphere for which the sum of the spherical distances from two given points is constant. Using spherical coordinates x and y, x for longitude and y for latitude, and assuming that the equator passes through the foci of the ellipse located symmetrically with respect to the first meridian, Fuss obtained for the spherical ellipse the following equation:

$$\tan y = \frac{\sqrt{(\sin^2 c - \sin^2 a)(\sin^2 c - \sin^2 x)}}{\sin c \cos c}.$$

If we go over to rectangular coordinates X, Y, Z, connected with the spherical coordinates x, y of the (unit) sphere by means of the equations

$$X = \sin x \cos y, \qquad Y = \sin y, \qquad Z = \cos x \cos y,$$

then

$$\tan y = \frac{Y}{\sqrt{1-Y^2}} = \frac{Y}{\sqrt{X^2+Z^2}}, \qquad \sin x = \frac{X}{\sqrt{1-Y^2}} = \frac{X}{\sqrt{X^2+Z^2}}.$$

If we put $\sin a = A$ and $\sin c = C$, then Fuss's equation takes the form

$$(C^2 - A^2)X^2 + C^2 Y^2 - \frac{C^2}{1-C^2}Z^2 = 0.$$

This shows that a spherical ellipse is the intersection of a sphere and a quadric cone whose vertex is the center of the sphere.

Chapter 2
The Theory of Parallels

Euclid's Theory of Parallels

The first systematic account of the theory of parallels to come down to us is contained in Euclid's *Elements [173]* dating from about 300 B.C. Euclid worked in Alexandria under Ptolemy I and was the head of the Museion, the most eminent scientific center of antiquity, which was founded at that time. Euclid's *Elements* is a revised version of a number of Greek works from the fifth and fourth centuries B.C., namely the *Elements* attributed to Hippocrates of Chios (books I–IV and XI), the arithmetic works of the Pythagoreans (books VII–IX), Eudoxus' theory of similarity and ratios (books V–VI) and his method of exhaustion (book XII), and Theaetetus' works on quadratic irrationalities (book X) and on regular polyhedra (book XIII). The *Elements* opens with a list of 23 definitions, many of which bear traces of ancient traditions. The last of these definitions deals with parallel lines:

> Parallel straight lines are straight lines which, being in the same plane and being produced indefinitely in both directions, do not meet one another in either direction *[173, vol. 1, p. 154]*.

There follow five "postulates." The first three are axioms bearing on geometric constructions with an ideal straightedge and ideal compass:

> Let the following be postulated:
>
> (1) To draw a straight line from any point to any point.
> (2) To produce a finite straight line continuously in a straight line.
> (3) To describe a circle with any center and distance *[173, vol. 1, p. 154]*.

The fourth postulate asserts

> That all right angles are equal to one another *[173, vol. 1, p. 154]*.

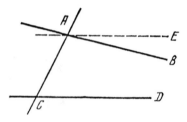

Figure 12

According to Ivan Nikolaevič Veselovskiĭ (1892–1977) *[613]* this axiom is meant to exclude spherical trigometry in which the right angles between a meridian and different parallels are different.

Euclid's fifth postulate asserts

> That, if a straight line falling on two straight lines makes the interior angles on the same side together less than two right angles, the two straight lines, if produced indefinitely, meet on that side on which the angles are together less than two right angles *[173, vol. 1, p. 155]*.

It follows from this postulate that there is at most one line passing through a point not on a given line that does not intersect it, that is, is parallel to it. In Figure 12 the angles *BAC* and *ACD* are together less than two right angles and the lines *AB* and *CD* meet, whereas the angles *EAC* and *ACD* are together equal to two right angles and the lines *AE* and *CD* are parallel.

Euclid goes on to state five "common notions," that is, axioms bearing on the comparison of magnitudes. They are as follows:

(1) Things which are equal to the same thing are also equal to one another.

(2) If equals be added to equals, the wholes are equal.

(3) If equals be subtracted from equals, the remainders are equal.

(4) Things which coincide with one another are equal to one another.

(5) The whole is greater than the part *[173, vol. 1, p. 155]*.

Euclid tries to prove as many theorems as possible without using the fifth postulate. The first 28 propositions of book I are so proved. Some of the first 26 of these propositions deal with the construction of an equilateral triangle on a given segment; the construction of a segment equal to a given segment and of an angle equal to a given angle; a theorem about the equality of triangles; a theorem about isosceles triangles; the bisection of a given angle and of a given segment; the drawing of a perpendicular straight line to a given infinite (that is, unbounded) straight line; theorems about adjacent and vertical angles; the theorem that an exterior angle in a triangle is greater than either of the interior angles not adjacent to it; the theorem that two angles in a triangle are less than two right angles; the theorems that in any triangle the greater side subtends the greater angle and the greater angle is subtended by the greater side; the theorem that in any triangle two sides are greater than the third; and the construction of a triangle with sides equal to three given segments. Propositions 27 and 28 are

If a straight line falling on two straight lines make the alternate angles equal to one another, the straight lines will be parallel to one another *[173, vol. 1, p. 307]*.

and

If a straight line falling on two straight lines make the exterior angle equal to the interior and opposite angle on the same side, or the interior angles on the same side equal to two right angles, the straight lines will be parallel to one another *[173, vol. 1, p. 309]*.

The fifth postulate (the parallel postulate) is first used in proposition 29, inverse to propositions 27 and 28:

29. A straight line falling on parallel straight lines makes the alternate angles equal to one another, the exterior angle equal to the interior and opposite angle, and the interior angles on the same side equal to two right angles *[173, vol. 1, p. 311]*.

In proposition 30 it is proved that straight lines parallel to the same straight line are also parallel to one another. Proposition 31 is the problem of drawing through a given point a straight line parallel to a given straight line. In proposition 32 it is proved that

In any triangle, if one of the sides be produced, the exterior angle is equal to the two interior and opposite angles, and the three interior angles of the triangle are equal to two right angles *[173, vol. 1, p. 318]*.

In proposition 33 it is proved that

The straight lines joining equal and parallel straight lines (at the extremities which are) in the same directions (respectively) are themselves also equal and parallel *[173, vol. 1, p. 322]*.

Thus in this proposition Euclid proves the existence of a parallelogram. (We note that this term appears here for the first time. Earlier in the book Euclid defines a rhombus and a rhomboid—a nonrectangular parallelogram that is not a rhombus.) In proposition 34 it is proved that in a parallelogram the opposite sides and angles are equal to one another, and the diameter bisects the area. Propositions 35 and 36 establish the equality of (areas of) parallelograms "on the same base and in the same parallels" and of parallelograms "on equal bases and in the same parallels."

In propositions 37 and 38 analogous results are proved for triangles "on the same base and in the same parallels" and for triangles "on equal bases and in the same parallels." The inverses of propositions 37 and 38 are proved in propositions 39 and 40.

In proposition 41 it is proved that a parallelogram having the same base with a triangle and being in the same parallels is double of the triangle.

Proposition 42 deals with the construction in a given rectilinear angle of a parallelogram equal to a given triangle. In proposition 39 it is proved that in any parallelogram the complements of the parallelograms about the diameter are equal to one another. Proposition 44 deals with the construction on a given straight line in a given rectilinear angle of a parallelogram equal to a given triangle. This construction is called the "application" (parabolē) of a parallelogram to a given line. Proposition 45 deals with the construction in a given rectilinear angle of a parallelogram equal to a given rectilinear figure (i.e., polygon). Proposition 46 deals with the construction of a square with a given side. Propositions 47 and 48 are, respectively, Pythagoras' theorem and its inverse.

Aristotle's Treatment of the Problem of Parallels

Owing to its relative complexity and scant intuitive appeal, the fifth postulate has given rise to a great many attempts to deduce it from the remaining axioms and thus prove it as a theorem. It seems that this postulate was absent from the works of mathematicians in the fourth century B.C. Be that as it may, Aristotle (384–322 B.C.), in his analysis of the logical error of *petitio principii* (the implicit use of an assertion equivalent to the one being proved), wrote in his *Prior Analytics* (Analytika protera):

> This is what happens with those who think they describe parallel lines, for they unconsciously assume things which it is not possible to demonstrate if parallels do not exist [29, vol. 1, p. 65ᵃ; 219, p.27].

To avoid this logical error one must explicitly assume the fifth postulate, as was done in Euclid's *Elements*, or an equivalent proposition.

It is possible that Aristotle stated such a proposition in a treatise that has not come down to us. In this connection we note that, in his commentaries on Euclid, ʿUmar Khayyām wrote:

> The cause of the error made by later scholars in the proof of this premise is that they did not take into account the principles borrowed from the Philosopher,

that is, Aristotle. Khayyām states five such principles:

    (I) Magnitudes are infinitely divisible, that is, they do not consist of indivisibles;

    (II) A straight line can be produced to infinity;

    (III) Any two intersecting straight lines open and diverge to the extent to which they move away from the vertex of the angle of intersection;

    (IV) Two converging lines intersect and it is impossible for the converging straight lines to diverge in the direction of convergence;

    (V) Of two unequal bounded magnitudes the smaller can be taken with such multiplicity that it exceeds the larger [272, pp. 119–120].

Principles I, II, and V are the following well-known assertions of Aristotle's *Physics* (Physika):

> For there are two senses in which length and time and generally anything continuous are called "infinite": they are called so either in respect of divisibility or in respect of their extremities *[29, vol. 2, p. 233ª];*

> Nothing that is continuous can be composed of indivisibles: e.g. a line cannot be composed of points, the line being continuous and the point indivisible *[29, vol. 2, p. 231ª];*

> Every finite magnitude is exhausted by means of any determinate quantity however small *[29, vol. 2, p. 206ᵇ].*

The last of the principles is the axiom of Eudoxus-Archimedes. Principle III is equivalent to an assertion of Aristotle found in his work *On the heavens* (Peri ouranon), commonly known under the Latin title *De caelo*:

> The body which moves in a circle must necessarily be finite in every respect, for the following reasons.... If the body so moving is infinite, the radii drawn from the centre will be infinite. But the space between infinite radii is infinite *[29, vol. 2, p. 271ᵇ].*

Only principle IV is not found in the known works of Aristotle. But it is possible that the medieval Eastern scholars were familiar with a work of Aristotle in which this principle was formulated. The principle consists of two assertions, each of which is equivalent to Euclid's fifth postulate.

Aristotle linked the problem of parallel lines to the question of the sum of the angles of a triangle. In *Prior analytics* Aristotle states:

> since it is not perhaps absurd that the same false result should follow from several hypotheses, e.g. that parallels meet, both on the assumption that the interior angle is greater than the exterior and on the assumption that a triangle contains more than two right angles *[29, vol. 1, p. 66ª].*

Clearly, the term "parallel lines" is used here not in Euclid's sense but in the sense of two lines that form equal opposite interior angles with a third or, in particular, in the sense of two lines perpendicular to a third.

Earlier we saw that proposition 10 of Menelaus' *On the sphere* proves that under certain conditions an interior angle of a spherical triangle is greater than an exterior angle, and proposition 11 proves that the sum of the angles of a spherical triangle exceeds two right angles. It is well known that any two great circles on a sphere, including circles that form equal opposite interior angles with a third circle and, in particular, circles perpendicular to a third circle, intersect. It is very likely that these facts of spherical geometry, possibly not supported by rigorous proofs, were known in Aristotle's time, and that in his example Aristotle had these facts in mind when referring to great circles on a sphere as "straight lines."

Aristotle regards triangles with angle sum equal to two right angles and those with angle sum different from two right angles as being of different kinds. In his *Metaphysics* (Meta ta physika) he states:

> E.g. if we suppose that the triangle does not change, we shall not suppose that at one time its angles are equal to two right angles while at another time they are not (for that would imply change) *[29, vol. 8, p. 1052ᵇ]*.

Later scholars no longer understood Aristotle's remarks on the theory of parallels. Thus the author of a scholium on Euclid's *Elements* wrote that

> It is impossible to find ... an isosceles right triangle whose hypotenuse is equal to the sides including the right angle *[172, vol. 5, p. 722]*,

and Maimonides (see p. 193) wrote that

> We do not attribute to God, may He be exalted, incapacity because He is unable ... to create a square whose diagonal is equal to its side *[349, p. 226]*.

We know that on a sphere there are a triangle and a square with the indicated properties.

Apparently, when he refers to triangles whose angle sum is not equal to two right angles, Aristotle has in mind spherical triangles. But he sometimes refers to an angle sum that is smaller than two right angles. Thus, for example, in his *Posterior analytics* (Analytika deutera) he states:

> Thus, as we maintain, to know a thing's nature is to know the reason why it is; and this is equally true of things in so far as they are said without qualification to *be* as opposed to being possessed of some attribute, and in so far as they are said to be possessed of some attribute such as [the angles in a triangle are] equal to two right angles, or greater or less *[29, vol. 1, p. 90ᵃ]*.

Aristotle does not investigate a geometry in which the angle sum in a triangle is less than two right angles. There are no simple models of such a geometry.[1]

The First Attempts in Antiquity to Prove the Parallel Postulate

The problem of the parallel postulate and the theory of parallel lines have been considered by many scholars over a period of 2000 years. It seems that the first work devoted to this question was Archimedes' lost treatise *On parallel lines* that appeared a few decades after Euclid's *Elements*. This title is one item in the list of Archimedes' works contained in the Arabic *Book*

---

[1] Concerning the problem of parallel lines in Aristotle see the papers of Imre Tóth *[588]*, Anna Evgen'evna Busurina *[85]*, and I. N. Veselovskiĭ *[613]*.

of bibliography of the sciences (Kitāb Fihrist al-ʿUlum) by Abū l-Faraj Muḥammad ibn al-Nadīm (d. 993) *[240, p. 266]*.

The fact that Ibn Qurra was the author of six of the seven extant Arabic translations of the works of Archimedes supports the assumption that he was familiar with, and may have translated into Arabic, Archimedes' treatise that Ibn al-Nadīm called *The book on parallel lines* (Kitāb al-khūṭūṭ al-mutawāziya). It is therefore possible that one of Ibn Qurra's preserved treatises on parallel lines considered later represents an edited version of Archimedes' treatise. Be that as it may, it is very likely that Archimedes used a definition of parallel lines different from Euclid's. Furthermore, given that metric relations played a greater role in Archimedes' geometry than in Euclid's, it is possible that Archimedes based his definition of parallel lines on distance. According to Proclus, the philosopher, astronomer, and mathematician Posidonius (ab. 135–50 B.C.), a native of Syria working in Rome, based his "proof" of the parallel axiom on just such a definition:

Parallel lines are lines in the same plane that come neither near nor apart, so that all the perpendiculars from the points of one of them to the other are equal *[440, p. 138]*.

This definition of parallel lines, based on the assumption that the locus of points on one side of a line and at an equal distance from it is a straight line, contains an assertion equivalent to the parallel axiom; in fact, in Lobačevskian geometry, in which the parallel axiom does not hold, the locus in question is not a straight line. Therefore the parallel postulate is a simple consequence of this definition of parallel lines.

Proclus' and Ptolemy's "Proofs" of the Fifth Postulate

Proclus *[440, pp. 282–288]* describes another "proof" of the parallel postulate due to Ptolemy. Ptolemy first "proved" that if two parallel lines are cut by a transversal, then the interior angles on the same side add up to two right angles—an assertion equivalent to the fifth postulate. His proof is by contradiction. Thus, suppose that the interior angles on one side of the transversal are together less than two right angles. Since the corresponding angles on the other side of the transversal are supplementary angles, they must add up to more than two right angles. But the lines on one side of the transversal are "no more parallel" than the same lines on its other side. Hence the false conclusion that if the interior angles on one side of the transversal add up to less than two right angles, then so do the interior angles on the other side of the transversal. The resulting "contradiction" proves Ptolemy's assertion. From it, again arguing by contradiction, he readily obtains the parallel postulate.

Notwithstanding his accounts of the "proofs" of the parallel postulate of Posidonius and Ptolemy, Proclus gives his own "proof" of that postulate

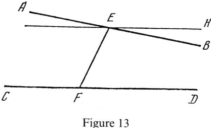

Figure 13

[440, pp. 290–292]. He considers lines *AB* and *CD* and a transversal *EF* that forms with these lines interior angles *BEF* and *EFD* on the same side that add up to less than two right angles (Figure 13). Through the point *E* he draws a line *EH* parallel to *CD* and argues that, since the distance between the points on the sides of the angle *BEH* can be made arbitrarily large by moving sufficiently far away from the vertex *E*, it is bound to exceed the distance between the parallels *CD* and *EH*. Consequently, the side *EB* of the angle *EBH* is bound to intersect the straight line *CD*. Proclus himself is guilty of the error of *petitio principii*, for his assumption that the distance between nonintersecting coplanar lines (he calls them parallel lines) is bounded is equivalent to the postulate which he wants to prove (in Lobačevskian geometry the distance between nonintersecting coplanar straight lines can be arbitrarily large).

Of the ancient theories of parallels that have not come down to us we mention those of Diodorus (first century B.C.) and Anthiniatus. Simplicius refers to them in these words:

> This postulate is not obvious but it is indispensable for a proof by means of lines. Anthiniatus and Diodorus proved many different propositions about it *[419, p. 154]*.

Arabic Translations of Byzantine Attempts to Prove the Fifth Postulate

There exist Arabic translations of two attempts, undertaken respectively by the Byzantine scholars Aghānīs (fifth century) and Simplicius (sixth century) to prove the parallel postulate. G. B. Petrosyan has advanced the thesis that Aghānīs is the Greek transcription of the name of *Aghan*, the Armenian advocate of enlightenment, connoisseur of Greek culture, and teacher of the famous Armenian scholars of the second half of the fifth century Ghazar Farpetzi and Vaghan Mamikonyan *[419, pp. 153–154]*. The extant version of Aghānīs' proof is due to Simplicius, who refers to him as his friend. We note that Simplicius was a student of Damaskius, himself a student of Proclus. Simplicius' version of Aghānīs' proof is found in the commentaries on Euclid's *Elements* by the 10th-century Iranian mathematician Abū l-ʿAbbās al-Fadl ibn Hātim al-Nayrīzī. The title of the chapter of al-Nayrīzī's commentaries devoted to this exposition is *Premises and propositions of Simplicius and Aghānīs indispensable for the twenty-ninth proposition of the first book [419, pp. 154–159].*[2] Here Simplicius points out that the proof of proposition 29 in book I uses the fifth postulate and says about this postulate that "it is not one of the generally accepted assertions." There follow the words of Simplicius quoted previously and his critique of Ptolemy's "proof." The critique is that whereas the proof is based on propositions 13, 15, and 16 of book I of the *Elements*, the postulate is not applied before proposition 29. Then Simplicius says that "As regards our friend Aghānīs, he does not see why this assertion should be taken as a postulate at a time when it requires a proof." Then Simplicius gives Aghānīs' "proof" of the fifth postulate in "its proper form."

The proof consists of four propositions inserted after proposition 26 in book I of the *Elements*. They are based on Posidonius' definition of parallels:

We define parallel lines as lines that are in the same plane, and such that if they are indefinitely produced in both directions, the distance between then stays the same; the distance is the shortest line joining them, the same is said of other distances *[419, p. 155].*

In proposition 1 it is proved that "if two lines are parallel, then the distance between them is perpendicular to each of them"; in proposition 2, that "if a straight line falling on two straight lines is perpendicular to each of them, then the lines are parallel and the perpendicular is the distance between them." As for propositions 3 and 4 *[419, pp. 155–156],* they are, respectively, propositions 29 and 27–28 of book I of Euclid's *Elements* (discussed previously). We note that while proving his proposition 3, Aghānīs proves the existence of a quadrilateral with four right angles—the key point of most medieval proofs of the parallel postulate. We note also that whereas Euclid's own proof of propositions 27 and 28 does not rely on the fifth postulate, Aghānīs' proof invokes his propositions that are based on an assertion equivalent to that postulate. After these four propositions (which he regards as propositions 27–30 of book I of the *Elements*) Aghānīs (to quote Simplicius) proves propositions 31–34 that coincide, respectively, with propositions 31, 34 (the part that asserts the equality of the opposite sides in a parallelogram), 30, and 33 in book I of the *Elements*. In proposition 35, Aghānīs uses the existence of a parallelogram (established in his proposition 34) to prove the parallel postulate in the case when the transversal is perpendicular to one of the two given lines (in this formulation, the statement of the postulate is that "a perpendicular line and an oblique line intersect"). Aghānīs considers lines *AB* and *CD* and a transversal *EG* perpendicular to *AB* (Figure 14a). From a point *F* on the line *CD* he drops a perpendicular *FI* to the transversal *EG*. Then

---

[2] A French translation of this chapter *[248a, pp. 127–136]* appeared in 1986.

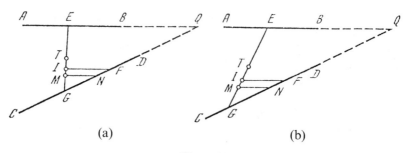

Figure 14

he bisects EG enough times to obtain a division point M between I and G (in Figure 14a the point M is obtained by dividing the segment EG into four equal parts), erects a perpendicular MN at M, extends it until it intersects the line CD, and constructs the segment GQ of the straight line CD that is the same multiple of the segment GN as the segment EG is of the segment MG. The rest of Aghānīs' construction is equivalent to the construction of a rectangle with side EG and diagonal GQ.

The Aghānīs construction is also applicable if the transversal is not perpendicular to either of the given straight lines. In those cases we have, in place of a rectangle, a parallelogram (Figure 14b). The fact that halving the segment EG enough times will always yield a segment GM smaller than the segment GI is a consequence of the Eudoxus-Archimedes axiom (mentioned previously) which asserts that given two unequal segments, areas, or volumes, there is always a natural number n such that n times the smaller magnitude exceeds the corresponding larger magnitude. This axiom was known to Eudoxus. It is given by Euclid as definition 4 of book V of the *Elements*, considerably later in the text than the theory of parallels presented in book I.

It is possible that the "proofs" of Posidonius and Archimedes consisted of the same stages and, in particular, also made use of the Eudoxus-Archimedes axiom. It may be that these "proofs" were more complete in the sense that they involved parallelograms rather than rectangles or that the general case of the fifth postulate was deduced from the case considered by Aghānīs.

Al-Nayrīzī does not give Simplicius' proof of the parallel axiom, but a short version of that proof is contained in a letter written by ʿAlam al-Dīn Qayṣar al-Ḥanafī (1178–1258) to Naṣīr al-Dīn al-Ṭūsī. In most cases, the manuscript versions of this letter were attached to the manuscripts of al-Ṭūsī's treatise on parallel lines (which will be discussed later). In that letter—published by A. I. Sabra *[494]*—al-Ḥanafī writes:

> One thing that may be proposed for your elevated consideration is what occurred to me regarding a proposition which Simplicius, in his commentary to the premises of the book of *Elements*, mentioned among lemmas of the famous proposition, which is this: If a straight line falling on two straight lines makes the two interior angles on one

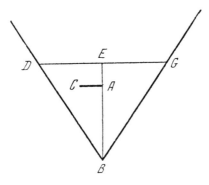

Figure 15

side equal to less than two right angles, then the two lines, if produced on that side, will meet.

He said that every angle can have infinitely many chords of increasing magnitudes each of which cuts off equal [segments] from the two lines containing that angle.

He used this in [the following]: If the line *AB* falls on the two lines *BD*, *AC*, and the angle *CAB* is a right angle, and the angle *ABD* (Figure 15) is acute, then the two lines *AC*, *BD* will meet on the side of *C*, *D*.

For he constructed on the point *B* of the line *AB* an angle *ABG* equal to the angle *ABD*. Then there are infinitely many chords of increasing magnitudes which subtend the angle *DBG*. And one of the chords will fall beyond point *A*—such as the chord *GED*. Therefore the two angles *A*, *E* being right angles, the line *AC*, if produced, will not meet the line *ED*. It will therefore meet the line *BD* *[494, pp. 8–9, 19–20]*.

What Simplicius' *petitio principii* comes down to is that he thinks that the fact that in a right angle one can draw infinitely many chords that cut off from its sides equal segments implies that for every interior point of the angle there exist chords with this property that fall "outside" the point in question. Although it is true that if the parallel axiom does not hold, that is, in Lobačevskian geometry, it is still possible to draw infinitely many chords in a right angle that cut off equal segments from its sides, it is also true that there are interior points of the angle that are "outside" all such chords; such points are the interior points of the angle on the other side of the line that is parallel in the sense of Lobačevskiĭ to both sides of the angle. This means that Simplicius' assumption is equivalent to the parallel axiom.

In the Bodleian Library at Oxford University there is a manuscript of commentaries on Euclid by Muḥyī al-Dīn al-Maghribī (d. ab. 1290), an associate of the Marāgha observatory, that contains a proof of the parallel postulate in four propositions. In the first of these propositions the author proves Simplicius' assertion on the existence of infinitely many "chords" of an angle (these are chords of arcs drawn with appropriate radii from the vertex of

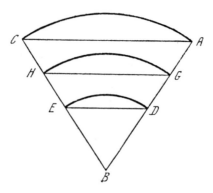

Figure 16

the angle as center; Figure 16). The second proposition is identical with the proposition of Simplicius quoted by al-Ḥanafī, and the third and fourth propositions are proofs of the parallel axiom in two special cases, namely, when a straight line falling on two straight lines makes acute angles with these lines, and when it makes an acute angle with one of them and an obtuse angle with the other. A. I. Sabra *[494, p. 7]* believes that the author of the first two propositions is undoubtedly Simplicius and the author of the last two is either Muḥyī al-Dīn al-Maghribī himself or another 13th-century author. We will return to this matter later.

We note that in his commentaries on Euclid, al-Nayrīzī attributes to Simplicius the following proposition:

if lines in the plane are such that the distances between them are unequal, then they are not parallel and, since they are not parallel, they meet *[419, p. 159]*.

This proposition is very similar to the principle attributed by Khayyām to Aristotle.

Al-Jawharī's Theory of Parallels

The first preserved attempt at a proof of the parallel axiom by a scholar of the medieval East was made by ʿAbbās ibn Saʿīd al-Jawharī, who flourished in Baghdad in the first half of the ninth century. Al-Jawharī's investigation was included in his *Improvement of the book the "Elements"* (Iṣlāḥ Kitāb al-Uṣūl) and was preserved in the treatise of Naṣīr al-Dīn al-Ṭūsī on parallel lines.

Al-Ṭūsī begins his exposition of al-Jawharī's attempt at a proof as follows:

As for al-Jawharī ... he is the author of *Improvement of the book the "Elements"* in which he introduced corrections in all assumptions and propositions of that book. In what he said on this question there was the following principle: if, given any two unequal lines, one takes away from the longer one half [of the shorter line], then half of the half many times, then the number of times can be such that of the half of the longer one there remains a line that is shorter than the shorter line. The first of his six propositions is the twenty-eighth [proposition of the book 1 of the *Elements*]. In the first proposition he set forth that which the author of the *Elements* set forth in the twenty-seventh [proposition], and added to it [one other proposition]; the last of these propositions is the thirty-third. Another proposition added by him is the twenty-third proposition of the *Elements* where it is shown that if one draws from any point three straight lines in different directions, then the three resulting angles are together equal to four right angles. The six propositions are added to the *Elements* after the twenty- seventh proposition *[593, p. 501]*.

The principle formulated by al-Jawharī is contained in proposition 1 of book X of Euclid's *Elements* (which is proved with the help of definition 4 of book V) that is, the Eudoxus-Archimedes axiom. Apparently, al-Jawharī added proposition 23 of book I of the *Elements* before Euclid's proposition 23 (which deals with the construction of a given angle at a given point of a given line), so that Euclid's propositions 23–26 are al-Jawharī's propositions 24–27. According to al-Ṭūsī, the six new propositions inserted by al-Jawharī after proposition 26 of the *Elements* are as follows.
Proposition 28:

If a straight line falling on two straight lines makes the [alternate] angles equal to one another, the straight lines will be parallel to one another, and if they are parallel then the distances from any point on one straight line to the corresponding point on the other straight line are the same.

Proposition 29:

If in a triangle two of its sides are halved and the midpoints are joined by a line, then [the third] side of the triangle is equal to two such lines.

Proposition 30:

In every angle one can draw infinitely many bases.

Proposition 31:

If we draw in an arbitrary angle a line passing through its vertex and intersecting a base of a triangle and lay off on each side of the angle a line equal to the side of the obtained triangle, then "this line [that is, the new base] cuts off from the line dividing the given angle a line equal to two lines [that is, twice the line] drawn from the given angle to the base of the triangle".

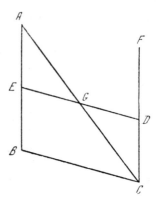

Figure 17

Proposition 32:

If we draw in an arbitrary angle a line passing through its vertex and choose on that line an arbitrary point, then "through this point we can draw in both directions a line which is a base of the given angle". Proposition 33 is a proof of the parallel postulate [593, pp. 501–507].

The first half of al-Jawharī's proposition 28 is the same as proposition 27 of the *Elements*. In the second half al-Jawharī attempts to prove that the equality of the alternate angles formed by a straight line falling on two other straight lines implies that the latter are equidistant; what he is in fact able to prove is that the two straight lines are symmetric with respect to the midpoint of the segment of the third straight line contained between them. It is this result that al-Jawharī uses to prove his proposition 29: he halves the sides AB and AC of the triangle ABC at E and G (Figure 17), draws the midline EG, constructs the angle ACF equal to the angle BAC, and extends the line EG to the point D on the side of CF of that angle. By proposition 28, the line CF is parallel to the line AB. Al-Jawharī proves the equality of triangles AGE and GCD which implies that $EG = GD$ and $ED = 2EG$. Therefore, in view of his proposition 28, which he supposes established, al-Jawharī concludes that the lines EB and CD are equidistant and that $ED = BC$, which proves the required equality $BC = 2EG$. Al-Jawharī's propositions 28 and 29 hold in Euclidean but not in Lobačevskian geometry. On the other hand, al-Jawharī's proposition 30 does not depend on the parallel postulate; we saw that it was used by Simplicius. Al-Jawharī's proposition 31 holds only in Euclidean geometry; its proof depends in a fundamental way on al-Jawharī's proposition 29. Al-Jawharī's proposition 32 also holds only in Euclidean geometry and coincides with the result used by Simplicius. Unlike Simplicius, al-Jawharī proves it by using his proposition 31—that is, by ultimately basing the argument on proposition 28, which is valid only in Euclidean geometry. With the help of his proposition 32 al-Jawharī gives, in proposition 33, the following proof of the parallel postulate.

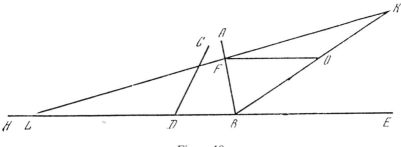

Figure 18

If one extends two lines from a line in the direction of angles less than
two right angles, then they intersect on that side. Example. The lines AB
and CD are drawn from the line BD in the direction of the angles ABD
and CDB that are less than two right angles (Figure 18). I claim that if
one extends them in their direction then they intersect. [Proof]. We
extend the line BD in its direction to the points E and H and lay off BF,
equal to BD, as proved in proposition 3. By assumption, angles ABD
and CDB are [together] less than two right angles. Hence angle ABE is
greater than angle CDB. At B on line BA we construct the angle ABK
equal to the angle CDB. Then angles ABD and ABK are [together] less
than two right angles, that is angle KBL. Through the point F we pass
the line KL—a base of angle KBL, as proved in proposition 32. Then,
as proved in proposition 16, the angle KFB, an exterior angle of the
triangle FBL, is greater than the interior angle FBL. Therefore we
construct at F on the line BF the angle BFO equal to the angle FBL. As
we know, angle KBA is equal to angle CDB. Therefore, angles BFO
and OBF are equal to angles ABD and CDB, respectively, and BF was
laid off equal to BD. If we superimpose BD on BF, equal to it, then
angle CDB is superimposed on angle OBF, equal to it, and angle ABD
is superimposed on angle BFO, equal to it. Hence lines BA and DC, if
extended in their directions, are superimposed on lines FO and BO
which intersect at the point O. This is what we wished to prove *[593,
pp. 507–508]*.

In spite of the crude logical error in the first proposition, al-Jawharī's proof
is of considerable interest, for he was the first to prove the possibility of
drawing a line through an arbitrary interior point of an angle that intersects
both of its sides and to use this result to prove the parallel postulate.

Thābit ibn Qurra's Theory of Parallels

Thābit ibn Qurra devoted two treatises to attempts to prove the parallel
postulate. One of them is called *Book on a proof of Euclid's well-known
postulate* (Maqāla fī burhān al-muṣādara al-mashhūra min Uqlīdis) or *The*

book whose theme is that if a straight line falling on two straight lines makes the angles on the same side together less than two right angles, the two straight lines, if produced indefinitely on that side, meet (Kitāb fī annahu idhā waqa'a Khaṭṭun mustaqīmum 'alā khaṭṭayn mustaqīmayn fa-ṣayyara al-zāwiyatayn allatayn fī jihatin wāhidatin aqalla min qā'imatayn fa-inna al-khaṭṭayn idhā ukhrija fī tilka al-jiha iltaqayā). *[493, pp. 28–32]*. The title of the second treatise is an abbreviation of the second title of the first treatise: *The book whose theme is that two lines drawn so as to form angles less than two right angles meet* (Maqāla fī anna al-khaṭṭay idhā ukhrijā 'alā zāwiyatayn aqall min qā'imatayn iltaqayā) *[493, pp. 19–17]*.

Proposition 1 of Ibn Qurra's first treatise is formulated in the same way as the first proposition of al-Jawharī. His proof also involves *petitio principii*, but its manner is less crude than al-Jawharī's. Ibn Qurra makes the naive assumption that if two lines diverge on one side then they necessarily converge on the other. This assertion is equivalent to Euclid's parallel postulate; in Lobačevskian geometry there are straight lines that diverge on both sides of their common perpendicular. Ibn Qurra formulates his first proposition as follows:

> If a straight line falls on two straight lines and the two alternate angles are equal to one another, then those two lines do not converge or diverge on either side *[493, p. 28]*.

In proposition 2 he proves the converse, which coincides with the first part of proposition 29 of book I of the *Elements*:

> If a straight line falls on two straight lines which do not converge or diverge on either side, then the two alternate angles are equal to one another *[493, p. 29]*.

The proof is by contradiction: the proposition that the angles are unequal is made to contradict what was "proved" in proposition 1. From these two propositions Ibn Qurra deduces in proposition 3 the existence of a parallelogram:

> If the extremities of two straight and equal lines which do not converge or diverge are joined by two straight lines, then these also are equal and do not converge or diverge *[493, p. 29]*.

In proposition 4 Ibn Qurra, like al-Jawharī, considers a midline of a triangle and shows that it is equal to half of the base and is parallel to it:

> In every triangle [if] two of the sides are each divided into two halves, and the two points at which they have been divided are joined by a straight line, then [this straight line] will be half the other [i.e., third] side and it does not converge with or diverge from it *[493, p. 30]*.

Ibn Qurra notes that using the same method one can prove that if one divides the sides of a triangle into an arbitrary number of parts and joins the division points by means of straight lines, then each of the resulting lines is the

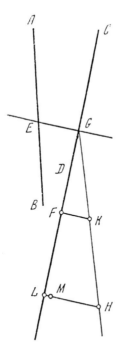

Figure 19

same part of the base as the segment it cuts off is of the side. Finally, in proposition 5, Ibn Qurra proves the parallel postulate. Using the axiom of Eudoxus-Archimedes, Ibn Qurra considers the lines AB and CD and the line EG falling on them such that the angles BEG and DGE are together less than two right angles (Figure 19) and claims that the lines AB and CD meet when extended in the direction of the points B and D. For proof, Ibn Qurra draws from the point G a line GH that "does not converge with or diverge from the line AB," chooses an arbitrary point F on GD and draws from F to GH the line FK that "becomes neither closer to nor more distant from EG." If $FK < EG$, then one lays off $FL = GF$ and $KH = GK$. Then the line LH is twice FK and also "does not converge with or diverge from the line EG." If $LH < EG$, then one continues the process until one obtains a line $\geq EG$. Suppose LH is such. On LH one lays off $MH = EG$. Then the lines GE and HM are equal and "do not converge or diverge." Hence the lines joining their ends are also equal and "do not converge or diverge." Therefore, says Ibn Qurra,

EB, if produced in a straight line on the side of B, will proceed to M ... thus it is necessary that it should meet a point on the line CD before it meets the point M. Therefore AB, CD, if produced on the side of B, D, meet. And that is what we wanted to prove [493, p. 31].

Ibn Qurra's second treatise begins with a long introduction that justifies the possibility of applying motion in geometry. As is well known, Aristotle condemned the use of motion in geometry and said that

the objects of mathematics, except those of astronomy, are of the class of things without movement [29, vol. 8, 989^b].

Euclid also tried to avoid using motion in his *Elements*. Ibn Qurra points out that measurement of magnitudes is itself impossible without their translation and superposition. He writes:

[Therefore] the principles of many demonstrations of those demonstrable first elements among geometrical propositions consist in the use of the said operation—I mean moving one of the two things to be measured by one another and pushing it from its place and transferring it in our imagination without changing its shape by movement so as to place it with its shape [unchanged] upon that which is to be measured by it.
Euclid was obliged to do the same thing in the demonstration of Proposition 4 of Book I of his work on *Elements*, and in the demonstration of Proposition 8 of [the same Book]; for these are two of the oldest elements, knowledge and demonstration of which are premised and taken as basis for other [propositions] [493, p. 20].

Ibn Qurra goes on to introduce "a certain simple motion," that is, a translation, and formulates the following "premiss":

Consequently, as a premiss, I have started from something known regarding the solid—which is this:
If any solid is imagined to move as a whole in one direction with one simple and straight movement, then every point in it will have a straight movement and will thus draw a straight line on which it will pass [493, p. 21].

Then Ibn Qurra proves the following proposition:

If two straight lines are in the same plane, and two straight lines equal to one another are drawn across them in such a way as to contain with one of the first two lines two equal angles on one side, then any two perpendiculars falling on that line from two points on the other will be equal.
Let the straight lines *AB*, *CD* be in the same plane; let the straight lines *AC*, *EF* be drawn across them and let them be equal; and let the angles *ACD*, *EFD* be equal.
I say that any two perpendiculars falling on the line *CD* from two points on the line *AB* will be equal (Figure 20) [493, p. 21].

For proof Ibn Qurra imagines "that a solid surrounds the line *AC* so as to be cut by *CG* which is part of [the line] *CD*" and imagines further that the

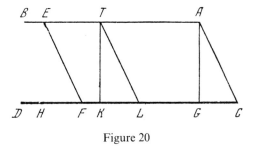

Figure 20

solid has moved as a whole from the side of C to that of D—this being a straight simple movement along CD.

Let analogues to the lines AC, CG be drawn in the solid and let them remain and preserve their shape in it. Thus the line that is drawn in the solid as an analogue to AB [sic; read AC] will not be situated along the solid's movement; but the other line drawn in the solid as an analogue to CG will be situated along that movement.

Therefore, the analogue drawn for the line CG will throughout the solid's movement pass along CD and will always be placed on it [493, pp. 21–22].

Ibn Qurra goes on to say:

If we imagine that the point C on the analogue drawn for the line AC has reached the point F as a result of the solid's movement, then the position of the analogue drawn in the solid for the line CG will be that of FH, for it has moved along CD.

But the angle EFH is equal to the angle ACG; therefore the analogue to the line AC in the solid will fall on FE when the point C on it comes to F.

And since AC is equal to FE, it coincides with it, the point A in AC falling on the point E in FE.

Therefore, the point A in the solid passes to the point E as a result of the movement of the solid, by its passage draws a straight line, for this is the case with every point in the solid. Therefore, the passage of the point A will be along the line AEB since there is no other straight line passing through the points A, E [493, p. 22].

Then Ibn Qurra marks off on the line AEB an arbitrary point T and drops from it a perpendicular TK to CD, considers separately the case when the angle ACD is a right angle and the case when it is not, and proves that in both cases the perpendicular from A to CD—which is AC in the first case and AG in the second—coincides with the perpendicular TK when the point A passes the position T.

Therefore, the perpendicular *TK* is equal to the perpendicular *AG*, and also equal to any perpendicular falling on *CD* from a point on the line *AB*. All these perpendiculars are therefore equal. And that is what we wanted to prove *[493, pp. 22–23]*.

Actually, this proposition uses kinematic considerations to prove the existence of a rectangle. The next four propositions also deal, basically, with the properties of rectangles, although the first two discuss isosceles trapezoids:

In any quadrilateral plane [figure] if two angles on one side are equal and the two sides joining that side are equal, then the remaining two angles are equal *[493, p. 23]*.

In any quadrilateral plane if two angles on one side are equal, and the two other angles are equal, then the two sides joining its first side are equal *[493, p. 23]*.

If any two straight lines are in the same plane, and if two perpendiculars drawn from two points on one of them to the other are equal, then they will also be perpendicular to the first line, and all perpendiculars falling from each one of the two lines upon the other—whatever the points they are drawn from—will be perpendicular to the companion line and will be equal among themselves and equal to the first two perpendiculars *[493, p. 24]*.

If any two straight lines are drawn from the extremities of a straight line in the same plane so as to contain with it two right angles, then any perpendicular drawn from a point on one of the two lines to the other will also be perpendicular to the first, and will be equal to the line from whose extremities the two lines have been drawn *[493, p. 25]*.

Ibn Qurra goes on to prove the following proposition:

If a straight line falling on two straight lines in the same plane is perpendicular to both, then any straight line cutting the two lines will make the alternate angles equal to one another and the exterior angle equal to the opposite interior angle *[493, pp. 25–26]*.

Finally, implicitly using the Eudoxus-Archimedes axiom, Ibn Qurra proves the parallel axiom.

From the extremities of the straight line *AB* let us draw the two straight lines *AC*, *BD* in the same plane, and let the angles *BAC*, *ABD* be together less than two right angles (Figure 21).

I say that the lines *AC*, *BD* meet when produced in the direction of *C*, *D*.

Demonstration: One of the angles *BAC*, *ABD* must be less than a right angle. Let it be the angle *ABD*.

Then from *A* let us draw *AE* perpendicular to *BD*, mark a point *F* on *AC* at random and from it draw *FG* perpendicular to *AE*.

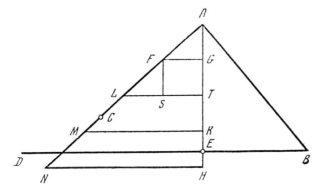

Figure 21

Then the lines *AG*, *AE* are finite, and *AE* is longer than *AG*, and therefore it is possible to multiply the smaller, viz. *AG*, until its multiple becomes greater than *AE*.

Let *AH* be the multiple which is greater than *AE*.

From *GH* let us cut off equals to the line *AG*, viz. *GT*, *TK*, *KH*.

And from the line *FC* let us cut off equals to the line *AF* as many times as *GT*, *TK*, *KH*, viz. *FL*, *LM*, *MN*. If *FC* is less than is sufficient we prolong it until it is enough.

I say that the line *AG* has cut the line *BD*.

Demonstration: From the point *T* let us draw *TS* perpendicular to *AE*, and from *F* we draw *FS* perpendicular to this perpendicular.

Then *FS* is also perpendicular to *FG* and equal to *GT*.

But *GT* was equal to *AG*.

Therefore, *FS* is equal to *A*.

And it is evident that the line *FL* will fall outside the (area) between *FS*, *GT*—for the angle *GFS* is right, and the angle *AFG* is less than a right angle since the angle *AGF* is right and there cannot be two right angles in the same triangle.

Again, the line *FG* has fallen on the two lines *AT*, *FS* in such a way as to be perpendicular to both, and the straight line *AC* has also fallen on them.

Therefore, the exterior angle *LFS* is equal to the interior and opposite angle *FAG*.

Therefore these two angles in the triangles *AFG*, *FLS* are equal.

But we have shown that their sides *AG*, *FS* are also equal, and side *AG* in one of the two triangles is likewise equal to the side *FL* in the other.

Therefore, the two bases are equal, and the remaining angles are respectively equal to one another.

And the angle *FSL* being equal to the angle *AGF*, and the angle *AGF* being a right angle, then the angle *FSL* is right.

But the angle *FST* was also right. Therefore the lines *TS, SL* are joined in such a way as to make one straight line. Therefore the line that joins the points *T, L* is the line *TSL* itself, and it is perpendicular to *AH*.

And likewise we show that the straight line *KM* joining the points *K, M* is perpendicular to *AH*, and that *HN* is perpendicular to *AH*. Therefore the angle *AHN* is a right angle.

But the angle *BEH* also was a right angle, being equal to the angle *AED*.

Therefore on the straight lines *HN, BD* a straight line *AEH* has fallen making the alternate angles equal.

The two lines are therefore parallel and will not meet even if produced indefinitely.

But one of them, *HN*, has been met by the line *AC* in *N*, and therefore *AC* has gone across to the other side of *BD*.

Therefore the line *AC* has met the line *BD*, cutting it and crossing it. And that is what we wanted to prove *[493, pp. 26–27]*.

Attempts to Prove the Parallel Postulate in the Tenth Century

Al-Nayrīzī, whose commentaries on Euclid contained an account of Aghānīs' attempt to prove the parallel postulate, was himself the author of a treatise devoted to such a proof—*Treatise on a proof of the well-known postulate of Euclid* (Risāla fī bayān al-muṣādara al-mashhūra li-Uqlīdis).

E. S. Grigoryan *[212a]* investigated the incomplete manuscript of this treatise kept in the Paris National Library (Ar. 2467/7). Another manuscript of this treatise, kept in the former Prussian State Library and described in its catalogue, was lost during the second world war (according to information given to the author by the management of the library). The beginning of this treatise is the same as that of the Paris manuscript. The complete text of this treatise was published in Arabic by Abū'l-Qāsim Qurbānī *[445, pp. 86–87]*. There is a Russian translation *[380]*. An analysis of the treatise is found in *[488, pp. 42–45]*.

Like the second treatise of Ibn Qurra, al-Nayrīzī's begins with a philosophical justification of the existence of equidistant lines, but the arguments of al-Nayrīzī, to the effect that "by its nature, equality precedes difference," are far less convincing than the kinematic argument of Ibn Qurra. Next come seven propositions:

The distances between equidistant straight lines are perpendicular to both straight lines;

Each of the perpendiculars to two equidistant straight lines is shortest;

Each of the perpendiculars to one of two equidistant straight lines is perpendicular to the other;

If there is a [common] perpendicular to two straight lines, then they are equidistant;

If one drops two perpendiculars EH and GF from one of two equidistant straight lines AB and CD to the other then the distance HF between their feet on the second straight line is equal to the distance EG between their feet on the first straight line;

If a straight line falls on two equidistant straight lines AB and CD, then the two interior angles on the same side are [together] equal to two right angles;

If a straight line AC falls on two straight lines AB and CD and if the sum of the two interior angles CAB and ACD on the same side is less than two right angles, then they meet on that side [380].

We see that al-Nayrīzī's first four propositions deal with common perpendiculars to two equidistant straight lines whose existence is al-Nayrīzī's point of departure; in proposition 5 a property of such perpendiculars is used to deduce the existence of a rectangle. In proposition 6 it is shown that if two equidistant straight lines are cut by a third, then the sum of the interior angles on the same side is equal to two right angles, and in proposition 7 the parallel postulate is proved. The whole line of al-Nayrīzī's argument is very close to that in the second treatise of Ibn Qurra; in particular, like Ibn Qurra, he uses the Eudoxus-Archimedes axiom in his proof of the parallel postulate.

The treatise of al-Nayrīzī is mentioned in Khayyām's commentaries on Euclid together with the treatises of Abū Ja'far al-Khāzin (d. ab. 965) and Abū 'Abdallāh al-Shannī (10th–11th century), who do not seem to be mentioned in other sources. Of these treatises Khayyām wrote as follows:

As for such later scholars as al-Khāzin, al-Shannī and al-Nayrīzī, who tried to prove this, none succeeded in giving a rigorous proof; each of them based himself on something that is no simpler an assumption than the one to be proved [272, pp. 114–115].

The treatise *On the intersection of two straight lines issuing from the ends of a straight line under angles that are [together] smaller than two right angles* (Fī iltiqā' al-khaṭṭayn al-mustaqīmayn al-khārijayn min ṭarafay khaṭṭin mustaqīmin 'ala aqall min zāwiyatayn qā'imatayn) by the Christian priest Yuḥannā ibn Yūsuf al-Hārith ibn al-Biṭrīq al-Qāṣṣ (d. ab. 980), dedicated to the sultan Ṣayf al-Dawla (d. 950), has not reached us. It is mentioned by Yuḥannā ibn Yūsuf in his *Book on rational and irrational magnitudes* (Maqāla fī l-maqādīr al-munṭaqa wa l-ṣumm)[3] and by al-Ḥanafī in his letter (quoted previously) to al-Ṭūsī, in which this treatise is mentioned in a list of treatises devoted to the theory of parallels together with one of the treatises of Ibn Qurra and the treatise of Ibn al-Haytham, to be discussed later.

---

[3] This treatise was studied by G. P. Matvievskaya [358, pp. 213–216].

Ibn Sīnā's Theory of Parallels

The features of the exposition of the theory of parallels common to Aghānīs, Ibn Qurra, and al-Nayrīzī also occur in the encyclopedic treatise of the eminent philosopher and physician Abū ʿAlī ibn Sīnā (980–1037), entitled *Book of knowledge* (Dānish-nāma) *[33; 540]*, who devotes to this theory the second section of the chapter on geometry. This chapter opens with the following words:

> Disjoint lines can be disposed so that the end of one is inclined toward the other; if extended in that direction they will intersect: they will not intersect if extended in the opposite direction *[540, p. 21]*.

We see that this statement is very close to the "principle" mentioned previously, which Khayyām ascribed to Aristotle.

Continuing, Ibn Sīnā states that

> disjoint lines ... can be disjoint so that the distances between their ends are equal. If one extends a perpendicular erected at one of these lines to the other, then this perpendicular will also be perpendicular to the other line. In fact, if it were not perpendicular to that other line, then one of the angles would be acute and the other obtuse, the ends on the side of the obtuse angle would be further apart and the ends on the side of the acute angle would be closer together *[540, pp. 21–22]*.

Two such lines are called parallel.

Ibn Sīnā's argument that a perpendicular to one of these lines is also perpendicular to the other is based on the refutation of the proposition that the second angle is acute or obtuse. In the sequel we will often encounter similar arguments in connection with two perpendiculars. Ibn Sīnā goes on to prove by contradiction that

> if a line intersects two other lines and two interior angles on the same side are [together] equal to two right angles, then the two lines are parallel *[540, p. 23]*,

and, finally, that

> if a line intersects two other lines and two interior angles on the same side are together less than two right angles then the two lines, if produced on that side, will intersect.
>
> This [says Ibn Sīnā] is because one of these lines is inclined toward the other and in consequence of this inclination they will intersect. In fact, if one of them did not incline toward the other, then they would be parallel, and if they were parallel, then, as we showed earlier, the two indicated angles would be equal to two right angles *[540, pp. 23–24]*.

Thus, essentially, Ibn Sīnā justifies the parallel axiom by means of the indicated principle. However, in defining parallel straight lines as equidistant, Ibn Sīnā assumes a proposition equivalent to the fifth postulate. This results in a remarkable simplification of his argument.

Ibn al-Haytham's Theory of Parallels

The famous Egyptian physicist, mathematician, and astronomer Abu ʿAlī Ibn al-Haytham (Alhazen, 965–1041) considered the theory of parallels in both of his commentaries on Euclid's *Elements*, namely, in his *Commentary on the premises to Euclid's book the "Elements"* (Sharḥ muṣādarāt kitāb Uqlīdis fī l-Uṣūl) and in his *Book on the resolution of doubts in Euclid's book of Elements* (Kitāb fī ḥall shukūk kitāb Uqlīdis fī l-Uṣul wa'sharḥ maʿānīhi). Only the second of these two works has been published *[237]*. The first one was studied by B. H. Sude in her thesis *[569]*.[4] The commentaries in the first book pertain to the introductory parts of the *Elements* and those in the second book to the propositions themselves. The key issue of Ibn al-Haytham's first work is the same as that of the second treatise of Ibn Qurra, that is, the proof of the existence of a rectangle. In turn, this proof is based on the proposition of the existence of equidistant straight lines, established with the help of kinematic arguments similar to those used by Ibn Qurra. The proof of the existence of a rectangle consists in the investigation of a quadrilateral with three right angles and of the three hypotheses concerning the fourth angle, and in the refutation of the hypotheses that the fourth angle is acute or obtuse. This quadrilateral and the three hypotheses pertaining to its fourth angle, of which the "right-angle hypothesis" holds in Euclidean geometry and the "acute-angle hypothesis" holds in Lobačevskian geometry, have played an important role in the history of non-Euclidean geometry. This quadrilateral is often referred to as the *"Lambert quadrilateral"*, in honor of the 18th-century mathematician J. H. Lambert who also considered it. After establishing the existence of a rectangle Ibn al-Haytham first proves the parallel postulate for the case of a perpendicular line and an oblique line, using, like Aghānīs before him, the Eudoxus-Archimedes axiom. Then he proves the parallel postulate for the case when the angles between the two straight lines and the transversal are acute and, finally, for the case when one of them is acute and the other obtuse. Ibn al-Haytham begins with a critique of Euclid's definition of parallel lines:

> Euclid stated: "Parallel straight lines are coplanar lines such that if produced indefinitely in both directions they do not intersect in either direction." Thus the lines are represented as coplanar and noninter-secting if produced in both directions; which means that the lines can be extended constantly and simultaneously in both directions. But it is

---

[4] In 1986 there appeared French translations of parts of these two books devoted to the theory of parallels *[248a, pp. 161–184]*.

impossible to imagine such constant increase that reaches no end; there is no way of imagining this, for whatever can be imagined is finite, and the lines dealt with here are depicted as lines of finite magnitude.

Ibn al-Haytham goes on to define parallel lines in a way that extends Ibn Qurra's definition:

> We further imagine a second bounded straight line that forms a right angle with the first in the same plane in which the first line is located. We further imagine this line to be moving with one of its ends along the first straight line in one direction. Its motion is one simple motion, that is, without change of motions, not made up of motion and rest, one motion, without bending. If in this translation, throughout the duration of the motion, the [translated] line remains perpendicular to the line in the plane, that is, the first line located in it, then, during the time when it is in motion, the end of that perpendicular line will describe a straight line perpendicular to it; and one can imagine this line as well as this kind of motion [274, p. 743].

In this way one can obtain two coplanar straight lines which when produced indefinitely in both directions do not intersect in either direction, the distances between these lines in both directions are always the same as they grow in either direction, and it is impossible that they should intersect in any place. Thus parallel lines exist and can be imagined in this way [274, p. 748].

Ibn al-Haytham proves the parallel postulate as follows:

> As for [the assertion that], if a straight line falling on two straight lines makes the interior angles on the same side [together] less than two right [angles], then the two straight lines, if produced indefinitely in that direction, will meet, this assertion is the innermost of all that we present here. It requires proof like most assertions. In the proof one should use those propositions in the book in whose proofs this premise is not used.... This premise must be preceded [by the following]: if one draws at the ends of a bounded straight line two straight lines that enclose together with the first line right angles, then each of the perpendiculars dropped from [the points] of one of these lines to the other is equal to the first line, that which encloses with these two lines right angles, and every perpendicular dropped from [the points] of one of the indicated lines to the other encloses with the line from which it is dropped a right angle.
>
> *Example.* Such is the line AB, from its ends A and B are drawn lines AC and BD so that each of the angles CAB and DBA is a right angle. Further, we take a point C on the line AC and drop from it the perpendicular CD to the line BD. I say that the line CD is equal to the line AB (Figure 22).

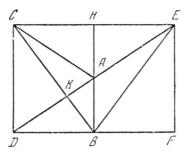

Figure 22

The proof of this, that is, that the opposite is impossible. Suppose, if possible, that they are unequal. If CD is unequal to AB then it is either greater than or smaller than it. Suppose it is greater than it. We extend the line CA in its direction toward A, let that be AE, we also extend BD in its direction toward B, let that be BF. We lay off [the line] AE equal to AC. We drop from the point E the perpendicular to the line BF, let that be EF. We draw the lines CB and BE. Since the line CA is equal to AE and the line AB is common, that is, the lines AC and AB are [respectively] equal to the lines AE and AB, the angles CAB and EAB are equal as two right angles, the base CB is equal to the base EB, and the triangle CAB is equal to the triangle EAB. Hence the base CB is equal to the base EB and the remaining angles [of the triangles] are equal [and in particular], the angle CBA is equal to the angle EBA. But the sum of the two angles ABD is equal to the sum of the two angles ABF [each being a right angle]. Hence the remaining angle CBD is equal [to the remaining angle] EBF. Angle CDB is equal to angle EFB as two right [angles]. Hence triangle CDB is equal to triangle EFB, for two angles of one of them are [respectively] equal to two angles of the other and the sides CB and BE of these triangles are equal. Hence the line CD is equal to EF. But [by assumption] CD is greater than AB. Therefore EF is also greater than AB. Imagine the line EF moving along the line FB so that during this motion the angle EFB stays a right angle throughout the time of the motion and EF is always perpendicular to [the line FB]. If, in the motion of the line EF, the point F coincides with the point B, then the line EF will be superposed on the line BA; the reason for this is that the angles EFB and ABD are equal, for each of them is a right angle. Now if the line EF is superposed on the line BA, the point E will be external to the line AB and the excess will be on the side of the point A for, as was shown, the line EF is greater than the line AB as a result of the equality of the angles EFB and ABD. Therefore, upon superposition on the line BA, [the line] EF will be the line BH. After that, the line BH will move in the direction BD where it will be equal to its original position. If in the process of motion of the line BH the

point *B* coincides with the point *D*, the line *BH* will be superposed on
the line *DC*, for the angles *HBF* and *CDB* are equal as two right [angles].
If the line *BH* is superposed on the line *DC*, the point *H* coincides with
the point *C*, for the line *HB* is the line *EF* and the line *EF* is equal to
the line *CD*, that is, it coincides with it when, in the process of motion
of the line *EF* along the line *FD*, the point *F* coincides with the point *D*
and the point *E* coincides with the point *C*. But when defining parallel
lines we showed that if any [straight line] moves in this way, then its
ends describe a straight line. Therefore, in the process of motion of
the line *EF* along the line *FB*, the point *E* describes a straight line. But
the line described by the point *E* is the line *EHC*. Thus the line *EHC* is
a straight line. But, by assumption, the line *EAC* is a straight line joining
the points *E* and *C*, and the line *EHC*, different from the line *EAC*, has
in common with it two points *E* and *C*. Since both of these lines are
straight lines, two lines bound a surface area, which is absurd. It follows
that our assumption that the line *CD* is greater than the line *AB* is
also absurd, that is, the line *CD* is not greater than the line *AB* *[274,
pp. 748–750]*.

In the same way it is proved that the line *CD* is not smaller than the line *AB*.
By a similar argument it is shown that every perpendicular dropped from the
line *AC* to the line *BD* is equal to the line *AB*.

Ibn al-Haytham first proves the parallel postulate for the case of a perpendicular and an oblique line. He begins by considering the lines *AC* and *EDG*
intersected by the straight line *BD* perpendicular to *EDG* (Figure 23). From

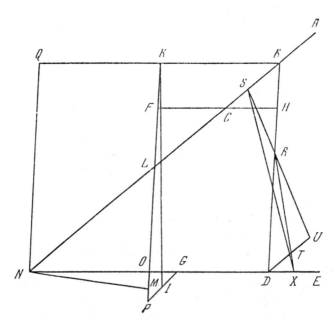

Figure 23

the point B Ibn al-Haytham draws the straight line BK perpendicular to BD, chooses an arbitrary point C on the straight line BC, and drops from it the perpendicular CH to the straight line BD. By a detailed analysis of all possible cases, and by using the fact that a straight line which cuts one side of a triangle and does not pass through any of its vertices must cut one of the two remaining sides (today this proposition is called "Pasch's axiom," for it was formulated as an axiom by M. Pasch at the end of the 19th century), Ibn al-Haytham proves that the foot H of the perpendicular CH to the line BD lies between B and D. Then the lines CH and AC are extended toward C and segments $CF = CH$ and $CL = BC$ are laid off on the extensions, the line CL is drawn, and the equality of the triangles LCF and BCH is proved, whence it is clear that CFL is a right angle. Further, the perpendicular FK to the line BK is dropped from the point F. In view of the premise, KFH is a right angle and $FK = HB$. It follows that LFK is a straight line and $LK = 2BH$. If LK is not greater than BD then this process is continued until one obtains a line greater than BD; in Ibn al-Haytham's diagram such a line is $NQ = MK = 4BH$. To justify the possibility of finding such a line Ibn al-Haytham argues that

for any two different lines, if the smaller one is doubled infinitely many times, its magnitude will become larger than the greater magnitude. This premise was not needed earlier in this proof but Euclid used it in his book in his other proofs *[274, p. 756]*.

Then Ibn al-Haytham lays off the line $KO = BD$ and proves that if the line DG is extended to the line KM then it cuts it at the point O; at this point a great many propositions are again stated and refuted. Noting further that by his premise KOD is a right angle and again making use of Pasch's axiom for the line DO and the triangle LMN, Ibn al-Haytham finds that the extension of the straight line EDG meets the side LN of that triangle, that is, the extension of the straight line AB.

Using the same diagram (Figure 23) Ibn al-Haytham proves the parallel postulate in the cases when the angle BDG is acute and when it is obtuse. In the first of these cases Ibn al-Haytham drops a perpendicular from B to DG, shows that it falls on the side of G, and reduces the problem to the previous case. In the second case he halves the line BD at R; drops the perpendicular RX from R to EDG, proves that it will fall on the side of E; constructs the angle RDT equal to the angle DBC; drops from R the perpendiculars RU and RS to the lines DT and BC, respectively; proves that the triangles RSB and RDU are equal and the perpendiculars RU and RS form a single line; draws the line SX; proves that it intersects the lines ABC and EDG at acute angles; and thus reduces the problem to the previous case.

Therefore [concludes Ibn al-Haytham] that perpendicular is the line CH and the point H is between the points B and D. [We extend the line] CH in its direction toward C and we also extend it in its direction toward H, we lay off [on the extension of the line CH] the line CF equal to [the line] CH, we lay off [on the extension of the line BC] the line CL equal to [the line] *BC*, and join *F* [to] *L*. Since the lines *BC* and *CL* are equal, and also the lines *CH* and *CF*, and the vertical angles *LCF* and *BCH* are equal, the base *LF* is equal to the base *BH* and each of the two other angles is equal to its corresponding one. Therefore the angle *CFL* is equal to the right angle *CHB*, that is, *CFL* is a right angle. From the point *F* we drop a perpendicular to the line *BK*, let this be *FK*. The line *FK* is equal to the line *HB* and, as was proved in the premise, *KFH* is a right angle. But *CFL* is also a right angle, so that the line *LFK* is a straight line. Since it was shown that the line *LF* is equal to the line *BH*, the line *LK* is equal to twice the line *LH*. If *LK* is not greater than the line *BD*, we extend the line *BL* in its direction toward *L* and we also extend the line *KL* in its direction toward *L*, we lay off the lines *LM* and *LN* and join *N* [to] *M*; here the lines *LM* and *LN* are [respectively] equal to the lines *BK* and *LK* and the vertical angles *NLM* and *BLK* are equal. Therefore the base *MN* is equal to the base *BK*, and the triangle *NLM* is equal to [the triangle] *BLK*, and the remaining angles [of the triangles] are equal to the remaining angles, each to the corresponding one. Therefore if a straight line falls on two straight lines and the interior angles on one of the two sides are together less than two right angles, then these two straight lines, when extended to that side, meet, and this is what we wished to prove *[274, pp. 758–762]*.

In his *Book on the resolution of doubts* Ibn al-Haytham refers to the fact that he proved the parallel postulate in the *Book of commentary on the premises* and notes that this postulate is equivalent to the assertion that two intersecting lines cannot be parallel to a third line (that is, it is not possible to pass through a point two different parallels to one line) and

this assertion is clearer to the senses and penetrates into the soul more than it *[237, p. 25]*.

Khayyām's Theory of Parallels

The first theory of parallel lines in which the proof of the parallel postulate is not based on *petitio principii* but on a more intuitive postulate is the theory of parallel lines of ʿUmar Khayyām, mathematician, astronomer, philosopher, and poet from the city of Nishāpūr (in Khorāsān), who worked in Samarkand, Bukhārā, Isfahān, and Marw. Especially popular is Khayyām's *Rubā ʿiyāt* in Persian.

It is the first book of his *Commentaries on the difficulties in the premises of Euclid's book* (Sharḥ mā ashkāla min muṣādarāt kitāb Uqlīdis) *[272, pp. 113–146]*[5] that deals with the theory of parallel lines. In the remaining two

---

[5] There is an incomplete English translation of Khayyām's *Commentaries* by Ali-Reza Amir Moéz *[277]*. In 1986 there appeared a French translation of the book devoted to the theory of parallels *[248a, pp. 185–199]*.

books of this work Khayyām expounds his studies of the theory of ratios, which we will consider later. Khayyām has no doubts of the truth of Euclid's parallel postulate but views it as less obvious than many of the propositions that Euclid thought needed proof, such as the theorem that equal central angles cut off equal arcs from equal circles. Khayyām refutes, as logically unsound, a number of the attempts of his predecessors to prove the parallel postulate. Khayyām includes in this category the previously mentioned attempts due to Heron, Eutocius, al-Shannī, and al-Nayrīzī. Khayyām also refutes Ibn al-Haytham's proof presented previously, which he criticizes for its use of the concept of motion. Like Aristotle, Khayyām rejected the application of motion in geometry.

Further, Khayyām formulates five "principles borrowed from the Philosopher" (Aristotle). We cited these principles previously and pointed out that the fourth is equivalent to the parallel postulate.

Khayyām proves first that two perpendiculars to the same line cannot intersect, for if they did then they would have to intersect in two points, one on each side of that line. This and the first assertion in Khayyām's principle IV imply that two perpendiculars to the same line cannot converge.

The second assertion of Khayyām's principle IV implies that these two perpendiculars also cannot diverge, for they would have to diverge on both sides of the line they are perpendicular to. It follows that two perpendiculars to a single line must be equidistant.

Khayyām goes on to prove eight propositions which, in his view, should be inserted in book I of Euclid's *Elements* in place of its proposition 29 with which Euclid starts the exposition of the theory of parallel lines based on his parallel axiom. Here Khayyām constructs a quadrilateral formed by two perpendiculars AC and BD of equal length erected at a line AB and segments AB and CD. Like the Ibn al-Haytham–Lambert quadrilateral, this quadrilateral has played an important role in the history of non-Euclidean geometry. It is often called the *Saccheri quadrilateral* for the 18th-century mathematician G. Saccheri who considered it anew. We note that the axis of symmetry of a Khayyām-Saccheri quadrilateral divides it into two Ibn al-Haytham–Lambert quadrilaterals.

Before setting forth his propositions Khayyām says:

We should now adopt the twenty-eight propositions of the book of *Elements*, for they do not involve this premise. But it requires a twenty-ninth proposition that embodies the law governing parallel lines. Therefore, let him who so desires replace the twenty-ninth proposition of book I with the first proposition of this book and thus include it ... in the content of the [former] book. Here we shall see the true "proof of why this is so" *[272, p. 120]*.

"The proof of why this is so" is a well-known term of Aristotelian logic opposite to the Aristotelian term "proof that it is so" *[29, vol. 1, p. 78ª]*.

In Khayyām's proposition I, which he calls the 29th proposition of Euclid's book I, two equal perpendiculars AC and BD are erected at the ends of a

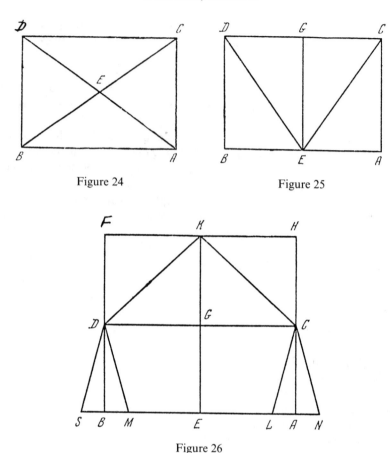

Figure 24   Figure 25

Figure 26

straight line AB, the upper ends of these perpendiculars are joined (Figure 24), and it is shown that the angle ACD is equal to the angle BDC. In proposition II (proposition 30 of the *Elements*) one erects, in the same quadrilateral $ABCD$, a perpendicular EG at the midpoint E of the line AB (Figure 25), and it is shown that $CG = GD$ and that EG is perpendicular to DC. The key role in Khayyām's proof is played by his proposition III (proposition 31 of the *Elements*):

> *Third proposition* (that is, proposition 31 of the *Elements*). We again consider the figure $ABDC$ (Figure 26). I claim that the angles ACD and BDC are equal.
>
> *Proof.* We halve AB at E, erect the perpendicular EG, extend it in its direction to G, lay off GK equal to GE, and draw HKF perpendicular to EK.
>
> Next we extend AC and BD. They intersect HKF at H and F, for AC and EK are parallel and the distance between parallel lines does not change and, if we produce indefinitely AC parallel to the line EK and

produce *HK* indefinitely parallel to the line *GC* then, obviously, they will intersect. We join *C* [to] *K* and *D* [to] *K*. Then, since the line *DG* is equal to *GC* and *GK* is common as well as perpendicular [to *DG* and *GC*], it follows that the bases *DK* and *KC* are equal and the angles *GCK* and *GDK* are equal. Therefore the angles *HCK* and *KDF* are also equal, the complementary angles *DKG* and *CKG* are equal and the remaining angles *KHC* and *KFD* are also equal. Therefore, since the line *DK* is equal to *KC*, it follows that *CH* is equal to *DF* and *HK* is equal to *KF*. If *ACD* and *BDC* are right angles, this is unavoidably true. If they are not right angles then each of them is either smaller than or greater than a right angle.

Assume first that they are smaller than a right angle. If we superpose the plane figure *CF* on the plane figure *CB* then *GK* is superposed on *GE* as well as *HF* on *AB*, and *HF* will be equal to the line *NS* for angle *HCG* is greater than angle *ACG* and line *HF* is greater than *AB*. In just the same way, if the two lines [*CH* and *DF*] are produced indefinitely, then each of the lines joining [them] will, progressively, be ever larger. Therefore, the lines *AC* and *BD* will diverge. In just the same way the lines *AC* and *BD* will diverge if produced in the opposite direction; this is proved in exactly the same way for, with regard to superposition, the circumstances on both sides are necessarily the same. Therefore two straight lines cut a straight [line] at right angles and, subsequently, on either side of that line the distance between them increases. But this is absurd in view of the axiom, if one envisages straightness. Therefore the two lines are a fixed distance apart. This follows from what was considered by the Philosopher.

Now suppose that each of them [the angles *ACD* and *BDC*] is greater than a right angle. Then upon superposition the line *HF* will be equal to *LM* and will be smaller than *AB*, just as will be all connecting lines, and these two lines [*CH* and *DF*] will converge. On the other side there will also be convergence for, with regard to superposition, the circumstances are identical on both sides. If you give it some thought you will follow it. But, according to what was said above, this is again absurd.

Thus the two lines [*AB* and *FH*] cannot be different, that is, they are equal. Since they are equal, the two angles must also be equal and thus are right angles. With a little thought you will follow it. Therefore, to avoid verbosity, we leave this question. He who will want to carry out a detailed proof will be able to do this without requiring our assistance.

The mistake of later [scholars] in the proof of this premise is due to the fact that they did not take this axiom into consideration even though its subject and predicate occurred correctly. Even persons of profound intuition and penetrating mind may fail to consider many axioms because they fail to conceive their subjects and predicates. But the primacy and truth of an assertion lie not only in the presentation of its subject and predicate; for the correctness or incorrectness of an assertion depends not on subject and predicate alone but only on the connection between them. Grasp this *[272, pp. 120–122]*.

We see that when he considers the acute- and obtuse-angle hypotheses Khayyām first folds the drawing along the line *CD* and shows that in the case of the acute angle hypothesis the segment *HF* coincides with the segment *NS* and in the case of the obtuse angle hypothesis *HF* coincides with the segment *LM*—that is, under the first hypothesis the upper base of the quadrilateral is greater than the lower one and its sides grow apart, whereas under the second hypothesis the upper base is smaller than the lower one and the sides of the quadrilateral come closer to one another. Further, upon folding the drawing along the line *AB* Khayyām sees that under the acute-angle hypothesis two perpendiculars to a line diverge on both sides of that line and under the obtuse-angle hypothesis those perpendiculars converge on both sides of that line. But this situation contradicts Khayyām's principle IV, that is, that two perpendiculars to the same line are a fixed distance apart. In other words, both the acute- and obtuse-angle hypotheses contradict this principle. This proves the existence of a rectangle.

Khayyām goes on to define the "distance between two arbitrary [straight] lines" as "the line that links them in such a way that the interior angles are equal." In this connection Khayyām argues as follows:

Two lines *AB* and *AC* intersect in a point *A* (Figure 27). I claim that they open and diverge indefinitely. To this end we draw a circle with center *A* at a distance *AB*. The distance between the two lines at their intersection with the circle is the line *BC*. We extend *AB* in its direction and draw the circle *ADE*. Then we extend *AC* in its direction to its intersection with the circle [*ADE*] at the point *E* and join *D* [to] *E*. Then *DE* is the distance between the two lines and the line *DE* is greater than *BC*; and if we imagine the meaning of circle, angle and straight line then, undoubtedly, this is an axiom *[272, p. 123]*.

We note that drawing arcs with center at the vertex of a right angle and their chords is reminiscent of the arguments by which Simplicius and al-

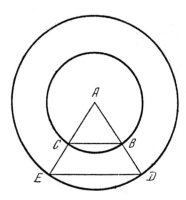

Figure 27

Jawharī proved the parallel axiom. It is possible that Khayyām adopted these arguments from these or other proofs of that axiom by his predecessors. We also note that the assertion we are considering is the third "principle borrowed from the Philosopher" (Aristotle) and it is a consequence of the axioms of Euclidean geometry that is independent of the parallel axiom.

After defining the distance between straight lines Khayyām poses the following question:

Given two straight lines *AB* and *DC* in a plane and, assuming a point *E* on *AB*, the distance between the point *E* and the line *DC* is the line *EG* and angle *E* is equal to angle *G* (Figure 28). But how is one to draw from the point *E* a line to *CD* such that the interior angles are equal? Correction of the foundations of geometry is a job for the geometer and not for the philosopher. Can one draw a line with this property? This question belongs to the art of the author of [philosophical] principles. We shall clarify this as follows. It is possible to draw from *E* countless lines that form at their ends countless angles that differ from one another in that one is greater or smaller than the other. But since at the two ends [of the connecting straight line] there are different [angles], one greater or smaller than the other, it follows from the infinite divisibility of magnitudes that the equality of two angles [*EGF* and *GEH*] is necessarily possible.

We lay off *EH* and *GF* equal to one another and join *H* [to] *F*. Then angle *H* is equal to [angle] *F*, as shown in the first case, so that *HF* is the distance. Hence if *HF* is greater than *EG*, the two lines diverge.

Next we lay off *HK* and *FL*, equal to one another, and join *K* [to] *L*. Then *KL* is the distance. But if *KL* is smaller than *HF*, the two lines converge in view of the axiom, for they diverged before. The same [contradiction] will also necessarily arise when they are equal.

If *HF* is smaller than *EG*, the two lines converge. By what we have

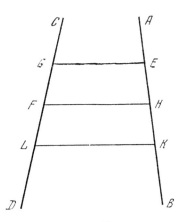

Figure 28

proved *KL* is necessarily smaller than *HF*, for otherwise, in view of the axiom, we obtain an absurdity.

It is thus clear that if two coplanar lines converge in a certain direction, it is impossible that they should diverge in that direction. The same is true if they diverge. This explanation is philosophical, not geometric. The example we gave is meant to make the presentation more visual and obvious for those without a keen intuition *[272, pp. 123–124]*.

We see that in proving the possibility of drawing a line that forms equal angles with two given lines Khayyām refers to the first "principle borrowed from the Philosopher," which in his case plays the role of a continuity principle. By "the axiom" Khayyām means the fourth principle, with which he replaced the parallel axiom.

In proposition IV Khayyām shows that in a rectangle the opposite sides are equal. In proposition V he shows that two perpendiculars to one line have the property that any perpendicular to one of them is their common perpendicular. According to his proposition VI, if two lines are parallel in the Euclidean sense, that is, if they do not intersect upon extension, then they constitute two perpendiculars to one line.

In proposition VII it is shown that a straight line falling on parallel straight lines makes the alternate angles equal to one another, the exterior angle equal to the interior and opposite angle, and the interior angles on the same side equal to two right angles. This is Euclid's proposition 29 of book I, but in proving it Khayyām relies on his propositions and not on Euclid's parallel axiom.

Finally, in proposition VIII Khayyām proves the parallel postulate in Euclid's formulation:

The *eighth proposition* (that is, proposition 36 of the *Elements*). The line *EG* is a straight line. From it are drawn two lines *EA* and *CG* such that the angles *AEG* and *CEG* are [together] less than two right angles (Figure 29). I claim that they intersect on the side of *A*.

Proof. We extend the two lines in their directions. Let the angle *AEG* be smaller than [the angle] *EGD*; we construct the angle *HEG* equal to [the angle] *EGD*. Then, as Euclid showed in proposition 27 of book I, the two lines *HEF* and *DGC* are parallel and the line *AE* that intersects [the line] *HF* will intersect the line *CD* on the side of *A*. This is what we wished to prove.

This is a true proof of the assertions on parallels according to its sense and purpose. These propositions should be added to the *Elements* in the order in which we have set them forth in this book. They follow from the principles of the First Philosophy.[6] We have included them here in spite of the fact that they go beyond the main point of this

---

[6] By *First Philosophy* Khayyām means Aristotle's philosophy (the philosophy of the First Philosopher); this was originally the title of Aristotle's *Metaphysics*.

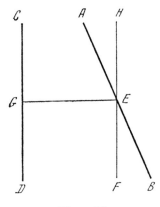

Figure 29

art, for we could not avoid it owing to the fact that this is a difficult question that has been treated by many people. Therefore we have added in the introduction the mentioned principles, for this art requires them as a sound philosophical basis and so as not to provoke suspicions and doubts in the minds of those who ponder it *[272, p. 127]*.

We see that Khayyām's proposition is close to Proclus' proof except that Khayyām justifies the constancy of the distance between the nonintersecting lines *EH* and *CD* whereas Proclus silently assumes its boundedness.

Ḥusām al-Dīn al-Ṣālār's Theory of Parallels

At the beginning of the 13th century, Ḥusām al-Dīn al-Ṣālār, author of a treatise on the complete quadrilateral cited by Naṣīr al-Dīn al-Ṭūsī *[590, pp. 23, 30]*, tried to improve Khayyām's proof. The title of al-Ṣālār's treatise is *Premises for the proof of the postulate, given by Euclid in the beginning of the first book, which refers to parallel lines* (Muqaddamāt li-tabyīn al-muṣādara allatī dhakarahā Uqlīdis fī Ṣadr al-maqāla al-ūlā fīmā yataʿallaqa bi-l-khuṭūṭ al-mutawāziya) *[499]*. Al-Ṣālār's proof consists of six "premises" and the proof of the parallel postulate itself.

Al-Ṣālār's *first "premise"* coincides with proposition I of Khayyām's proof. Here al-Ṣālār, like Khayyām, assumes that the upper angle of his Khayyām-Saccheri quadrilateral can be acute as well as obtuse. Also, al-Ṣālār says that

the distance between two lines, or the distance between two points on them, is defined in terms of the magnitude of the line joining the two lines and forming with them two equal angles *[499, p. 285]*

that is, al-Ṣālār mentions the term *distance between two lines* in the same sense as Khayyām.

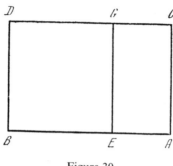

Figure 30

Next comes al-Sālār's second "premise," which is the core of his proof.

Second premise. For every straight line from whose ends are drawn two straight lines that stand straight on it, that is, do not incline to either side, like the lines AC and BD drawn from the ends of the line AB in the indicated manner (Figure 30), that is, are perpendicular to it, [it is true that] however far they go from the points of issue they do not tend to come together or move apart.

While this is obvious and easy to understand we give a proof, namely, we draw from a point E between the points A and B a line EG which also stands on that line like the two earlier lines. If the two perpendiculars issuing from the ends of the given line tend to come together then, the position of EG relative to each of the lines AC and BD being the same as their mutual position, the line EG necessarily tends to come near each of them. On the other hand, if they tend to move apart, then it too will tend to move away from each of them. Obviously, such a situation cannot develop for, if a line ends up between two other lines then it is not possible that it should tend to come near one of them without tending to move away from the other, or that it should tend to move away from one of them without tending to come near to the other.

From this we learn the proof that a line joining the ends of two equal perpendiculars issuing from the ends of a given line must necessarily be equal to the given line, just as the line CD, joining [the ends of] the equal perpendiculars AC and BD issuing from the ends of the line AB, is necessarily equal to AB: if CD were unequal to AB then it would be greater or smaller than it. If it is greater then the two lines tend to move apart, and if it is smaller, then the two lines tend to come together. But we know already that under these circumstances the distance between them is constantly in the same state, it neither increases nor decreases [499, p. 286].

The assertion that two perpendiculars to a straight line do not come together or move apart was proved by Khayyām in his proposition III: he refuted the proposition that the upper angles of a Khayyām-Saccheri quadrilateral are either obtuse or acute with the help of the fourth "principle borrowed from the Philosopher." Al-Sālār replaces Khayyām's sophisticated proof with a crude and defective argument in which one first supposes that two perpendiculars come together or move apart and then one considers a third perpendicular located between them. The author finds a contradiction between the notion that, in that case, the third perpendicular must simultaneously come near the boundary perpendiculars or move away from them and the notion that if a straight line between two perpendiculars comes near one of them then it must move away from the other. Actually in the latter assertion it is tacitly assumed that the boundary perpendiculars are a fixed distance apart. Al-Sālār goes on to say:

If this premise is established, it suggests to us the *third premise*: if the two angles *A* and *B* are not themselves right angles but are equal to them [in sum], then the position of these two lines is the same as the position of the [straight lines] mentioned earlier, that is, they never come near or move apart *[499, p. 286]*.

This "premise" of al-Sālār is identical with one of the assertions of Khayyām's seventh proposition.

Al-Sālār's *fourth "premise"*, to the effect that

A line joining the ends of two equal perpendiculars drawn from the ends of a straight line is the boundary of two right angles *[499, p. 287]*

coincides with proposition III of Khayyām but is proved by using the notion that the sides of a Khayyām-Saccheri quadrilateral are equidistant straight lines. Al-Sālār's *fifth "premise"* coincides with proposition IV of Khayyām. Al-Sālār's *sixth "premise"* is very interesting. Its statement is that

When two lines diverging from a point and bounding a right or nonright angle are indefinitely extended then the distance between them will exceed a multiple of any distance and of [any] given magnitude [when the multiple is increased] indefinitely *[499, p. 288]*.

It coincides with assertion III of the "principles borrowed from the Philosopher," accepted without proof by Khayyām.

The proof of this "premise," which consists in inserting a triangle in the angle, doubling its sides, and proving that the base of the new triangle is twice as large as the base of the original triangle, is very close to the argument in propositions 31 and 32 of al-Jawharī.

Al-Sālār's proof of the parallel postulate is basically the same as the proof of Khayyām's proposition VIII. Al-Sālār's *sixth "premise"* shows that, beginning with Khayyām's proof and trying to improve it, he proved the third "principle borrowed from the Philosopher" that Khayyām accepted without proof. However, having misunderstood the proof of Khayyām's fundamental proposition III, al-Sālār replaced it with a faulty argument in which he assumed what he tried to prove.

Naṣir al-Dīn al-Ṭūsī's Theory of Parallels

On a number of occasions we have mentioned Naṣīr al-Dīn al-Ṭūsī's work devoted to the theory of parallel lines. This work is the *Treatise that heals the doubt raised by parallel lines* (Al-risāla al-shāfiya 'an al-shakk fī l-khuṭūṭ) *[593].*[7] It begins with an exposition and critique of the theories of parallel lines of Ibn al-Haytham, al-Jawharī, and Khayyām. At the time when he wrote his treatise al-Ṭūsī was in the "state of Assassins" and was isolated from the outside world (the treatise was finished not later than 1251, at which time 'Alam al-Dīn al-Hanafī, with whom al-Ṭūsī corresponded about it, died; al-Ṭūsī remained in the state of Assassins until it was overrun by the Mongols in 1256) with no access to many relevant mathematical works. In particular, although familiar with Ibn al-Haytham's *Book on the resolution of doubts in Euclid's book the "Elements"*, he did not have a copy of his *Commentaries on the introductions of Euclid's book the "Elements"*, which contained a proof of the parallel postulate. That is why he took Ibn al-Haytham's statement that the parallel postulate is equivalent to the impossibility of the existence of two intersecting straight lines parallel to a third and that the latter assertion is more intuitive than the parallel postulate to mean that he (that is, Ibn al-Haytham) had not tried to prove the parallel postulate but only to replace it with a more intuitive assertion. As for Khayyām's geometric treatise, al-Ṭūsī quotes not all of the material relating to parallel lines but only the eight propositions that contain the proof of the parallel postulate and fails to notice that Khayyām's proof is based on the fourth "principle borrowed from the Philosopher," which consists of two assertions equivalent to the parallel postulate. Al-Ṭūsī thinks that, by using the assertion that the distance between two intersecting lines increases beyond all bounds, that is, the third "principle borrowed from the Philosopher" (the principle proved by al-Sālar), Khayyām had made a logical error. As pointed out earlier, this assertion is independent of the parallel axiom, so that al-Ṭūsī's reproach is completely unfounded. All that *is* correct is al-Ṭūsī's critique of the "proof" of al-Jawharī.

Al-Ṭūsī goes on to give his proof of the parallel postulate:

> As for the approaches by means of which I investigated this after studying the words of these scholars, we will set forth our discourse in seven propositions, two of which are taken from al-Khayyām's propositions; they are our second and fourth which are his first and fourth. Let the beginning of the book the *Elements*, the twenty-eight propositions that do not include a doubtful postulate, remain unchanged, and then we will add these propositions *[593, p. 511].*

In proposition I al-Ṭūsī proves that

> The shortest one of the lines drawn from a point to any line whose ends are not bounded, called the distance from that point to that line, is the perpendicular dropped from the point to the line *[593, p. 511].*

---

[7] A French translation of this treatise was published in 1986 *[248a, pp. 137–144, 201–226].*

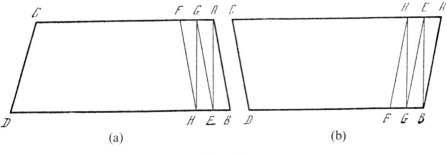

(a)   (b)

Figure 31

Al-Ṭūsī's proposition II is that

If one erects two equal perpendiculars to a straight line and joins their ends by means of a straight line then they form equal angles [with the latter] *[593, p. 512]*.

This proposition coincides with proposition I of Khayyām's proof.

The core of al-Ṭūsī's proof is proposition III, which coincides in its formulation with Khayyām's proposition III but differs in its proof.

Proposition III. If one erects two equal perpendiculars and joins their ends by means of a straight line then the angles formed by them are right angles.

Example. Equal perpendiculars *AB* and *CD* are erected on the line *BD*, and their ends are connected by means of the line *AC* (Figure 31a and 31b). I claim that the equal angles *BAC* and *DCA* are right angles.

Proof. If they are not right angles then they are obtuse or acute. First we assume that they are obtuse and we erect in the first drawing (Figure 31a) at the point *A* a perpendicular *AE* to the line *AC* as proved in proposition II. It necessarily falls between the lines *AB* and *CD* and, as proved in proposition 16, the angle *AED*, an exterior angle of the triangle *ABE*, is greater than the interior right angle. Therefore it is also obtuse. Now we erect at the point *E* a perpendicular *EG* to the line *BD*. It falls between the lines *AE* and *CD*, and the angle *EGC*, an exterior angle of the triangle *EAG*, is greater than the interior right angle *A*. Therefore it is also obtuse. Further, we erect at the point *G* a perpendicular *GH*, again to the line *AC*, and in this order we will indefinitely continue to erect perpendiculars. Then the perpendiculars drawn from points located on the line *AC* at a right angle to the line *BD*, [that is,] the perpendiculars *AB*, *GE*, *FH*, successively increase in length. The shortest of them is the perpendicular *AB* which subtends the acute angle *AEB* in the triangle *AEB* and is therefore shorter than *AE* which subtends the right angle *ABE*; this follows from proposition 19. *AE*, which subtends the acute angle *AGE* in the triangle *AEG*, is shorter than *GE* which subtends the right angle *EAG*. Therefore *AB* is also shorter than *GE*. In just this way it is shown that *GE* is shorter than *FH* and *FG* is shorter than that which follows it. Thus the perpendiculars closer to *AB* will be shorter and the distances between the points that are the feet of the perpendiculars dropped from the points of the line *AC* to the line *BD* successively increase in the direction of *C*, so that the lines *AC* and *BD* diverge in the direction of *C* and converge in the direction of *A*. But the angle *DCA* is also obtuse, for it is equal to the angle *BAC* by the previous proposition. Thus we prove, as before, that the lines *CA* and *DB* diverge in the direction of *A* and converge in the direction of *C*. But this is absurd. Hence the angles *BAC* and *DCA* are not obtuse.

If, on the other hand, these angles are acute, then in the second drawing (Figure 31b) we will drop from the point *B* a perpendicular *BE* to the line *AC*, as proved in proposition 12. It necessarily falls between the lines *AB* and *CD*, for the angle *A* is acute and it is impossible that it falls outside these lines. In the right triangle *AEB* the angle *ABE* is acute, so that the angle *EBD* which, together with the angle *ABE*, forms the right angle *ABD* is also acute. Next we drop from the point *B* a perpendicular *EG* to the line *BD*. It falls between the lines *AB* and *CD* and the angle *GEC* is acute. Then we drop from the point *G* a perpendicular *GH* again to the line *AC* and in this order we will indefinitely continue to drop perpendiculars. Then the perpendiculars drawn from the points on the line *AC* at right angles to the line *BD*, the perpendiculars *AB*, *EG*, *HF*, successively decrease in length. The longest one of them is the perpendicular *AB*. In this way it is shown that the lines *AC* and *BD* converge in the direction of *C* and diverge in the direction of *A*. But the angle *DCA* is also acute, for it is equal to the angle *BAC* by the previous proposition. Hence, as before, it is shown that the lines *CA* and *DB* converge in the direction of *A* and diverge in the direction of *C*. But this is absurd. Hence the angles *BAC* and *DCA* are not acute, and since they are not obtuse either, they are right angles. And this is what we wished to prove *[593, pp. 512–514]*.

Here al-Ṭūsī, like Khayyām before him, considers Khayyām-Saccheri quadrilaterals and three hypotheses about their upper angles. While refuting the obtuse- and acute-angle hypotheses, al-Ṭūsī shows that in the first case the perpendiculars he erects increase from the edge of the base toward its midpoint and in the second case the perpendiculars he drops decrease from the edge of the base toward its midpoint; and he thinks that this contradicts the symmetry of these quadrilaterals relative to the perpendiculars joining the midpoints of their lower and upper bases. In fact, all perpendiculars constructed by al-Ṭūsī turn out to lie on one side of the axis of symmetry of the quadrilateral; this shows that in the case of the obtuse-angle hypothesis the bases of a Khayyām-Saccheri quadrilateral come together on both sides of its axis of symmetry, and in the case of the acute-angle hypothesis they move apart, which is what Khayyām proved for two perpendiculars and one straight line.

Al-Ṭūsī's proposition IV that

Every two opposite sides of a right-angled quadrilateral are equal *[593, p. 515]*

coincides with proposition IV of Khayyām's proof. Al-Ṭūsī's proposition V, that

If a straight line falls on two perpendiculars erected in an arbitrary way on another straight line then the alternate angles are equal, each exterior angle is equal to the [corresponding] interior opposite angle, and interior angles on the same side are equal to two right angles *[593, p. 515]*

coincides with proposition VII of Khayyām's proof. In proposition VI al-Ṭūsī gives a proof of the parallel postulate for the case of a perpendicular and an oblique line that is very close to the second proof given by ibn Qurra and to the proof given by Ibn al-Haytham. In proposition VII al-Ṭūsī gives a very original proof of the general case of the parallel postulate:

Proposition VII containing the proof of the postulate: If a straight line falling on two straight lines makes the interior angles on the same side [together] less than two right angles, the two straight lines, if produced indefinitely, meet on that side.

Example. The line *AB* falls on the lines *CD* and *EG* and forms angles *CHF* and *EFH* that are together less than two right angles (Figure 32). I claim that the lines *CD* and *EG*, if extended in the direction of *C*, will meet.

Proof. If one of the angles *CHF* and *EFH* is a right angle, then the other angle is necessarily acute. Therefore one of the lines *CD* and *EF* intersects the line *AB* at an angle that is not a right angle and the other is perpendicular to it, so that if we extend them, they will meet on the side of the acute [angle], as was proved in the previous proposition.

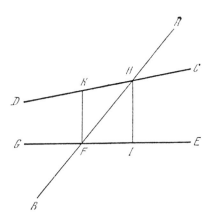

Figure 32

If one of the angles is obtuse, let it be the angle CHF, we erect at H a perpendicular HI to the line CD, as shown in proposition 11, and from F we drop to it a perpendicular FK as shown in proposition 12. Then we claim that, since the angles CHF and EFH are together less than two right angles and CHI is a right angle, the angles IHF and HFI are together less than a right angle. But, as was shown in the fifth of these propositions, the angles IHF and HFK, as alternate angles formed by the line AB falling on two perpendiculars IH and FK, are equal. Hence the whole angle KFI is less than a right angle, that is, it is acute. Therefore the lines KF and EF intersect at an angle that is not a right angle and the line HK is perpendicular to one of them, namely to the line KF. Therefore, as shown in the previous proposition, the lines CK and EF will meet if extended in the direction of C and E.

If both angles are acute then we will erect at the point F a perpendicular FK to the line GE, as shown in proposition 11, and drop to it from the point H a perpendicular HI, as shown in proposition 12. Then EFK is a right angle, and the angles KFH and FHI are equal as alternate angles formed by the line AB falling on two perpendiculars HI and KF, as proved in the fifth of these propositions. Hence the angles FHI and HFI are together equal to a right angle. Since, by assumption, the angles EFH and CHF are less than two right angles, the angle IHC is less than a right angle, that is, it is acute. Therefore the lines IH and CH intersect at an angle that is not a right angle and EI is perpendicular to one of them, namely to IH. Therefore [the lines] CD and EG intersect when extended in the direction of C and E, as shown in the previous proposition. And this is what had to be proved *[593, pp. 519–520]*.

Al-Ṭūsī continues with a variant of the proof of the parallel postulate in which he replaces his propositions VI and VII with other propositions (but keeps proposition VIII).

Instead of proposition VI. In every acute rectilinear angle, if one lays off equal lines on one side and drops from the division points perpendiculars to the other side, then the lines cut off from that side by the feet of the perpendiculars are also equal.

Example. In the acute angle BAC equal lines AD, DE and EG are laid off on AB and from their [ends] perpendiculars DH, EF and GI are dropped to the line AC (Figure 33). I claim that the lines AH, HF and FI cut off by the feet of the perpendiculars are also equal.

Proof. At the point D on the line ED we construct an angle EDK equal to the angle A, as shown in proposition 23. Then in the triangles AHD and DKE the angles A and D are equal; the angles D and E, one exterior and the other interior, formed by the line AE falling on the two perpendiculars DH and EF are equal, as shown in the fifth of these propositions; and the sides AD and DE are equal. Hence these two triangles are equal, the side AH is equal to the side DK and the right

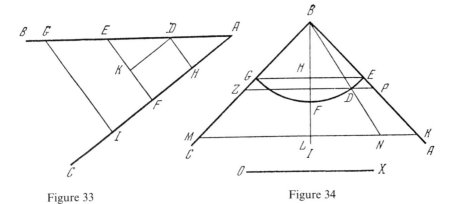

Figure 33   Figure 34

angle *H* is equal to the angle *K*, as shown in proposition 26. Therefore the plane figure *DHFK* is a right-angled quadrilateral and its opposite sides *DK* and *HF* are equal, as was shown in the fourth of these propositions. Therefore the line *AH*, equal to *DK*, is also equal to *HF*. In the same way one shows that *HF* is equal to *FI*. And this is what we wished to prove.

Instead of proposition VII. If one chooses a point between the sides of a right angle then it is possible to join them by means of a line passing through that point.

Example. In the right angle *ABC* one chooses a point *D* between the sides *AB* and *BC* (Figure 34). I claim that it is possible to join the sides *AB* and *BC* by means of a straight line passing through the point *D*.

Proof. From *B* as center we draw at a distance *BD* the arc *EDG* passing through the point *D*. We draw the chord *EG* and halve the angle *EBG* by means of the line *BI*, as shown in proposition 9. Then the triangles *EBH* and *GBH* are equal, for the sides *EB* and *BH* are equal to the sides *GB* and *BH* and the angles *B* are equal. Therefore the sides *EH* and *HG* and the angles *H* are also equal, as proved in proposition 4. Therefore *EH* is perpendicular to *BH*. We extend *BH* until it intersects the arc *EDG* at the point *F*. We repeat the line *BH* until the sum of these lines exceeds the line *BF*; let *OX* be that sum. On *BA* we lay off as many lines, each equal to the line *BE*, as the number of times *OX* is a multiple of *BH*; these are *BE*, *EK*. From the ends of these lines we drop perpendiculars to the line *BH*; these are the perpendiculars *EH*, *KL*. As proved in the previous proposition, these perpendiculars will cut off from the line *BH* equal lines *BH*, *HL*. But their sum, equal to the line *OX*, is longer than the line *BF*, so that the end of the perpendicular *KL* on the line *BI*, that is, the point *L*, is outside the line *BF*. Further, we lay off on *BC* a line *BM* equal to *BK* and join *M* [to] *L*. Then the triangles *BKL* and *BML* are equal, for they have a common side *BL*, equal sides *BK* and *BM*, and equal angles *B*, as shown in proposition 4.

Therefore the angle *MLB* is equal to the right angle *KLB* and, in view of proposition 14, the lines *KL* and *LM* combine in direction into a single line. Next we join *B* [to] *D*, extend it to *N* and construct at *D* on the line *DN* an angle *PDN* equal to the angle *DNL*, as shown in proposition 23. Then the lines *PD* and *NM* are parallel, as before, in view of the equality of alternate [angles], that is, the angles *PDN* and *DNB*, as proved in proposition 26. We extend *PD* until it passes outside the triangle *BKM* at the points *P* and *Z*. The line *PZ* joins the sides *AB* and *BC* and passes through the given point *D*. And this is what we wished to prove *[593, pp. 520–523]*.

Al-Ṭūsī's proposition VIII is the same as the last proposition in the proof given by al-Jawharī.

The correspondence between al-Ṭūsī and al-Ḥanafī referred to previously shows that al-Ṭūsī discussed his treatise on parallel lines with other scholars. It seems that, influenced by criticism that followed his inclusion of both variants of his proof of the parallel postulate in his *Exposition of Euclid* (Taḥrīr Uqlīdis) *[596]*,[8] in distinction to his treatise, al-Ṭūsī put the following remark after the statement of that postulate:

> I say that the latter assertion is not an axiom and can be proved only in the geometric science. It is best not to talk about this in the introduction, and I will prove this in the appropriate place. Instead of it I put the following assertion: if coplanar lines converge in one direction they cannot diverge in that direction, provided that they do not meet *[353, p. 13; 596, p. 4]*.

Attempt at a Proof of the Parallel Postulate Attributed to al-Ṭūsī

The exposition of the theory of parallels referred to by J. Wallis and G. Saccheri in their respective papers on works on the parallel postulate and called by them "the proof of Naṣīr al-Dīn al-Ṭūsī" is not the same as the proof given previously. Wallis and Saccheri had in mind the *Book of exposition of the "Elements" of Euclid authored by Khwāja Naṣīr al-Dīn al-Ṭūsī* (Kitāb taḥrīr Uṣūl li Uqlīdis min ta' līf khwāja Naṣīr al-Dīn al-Ṭūsī) *[175]*[9] issued in Arabic with a Latin front page (Euclidis Elementorum geometricorum libri tredecim ex traditione doctissimi Nasiridini Tusini) in Rome in 1594. *Exposition of the "Elements" of Euclid* differs significantly from *Exposition of*

---

[8] A Russian translation of the proof of the parallel postulate by Gabibulla Jafar-kulu oglu Mamedbeĭli (1914–1982) is given in *[353, pp. 13–32]*.

[9] A Russian translation of the proof of the parallel postulate by G. J. Mamedbeĭli is given in *[353, pp. 22–32]* and in V. F. Kagan's book *[255, pp. 119–121]*. A French translation is given in *[248a, pp. 233–241]*.

Euclid. For one thing, *Exposition of the "Elements" of Euclid* is an exposition of just the 13 books of the *Elements* written by Euclid himself, whereas *Exposition of Euclid* is an exposition of all 15 books traditionally comprising the text of the *Elements*. For another thing, in the *Exposition of the "Elements" of Euclid* the text has been fundamentally reworked. An important difference between the two "expositions" is that there are very many preserved manuscripts of the *Exposition of Euclid* (the author knows of 96 such manuscripts), whereas there are only two copies, one complete and one incomplete, of the *Exposition of the "Elements" of Euclid*, preserved in the Medici Library in Florence. The Roman edition is based on the complete Florentine manuscript. The fact that the significantly more perfect variant of the exposition of Euclid has been disseminated far less widely than the less perfect one has given rise to doubts about al-Ṭūsī's authorship of the *Exposition of the "Elements" of Euclid*. This doubt was further substantiated after A. I. Sabra *[494, p. 15]* established that, as indicated in the Florentine manuscript, the *Exposition of the "Elements" of Euclid* was written in 1298, that is, after al-Ṭūsī's death. This is why J. Murdoch *[379]* refers to the author of the *Exposition of the "Elements" of Euclid* as "pseudo-Ṭūsī."[10] However, a bibliography of the works of al-Ṭūsī prepared by his student Niẓām al-Dīn al-Nayshābūrī, recently discovered by H. Tllašev *[587]* and kept in the Taškent Institute for Oriental Studies, lists three works of al-Ṭūsī devoted to commentaries on Euclid: number 9 is *Exposition of Euclid*; number 24 is *Treatise on Euclid* (Risāla fī-l-Uqlīdis), undoubtedly short for *Treatise healing doubts about parallel lines*; and number 27 is *Remarks on Euclid* (Ḥawāshī ʿalā Uqlīdis). The last work has not survived. It is possible that the *Exposition of the "Elements" of Euclid* was written after al-Ṭūsī's death by one of his students (very likely by his son, Ṣadr al-Dīn, whose full name was Ṣadr al-Dīn ibn khwāja Naṣīr al-Dīn, and who became the director of the Marāgha observatory, founded by al-Ṭūsī, after his father's death) with due regard to the *Remarks on Euclid*. The *Exposition of the "Elements" of Euclid* contains a new and very original proof of the parallel axiom:

First premise. If on any two coplanar straight lines such as the lines *AB* and *CD* there fall straight lines such as the lines *EG, HF, KL, MN* and *XO*, each of which is perpendicular to the line *CD* and intersects the line *AB* at acute and obtuse angles such that all angles directed toward *BD* are acute and [all angles] directed toward *AC* are obtuse (Figure 35), then I claim that the lines *AB* and *CD* come steadily closer together in the direction of *BD* until they meet and move steadily apart in the direction of *AC*, that is, the perpendicular *EG* is greater than the perpendicular *HF*, the latter [is greater] than the perpendicular *KL*, the latter [is greater] than the perpendicular *MN*, and the latter

---

[10] The authorship of this work has also been discussed by B. A. Rosenfeld, A. Kubesov, G. Sobirov and A. Ahmedov *[467, 481, 10]*.

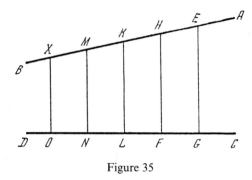

Figure 35

[is greater] than the perpendicular XO, whereas the perpendicular XO is smaller than the perpendicular MN, the latter [is smaller] than the perpendicular KL, [and so on,] to the last. In much the same way, if straight lines falling on two straight lines perpendicular to one of them increase when we take them in one of the two directions of the two lines and decrease when we take them in the other of the two directions of the two lines, that is, if two straight lines move steadily apart in the direction of increasing perpendiculars and come together in the opposite direction, that is, in the direction of decreasing perpendiculars, until the two lines meet, then each of the straight lines perpendicular to one of these two straight lines intersects that straight line at a right angle and that line is not oblique with respect to any of the perpendiculars; but each of these perpendiculars will intersect the second of the two straight lines at two angles one of which is acute and the other obtuse, and all acute angles—in the direction in which the two lines come together, and all obtuse angles—in the direction in which they come apart, and that line inclines to each perpendicular in the direction of decreasing separation and declines from each of them in the direction of increasing separation. These two assertions are obvious, and in view of their obviousness both were used by some ancient, as well as later, geometers.

   Second premise. If two straight lines erected at the ends of a straight line perpendicular to it are equal and their ends are joined by a straight line then each of the angles formed by the perpendiculars and that straight line is a right angle. . .

   Third premise. In every triangle with rectilinear sides the three angles [are equal to] two right angles *[175, pp. 28–30]*.

The second premise is proved on the basis of the first and second assertions of the first premise, and the third on the basis of the existence of a rectangle proved in the second premise. The latter premise is first proved for a right triangle and then, by the subdivision of an obtuse-angled triangle into two right triangles and the complementation of an acute-angled triangle, for the general case. Here the first premise plays the role of an axiom. However, unlike in al-Ṭūsī's *Exposition of Euclid*, this axiom, obviously meant as a replacement for Euclid's axiom, is incorrectly stated: taken literally, this assertion is independent of Euclid's parallel axiom. However, whenever the author refers to the "first proposition" he invokes not this assertion but some assertion close to the axiom in al-Ṭūsī's *Exposition of Euclid*. In the "second premise," in which the author proves the existence of a rectangle, he actually considers an Ibn al-Haytham–Lambert quadrilateral and three hypotheses concerning its angle *D* and refutes the acute- and obtuse-angle hypotheses by using the "first premise" interpreted in the indicated sense. Nevertheless, the quadrilateral in the "second premise" can also be looked at as a Khayyām-Saccheri quadrilateral, which is, apparently, what Saccheri himself thought. In the "third premise," the existence of a rectangle is used for the first time in the history of geometry to prove that the angle sum in a triangle is two right angles. In the formulation of this premise it is stressed that one deals with rectilinear triangles; the author was undoubtedly aware that the angle sum of a spherical triangle is greater than two right angles.

Next the author proves the parallel postulate. He does this first for the case of a perpendicular and an oblique line and then for the two remaining cases. In the first case, his proof of the parallel postulate, based on the existence of a rectangle, differs little from the traditional proof used by Ibn Qurra and Ibn al-Haytham. In the remaining two cases the proofs are also similar to those of Ibn al-Haytham, but there are occasional touches of originality. We quote these proofs next.

As for the second case, when each of the angles *BEC* and *DCE* is acute (Figure 36), the fact that angle *DCE* is acute implies, by the thirteenth poposition, that angle *DCG* is obtuse. By the eleventh proposition, we erect at the point *C* a perpendicular *HC* to the line *EC* in

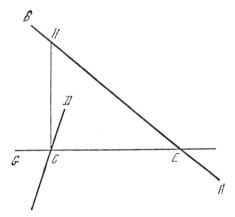

Figure 36

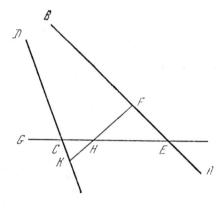

Figure 37

the direction of *D*; it falls between the sides *DC* and *CG*. Therefore if we extend it toward *H* in its direction, it will meet the line *AB* in view of the previous proposition. It will meet it at the point *H*. Therefore if we extend the line *DC* in its direction toward *D* it will meet the line *AB* between the points *E* and *H*. This is obvious in view of the impossibility of two straight lines bounding a plane figure.

As for the third case, when the angle *BEC* is acute and the angle *DCE* is obtuse (Figure 37), since the angles *BEC* and *DCE* are [together] less than two right angles and, by the thirteenth proposition, the angle *DCE* together with the adjacent angle equal two right angles, the angle *C*, adjacent to the angle *DCE*, is greater than the angle *BEC*. We mark off on the line *EC* an arbitrary point *H* and drop from it (by the twelfth proposition) a perpendicular *HF* to the line *EB*. It is clear that it will not fall on the point *E*. It will not fall on the line *AE* either, for if that were so then two angles in a triangle would be greater than two right angles, whereas in the seventeenth proposition it was shown that they are less than two right angles, which is absurd. Suppose it falls on the point *F*. We extend the line *FH* in its direction toward *H* up to *K*. Since, by the seventeenth proposition, angles *HFE* and *EHF* are together less than two right angles and, by the fifteenth proposition, the acute angle *EHF* is equal to the angle *CHK*, the angle *C*, adjacent to the angle *DCE*, is smaller than a right angle. Therefore each of the angles *CHK* and *C* adjacent to the angle *DCE* is acute. Therefore the lines *HK* and *DC*, if extended toward *K*, meet, by the second of the two preceding propositions. Suppose they meet at the point *K*. Since, by the fifteenth proposition, the angles *EHF* and *CHK* are equal, and the angle *C* is greater than the angle *HEF*, the right angle *EFH* is greater than the angle *HKC*, since the three angles of every triangle with rectilinear sides are equal to two right angles by the third premise; therefore *HKC* is an acute angle. But angle *BFK* is a right angle by the thirteenth proposition.

Therefore, if we extend the lines AB and CD toward BD, they meet, by the first of these premises, on that side of the line $[FK]$ where it falls at [angles less than] two right angles. And this is what we wished to prove *[175, pp. 32–33]*.

Although the last proof is rather close to the proof of Ibn al-Haytham, what is special about it is the use of the equality of the angle sum in a triangle to two right angles.

The incorrect formulation of the "first premise" of this proof makes it very unlikely that al-Ṭūsī was the author of the whole proof as well as of all of the *Exposition of the "Elements" of Euclid.* What is very likely is that the most original parts of this proof—the "third premise" and the proof of the parallel axiom in the last two cases—have been borrowed from the *Remarks on Euclid* or from another work written by al-Ṭūsī at the end of his life.

The Theories of Parallels of al-Ḥanafī and al-Abhārī

In the previous sections we presented three proofs of the parallel axiom due, respectively, to al-Sālār, al-Ṭūsī, and the author of the *Exposition of the "Elements" of Euclid.* These are not the only 13th-century works devoted to the problem of parallel lines. We have already mentioned the letter of ʿAlam al-Dīn al-Ḥanafī to Naṣīr al-Dīn al-Ṭūsī about his treatise on parallel lines and quoted from it the beginning of Simplicius' attempt at a proof. Al-Ḥanafī's critique of this attempt, which follows his exposition, is of great interest:

> But we may assume that, from the very beginning, the line BD will deviate from the direction of the line BG and every chord that subtends the angle GBD will fall between the points A and B if AB is infinitely divisible *[494, pp. 8–9, 19]*.

We see that al-Ḥanafī points to the very case which takes place in Lobačevskian geometry when all "chords" subtending the angle GBD pass below a certain interior point of that angle. He does this because he is aware that although Simplicius' proof depends on the impossibility of this case, this impossibility does not follow from the axioms on which his proof is based.

The attempt to prove the parallel axiom that enjoyed the greatest popularity in the 13th as well as in subsequent centuries was due to Athīr al-Dīn al-Mufaḍḍal ibn ʿUmar al-Abhārī, known also as al-Abahrī (d. 1263), a native of Abhār in Jibal and a student of the same Kamāl al-Dīn ibn Yūnis as Naṣīr al-Dīn al-Ṭūsī, who worked in Mosul and Arbil (Iraq). His attempt to prove the parallel axiom was set forth in his reworked version of Euclid's *Elements*, which has come down to us under the name of *Improvement of the "Elements"* (Iṣlāḥ al-Ustuquṣāt) and *Improvement of the "Elements" of Euclid* (Iṣlāḥ Uṣūl Uqlidis). Single chapters from this book, including the one on parallel lines, were reproduced in the very popular *Propositions of the foundation* (Ashkāl al-ta'sīs) of Shams al-Dīn Muḥammad Ashraf al-Ḥusaynī al-Samarqandī, a native of Samarkand who lived in the second half of the 13th century and worked as a scientist at the Marāgha observatory of al-Ṭūsī. The most popular version of this book, reprinted a number of times in Istanbul (e.g., *[501]*,) is the one with commentaries by the eminent Samarkand mathematician and astronomer of the 15th century Qādī-Zada al-Rūmī and by Muḥammad al-Hādī. The chapter of al-Samarqandi's book containing al-Abhārī's attempt to prove the parallel axiom has been printed a number of times but without indication of al-Abhārī's authorship.[11]

Al-Abhārī's proof begins as follows:

This is the place for the promised proof of the well-known postulate. The philosopher Athīr al-Dīn al-Abhārī said: if an angle *ABC* is halved by a line *BH* then it is possible to draw in that angle infinitely many chords in such a way that they are located one under the other and each of them is the base of an isosceles triangle *[500, p. 598]*.

This proposition is the same as Simplicius' proposition 1 and al-Jawharī's proposition 30, and al-Abhārī's proof differs little from other proofs of these propositions. Al-Abhārī's proof of the parallel postulate for the case of a perpendicular and an oblique line is the same as the proof of Simplicius' proposition 2 (al-Abhārī's figure differs from Figure 15 only by the use of different letters). Unlike Simplicius, al-Abhārī gives proofs of the two remaining cases of the parallel postulate. The case when the transversal makes two acute angles with the two straight lines is proved in the same way as the first case by using an analogous figure . The proof of the case when the transversal makes an acute and an obtuse angle with the two straight lines is the same as Ibn al-Haytham's proof of this case.

Al-Maghribī's Theory of Parallels

Al-Maghribī's proof, contained in the manuscript of the Bodleian Library, is very close to al-Abhārī's.

Muḥyī al-Dīn Yaḥyā ibn Abī l-Shukr al-Maghribī, also known as al-Andalusī, was born in Muslim Spain (al-Andalus) or in northwestern Africa (Maghrib). While in Syria in 1260 he was taken prisoner by the Mongols, who brought him to the Marāgha astronomical observatory of Naṣīr al-Dīn al-Ṭūsī; he was there during the last years of his life.

The first two of the four propositions of the manuscript just mentioned

---

[11] The Turkish and French translations of the proof of the parallel postulate have been published by H. Dilgan *[149, 150]*. For a Russian translation see *[500]*. It was A. I. Sabra who called our attention to the fact that the proof is al-Abhārī's.

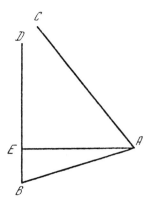

Figure 38

were discussed previously. Apart from notational differences, the fourth one is the same as al-Abhārī's proposition IV. The third proposition is the same in formulation as al-Abhārī's proposition III but differs from it in the proof, which is closer to the penultimate proposition in the proof of the *Exposition of the "Elements" of Euclid*. The statement of the third proposition in al-Maghribī's work cited previously is as follows:[12]

> Let each of the angles *A*, *B* be acute (Figure 38); then the lines *AC*, *BD*, if produced, will meet on the side of *C*, *D*.
>
> We draw *AE* perpendicular to *BD*.
>
> Then, any angle *E* being right and the angle *EAC* acute, the lines *AC*, *BD* will meet if produced *[494, pp. 11, 20]*.

Al-Maghribī was also the author of a work whose title, *Exposition of the "Elements" of Euclid* (Taḥrīr Uṣūl Uqlīdis), is the same as that of the work attributed to al-Ṭūsī. This work of Maghribī contains yet another variant of a proof of the parallel postulate. In the introduction to this work al-Maghribī defines parallel lines as follows:

> Parallel straight lines are those that lie in the same plane surface and are such that if a straight line falls on any two of them at random it makes the two angles on one side equal to two right angles *[494, p. 17]*.

This definition implies that every perpendicular to one of two such straight lines is a common perpendicular. In particular, this implies the existence of rectangles. Also, if two such straight lines are intersected by two other straight lines at equal angles then we obtain a parallelogram; that is, parallelograms exist. From the existence of rectangles it follows easily that lines parallel in al-Maghribī's sense are equidistant. It is clear that al-Maghribī's

---

[12] In what follows, the English translation of A. I. Sabra *[494, p. 11]* was corrected according to the Arabic text publshed by him *[494, p. 20]*.

definition includes a definition equivalent to the parallel postulate (it excludes Lobačevskian and elliptic geometry).

We quote al-Maghribī's proof (the numbers in parentheses are numbers of propositions in his book; for the most part, they coincide with the numbers of the corresponding propositions of Euclid). The abbreviation (*Post.*) refers to a postulate, an axiom, or a definition in the introduction to al-Maghribī's book.

Premise: If a line falling on two straight lines makes the two angles on one of the two sides less than two right angles, then the two lines, if produced indefinitely on that side, meet.

Example: The line AC has fallen on the two lines AB, CD, making the two angles BAC, DCA less than two right angles (Figure 39).

I say that if the two lines are indefinitely produced, they meet.

Demonstration: If one of the two angles is right, we complete the demonstration as will be shown.

But if not, we produce DC indefinitely and let fall on it the perpendicular AE (12) which we then produce indefinitely on the side of F.

Then, on the point A of the line AE, we construct the angle KAE equal to the angle BAE (23);

and we indefinitely produce the lines AB, AK in the directions of B, K (Post.) [i.e. Eucl. Post. 2 = Maghribī's Post. 2];

we mark on AB the point L and cut off AM equal to AL (3) and join ML.

Then, because the angles KAE, BAE are acute (Post.), the line ML cuts AF in N;

and since the two sides AL, AM are equal, and the side AN is common, and the two angles at A are equal, then the two angles N in the two triangles ANL, ANM are right (4 and Post.).

Now if point N lies between the points E, F, we complete the construction;

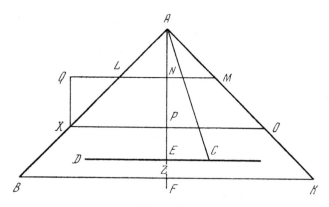

Figure 39

otherwise, we cut off the lines MO, LX equal to AM, AL (3), and join XO which cuts AF in P.

We then prove as before that the two angles F in the two triangles APO, APS are right angles (4).

Now if point P falls between the points E, F, that will be sufficient for us;

otherwise, we produce the line NL indefinitely and from point X draw XQ perpendicular to it (12).

Then, since the angle ANL is right, and the angle LQX is right, and the angles ALN, XLQ are equal (15), and the lines AL, LX are equal, then the lines AN, QX are equal (26).

Further, since the angle QNF is right, and the angle NFX is also right, as we showed, and the angle Q is right,

then the surface FQ is a parallelogram,

and, therefore, the lines QX, NF are equal (34).

We continue with this construction, cutting off from AT multiples (amthāl) of the line AN until we reach a multiple (of AN) greater than AE;

let it [this multiple] be AZ and let the line which has cut off AZ from AF be the line KZ.

Then, the angles at Z are right—as we showed for the angle ANL, and the angles at E are also right,

therefore, ED does not meet ZB (28), nor does it meet AE (Post.),

therefore, it necessarily meets AB.

It is evident from this and from kh [Eucl. I. 28] that every two lines in a plane surface are either secant or parallel;

for if a straight line falls on them, making the two angles on one of the two sides less than two right angles, they meet;

but if it makes (the angles) equal to two right angles, then the two lines are parallel.

Thus the doubt concerning this question has been removed by virtue of what we have posited as an emendation of the introduction *[494, pp. 15–17, 21–24]*.

Al-Maghribī's references to the Post[ulates] pertain, respectively, to the following: (1) postulate II (the possibility of unlimited extension of a line); (2) the definition of an acute angle as an angle smaller than a right angle; (3) the definition of a right angle as an angle equal to the adjacent angle; (4) the axiom of Eudoxus-Archimedes; (5) the axiom that two straight lines cannot bound a plane figure. We see that, on the one hand, al-Maghribī's proof of the "premise" exhibits traces of proofs derived from Simplicius' proof and, on the other hand, it is analogous to proofs based on a definition of parallel lines that contains an assertion equivalent to the parallel postulate, and this makes it possible to prove the existence of a rectangle. Like Ibn al-Haytham and al-Ṭūsī, al-Maghribī uses the Eudoxus-Archimedes axiom as well as an argument equivalent to Pasch's axiom.

An interesting attempt to prove the parallel postulate was made by a student of Naṣīr al-Dīn al-Ṭūsī, Quṭb al-Dīn al-Shīrāzī (1236–1311) in his encyclopedia *Pearls of the crown for decoration of Dubāj* (Dūrra al-tāj li ghurrat al-Dubāj) *[528; 488, pp. 107–110]*.

The Theory of Parallel Lines of Levi ben Gerson and Alfonso

The first three attempts in medieval Europe to prove Euclid's parallel postulate date back to the 13th and 14th centuries. The first of these attempts is due to the Polish scholar Vitello (about 1230–after 1275) and is found in proposition 14 of book I of his *Perspective* (Perspectiva *[604]*). Vitello's proof was studied by Sabetai Unguru *[598]*. The second and third attempts are due to scholars who wrote in Hebrew and were directly influenced by Arabic works, Hebrew translations of which were, at the time, widespread in southern France. One of these proofs is due to Levi ben Gerson (1288–1344), who was born in Bagnols in southern France and worked in the southern French cities of Orange and Avignon. Levi ben Gerson was a well-known Jewish religious philosopher and an author of a number of Hebrew mathematical works, including *Commentaries to the introductions of Euclid's book* (Beyur ptikhat sefer Iqlidus) *[201]* and a treatise *On sines, chords and arcs*. The astronomical Book V of his philosophical work *Wars of the Lord* (Milhamot ha-Adonai) was published by Bernard R. Goldstein *[208]*.

Levi's proof of the parallel postulate is set forth in his *Commentaries*, written under the obvious influence of the commentaries of scholars of the Near and Middle East, above all Ibn al-Haytham.

Unlike Ibn al-Haytham, who in his "proof" is guilty of *petitio principii*, Levi, like Khayyām, precedes his proof with "two well known premises." One of them is the "Eudoxus-Archimedes axiom," also used by Ibn al-Haytham, and the other is that

a line which is inclined comes closer to the side where a right angle is formed *[201, p. 764]*.

Here, by "a line which is inclined" is meant a straight line for which, relative to some other straight line and some straight line falling on these two, the parallel postulate holds. Before formulating his premises Levi notes that

If a straight line falls on two straight lines and the interior angles on the same side are less than two right angles, every straight line falling one them on that side forms interior angles less than two right angles *[201, p. 765]*.

Thus there is no doubt that what is intended in the second premise is that "a line which is inclined" approaches the given straight line throughout their extent. This assertion is equivalent to Euclid's parallel axioms but is more intuitive.

In proposition 1 Levi shows that

there can be no rectilinear quadrilateral all of whose angles are obtuse or acute" *[201, p. 765]*.

Like Ibn al-Haytham and Khayyām, Levi considers the three hypotheses, respectively involving acute, obtuse, and right angles, and refutes the first two. Just as the Khayyām isosceles quadrilateral with two right angles can be obtained from the Ibn al-Haytham quadrilateral with three right angles by reflecting the latter in a side bordering on two right angles, so too we obtain the quadrilateral considered by Levi by reflecting an isosceles quadrilateral with two right angles in its base bordering on the two right angles. The first two hypotheses are refuted with the aid of the second premise.

Of the consequences of the right-angle hypothesis found by Levi we note the theorem that

if one extends one of the [equal] sides of an isosceles triangle from their point of intersection in its direction by the same magnitude and draws a base, then it forms a right angle with the first base *[201, p. 766]*.

Levi extends the side *AB* of an isosceles triangle *ABC* by a distance *BG = BA* (Figure 40) and shows that the line *CG* is perpendicular to *AC*. For proof, he constructs the altitude *BD* of the triangle *ABC*, extends it to *E* by an amount *BE = BD*, and proves that the quadrilateral *CDEG* is a rectangle. In the course of the proof he refutes the assumptions *CG = KL > ED* and *CG = MN < ED* corresponding to the hypotheses of the acute and obtuse angles. Another consequence of the right-angle hypothesis established by Levi is that the angle sum in a triangle is equal to two right angles.

The parallel postulate is proved first for the case when the straight lines *AB* and *CD* are intersected by a line *AG* (Figure 41) perpendicular to the line *CD*. From the point *H* on the line *AB* Levi drops a perpendicular to *AG* and (successively) doubles the segments *AH* and *AF* until the segment *AL* obtained by doubling the segment *AF* exceeds the segment *AG*. Then the straight line

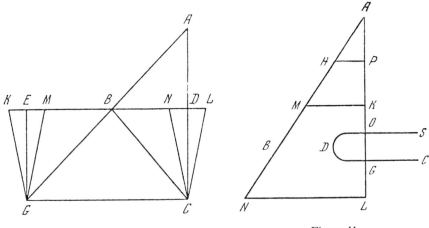

Figure 40   Figure 41

CD intersects the side *AL* of the resulting triangle *ALN* and, since it cannot intersect the side *LN* or intersect a second time the side *AL*, it necessarily intersects the side *AN*, that is, the line *AB*. In this proof Levi uses his first premise, that is, the Eudoxus-Archimedes axiom, and, implicitly, Pasch's axiom. What is refuted in his diagram (see Figure 41) is the assumption that the straight line *[CD]* intersects the side *AL* of the triangle *ALN* a second time. This proof is very similar to Ibn al-Haytham's proof. Like Ibn al-Haytham, Levi goes on to prove the parallel postulate for the case when two straight lines are intersected by a third at two acute angles and at an acute and obtuse angles.

The second proof of Euclid's parallel postulate, written in a Hebrew dialect used by Spanish Jews, is due to a certain Alfonso. Since its author wrote in Hebrew and had a Christian name, he must have been a baptized Jew. Of the many baptized Jews living in Spain and named Alfonso the most likely author of this treatise is Alfonso de Valladolid (1270–1346), known primarily as a physician and author of polemical works on religious topics as well as a cònnoisséur of calendars. Alfonso's treatise *Rectifier of the curved* (Meyashēr ʿāqom) *[17]*[10] is more the work of a philosopher than of a mathematician. Basically, the treatise deals with the quadrature of a circle and with other problems involving infinitesimal considerations. Here the question of the parallel postulate is an example of a fruitful application of motion in geometry. Alfonso begins by criticizing Aghānīs' proof, which he takes to be the proof of al-Nayrīzī, and then gives his own proof in six propositions. In the first proposition he proves that if we extend a median in a triangle by an amount equal to that median then we obtain a quadrilateral whose opposite sides and angles are equal. This is a theorem of absolute geometry. In the second proposition he applies the previous construction to a right triangle and, using the "simple motion" of Ibn Qurra and Ibn al-Haytham, shows that the resulting quadrilateral is a rectangle. In propositions 3–5 Alfonso relies on the "proved" existence of a rectangle to establish various facts implied by the parallel postulate. They are that the two acute angles in a right triangle add up to a right angle; that the "right-angle hypothesis" holds for a Khayyām quadrilateral and that its upper and lower bases are equal; and (the theorem) that the perpendicular dropped from the midpoint of the hypotenuse of a right triangle to a leg halves it. In proposition 6, Alfonso, using the Eudoxus-Archimedes axiom, deduces the parallel postulate from the previous propositions *[17, pp. 54–64]*.

The Theory of Parallels in the 16th Century

The first Latin work devoted to a study of the theory of parallel lines was *The astronomical Mirror which terminates the human intellect in every science*

---

[13] S. Ya. Lur'e *[343, p. 20]* was the first historian of science to call attention to the manuscript of *[17]*.

(Speculum astronomicum terminans intellectum humanum in omni scientia. Venice, 1507). Its author, born in the city of Zadar (now in Yugoslavia), was Federik Bartolaćić Grisogono (1472–1538), who worked in Italy. This work was recently studied by Ernest Stipanić *[561]*.

In the 9th chapter, entitled "On parallel lines," Grisogono criticizes "many mathematicians, ancient, Arabic and Latin," who tried to prove Euclid's parallel postulate starting with the definition of parallel lines as equidistant. Although admitting that he has not yet found a solution, Grisogono expresses an interesting thought:

> I imagined a way of drawing on one surface two nonequidistant straight lines which can be indefinitely prolonged and yet may never meet *[561, p. 371]*.

E. Stipanić renders the Latin *superficies*—surface—as the Serbo-Croat *ravan*—plane—and writes that the lines mentioned by Grisogono occur in the Lobačevskian plane. The fact is that nowhere does Grisogono mention that his surface is flat, and it is conceivable that he had in mind the property of rectilinear generators of a one-sheeted hyperboloid. Be that as it may, Grisogono's reflections are extremely interesting.

We mentioned a number of authors who wrote in Greek and Arabic and who tried to prove the parallel postulate from the definition of parallel lines mentioned by Grisogono. It appears that Grisogono was familiar with some of these "proofs." But we know only Vitello's work in Latin dealing with this question and written before 1507, and Grisogono's evidence about the existence of many such works is very interesting. The first Latin exposition of a "proof" of the parallel postulate that we know of is the exposition of Proclus' "proof" in the Latin translation of Euclid's *Elements* by the Italian mathematician Federigo Commandino (1509–1575) *[174]*. published in 1572.

An original "proof" of the parallel postulate is due to Christopher Clavius (Schlüssel, 1537–1612), a German born in Bamberg who worked in Rome and took part there in the elaboration of the Gregorian calendar. Clavius' proof is found in an exposition of Euclid's *Elements [117]* first published in 1574.

Clavius' proof is based on the theorem that

> A line each of whose points is at the same distance from a coplanar straight line is a straight line *[117, p. 50]*.

Clavius proves this from Euclid's definition of a straight line as "a line which lies evenly with the points on itself." Clavius thinks that since all the points of the line he considers are at the same distance from the straight line, this line "lies evenly" with the points on it. Clavius refutes the possibility that this line is a circle, for a line equidistant from a circle is itself a circle, which means that if the line under consideration were a circle then the original line would likewise be a circle. Clavius deems it obvious that the only lines which "lie evenly with the points on themselves," that is, essentially, can be superposed on themselves, are a straight line and a circle. In Lobačevskian geometry such a line, other than a straight line and a circle, is an equidistant curve, which, in that geometry, is the locus of points equidistant from a straight line.

Clavius' starting position is very similar to Ibn al-Haytham's. Clavius does not mention Ibn al-Haytham by name but says that

> I learned that this was also done in a certain Arabic Euclid but I never had the opportunity to read this proof in spite of the fact that I insistently, and on a number of occasions, asked this of the one who owns this Arabic Euclid *[117, p. 50]*.

Apparently, Clavius had only second-hand knowledge of Ibn al-Haytham's proof. The following propositions of Clavius are also similar to those of Ibn al-Haytham:

> If a straight line moves along another straight line, always forming with its end a right angle, then its other end traces a straight line;
>
> If two equal perpendiculars are erected on a straight line and their endpoints are joined by a straight line then the perpendicular dropped from an arbitrary point of that straight line to the first straight line is equal to the first perpendicular *[117, pp. 51–52]*.

In the next proposition Clavius considers a Khayyām-Saccheri quadrilateral and shows that its two upper angles are equal. Unlike al-Haytham and Khayyām, Clavius does not consider the three hypotheses but erects a perpendicular EF at the midpoint of the lower base of the quadrilateral $ABCD$ with two equal angles (Figure 42). By the previous proposition, this perpendicular is equal to the sides CA and DB of the quadrilateral with two equal angles. After proving that the upper angles of the quadrilateral $ABCD$ are right angles Clavius notes that the same holds for the quadrilaterals $AEFC$ and $EBDF$, whence it follows that all upper angles of these quadrilaterals are equal to one another. Since the angles CFE and DFE are adjacent, all of these angles are right angles.

Having proved the existence of a rectangle, Clavius "proves" the parallel postulate. He proceeds like Ibn al-Haytham, al-Ṭūsī, and Levi ben Gerson in

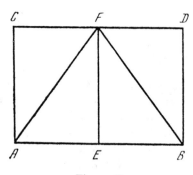

Figure 42

that he deals first with the case of a perpendicular and an oblique straight line and then with the general case.

Cataldi's "Proof" of the Parallel Postulate

At the very beginning of the 17th century there appeared *A small work on equidistant and nonequidistant straight lines* (Operetta delle linee rette equidistanti et non equidistanti. Bologna, 1603) *[98]* and *A supplement to a small work on equidistant and nonequidistant lines* (Aggiunta all' Operetta delle linee equidistanti et non equidistanti. Bologna, 1604) *[99]* by the Italian mathematician Pietro Antonio Cataldi (1548–1626), famous for his discovery of continued fractions. Cataldi's first work contains two definitions and propositions 1–15, and his second work propositions 16–33 and the concluding proposition in which he "proves" the parallel postulate. Definition 1 defines the distance from a point to a line, and definition 2 defines equidistant and nonequidistant lines. From the assumed existence of equidistant lines Cataldi deduces in his first work a number of assertions from which it is already possible to prove the parallel postulate. One such assertion is proposition 2,

If two straight lines are equidistant then the lines drawn perpendicularly from the first to the second are perpendicular to the first line *[98, p. 3]*,

in which he proves the existence of a rectangle. Another is one of the consequences of proposition 10:

If we add the three angles in any triangle then the sum is two right angles *[98, p. 16]*.

Cataldi's proof of the parallel postulate is based on proposition 30:

If two nonequidistant straight lines converge in one direction, then it is unavoidable that, when extended in the direction in which they come closer, they should ultimately intersect *[99, p. 59]*.

We note that the formulation of this assertion is the same as that of the "principle" attributed by Khayyām to Aristotle.

"Proofs" of the Parallel Postulate by Borelli and by Vitale Giordano

There are two more 17th-century "proofs" of the parallel postulate due to Italian mathematicians. In his *Euclid restored* (Euclides restitutus. Pisa, 1658) *[73]*, Giovanni Alfonso Borelli (1608–1679) proceeds in his proof from the following "axiom 14":

If a straight line which remains always in the same plane as a second straight line, moves so that one end always touches this line, and during the whole displacement the first remains continually perpendicular to the second, then the other end, as it moves, will describe a straight line *[73, p. 32; 71, p. 13]*.

Since the phrase "is transferred across" means transfer at a constant angle to the given line, Borelli's assertion is essentially the same as those of Ibn Qurra and Ibn al-Haytham. Like these Arab mathematicians before him, Borelli tries to justify this assertion by means of kinematic considerations. He proves that two perpendiculars to the same straight line are equidistant, defines parallel straight lines as equidistant lines, and deduces the parallel postulate from the existence of a rectangle.

A similar "proof" of the parallel postulate is found in *Euclid restored, or restored and simplified ancient elements of geometry in* 15 *books* (Euclide restituto, overo gli antichi elementi geometrici ristaurati, e facilitati libri XV. Rome, 1680) *[203]* of Vitale Giordano (1633–1711). Giordano "proves" that the locus of points equidistant from a straight line is a straight line by means of the following lemma:

If two points, *A*, *C* upon a curve, whose concavity is towards *X*, are joined by the straight line *AC*, and perpendiculars are drawn from the infinite number of points of the arc *AC* upon any straight line, then these perpendiculars cannot be equal to each other *[203, p. 4; 71, p. 14]*.

Giordano considers the straight line *GF* and joins the ends of the equal perpendiculars *AG* and *DBF* by means of the straight line *AC* (Figure 43). Of course, the line *ABC* is not equidistant from the straight line *GF*, but in "proving" that the locus of points equidistant from a straight line is a straight line Giordano applies this lemma to lines for which the relations between the straight line *GF* and the line *ABC* do not hold. Further, Giordano considers a Khayyām quadrilateral and proves that its upper angles are equal and from the existence of a rectangle deduces the parallel postulate.

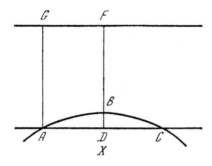

Figure 43

Wallis's Theory of Parallel Lines

John Wallis (1616–1703), one of the greatest English mathematicians of the 17th century, also wrote a treatise dealing with the parallel postulate and related matters, under the title *On the fifth postulate and fifth definition in book 6 of Euclid*; *a geometric discourse* (*De postulato quinto et definitione quinta lib. 6 Euclidis*; *disceptatio geometrica*. Oxford, 1693) *[617, vol. 2, pp. 665–678]*. The treatise consists of three parts. The first part deals with the so-called fifth definition of book VI of the *Elements*, a definition of composite ratio which is a later insertion. The second part is Edward Pococke's translation from the Arabic of the "proof" of the parallel postulate taken from the previously mentioned 1594 Roman edition of the *Exposition of the* "*Elements*" *of Euclid* attributed to al-Ṭūsī. The third part is Wallis's proof of the parallel postulate. It rests on the following postulate:

> Finally (supposing the nature of ratio and of the science of similar figures already known), I take the following as a common notion: to every figure there exists a similar figure of arbitrary magnitude *[617, vol. 2, p. 676]*.

Wallis justifies the naturalness of this common notion, that is, axiom, with the argument that Euclid's postulate III ("to describe a circle with any center and distance") is a special case of this "common notion." In order to prove that two straight lines *AB* and *CD* forming with a straight line *AC* angles whose sum is less than two right angles meet (Figure 44), Wallis moves the straight line *AB* along the straight line *AC* so that the angle *CAB* remains constant. If one comes sufficiently close to the point *C*, then the straight line will occupy a position *αβ* in which it intersects the straight line *CD* at the point *π*. In view of Wallis's assumption, there is a triangle *ACP* similar to the resulting triangle *αCπ*; the required point is the point *P*.

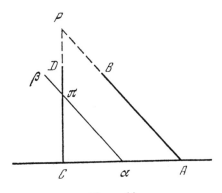

Figure 44

Saccheri's Theory of Parallel Lines

In the first half of the 18th century the Italian Girolamo Saccheri (1667–1733) made what turned out to be an important attempt to prove the parallel postulate. He did this in a work entitled *Euclid cleared of every flaw or A geometrical attempt to establish the very first elements of all geometry* (Euclides ab omni naevo vindicatus sive Conatus geometricus quo stabiliuntur prima ipsa universae geometriae principia. Milan, 1733) *[495; 168, pp. 45–135]*. Saccheri criticizes Wallis's proof of the parallel postulate as well as the proof attributed to al-Ṭūsī. In his own proof Saccheri considers the same isosceles quadrilateral with two right angles as Khayyām and al-Ṭūsī (we have already mentioned that this quadrilateral is often referred to as a "Saccheri quadrilateral") and states the same three hypotheses about its upper angles as Khayyām and al-Ṭūsī.

Saccheri refutes the obtuse-angle hypothesis by showing that under this hypothesis, as well as under the right-angle hypothesis, the parallel postulate holds. From this he makes the deduction that under the obtuse-angle hypothesis we must have the usual geometry in which, "as is clear to all geometers," the right-angle hypothesis holds. This being so,

> The hypothesis of obtuse angle is completely false, because it destroys itself *[495, p. 59]*.

The sense of this proof is that since the remaining axioms of Euclidean geometry and the parallel postulate imply the right-angle hypothesis, the obtuse-angle hypothesis, under which the parallel postulate holds, contradicts the remaining axioms of Euclidean geometry.

Now Saccheri sets about refuting the acute-angle hypothesis. Here Saccheri penetrates far deeper into Lobačevskian geometry, in which this hypothesis holds, than his predecessors. He shows that under the acute-angle hypothesis two straight lines intersect, or have a common perpendicular on each side of which they diverge, or diverge in one direction and come asymptotically close to one another in the other direction. In the latter case, Saccheri concludes that these straight lines must have a common point and a common perpendicular at infinity. The sense of this deduction is that if one drops perpendiculars from the points of one of these straight lines to the other, then, as the point tends to infinity, the length of the perpendicular tends to zero and the angle between the first straight line and the perpendicular tends to a right angle. But Saccheri envisages the common point and common perpendicular at infinity as an ordinary common point and an ordinary common perpendicular and deduced that in the third case the two straight lines touch at infinity. From this he concludes that

> The hypothesis of acute angle is absolutely false; because repugnant to the nature of the straight line *[495, p. 173]*.

Not satisfied with this proof Saccheri considers the locus of points in the plane equidistant from a straight line. Unlike his predecessors, he is aware that under the acute-angle hypothesis this line is neither a straight line nor a circle. In computing the length of an arc of this curve by means of infinitesimals Saccheri makes a mistake and concludes that the required length is equal to the distance between the feet of the perpendiculars dropped from the ends of the arc to the straight line. On the other hand, Saccheri has shown that the perpendiculars move apart, so that the distance between the ends of the arc, to say nothing of the arc length, is greater than the distance between the feet of the perpendiculars. Having found this contradiction Saccheri again declares that

The hypothesis of acute angle is absolutely false, because it destroys itself *[495, p. 225]*.

Saccheri's concluding remark, however, shows that he is not entirely satisfied:

It is well to consider here a notable difference between the foregoing refutations of the two hypotheses. For in regard to the hypothesis of obtuse angle the thing is clearer than midday light. . . .
But on the contrary I do not attain to proving the falsity of the other hypothesis, that of acute angle, without previously proving that the line, all of whose points are equidistant from an assumed straight line lying in the same plane with it is equal to this straight line *[495, p. 233]*.

Notwithstanding the falsity of his deductions, Saccheri's investigations of geometry under the acute-angle hypothesis were an important step on the road to the discovery of non-Euclidean geometry.

We note that the existence of a rectangle, proved by Saccheri and many of his predecessors, was explicitly adopted as a foundation for an exposition of the theory of parallels by the eminent 18th-century mathematician Alexis Claude Clairaut (1713–1765) in his *Elements of geometry* (Éléments de géométrie. Paris, 174). Clairaut justifies the existence of a rectangle by the "form of houses, gardens, rooms, walls" *[115, p. 4]*.

Lambert's Theory of Parallels

In the second half of the 18th century, Johann Heinrich Lambert (1728–1777), born at Mulhouse in Alsace, one of the greatest mathematicians of the century, wrote a special treatise under the title *Theory of parallel lines* (Theorie der Parallellinien. Leipzig, 1786) *[168, pp. 152–207]*. Lambert worked in Munich and in Berlin and is known for his proofs of the irrationality of *e* and π. In the introductory part of his treatise Lambert writes:

This work deals with the difficulty encountered in the very beginnings of geometry and which, from the time of Euclid, has been a source of discomfort for those who do not just blindly follow the teachings of others but look for a basis for their convictions and do not wish to give up the least bit of the rigor found in most proofs. This difficulty immediately confronts every reader of Euclid's *Elements*, for it is concealed not in his propositions but in the axioms with which he prefaced the first book *[168, p. 152]*.

Then Lambert formulates the 11th axiom (the parallel postulate) and comments:

> Undoubtedly, this basic assertion is far less clear and obvious than the others. Not only does it naturally give the impression that it should be proved, but to some extent it makes the reader feel that he is capable of giving a proof, or that he should give it.
>
> However, to the extent to which I understand this matter, this is just a *first* impression. He who reads Euclid further is bound to be amazed not only at the thoroughness and rigor of his proofs but also at the well-known delightful simplicity of his exposition. This being so, he will marvel all the more at the position of the 11th axiom when he finds out that Euclid proved propositions that could far more easily be left unproved *[168, p. 153]*.

Then Lambert considers the same quadrilateral as Ibn al-Haytham (we have already mentioned that this quadrilateral is often called the *Lambert quadrilateral*) and the same hypotheses of a right, obtuse, and acute angles.

Like Saccheri, Lambert refutes the obtuse-angle hypothesis. He does this by showing that under the obtuse-angle hypothesis two perpendiculars to the same straight line must intersect. This does not contradict the parallel postulate, but it contradicts the remaining axioms of Euclidean geometry. Lambert notes that the obtuse-angle hypothesis holds on a sphere if we regard its great circles as straight lines.

Going over to the acute-angle hypothesis, Lambert proves even more assertions that hold in Lobačevskian geometry than had Saccheri. In particular, he finds that under the acute-angle hypothesis the sum of the angles in a triangle is less than two right angles. Comparing this fact with the theorem that under the obtuse-angle hypothesis the angle sum in a triangle is more than two right angles, Lambert says:

> It is easy to see that under the third hypothesis one can go even further and that analogous, but diametrically opposite, consequences can also be found under the second hypothesis. But, for the most part, I looked for such consequences under the third hypothesis in order to see if contradictions might not come to light. From all this it is clear that it is no easy matter to refute this hypothesis. I will cite some more consequences without considering to what extent they can be extended, *mutatis mutandis*, under the second hypothesis.

The most striking of these consequences is that *under the third hypothesis we would have an absolute measure of length for every line, of area for every surface and of volume for every physical space.* This refutes an assertion that some unwisely hold to be an axiom of geometry, for until now no one has doubted that there is no absolute measure whatsoever. There is something exquisite about this consequence, something that makes one wish that the third hypothesis be true!

In spite of this gain I would not want it to be so, for this would result in countless inconveniences. Trigonometric tables would be infinitely large, similarity and proportionality of figures would be entirely absent, no figure could be imagined in any but its absolute magnitude, astronomers would have a hard time, and so on.

But all these are arguments dictated by love and hate, which must have no place either in geometry or in science as a whole.

To come back to the third hypothesis. As we have just seen, under this hypothesis the sum of the three angles in every triangle is less than 180 degrees, or two right angles. But the difference up to 180 degrees increases like the area of the triangle; this can be expressed thus: if one of two triangles has an area greater than the other then the first has an angle sum smaller than the second. . . .

I will add just the following remark. Entirely analogous theorems hold under the second hypothesis except that under it the angle sum in every triangle is greater than 180 degrees. The excess is always proportional to the area of the triangle.

I think it remarkable that the second hypothesis holds if instead of a plane triangle we take a spherical one, for its angle sum is greater than 180 degrees and the excess is also proportional to the area of the triangle.

What strikes me as even more remarkable is that what I have said here about spherical triangles can be proved independently of the difficulty posed by parallel lines and upon assuming solely the axiom that every plane through the center of a sphere divides it into two equal parts.

From this I should almost conclude that the third hypothesis holds on some imaginary sphere. At least there must be something that accounts for the fact that, unlike the second hypothesis, it has for so long resisted refutation on planes *[168, pp. 200–203]*.

In spite of these sentiments Lambert also "disproves" the acute-angle hypothesis. To this end, he erects perpendiculars of equal length to a line at equidistant points of this line, joins their ends, and obtains a broken line. He states that the vertices of this broken line lie on a circular arc and obtains a contradiction. (In hyperbolic geometry these vertices lie on an equidistant curve.) This spurious contradiction "disproved" the acute-angle hypothesis. Lambert's error was first brought to light by Boris Lukič Laptev (b. 1905) *[305]*.

Bertrand's "Proof" of the Parallel Postulate

In 1778 the Swiss mathematician Louis Bertrand (1731–1812), a student of Euler's, published a clever "proof" of the parallel postulate. This "proof" appeared in the second volume of his *New exposition of the elementary part of mathematics* (Development nouveau de la partie élémentaire des mathématique. Geneva, 1778) and was reissued in his *Elements of geometry* (Éléments de géométrie. Paris, 1812) *[51]*. The "proof" is based on operations with infinitely large magnitudes and consists in the following: suppose that the lines *LC* and *KA* (Figure 45) form with the line *KL* interior angles *AKL* and *CLK* whose sum is less than two right angles. Then there exists a straight line *LM* that forms with *LC* an angle *CLM* such that the sum of the three angles *AKL*, *CLK*, and *CLM* is equal to two right angles. Hence if the straight line *LC* did not intersect the straight line *KA* then the angle *MLC* would be contained inside the strip *MLKA*. But this strip is contained in the plane "infinitely many times," whereas the angle *MLC* is contained in it only as many times as the arc *MC* is contained in the circle with center *L* and radius *LM*. From this Bertrand concludes that the angle *MLC* cannot be entirely contained in the strip *MLKA*. But then its side *LC* must leave that strip, and so intersects *KA*.

The fact is that neither the whole plane nor its part bounded by the sides of the angle can be considered as a magnitude that admits of numerical comparisons; for example, if we move the angle in the plane into the domain bounded by its sides and argue like Bertrand, then we obtain an absurd equality between the "magnitude of the angle" and its sum with the "magnitude" of the domain bounded by the first and second positions of the angle. And yet even serious mathematicians of the beginning of the 19th century found the intuitive imagery of Bertrand's argument convincing, and A. Crelle, the editor of the "Journal fur die reine und angewandte Mathematik," who in 1834 published one of Lobačevskiĭ's papers on the geometry he discovered, published in 1835 a modified version of Bertrand's proof.

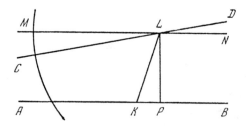

Figure 45

"Proofs" of the Parallel Postulate by Legendre and Gur'ev

At the end of the 18th and in the beginning of the 19th century the great French mathematician Adrien Marie Legendre (1752–1833) made a number of attempts to prove the parallel postulate in his *Elements of geometry* (Éléments de géométrie. Paris, 1794–1823) *[309–311]*—a textbook of elementary geometry that continued the tradition of A. C. Clairaut's textbook with the same title.

In the first edition of the *Elements of geometry* Legendre gave the following proof of the parallel postulate. Suppose that the straight line *BD* is perpendicular to the straight line *AB* and the straight line *AC* forms with it an acute angle *BAC* (Figure 46). Then the foot *G* of the perpendicular *FG* dropped from some point *F* on the line *AC* cannot coincide with the point *A* and cannot end up on the extension *AL* of the line *AB* on the other side of *A*. The first possibility cannot occur because the angle *BAF* is acute. The second possibility cannot occur because, if the point *G* coincided with a point *H* on the line *AL*, then the perpendicular *FH* would intersect the perpendicular *AE* erected at the point *A* at some point *K* and we would have two perpendiculars from that point to the straight line *AL*. It follows that the foot *G* of the perpendicular is on the line *AJ*. Similarly the foot *M* of the perpendicular dropped from the point *C* to the line *AB* cannot coincide with the point *G* or fall on the line *GL*, and the foot *N* of the perpendicular *PN*, dropped from some point *P* of the extension of the line *AC*, cannot coincide with the point *M* or fall on the line *ML*, and so on. Also, if a point on the straight line *AC* moves away from *A* then the foot of the perpendicular from that point to the straight line *AB* also moves away from *A*. None of the feet of these perpendiculars can be a last one, for the assumption that the foot *N*, say, is last contradicts the fact that

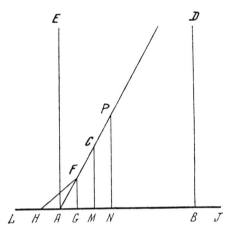

Figure 46

there are points on the straight line AC that are farther away from A then the corresponding point P and the feet of the perpendiculars from such points are farther away from the point A than N. From this Legendre concluded that the distances from A of the feet of the perpendiculars dropped from the points on the straight line AC to the straight line AB can be arbitrarily large and therefore one of them coincides with the point B. But then the perpendicular BD must also have been dropped to the straight line AB from some point on the straight line AC, and it is this very point that is the point of intersection of the straight lines AC and AB. Thus Legendre "proved" that a perpendicular and an oblique line must intersect and from this it is not difficult to deduce the general case of the parallel postulate.

The mistake in this argument was quickly brought to light by the Russian academician Semen Emel'yanovič Gur'ev (1746–1813) in *An attempt to perfect the elements of geometry* (Opyt ob usoveršenii elementov geometrii. Petersburg, 1798) *[213]*. Gur'ev pointed out that just as the monotonic increase of the partial sums of a convergent series of positive terms does not imply that these partial sums can exceed the sum of the series, so too the monotonic increase of the distances of the feet of the perpendiculars from the point A does not at all imply that these distances can be made arbitrarily large. In this connection we note that the same kind of "convergence" of the feet of perpendiculars is involved in al-Ṭūsī's attempt to refute the hypotheses of the obtuse and acute angles in a Khayyām-Saccheri quadrilateral. Legendre himself was dissatisfied with his proof and in the third edition of the *Elements of geometry* proposed a new "proof" based on the proposition that the angle sum in a triangle is equal to two right angles. To prove this proposition, Legendre shows first that the angle sum in question is not greater than two right angles. Suppose it is; specifically, suppose the angle sum in the triangle ABC is greater than $2d$ (Figure 47). On the extension of the side AC Legendre constructs triangles $CDE, \ldots, NOP$ congruent to the triangle ABC and joins the vertices $B, D, \ldots, O$ by means of straight lines. The sum of the angles ACB, BCD, and DCE is $2d$. Since, by construction, angles DCE and CAB are equal and the angle sum of triangle ABC is greater than $2d$, it follows that angle BCD is smaller than angle ABC. Since the sides AB, BC of triangle ABC are respectively equal to the sides CD, CB of triangle BCD, it follows that $AC > BD$. Let $AC - BD = \delta$. If there are n triangles $ABC, CDE, \ldots, NOP$,

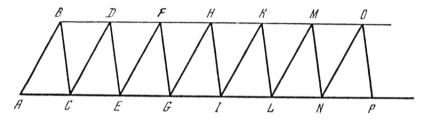

Figure 47

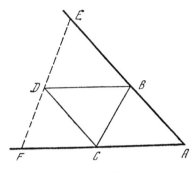

Figure 48

then the length of the segment AP is $n \cdot AC$ and the length of the polygonal line $ABDF, \ldots, OP$ is $AB + BC + (n-1)BD = AB + BC + AP - BD - n\delta$ (since $AC - BD = \delta$). Therefore the difference between the polygonal line and the segment AP is $AB + BC - BD - n\delta$, which can be made negative by taking n sufficiently large. Then the polygonal line would be shorter than the segment joining its endpoints, which is impossible.

Then Legendre tries to prove that the angle sum in a triangle cannot be less than $2d$. He assumes that the angle sum in the triangle ABC is smaller than $2d$ (Figure 48) and argues as follows. Let A be the smallest angle of triangle ABC. On the side of ABC opposite to A one constructs angles $DBC = ACB$ and $DCB = ABC$. The triangles BCD and ABC are congruent. One draws a line through D that intersects the sides of angle A in points E and F. By what has been proved, the angle sum in each of the triangles BDE and CDF is $\leq 2d$. The angle sum in each of the congruent triangles ABC and BCD is equal to $2d - \delta$ for some $\delta > 0$. Hence the angle sum in triangle AEF, which is 6d less than the sum of the angles in all four triangles, is less than $2d - 2\delta$. Continuing this process one obtains a sequence of triangles with angle sum, respectively, less than $2d - 4\delta$, $2d - 8\delta$, and so on. In this way one obtains a triangle with negative angle sum, which is absurd.

In this proof Legendre used the assertion that given a point in the interior of an acute angle one can always pass through it a straight line that meets both of its sides. Before Legendre, Simplicius, al-Jawharī, al-Abharī, and al-Ṭūsī all used this assertion to derive the parallel postulate.

Upon locating the error in his proof Legendre, in the 12th edition of *Elements of geometry*, proposed yet another proof of the proposition that the angle sum in a triangle is equal to $2d$. Thus consider the triangle ABC (Figure 49), in which AB is the largest side and BC the smallest, so that ACB is the largest angle and BAC the smallest. Let I be the midpoint of BC. On the ray AI lay off the segment $AC' = AB$ and on the ray AB lay off $AK = AI$ and $AB' = 2AI$. Then the triangles AKC' and $B'KC'$ are, respectively, congruent to the triangles ABI and AIC, and the angles A', B', C' of the triangle $AB'C'$ are connected with the angles A, B, C of the triangle ABC

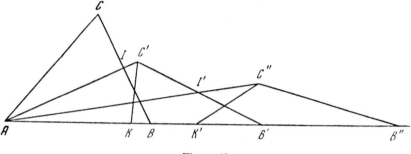

Figure 49

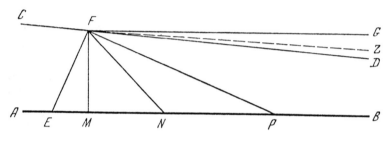

Figure 50

by means of the relations $C' = B + C$ and $A = A' + B'$, so that $A + B + C = A' + B' + C'$; since $AC < AB$, the angle A' is smaller than $\frac{1}{2}A$. Similarly, construct triangles $AB''C''$, $AB'''C'''$, and so on, whose angle sums remain constant and whose angles A'', A''', ... are, respectively, less than $\frac{1}{4}A$, $\frac{1}{8}A$, ... By repeating this step sufficiently many times one obtains a triangle whose angle at the vertex A is less than any preassigned number.

Now Legendre concludes that in the limit this angle will be equal to 0 and all vertices of the triangle will lie on the same line. But then one of the two remaining angles will be 0 and the other $2d$. Since in the process the angle sum remains unchanged, Legendre concludes that the angle sum must have been equal to $2d$ to begin with.

Here too Legendre made a mistake. In hyperbolic geometry the area of a triangle is determined by its angle sum. Hence a variable triangle with constant angle sum has constant area. But this means that its vertices cannot tend to three collinear points.

Legendre's argument leading from the assertion that the angle in a triangle is $2d$ to the parallel postulate is as follows. If the straight lines AB and CD form with the straight line EF angles BEF and DFE whose sum is less than $2d$ (Figure 50), then one draws the straight line FG at an angle EFG which supplements the angle BEF to $2d$. Then angle EFD is less than angle EFG. Next one draws an arbitrary straight line FM that intersects the straight line AB at a point M, lays off on the straight line AB the segment $MN = FM$, and

draws the line *FN*; in the isosceles triangle *FMN* each base angle equals half
the angle *FME* = *MFG*, so that the angle *NFG* is also equal to half the angle
MFG. Then one lays off on the straight line *AB* a segment *NP* = *FN*; in
the isosceles triangle *FNP* each of the base angles equals half the angle *FNE*,
that is, a quarter of the angle *MFG*, so that the angle *PFG* is also equal to
a quarter of the angle *MFG*. By continuing this process one obtains a sequence
of isosceles triangles whose bases make with the straight line *FG* angles that
form a geometric progression with multiplier $\frac{1}{2}$. This means that after
sufficiently many steps one obtains an angle that is less than any preassigned
magnitude and, in particular, less than the angle *DFG*. By then the line *FD*
will have ended up inside one of these triangles and therefore will have
intersected its side opposite the angle with vertex at *F*, that is, the straight
line *AB*.

Mistakes notwithstanding, the investigations of Legendre, like those of
Saccheri before him, have played an important role in the history of non-
Euclidean geometry. A particularly helpful factor was the wide dissemination
of Legendre's *Elements of geometry*.

We recall the name of S. E. Gur'ev, who first found the error in the first
edition of Legendre's *Elements of geometry*. In his *Attempt to perfect the
elements of geometry* Gur'ev also proposed a proof of the parallel postulate.
Gur'ev considers first the case when the straight lines *AC* and *BD* are cut by
the straight line *AB* perpendicular to *BD* (Figure 51). By dropping perpendiculars from the points on the straight line *AC* to the straight line *AB* one
obtains on the latter points such that perpendiculars erected at these points
intersect *AC*. If the perpendicular *BD* does not intersect *AC* then

among the perpendiculars erected on *AB* from *A* to *Z* there is one which
intersects *AC* and others that do not intersect *AC*.

Gur'ev concludes that

there is a common limit where certain perpendiculars end and others
begin, for without such a limit all perpendiculars would intersect *AC* ...

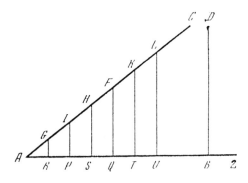

Figure 51

and thus such a limit is in order; I say that it does not exist, for no matter where we suppose it to be there are always perpendiculars that go beyond this limit and intersect AC: indeed, let the perpendicular KT be that limit. By taking on AC produced a point L beyond K and dropping from it a perpendicular LU you will find that there are a great many perpendiculars to AB erected between T and U that intersect AC and pass the supposed limit TK [213, pp. 237–238].

From the fact that there is no last perpendicular that intersects AC, Gur'ev draws the completely false conclusion that there is no first perpendicular among the perpendiculars that do not intersect AC; it is precisely this first perpendicular that is the "limit" Gur'ev has in mind. Having "proved" that the perpendicular and oblique line must intersect, Gur'ev easily proves the parallel postulate in the case when the transversal makes two acute angles with the two given straight lines as well as in the case when it makes with them an acute and an obtuse angle.

Farkas Bolyai's Theory of Parallels

In the first half of the 19th century there appeared several "proofs" of the parallel postulate by the Hungarian mathematician Farkas Bolyai (read: Farkash Boyai) (1775–1856). He was born in the small town of Bolya in Transylvania, studied with Gauss at the university of Göttingen, and became a professor of mathematics in the Reformed college at Maros-Vasarhely (now Tîrgu-Mureş in Rumania). While in Göttingen, Bolyai became interested in the theory of parallel lines and published *The theory of parallels* (Theoria parallelarum. Maros-Vasarhely, 1804) [67], in which he tried to prove the existence of equidistant straight lines. He sent this work to Gauss, who pointed out his error. After this, F. Bolyai made a number of attempts to prove the parallel postulate.

Under Farkas Bolyai's influence, his son János (read: Yanosh), subsequently one of the creators of non-Euclidean geometry, took an interest in the theory of parallels. Farkas Bolyai tried to dissuade János from the study of this theory. Mindful of his recurrent failures to prove the parallel postulate he wrote to his son:

I entreat you, leave the doctrine of parallel lines alone; you should fear it like a sensual passion; it will deprive you of health, leisure and peace—it will destroy all joy in your life. These gloomy shadows can swallow up a thousand Newtonian towers and never will there be light on earth; never will the unhappy human race reach absolute truth—not even in geometry [137, p. 9].

F. Bolyai's major work was a survey of attempts to prove the parallel postulate, with explanations of the hypotheses on which they were based.

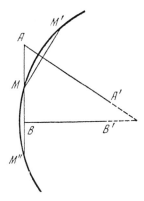

Figure 52

Its title is *An attempt to introduce young students to the elements of pure mathematics, both elementary and advanced, by means of a specially devised intuitive method* (Tentamen juventutem studiosam in elementa matheseos purae, elementaris ac sublimioris, methodo intuitiva, evidentiaque huic propria introducendi. Maros-Vasarhely, 1832) *[68]*. The most interesting proof of the parallel postulate is set forth in F. Bolyai's book *A short sketch of an attempt to* (1) *present arithmetic in a logically rigorous manner,* (2) *precisely define the concepts of geometry* (Kurzes Grundriss eines Versuches (1) die Arithmetik logisch-streng darzustellen, (2) in der Geometrie die Begriffe scharf zu bestimmen. Maros-Vasarhely, 1851) *[69]*.

The proof is based on the postulate that

Three points not on one line are always on some circle *[69, p. 246]*.

This assertion is equivalent to the parallel postulate. (In fact, in Lobačevskian geometry three noncollinear points determine a circle, an equidistant curve, or a horocycle.) F. Bolyai deduces the parallel postulate from this assertion as follows: Let AA' and BB' be two straight lines, one of which is perpendicular to and the other oblique relative to the straight line AB (Figure 52). Choose a point M on the segment AB and construct points M' and M'' symmetric to M relative to AA' and BB'. By F. Bolyai's postulate, since these three points are not collinear, they are concyclic. Hence the lines AA' and BB' intersect at the center of this circle. From the fact that a perpendicular and an oblique line intersect, F. Bolyai proves the parallel postulate in general.

This proof was published by F. Bolyai after the discovery of non-Euclidean geometry by his son and by N. I. Lobačevskiĭ but, regrettably, F. Bolyai failed to understand this discovery.[14]

---

[14] Many proofs of the parallel postulate made before and after the discovery of non-Euclidean geometry are analyzed in the book *[475a]* of J.-C. Pont, published in 1986.

Chapter 3
Geometric Transformations

Application of the Concept of Motion in Geometry in Antiquity and in the Middle Ages

On a number of occasions we have come across the use of motions in geometry. In book I Euclid uses superposition in propositions 7 and 8 (theorems on the congruence of triangles) and later relies on these propositions. Although his definition of a circle (definition 15, book I) *[173, vol. I, p. 153]* does not involve the concept of motion, his definitions of a sphere, a circular cone, and a cylinder (definitions 14, 18, and 21 in book XI) do *[173, vol. 3, pp. 261–262]*. The motions involved are, respectively, the rotation of a semicircle about its diameter, of a right triangle about a leg, and of a rectangle about a side. The use of motions in these definitions seems to reflect an older tradition. In fact, we saw that in Theodosius' later *Sphaerica [578, p.1]* a sphere is defined without the use of motions, in a manner analogous to Euclid's definition of a circle.

That motions were extensively used in geometry before Euclid is apparent from, say, the formulations of the theorems of Thales (sixth century B.C.), the first Greek scholar credited with proving theorems. In his commentaries on Euclid, Proclus pointed out that Eudemus, in his *History of geometry*, attributed to Thales the proofs of the following theorems: That a circle is divided into equal parts by its diameter; that the base angles in an isosceles triangle are equal; that when two straight lines intersect, angles are equal; that two triangles having a side and two angles, respectively, equal are themselves equal *[440, pp. 124–125, 195, 233, 275]*. These proofs were not based on axioms and other theorems for, at that time, there were neither axioms nor other theorems. We see that Thales' theorems concerned the congruence of semicircles, angles, and triangles. Doubtless Thales proved these theorems by folding drawings or by other means of superposing figures. Thales referred to congruent figures as similar; after the formulation of the theorem on the angles in an isosceles triangle Proclus notes that

in ancient fashion, he [Thales] called these angles not equal but similar *[440, p. 195]*.

It appears that the term "equal" for figures of the same size is due to the Pythagoreans, who thought that such figures consisted of equal numbers of points. Later the term "similar figures" acquired the modern meaning, and Euclid and his followers called congruent figures "similar and equal."

Motions were used systematically by the Pythagoreans, who regarded lines as traces of moving points and surfaces as traces of moving lines. Aristotle debated this view in his treatise *On the soul* (Peri psychēs), known as *De anima*, in these words:

Since they say a moving line generates a surface and a moving point a line, the movements of the psychic units must be lines (for a point is a unit having position, and the number of the soul is, of course, somewhere and has position) *[29, vol. 3, p. 409ᵃ]*.

Later, the Pythagorean Archytas *[26, vol. 3, pp. 98–111]* solved the classical Delian problem of doubling a cube with side a by considering (in modern notation) the point of intersection of the cylinder $x^2 + y^2 = 2ax$, the cone $x^2 + y^2 + z^2 = 4x^2$, and the torus $x^2 + y^2 + z^2 = 2a\sqrt{x^2 + y^2}$, obtained by rotating the circle $x^2 + z^2 = 2ax$ about the z-axis. (The coordinates x, y, z of the point of intersection satisfy the proportions

$$\frac{2a}{\sqrt{x^2 + y^2 + z^2}} = \frac{\sqrt{x^2 + y^2 + z^2}}{\sqrt{x^2 + y^2}} = \frac{\sqrt{x^2 + y^2}}{a}.$$

The segment joining the origin to this point is the required side of the doubled cube.)

Aristotle condemned the use of motions in geometry. In his view, "The objects of mathematics are without movement" *[29, vol. 8, p. 989ᵇ]*. This view derives from the fact that Aristotle regarded mathematical objects as abstractions of physical objects. In his *Metaphysics* Aristotle states:

As the mathematician investigates abstractions (for before beginning his investigation he strips off all the sensible qualities, e.g. weight and lightness, hardness and its contrary, and also heat and cold and the other sensible contrarieties, and leaves only the quantitative and continuous, sometimes in one, sometimes in two, sometimes in three dimensions *[29, vol. 8, p. 1061ᵃ]*.

Aristotle regarded a surface as more abstract than a solid (for it is devoid not only of "sensible properties" but also of thickness), a line as more abstract than a surface (for it is devoid of breadth), and a point as more abstract than a line (for it is devoid of length). Hence a line cannot consist of points, a surface of lines, or a solid of surfaces. To repeat a quotation from Aristotle:

Nothing that is continuous can be composed of indivisibles: e.g., a line cannot be composed of points, the line being continuous and the point indivisible [29, vol. 2, p. 231ª].

It follows that one cannot obtain a line by moving a point, a surface by moving a line, and a solid by moving a surface. This explains why Euclid made every effort to avoid using motions and superpositions. When (as in the definitions in book XI) he used motions without having to do so, Euclid followed an older tradition.

We saw that Ibn Qurra was critical of Aristotle's position and extensively used motions in geometry. The same is true of Ibn al-Haytham. On the other hand, Khayyām shared Aristotle's view and criticized Ibn al-Haytham in his commentaries on Euclid:

> What is the connection between geometry and motion, and what is meant by motion? According to scholars, there is no doubt that a line can only exist on a surface, and a surface in a solid, that is, a line can exist only in a solid and cannot precede a surface. How could it move apart from its object? How could a line be the result of the motion of a point if it precedes a point by its essence and by its existence? [272, p. 115].

Nevertheless, in his proof of the fifth postulate, Khayyām—as we saw—more than once resorted to the folding of figures.

As a rule, later Near Eastern, Middle Eastern, and West European scholars systematically used motions in their geometric works.

Geometric Transformations in the Works of Archimedes

The earliest geometric transformations more complex than motions are *axial affinities* and *central dilatations* or *homotheties*. An axial affinity with *axis a* and *ratio k* maps each point A into a point A' such that the lines AA' are parallel but not parallel to a and—with A_0 defined as the point of intersection of the lines a and AA'—the ratio A_0A'/A_0A has same value k for all points A. Here we assume that if $k \neq -1$, then $k > 0$. An axial affinity is called *right* if AA' is perpendicular to a and *skew* otherwise. An axial affinity with $k = -1$ is called an *affine reflection*. A central dilatation with *center A_0* and *ratio k* maps each point A into a point A' such that the ratio A_0A'/A_0A has the same value k for all points A. A central dilatation with $k = -1$ is called a "half-turn" or a "reflection in a point." Figure 53a depicts an axial affinity with axis a and ratio 2, and Figure 53b a central dilatation with center A_0 and ratio 2.

If the x-axis of a rectangular or skew coordinate system coincides with the axis of an axial affinity and the y-axis with one of the lines AA', then its analytic description is

$$x' = x, \qquad y' = ky. \tag{3.1}$$

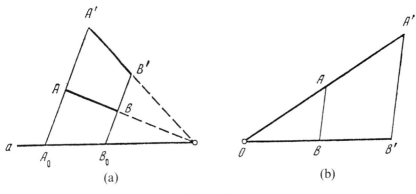

Figure 53

Similarly, if the origin of a rectangular or skew coordinate system coincides with the center of a central dilatation, then its analytic description is

$$x' = kx, \qquad y' = ky. \tag{3.2}$$

It seems that axial affinities first appeared in Archimedes' treatise *On conoids and spheroids* (Peri kōnoeideōn kai sphairoeideōn *[25, pp. 99–150]* dealing with the computation of the volumes of segments of ellipsoids of rotation (*spheroids*), two-sheeted hyperboloids of rotation (*obtuse-angled conoids*), and paraboloids of rotation (*right-angled conoids*). The names of the conoids reflect the fact that Archimedes used the pre-Apollonian terms *section of an acute-angled, right-angled*, and *obtuse-angled cone* for an ellipse, parabola, and hyperbola. Proposition 4 of this treatise states that

The area of any ellipse is to that of the auxiliary circle as the minor axis to the major *[25, p. 113]*.

Archimedes reasons as follows: Let *BD* be the smaller axis of the ellipse and *EG* the diameter of the circle *AECG* constructed on the larger axis of the ellipse as diameter, and such that *EG* and *BD* lie on the same straight line. Let *Z* (Figure 54) be the circle whose area is to the area of the circle *AECG* as *BD* is to *EG*. Then "the circle *Z* is equal to the ellipse." Assuming that *Z* is larger than the ellipse, Archimedes inscribes in the circle *Z* a polygon with an even number of sides that is larger than the ellipse, inscribes a similar polygon in the circle *AECG*, joins the pairs of vertices symmetric with respect to the larger axis of the ellipse by means of straight lines, marks off the points in which these lines meet the ellipse, and shows that the resulting polygon is to the polygon inscribed in the circle as *BD* to *EG*. Also, this polygon is in that same ratio to the polygon inscribed in the circle *Z*. But then the polygon inscribed in the ellipse is equal to the polygon inscribed in the circle *Z*, contradicting the fact that the latter polygon is larger than the ellipse. The assumption that the circle *Z* is smaller than the ellipse is disproved by a similar argument.

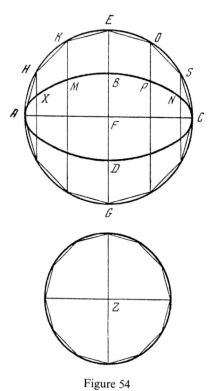

Figure 54

We see that Archimedes uses a right axial affinity with ratio equal to the ratio of the axes of the ellipse and shows that the ratio of the areas of his inscribed polygons is equal to the ratio of the areas of the figures obtained from these polygons by letting the number of sides tend to infinity. From this it follows that the area of an ellipse with semiaxes a and b is πab and the radius of the circle Z is $\sqrt{ab}$.

Inversions in the Works of Apollonius

It seems that central dilatations appeared first in Apollonius' treatise *On plane loci* (Peri topoi epiphanoi), which we know about through Pappus' *Mathematical collection* (Synagōgē mathēmatikē) *[404]* of the third century A.D. In the same work Apollonius also considers *reflections in circles* or *inversions*. An inversion in a circle with center O and radius R maps a point A other than O to a point A' on the ray OA such that $OA \cdot OA'$ is a constant. If that constant is R^2, then the points of the circle are mapped onto themselves, the points of the interior are mapped to the points of the exterior, and conversely. The analytic description of the inversion in the circle $x^2 + y^2 = R^2$ is given by

$$x' = \frac{R^2 x}{x^2 + y^2}, \qquad y' = \frac{R^2 y}{x^2 + y^2}. \qquad (3.3)$$

Whereas a central reflection maps lines into lines and circles into circles, an inversion maps the lines through O into themselves, the remaining lines into circles, circles through O into lines, and the remaining circles to circles. To make an inversion a bijective (one-one) mapping, we supplement the plane with a *single* point at infinity which we define as the image of O, and conversely. Also, we think of lines as circles passing through the point at infinity. Apollonius knew the properties of central dilatations and inversions just listed; in fact, the term "plane loci" referred to lines and circles (curves that can be drawn with ruler and compass). Referring to Apollonius' *On plane loci*, Pappus describes central dilatations, inversions, and their combinations with plane motions in these words:

If two straight lines are drawn from a single point, or from two points along a single straight line, or parallel to each other, or form a given angle and are in a given ratio, or contain a given rectangle, and if the end of one of these straight lines describes a plane locus, then the end of the other straight line also describes a plane locus of one kind or another, disposed in a similar manner with respect to the straight line or disposed in an opposite manner *[404, vol. 2, pp. 663–665]*.

We obtain a central dilatation if two "straight lines" (that is, two rectilinear segments) are drawn from a single point along a single straight line and their ratio is constant, and an inversion if their product is constant (they "contain a given rectangle"). If the lines are parallel, then we obtain, respectively, a central dilatation or an inversion followed by a *translation*. If the lines are drawn from a single point and form a fixed angle, then the transformation is a central dilatation or inversion followed by a *rotation* through this angle. Thus, in general, the transformation is a central dilatation followed by an arbitrary motion (and thus an arbitrary similitude), or an inversion followed by an arbitrary motion.

The quantitative relations underlying inversion in *all* conic sections are found in propositions 33, 35, and 37, book I, of Apollonius' fundamental *Treatise on conic sections* (Kōnika). We quote them below. (Recall that an *ordinate to a diameter* is a semichord conjugate to it.)

[I. 37]

In a hyperbola, an ellipse, or a circle, if QV be an ordinate to the diameter PP', and the tangent at Q meet PP' in T, then

$$CV \cdot CT = CP^2$$

[23, p. 66; 24, p. 28] (see Figures 55a, 55b, and 55c).

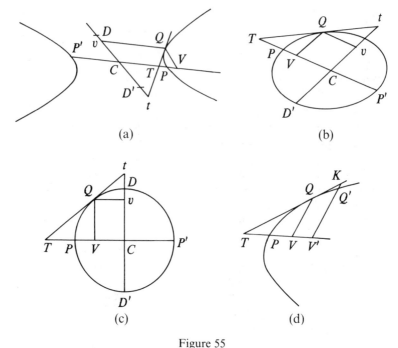

(a)   (b)

(c)   (d)

Figure 55

[I. 33, 35]

If a point T be taken on the diameter of a parabola outside the curve and such that $TP = PV$, where V is the foot of the ordinate from Q to the diameter PV, the line TQ will touch the parabola.

Conversely, if the tangent at Q meet the diameter produced outside the curve in the point T, $TP = PV$ *[23, p. 64; 24, p. 26]* (see Figure 55d).

These propositions play the role of equations of tangents to conics. The inversions implicit in these propositions map a point of the plane to the point of intersection of the diameter of the conic through the first point and its polar. All these inversions are *birational* transformations (see pp. 140–142 and the paper *[484]*).

Projections and Perspective

A number of projections were used in ancient Greece. In his *Ten books on architecture* (De architectura libri decem) *[605; 606]*, Vitruvius, the famous Roman architect of the first century A.D., lists three projections used by architects:

> Arrangement, however, is the fit assemblage of details, and, arising from this assemblage, the elegant effect of the work and its dimensions, along

with a certain quality or character. The kinds of the arrangement (which in Greek are called *ideae*) are these: ichnography (plan); orthography (elevation); scenography (perspective). Ichnography (plan) demands the competent use of compass and rule; by these plans are laid out upon the sites provided. Orthography (elevation), however, is the vertical image of the front, and a figure slightly tinted to show the lines of the future work. Scenography (perspective) also is the shading of the front and the retreating sides, and the correspondence of all lines to the vanishing point, which is the centre of a circle. These three (plan, elevation and perspective) arise from imagination and invention *[605, pp. 25, 27]*.

The first of these terms is made up of the words *ichnos*—trace—and *grapho*—I write—and denotes the construction of the horizontal projection of the building. The second term, derived from *orthos*—standing straight—denotes the construction of the frontal projection. The third term, derived from *skēnē*—stage—denotes the perspective representation of a locality on theatrical decorations. We may assume that these three projections were known to the Greeks centuries before Vitruvius. At any rate, the Roman poet-philosopher Titus Lucretius Carus (98–55 B.C.) described perspective representations in his philosophical poem *On the nature of things* (De rerum natura).

Though a colonnade runs on straight-set lines all the way, and stands resting on equal columns from end to end, yet when its whole length is seen from the top end, little by little it contracts to the pointed head of a narrow cone, joining roof with floor, and all the right hand with the left, until it has brought all together into the point of a cone that passes out of sight *[342, p. 385]*.

Claudius Ptolemy's *On projection* (Peri analēmmatos), usually referred to as *Analemma [443, vol. 1, pp. 187–223]*, deals with the orthogonal projection of the celestial sphere on the plane of the horizon and uses it to solve problems of spherical astronomy.

In book VII of Pappus' *Mathematical collection* (which also contains the excerpt from Apollonius' *On plane loci* quoted previously) there are geometric theorems on central projections and perspective. They are *Lemmas on the second porism in book I of "Porisms"*, referring to Euclid's lost work *Porisms*; literally, *porisma* means "acquisition, extraction." It is possible that Euclid's *Porisms* also dealt with projections and perspective.[1]

In lemma 3 (proposition 129) of book VII of the *Mathematical Collection* Pappus states:

If three straight lines AB, CA, DA are intersected by two straight lines FB and FE issuing from a single point, then I assert that [the rectangle] on FB, DC is to [the rectangle] on FD, BC as [the rectangle] on FE, HG is to [the rectangle] on FH, GE (Figure 56).

---

[1] In connection with Euclid's *Porisms* see Michel Chasles' works *[107]* and *[108]*.

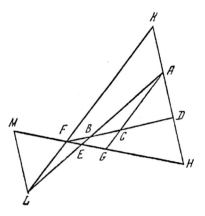

Figure 56

Through [the point] F draw [the straight line] KL parallel to the straight line GCA. Then DA and AB meet it in the points K and L. On the other hand, draw through L [the straight line] LM parallel to DA. It meets EF at M. Therefore EG is to GA as EF to FL, AG is to GH as FL to FM for, in view of parallelism, both ratios are the same as the ratio of FK to FH. Hence, "by equality," EG is to GH as EF to FM. Hence [the rectangle] on FE, HG is equal to [the rectangle] on EG, FM.

Consider another [rectangle] on EG, FH. [The rectangle] on EF, HG is to [the rectangle] on EG, HF as [the rectangle] on EG, FM is to [the rectangle] on EG, HF, that is, as FM is to FH or LF to FK. For the same reasons [the rectangle] on FD, BC is to [the rectangle] on FB, CD as KF is to FL. Therefore, upon "inverting," we see that [the rectangle] on FB, CD is to [the rectangle] on FD, BC as LF is to FK. But we showed that [the rectangle] on EF, HG is to the rectangle on EG, HF as LF is to FK. Hence [the rectangle] on EF, HG is to [the rectangle] on EF, HG is to [the rectangle] on EG, HF as [the rectangle] on FB, CD is to [the rectangle] on FD, BC [404, vol. 2, pp. 868–870].

Using modern notation we can write this proposition in the form of the proportion

$$\frac{FB \cdot DC}{FD \cdot BC} = \frac{FE \cdot HG}{FH \cdot GE},$$

or

$$\frac{FB}{FD} \bigg/ \frac{CB}{CD} = \frac{FE}{FH} \bigg/ \frac{GE}{HG}.$$

If FB, FD, CB, and CD are directed collinear segments then we call the ratio $\dfrac{FB}{FD} \bigg/ \dfrac{CB}{CD}$ a *cross ratio*; a cross ratio is the ratio of two simple ratios. We view the oriented lengths of segments, their simple ratios, and their cross ratios as real numbers. The cross ratio

$$(AB, CD) = \frac{AC}{CB} \Big/ \frac{AD}{DB} \qquad (3.4)$$

is positive if the pairs of points A, B and C, D do not separate each other (the circles on AB and CD as diameters do not intersect) and negative if they do (the circles in question intersect). Using modern notation, we can write Pappus' proposition in the form

$$(FC, BD) = (FG, EH)$$

Since the four points F, E, G, H are obtained from the four points F, B, C, D by projection from the point A, Pappus' theorem is seen to be a special case of the general theorem to the effect that *the cross ratio of four points is invariant under projection.* Following Euclid, Pappus called the proportion $a:c = A:C$, obtained from the proportions $a:b = A:B$ and $b:c = B:C$, a proportion "by equality," and the ratio $b:a$ the result of "inverting" the ratio $a:b$.

In lemma 5 (proposition 131) Pappus proves the following theorem:

Given the figure $ABCDEFGH$, AD is to DC as AB to BC. And if AD is to DC as AB to BC, then I claim that the line passing through the points A, H, F is a straight line (Figure 57).

Pass throught H [the straight line] KL parallel to AB. Then AD is to DC as AB to BC. Now AD is to DC as KL to LH, and AB is to BC as KH to HM. Hence KL is to LH as KH to HM and, finally, LH is to LM as KL to LH, that is, as AD to DC. "By interchanging," we find that AD is to HL as CD to LM, that is, as DE to KL. Since HL is parallel to AD, it follows that the line passing through A, H, F is a straight line, which is what was to be proved *[404, vol. 2, pp. 872–874].*

Comments are in order. The figure $ABCDEFGH$ consists of the complete quadrilateral $AHCFEG$ and its diagonals EHB, GFD, and ABC. Essentially,

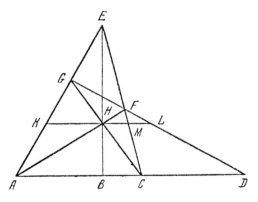

Figure 57

the first claim made in this proposition is that two diagonals of a complete quadrilateral intersect the third one in points B and D such that

$$(AC, BD) = -1$$

(we say that the points A, C divide the points B, D *harmonically*, or that the points A, C, B, D form a *harmonic tetrad*). The second claim is the inverse theorem, which asserts that every harmonic tetrad can be obtained in the indicated manner from a complete quadrilateral.

As for the missing proof of the first assertion, there is no doubt that what Pappus has in mind when he states that "Then AD is to DC as AB to BC" is that, in view of his lemma 3, the cross ratio (AC, BD) is equal to the cross ratio of the four points on the straight line GFD obtained by projecting the points A, B, C, D from E—that is, the points G, F, the point of the intersection of the straight lines GFD and EHB, and the point D. In turn, *their* cross ratio is equal to the cross ratio (CA, BD) of the points obtained by projecting them from H to the line $ABCD$. Since interchanging the points of a pair in a cross ratio inverts its value, it follows that, in our case, $(AC, BD) = \pm 1$ (for Pappus the value is 1; for us it is -1, since the pairs A, C and B, D separate each other). To "interchange" a ratio is to change $a:b = c:d$ to $a:c = b:d$.

In lemma 10 (proposition 136) Pappus proves the following remarkable theorem:

From [a point] F draw two straight lines DF and FE to two straight lines BAE and DAH. Suppose that [the rectangle] on DF, BC is to [the rectangle] on DC, BF as [the rectangle] on FH, GE is to [the rectangle] on FE, GH. I claim that the line passing through C, A, G is a straight line (Figure 58).

Draw [the straight line] KL parallel to CA and let K and L be the points in which it meets AB and AD. Through L draw [the straight line] LM parallel to AD and extend [the straight line] EF to M. Finally, draw through K [the straight line] KN parallel to AB and extend DF to N.

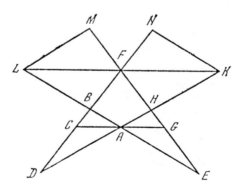

Figure 58

By parallelism, *DF* is to *FN* as *DC* to *CB*, so that [the rectangle] on *DF*, *CB* is equal to [the rectangle] on *DC*, *FN*. Now take another [rectangle on *DC*, *BF*. Then [the rectangle] on *DF*, *BC* is to [the rectangle] on *DC*, *BF* as [the rectangle] on *CD*, *FN* to [the rectangle] on *DC*, *BF*, that is, as *FN* to *BF*. But, by hypothesis, [the rectangle] on *FD*, *BC* is to [the rectangle] on *DC*, *BF* as [the rectangle] on *FH*, *GE* to [the rectangle] on *FE*, *GH*. By parallelism, *FN* is to *FB* as *KF* to *FL* and as *HF* to *FM*. It follows that [the rectangle] on *FH*, *GE* is to [the rectangle] on *FM*, *GE* as [the rectangle] on *FH*, *GE* to [the rectangle] on *FE*, *GH*. But then [the rectangle] on *FE*, *GH* is equal to [the rectangle] on *FM*, *GE*, and *FM* is to *FE* as *HG* to *GE*. Therefore, by "adjoining" and "interchanging," we find that *ME* is to *EH* as *FE* to *EG*. But *ME* is to *EH* as *LE* to *EA*, and *LE* is to *EA* as *FE* to *EG*. This implies that *AG* is parallel to *KL*. But the straight line *CA* is also parallel to it. Hence [the line] *AG* is a straight line, which is what was to be proved *[404, vol. 2, pp. 888–890]*.

Nowadays a correspondence between two straight lines such that for all corresponding quadruples *A*, *B*, *C*, *D* and *A'*, *B'*, *C'*, *D'*, we have the equality

$$(AC, BD) = (A'C', B'D')$$

is called a *projective correspondence*. Then the sets of points of these lines are called *projective ranges*. Since ordinary projection of a line onto a line preserves cross ratios, it is an instance of a projective correspondence. In that case the correspondence is called *perspective*, and the sets of points of these lines are called *perspective ranges*. It is clear that under a perspective correspondence the point of intersection of the lines corresponds to itself. In this proposition Pappus proves the converse theorem that *if the point of intersection of two projective ranges corresponds to itself, then the two ranges are perspective*.

Pappus used this proposition to prove—in lemmas 12 and 13 (propositions 138 and 139)—his famous theorem that the points of intersection of opposite sides of a hexagon inscribed in a pair of (parallel or intersecting) straight lines are collinear.

These theorems of Pappus are theorems of projective geometry, which emerged as a distinct geometric discipline only between the 17th and 19th centuries.

Stereographic Projection

One of the most important projections—applied already in antiquity—is stereographic projection. This is the projection of a sphere from one of its points (the *pole*) to the plane tangent to the sphere at the point antipodal to the pole—the *base plane*—or to a plane parallel to that plane (Figure 59). Stereographic projection has three remarkable properties: (a) Circles through

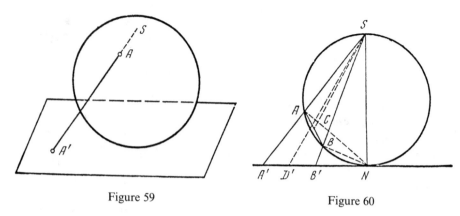

Figure 59   Figure 60

the pole are mapped onto straight lines and all other circles are mapped onto circles; (b) the angle between two curves on the sphere (that is, the angle between the tangents to the curves at their intersection point) is equal to the angle between the image curves; (c) if the sphere is rotated through some angle about the diameter passing through the pole, then this rotation induces a rotation through the same angle of the base plane about the point of tangency.

Property (a) of stereographic projection can be proved by using proposition 5 of book I of Apollonius' *Conics [23, pp. 9–10]*. In this proposition Apollonius proves that, in addition to the family of circular sections by planes parallel to its base, an oblique circular cone has another family of circular sections. The following is a simple proof of this proposition. Drop a perpendicular from the vertex of the cone to its base plane. This line and the line joining the vertex to the center of the base determine a plane of symmetry of the cone. An oblique cone has a second plane of symmetry, perpendicular to the first plane and passing through the bisector of the angle in which the cone meets the first plane (the planes perpendicular to these two planes of symmetry meet the cone in ellipses, and an oblique circular cone may be thought of as a right elliptical cone whose bases are these ellipses). Upon reflection in the first plane of symmetry the circular sections in the first family are mapped onto themselves, and upon reflection in the second plane of symmetry they are mapped onto the circular sections in the second family.

If the first plane of symmetry meets an oblique cone in the triangle SAB (Figure 60), then the circular sections in the second family meet this plane in straight lines parallel to a straight line $A'B'$ such that the angle $SA'B'$ is equal to the angle SBA, and the angle $SB'A'$ is equal to the angle SAB. Now, under stereographic projection, the circle with diameter AB on the sphere is projected to a section of the cone under consideration whose diameter $A'B'$ is such that the angle equalities just mentioned hold. (The similarity of the right triangles SAN and $SA'N$ with common acute angle S implies the proportion $SA/SN = SN/SA'$, that is, $SA \cdot SA' = (SN)^2$. Similarly, $SB \cdot SB' = (SN)^2$. Hence $SA/SB = SB'/SA'$. But then the triangles SAB and $SB'A'$ are similar and the asserted angle equalities hold.)

The earliest references to stereographic projections that have come down to our time are found in Vitruvius' *Ten books on architecture* (see previous discussion) and in Ptolemy's *Representation of the sphere in the plane* (Aplōsis epiphaneias sphairas), usually referred to as *Planisphaerium [443, pp. 225–229]*. (See also *[156]*).

Vitruvius described an astronomical instrument called a *spider* or *arachne* (*arachné*—spider); he writes that

> Berosus the Chaldean is said to have invented the semicircular dial hollowed out of a square block and cut according to the latitude; Aristarchus of Samos, the Bowl or Hemisphere, as it is said, also the Disk on a level surface; the astronomer Eudoxus, or as some say Apollonius, the Spider (arachne) *[605, pp. 255–256; 606, p. 320; 614, pp. 326–327]*.

On the arachne

> the hours are to be indicated cross-wise on a small column, in accordance with the analemma. The lines of the month are also to be marked on a column. . . . An analemma is described, and the hours are marked with bronze rods, beginning from a centre on the clock face. On this circles are described which limit the spaces of the months. Behind these rods there is a drum, on which the firmament and zodiac are drawn and figured: the drawing being figured with the twelve celestial signs *[605, p. 261; 606, p. 322; 614, p. 339]*.

Vitruvius' commentator Daniele Barbaro (1513–1570) describes the principle of Vitruvius' analemma (*analēmma*—survey, projection) thus:

> [an] analemma is projected from the pole of the sphere onto a plane. To project the sphere onto the plane [by means of an analemma] is to describe in the plane all circles and all [zodiacal] signs that are on the sphere. Thus all that is on the sphere is represented in the plane according to the same optical mode as in the making of a table of the astrolabe *[606, p. 322; 614, p. 339]*.

That is, the *analemma* is a stereographic projection. The drum bearing a representation of the ecliptic and some fixed stars was rotated. Before it were fixed wires representing hour lines. The drum was set by means of instruments that measured the altitudes of stars and was rotated by a hydraulic drive. In the *Planisphaerium*, the exposition of the stereographic projection of a sphere on a plane involves the so-called spider in a horoscopic instrument *[156, p. 271]*. A horoscopic instrument (*hōroskopon organon*) was an instrument for determining time (*hōroskopos*—time indicator). Later the word *horoscope* came to denote the point of intersection of the ecliptic and the eastern part of the horizon determined by means of this instrument, and, still later, the astrological prediction largely based on the location of the *horoscope* point. It seems that Theon, in his *Memoir on a small astrolabe* (Eis ton mikron astrolabon hypomnēma), was the first to combine the annular measuring instrument, the drum (*tympanum*), and the *spider* that modeled the celestial sphere, in a single compact instrument. The title of Theon's work is mentioned by the 10th-century Byzantine historian Suida *[570, v. 2, p. 702]*, and the description of the instrument is found in the work of the 9th-century Arab historian Aḥmad al-Yaʻqūbī *[280, pp. 23–25; 383, pp. 242–245]*, who referred to it as "[The instrument] possessing tympanums" ("Dhāt al-Ṣafāʼiḥ"). Al-Yaʻqūbī attributed this treatise, as well as other works of Theon mentioned by him, to Ptolemy. Ptolemy himself used the term *astrolabe* (*astrolabon organon*), literally "instrument for catching stars") to denote an armillary sphere (from the Latin *armilla*—ring)—an arrangement rings—that combined the functions of the annular and modeling instruments *[442, vol. 1, pp. 217–219]*. The first works on the astrolabe—in the sense of a "small astrolabe"—to come down to our time are the treatise *On the construction and use of the astrolabe* (Peri ton astrolabon chrēseōs kai kataskēnēs) of the 6th-century Alexandrian Christian philosopher and mathematician Joannes Philoponus *[574, vol. 9, pp. 341–367]*, and the *Treatise on the astrolabe* (Skolion demettul astrolabon) of the 7th-century Syrian bishop Severus Sebokt *[382]*. This instrument was widely known in the medieval East under the name *asṭurlāb*, so that in the Middle Ages stereographic projection was called *astrolabe projection* (*tasṭīḥ al-asṭurlāb*). The term *stereographic projection* (from *stereon*—solid) was introduced by François D'Aguillon (1566–1617) in his *Six books on optics* (Opticorum Libri VI. Antwerp, 1613) *[133]*.

In these ancient treatises properties (a), (b), and (c) of stereographic projection were used but not proved. The earliest extant exposition of the theory of stereographic projection with a complete proof of property (a) is found in *Book on the construction of the astrolabe* (Kitāb ṣanʻat al-asṭurlāb) (see *[474, 522]* by the ninth-century scholar Aḥmad al-Farghānī, a native of Farghāna who worked in Baghdad. The first chapter, devoted to stereographic projection, is called *Survey of the geometric propositions from which the form of the astrolabe is deduced*. It contains proofs of three theorems: the theorem on the equality of the angles $SA'B'$ and SBA and the angles $SB'A'$ and SAB (see Figure 60), the theorem on property (a) of stereographic projection, and the theorem that under stereographic projection the center of a circle is not mapped onto the center of the image circle. If, in Figure 60, C is the midpoint of AB and D' is the midpoint of $A'B'$ (and thus C and D' are the centers of corresponding circles) then the angle $A'SD'$ is equal to the angle BSC. Al-Farghānī's proof of property (a) is very close to the proof of proposition 5 of book I of Apollonius' *Conics*. The later chapters of the treatise deal with the construction of an astrolabe.

The astrolabe we are discussing combines an annular measuring instrument on the *back side* and a modeling instrument on the *front side*. The annular instrument of the astrolabe is a disk, on one side of which rotates an *alidad*—a ruler with two diopters. In use, the astrolabe is suspended in a vertical plane passing through the star, the alidad is aligned with it, and then its hand points to the star's altitude on the degree scale on the rim of the astrolabe. The second coordinate of the star is determined by means of the modeling instrument of the astrolabe, consisting of a fixed *tympanum* and a carved disk—the *spider*—rotating about the center of the tympanum. On the tympanum are represented the stereographic projections of the circles on the celestial sphere that do not change during its apparent diurnal motion: the celestial equator, the tropics of Cancer and Capricorn, the horizon and its parallels—the almucantarats, zenith points, and verticals—great circles passing through the zenith and nadir. In view of property (a), all these circles on the sphere are represented on the tympanum by circular arcs or straight line segments. Since the southern pole of the celestial sphere is usually taken as the pole of the projection, the equator and the tropics are represented on the tympanum by means of concentric circles. The tympanum is usually cut off at the circle representing the tropic of Capricorn (Figure 61). Since in a locality with geographic latitude φ the celestial equator forms with the horizon an angle of $90° - \varphi$, it follows from property (b) that the horizon is represented by a circle which cuts the representation of the equator in two antipodal points at an angle of $90° - \varphi$. The almucantarats are represented by circles which, together with the representation of the horizon, form a pencil of circles that are the loci of points for which the ratio of the distances to the points representing the zenith and nadir is constant. Verticals are represented by circles passing through the representation of the zenith and perpendicular to the representation of the horizon. The almucantarats and verticals form on the tympanum the *spiderweb* over which the spider moves. On the spider are represented the

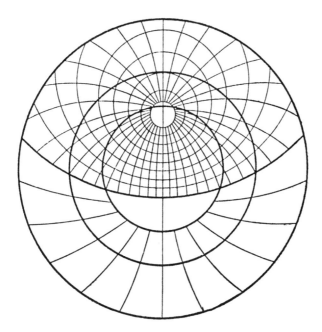

Figure 61

Figure 62

ecliptic and the brightest stars that rotate during the apparent diurnal motion of the celestial sphere. The ecliptic is represented by a circle tangent to the representations of the tropics. On the ecliptic are shown the twelve zodiacal constellations through each of which the Sun passes in the course of a month as well as further subdivisions of these sections that make it possible to determine the location of the Sun on every day of each month. Stars are represented by spikes issuing from the rim of the spider and from the representation of the ecliptic (Figure 62).

By means of the astrolabe one could determine the azimuth of just those heavenly bodies that were represented on its "spider," that is, the Sun and the other stars on it. After determining the altitude of the Sun or star by means of the alidade, one turned the astrolabe over—so that the tympanum was on top—and rotated the spider through an angle such that the representation of the celestial body fell on the almucantarat with the same altitude. Here one used property (c) of stereographic projection which implies that the diurnal rotation of the celestial sphere is represented by a rotation of the spider. After rotating the spider one obtained an accurate representation of the celestial sphere on the plane at the corresponding moment. At that moment the azimuth of the celestial body was given by the angle between the vertical passing through that body and some initial vertical. The angle of rotation of the spider determined the exact time that had passed from the beginning of the day or night. In terms of astronomical hours,[2] the position of the spider

---

[2] Astrolabes could also be used to determine time in so-called temporal hours (see chapter 1). Then one used hour lines on tympanums similar to the hour lines on the *arachne* described by Vitruvius.

corresponding to this angle of rotation was such that the Sun was on the horizon. The astrolabe with a "spider," just described, can be viewed as a nomogram with a transparent chart *[542]*.

Mathematicians in the mediaeval East tried to use other geometric transformations for constructing astrolabes. Thus Abū Ḥāmid al-Ṣaghānī, a native of Saghāniān (d. 990), in his *Book on the perfect projection onto the plane* (Kitāb fī al-tasṭīḥ al-tamm) (see [266]) suggested replacing stereographic projection of the sphere onto the plane from one of its poles by a projection from an arbitrary point on the axis. (In such a projection circles on the sphere are mapped onto conics). In his book *Exhaustion of the ways of constructing [an] astrolabe* (Istiʿāb al-wujūh al-mumkina fī ṣanʿat al-asṭurlāb) (see *[483, pp. 152–156, 162–166, 168–172]*), Abū l-Rayḥān al-Bīrūnī, after describing many ways of constructing astrolabes—including al-Ṣaghānī's "perfect projection"—suggested as a basis for the construction of astrolabes "cylindrical projection," that is, orthogonal projection along the axis—the limiting case of al-Ṣaghānī's projection when the center of projection recedes to infinity. In his *Chronology of Ancient Nations* (Al-Athār al-Bāqiya min al-qurūn al-khāliya) *[58; 60, vol. 1]*, written about 1000, al-Bīrūnī has this to say about these methods of projection:

'Abû-Ḥâmid Alṣaghânî has transferred the tops of the cones from the two poles, and has placed them inside or outside the globe in a straight line with the axis. In consequence the cones represent themselves as straight lines and circles, as ellipses, parabolas, and hyperbolas, as he (Abû-Ḥâmid) wants to have them. However, people have not been in a hurry to adopt such a curious plane. (This is the central projection, or the general perspective projection.)

Another kind of projection is what I have called *the cylindrical projection* (orthographic projection), which I do not find mentioned by any former mathematician. It is carried out in this way: You draw through the circles and lines of the globe lines and planes parallel to the axis. So you get in the day-plane straight lines, circles, and ellipses (no parabolas and hyperbolas). All this is explained in my book [*Exhaustion of the ways of constructing [an] astrolabe*], which gives a complete representation of all possible methods of the construction of the astrolabe *[58, p. 357–358; 60, vol. 1. pp. 407–408]*.

Stereographic projection is also used to project the surface of the earth onto a plane, that is, for making maps. Property (b) tells us that on such maps angles between lines are true. This fact makes such maps especially useful for seamen, for the angle through which the ship's steering wheel is rotated is equal to the corresponding angle on the map. Al-Bīrūnī's *Treatise on projection of constellations and on the representation of countries on a map* or *Cartography* (Risāla fī tasṭīḥ al-ṣuwar wa-tabṭīḥ al-kuwar) *[457; 11]* (partially set forth in his *Chronology of Ancient Nations [58, pp. 357–365; 60, vol. 1, pp. 407–413]*), are devoted to the application of stereographic projection to mapmaking.

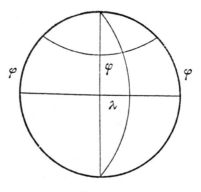

Figure 63

Both treatises also describe a projection, discovered by al-Bīrūnī and now known as *globular*,[3] of a sphere onto a plane.

In globular projection a hemisphere is mapped onto a circle whose circumference is divided into 360° and whose horizontal and vertical diameters are each divided into 180 parts. To represent a point on the sphere with longitude λ and latitude φ one lays off, beginning at center of the circle, λ divisions on the horizontal diameter and draws a circular arc through the resulting point on the horizontal diameter and the endpoints of the vertical diameter. Then one lays off, beginning at the center of the circle, φ divisions on the vertical diameter and, beginning at the endpoints of the horizontal diameter, φ degrees on the circumference and draws a circular arc through the three resulting points. The required point is the intersection of the two circular arcs (Figure 63). The resulting representation resembles the stereographic projection of a hemisphere, that is, the stereographic projection of a sphere cut off at the circle representing the equator, except for the nonuniformity of the λ and φ scales on the horizontal and vertical diameters.

A remarkable application of stereographic projection was proposed by Abū Ishāq Ibrāhīm ibn Yaḥyā al-Naqqāsh al-Zarqālī al-Qurṭūbī (c. 1030–c. 1090), a native of Cordoba, in his *Book of operations with the zīj of a tympanum* (Kitāb al-ʿamal bi-l-ṣafīḥa al-zījiyya) (see *[18, pp. 135–237; 34]*). In the medieval East, zījes were astronomical works consisting of large numbers of tables, including tables for the transition from one of the three systems of spherical coordinates on the celestial sphere—horizontal, equatorial, and ecliptic—to another.[4] Al-Zarqālī called his invention a *zīj of a*

---

[3] This projection was rediscovered by Nicolosi of Paterno in 1624 and by A. Arrowsmith (1750–1823) in about 1804. It was used to construct an astrolabe by Philippe de Lahire (1640–1717); see *[365, p. 21]*.

[4] The system of horizontal coordinates (altitude h and azimuth A) and one of the systems of equatorial coordinates (right ascension α and declination δ) are not involved in the diurnal rotation of the celestial sphere. The system of ecliptic coordinates (longitude λ and latitude β) and the second system of equatorial coordinates (hour angle t and declination δ) are involved in that rotation.

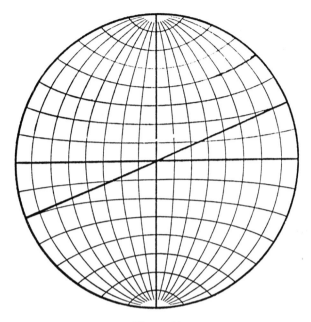

Figure 64

tympanum because geometric operations performed on his instrument dispensed with the need for transition tables. Essentially, the *zīj of a tympanum* is an astrolabe, whose special features earned it the name *universal astrolabe.*[5] Whereas the usual astrolabes are based on stereographic projection of the celestial sphere from a celestial pole, the *zīj of a tympanum* is based on stereographic projection from a solstice point (a point of intersection of the celestial equator and the ecliptic). Therefore, on al-Zarqālī's tympanum, the celestial equator and ecliptic are represented by straight lines intersecting at its center and forming an angle equal to the angle between the ecliptic and the equator. On this tympanum are also represented parallels to the celestial equator and meridians, that is, great circles passing through the celestial poles (Figure 64). The tympanum is supplied with a ruler that can rotate about its center. A cursor, perpendicular to the ruler, can slide along it. A finger is attached to the end of the cursor by means of two hinges, and its end can be placed at any point on the tympanum. To go from ecliptic coordinates (longitude λ and latitude β) to equatorial coordinates (right ascension α and declination δ) one must align the ruler with the representation of the celestial equator; place the finger at the point representing the point of the celestial sphere whose equatorial coordinates α, δ are numerically equal to the given ecliptic coordinates α, β; and rotate ruler, cursor, and finger—as a whole—until the ruler coincides with the representation of the ecliptic. Then the finger indicates

---

[5] The Dutch scholar Gemma Frisius (1508–1555) described this astrolabe in *[200]* and called it "catholic" (*katholikos*—universal).

the representation of the required point. The arcs passing through this point represent a parallel and a meridian and yield its coordinates α and δ. From the horizontal coordinates one goes over not to the equatorial coordinates α, δ, linked in a fixed way to the ecliptic, but to the equatorial coordinates t (hour angle), δ, linked in a fixed way with the horizon. In this case the same tympanum is viewed as the projection of the celestial sphere from a point of intersection of the horizon and the equator onto the plane of the celestial meridian, that is, the meridian passing through the zenith and nadir. Again, the ruler is aligned with the representation of the celestial equator; the finger is placed at the point representing the point of the celestial sphere whose equatorial coordinates t, δ are numerically equal to the horizontal coordinates A, h; and ruler, cursor, and finger are rotated—as a whole—through the complement of the latitude φ of the locality. This aligns the ruler with the representation of the horizon. Then the finger indicates the representation of the required point under the new projection and the circles passing through it—representing a parallel and a meridian—yield its t, δ coordinates.

Al-Zarqālī's astrolabe also represents a nomogram with a transparent chart (see *[573]*). Since it is supplied with an alidad for measuring the altitudes of stars, one can perform on it all the operations one can perform on an ordinary astrolabe. The advantage it offers is that its tympanum suits all geographic latitudes (hence the name *universal astrolabe*).

Of the works of more recent scholars on the use of stereographic projection in mapmaking we mention two papers of Euler's: *On the representation of a spherical surface in the plane* (De repraesentatione superficiei sphaericae super plano. Petersburg, 1778), and *On geographic projection of a spherical surface* (De projectione geographica superficiei sphaericae. Petersburg, 1778) *[176, vol. 28, pp. 228–235, 133–141]*. In those papers Euler posed the problem of the most general angle-preserving mapping of the sphere onto the plane. To solve this problem Euler used a stereographic projection of the sphere onto the plane that mapped the point on the sphere with latitude v and longitude t onto the point of the plane determined by the complex number

$$z = \tan v/2(\cos t + i\sin t),\qquad(3.5)$$

and then applied to the plane a conformal (that is, angle-preserving) mapping using a complex analytic function.

Affine Transformations in the Works of Ibn Qurra and Ibn Sinān

Central dilatations and axial affinities are special cases of *affine transformations*, the most general transformations of the plane in which straight lines are mapped onto straight lines. Relative to rectangular, skew, and *affine coordinates* (in the latter the units on the coordinate axes may be different) affine transformations are described by means of equations of the form

$$x' = Ax + By + a, \qquad y' = Cx + Dy + b, \qquad (3.6)$$

with nonzero determinant $\begin{vmatrix} A & B \\ C & D \end{vmatrix}$.

Since they are one-to-one mappings, affine transformations preserve the parallelism of straight lines. It is easy to show that these mappings also preserve simple ratios of segments on a line or on parallel lines.

The mapping (3.6) multiplies areas by $\begin{vmatrix} A & B \\ C & D \end{vmatrix}$. If this multiplier is ± 1, then we call the mapping *equiaffine*.

General equiaffine mappings first turn up in Thābit ibn Qurra's *Book on sections of a cylinder and its surface* (Kitāb quṭūʿ al-usṭuwāna wa-basīṭhā) *[267; 492, pp. 196–236]*. After proving that the area of an ellipse with semiaxes a and b is equal to the area of a circle of radius $\sqrt{ab}$, Ibn Qurra proves the following result (proposition 17):

Every segment of an ellipse is equal to a segment of a circle of the same area such that if we drop two perpendiculars from the endpoints of its base to a diameter of the circle, then each of them is to the diameter as the corresponding perpendicular, dropped from an endpoint of the base of the segment of the ellipse on one of its axes, is to the other axis; provided that the segments are both smaller or both greater [than half the ellipse or circle], the position of the center of the ellipse relative to the perpendiculars is the same as the position of the center of the circle relative to its perpendiculars, and the position of the feet of the perpendiculars of the ellipse on its axis is the same as the position of the feet of the perpendiculars of the circle on its diameter *[267, p. 69]*.

The provisions made by Ibn Qurra for the coincidence of the position of the centers and the feet of the perpendiculars (he considers eight cases of location of the segments of the ellipse and circle) guarantee equality of the signs of the corresponding oriented segments and make his mapping of the ellipse onto the circle affine, and the equality of the areas of the ellipse and the circle makes it equiaffine.

In his theorem Ibn Qurra proves that an equiaffine transformation maps any segment of an ellipse onto a segment of a circle of equal area. The proof is by the ancient method of exhaustion.

General affine transformations first occur in the *Book of measuring the parabola* (Kitāb fī misāhat al-qaṭʿ al mukāfī) of Ibn Qurra's grandson Ibrāhīm ibn Sinān ibn Thābit (908–946) *[242, pp. 53–66]* (see also *[482]*). Ibn Sinān's treatise consists of four propositions. The following result is proved in proposition 1:

If *ABCDE* is a multiangled figure and *GHJIK* is another multiangled figure (Figure 65), and if lines *BL* and *CM* are drawn parallel to [the line]

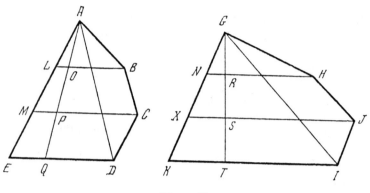

Figure 65

DE and [lines] HN and JX are drawn parallel to the line IK so that the lines AL, LM and ME are to one another as the lines GN, NX and XK and the lines BL, CM and DE are to one another as the lines HN, JX and IK, and if the lines AD and IG are drawn, then the triangle ADE is to the triangle GIK as the figure $ABCDE$ is to the figure $GHJIK$ [242, pp. 57–58; 482, p. 179].

The assumptions that the ratios of the segments AL, LM, and ME are the same as those of GN, NX, and XK, and the ratios of the segments BL, CM, and ME are the same as those of HN, JX, and IK guarantee that the polygon $ABCDE$ is mapped onto the polygon $GHJIK$ by the same affine transformation that maps the triangle ADE onto the triangle GIK.

This theorem tells us that affine transformation preserve the ratio of the areas of polygons. In proposition 2 Ibn Sinān extends this result by means of the method of exhaustion to segments of a parabola:

One of two arbitrary segments of a parabola is to the other as the triangle, whose base is the base [of the segment] and whose vertex is its vertex, is to the triangle constructed in the same way in the other segment [242, p. 59; 482, p. 179].

The vertex of a segment of a parabola is the point of intersection of the parabola and the diameter conjugate to the base of the segment. Figure 66 shows a segment of a parabola ABC and the triangle ABC inscribed in it in the indicated manner. In proposition 3 Ibn Sinān proves a theorem known to Archimedes, that

every segment of a parabola is to the triangle with the same base and vertex as four is to three [242, p. 62; 482, p. 179].

The proof consists in comparing the parabolic segment ABC with the segment BGC (see Figure 66)—one of the two small segments by which the parabolic segment exceeds the triangle. Ibn Sinān proves that the areas of the

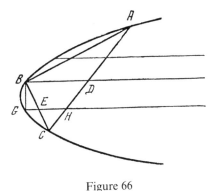

Figure 66

triangles ABC and BGC inscribed in the corresponding segments are in the ratio $8:1$. By proposition 2, this implies that this is also the ratio of the segments. Therefore the area of the segment is to the area of the two small segments as $4:1$, and hence to the area of the inscribed triangle as $4:3$. This enables him to prove in proposition 4 that the ratio of two segments P and Q of a parabola with parallel bases and diameters s and t is related to the ratio of the diameters by the equality $P:Q = (s:t)^{3/2}$.

The Point at Infinity in the Works of Kepler

Affine transformations are a special case of the more general *projective transformations*. To define projective transformations of the plane we must add to the plane *points at infinity*, one for each pencil of parallel lines. The need for this arises in connection with the central projection of a plane to a plane, in which some points of the first plane have no images and some points of the second plane have no preimages. To ensure the one-one character of such a projection it is necessary to supplement each plane with points such that there is a one-to-one continuous correspondence between the points of the extended plane and the bundle of straight lines through the center of projection. The straight lines of the bundle that intersect the plane correspond to the points of intersection with the plane, and the straight lines parallel to the plane represent the new points. They are called *points at infinity* for, when a straight line of the bundle that intersects the plane comes closer to a straight line parallel to it, its point of intersection with the plane recedes to infinity.

When a plane is projected to a plane that is not parallel to it, the parallel lines of the first plane are mapped to intersecting lines in the second plane. An example of such a projection is furnished by the projective representation of the horizontal plane in a vertical *picture plane*, where the images of parallel lines intersect at the *center of vision*.

A plane supplemented with points at infinity is called a *projective plane*.

Projective transformations (*collineations*) of the plane are one-one transformations of the *projective* plane that map lines onto lines; that is why projective transformations can map parallel lines onto intersecting lines.

Collineations are described in affine coordinates by equations of the form

$$x' = \frac{Ax + By + a}{Ex + Fy + c}, \qquad y' = \frac{Cx + Dy + b}{Ex + Fy + c}. \tag{3.7}$$

The concept of a point at infinity is first mentioned explicitly in the *Optical part of astronomy* (Astronomiae pars optica. Frankfurt am Main, 1604) *[271, vol. 2]* of the great astronomer and mathematician Johann Kepler (1571–1630). Its subtitle—*Supplement to Vitello* (Ad Vitellionem paralipomena)—shows that its was regarded as a development of the previously mentioned *Perspective* of the 13th-century Polish physicist Vitello *[604]*, itself an elaboration of the *Book of Optics* of Ibn al-Haytham *[19]*. Almost simultaneously with Kepler's book there appeared the book on optics by d'Aguillon discussed earlier. In it d'Aguillon considered not only stereographic projection but also orthogonal and general central projections, which he called *orthography* and *scenography*—terms he borrowed from Vitruvius (under whose influence he created the term *stereography*). Both books on optics, Kepler's and d'Aguillon's, were prepared by the many works on perspective that appeared in the 14th and 15th centuries. Of these we mention, in the first place, the treatise *On painting* (Della pittura. Florence, 1435) by Leon Battista Alberti (1404–1472) and *On perspective in painting* (De perspectiva pingendi. Rome, ab. 1480) by Piero della Francesca (1416–1492).

The first of these works develops a method of representing a row of equal and parallel segments as parallel segments contained by two lines intersecting on the line of the horizon, and the second describes the construction of the perspective representation of an object based on its vertical and horizontal projections.

An important role in the history of perspective was also played by Leonardo da Vinci's (1452–1519) *Treatise on painting* (Il trattato della pittura, published posthumously in 1651) and by Albrecht Dürer's *Instruction in measurement with compass and ruler* (Unterweysung der Messung mit Zirckel und Richtscheyt. Nürnberg, 1525), and *On human proportion* (Von menschlicher Proportion. Nürnberg, 1528). Both of these great artists were deeply occupied with geometric questions and, in particular, geometric transformations, and many pages of Leonardo's notebooks and Dürer's books are devoted to these questions.

In the *Optical part of astronomy*, in the chapter "*On conics*", Kepler notes that a section of a cone by a plane can be a straight line, a circle, a parabola, a hyperbola, or an ellipse; also, that

a straight line goes over into a parabola through infinite hyperbolas, and further through infinite ellipses into a circle (Figure 67),

and that

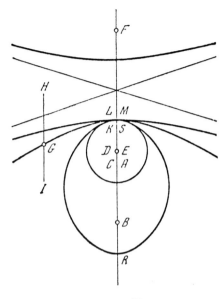

Figure 67

the most obtuse hyperbola is a straight line, and the most acute, a parabola; the most acute ellipse is a parabola, and the most obtuse, a circle.

Further, Kepler introduces the foci of conics—points such that

straight lines drawn from these points to the point of tangency of the tangent to the conic form with it equal angles.

"Because of [the doctrine of] light", says Kepler, "and with eyes intent on Mechanics we shall call these points 'foci.'"

He writes further:

A circle has one focus A, which is its center. An ellipse has two foci B and C equidistant from the center of the figure, and the more acute the ellipse, the further apart the foci. A parabola has just one focus D inside the figure, and the second must be imagined on the axis of the conic, inside or outside of it, at an infinite distance from the first, so that the line HG or IG from the invisible focus to any point of the conic is parallel to the axis DK. In the case of an hyperbola, the more obtuse the hyperbola the closer the outer focus F to the inner focus E. Also, the focus outside one of opposite conics is inside the other, and conversely [271, vol. 2, p. 90].

Kepler introduced the term *focus*—"fire, hearth"—under the influence of the term *ignition place*, used by Ibn al-Haytham and other Eastern writers on optics for the focus of a parabola, often called in the East an "incendiary mirror." Kepler's "directing the eyes to mechanics" shows that when he wrote this work he already knew of the importance of the foci of an ellipse in celestial mechanics—an importance bestowed on them by the first of Kepler's laws of planetary motion, published in his *New astronomy* (Astronomia nova) *[271, vol. 3]*. What is important to us at this point is that Kepler was aware that "the invisible focus of the parabola" closes the straight line *DK* and all straight lines parallel to it.

Projective Transformations in the Works of Desargues

Projective transformations were first systematically investigated by the French engineer, architect, and geometer Girard Desargues (1591–1661) in his *Rough draft of an attempt to deal with the outcome of a meeting of a cone with a plane* (Brouillon project d'une atteinte aux événemens des recontres du cone avec un plan. Paris, 1639) *[142]*.[6] Desargues added to the plane a line at infinity and viewed hyperbolas as closed curves that intersect it in two points, and parabolas as closed curves that touch it. Also, he thought of the asymptotes of a hyperbola as touching it at its points at infinity. Desargues studied cross ratios of quadruples of points (considered earlier by Pappus) and projective transformations of lines that preserve cross ratios (these transformations are called *projective*, for they arise as projective transformations of the plane that map the line onto itself).

We consider the most important of Desargues' transformations. Desargues used the term *tree* for a straight line with several given pairs of points such that the products of the lengths of the segments that begin at a point common to all pairs—the "trunk"—and end at the points of a pair is constant. He called the points of a pair *knots*; a segment joining the "trunk" to a "knot" was called a *branch*, segments between "knots" were termed *shoots*, and corresponding "branches" were said to be *paired*. This botanical terminology was probably created by analogy with the arithmetic and algebraic term *root*. Considering a "tree" with "trunk" *A* and pairs of corresponding "branches" *AC* and *AG, AF* and *AD, AB* and *AH*, that is, assuming that the equalities

$$AC \cdot AG = AF \cdot AD = AB \cdot AH$$

hold, assuming that "the trunk *A* is free relative to both branches of each pair" (that is, that *A* lies outside the segments *CG, DF*, and *BF* (Figure 68)), and calling products of segments *rectangles* on these segments, Desargues wrote:

In view of the equality of the rectangles on the two branches of each of the three pairs *AB, AH; AC, AG; AD, AF*, the four branches *AG, AF, AD, AC* are pairwise proportional. It follows that *GD* is to *CF* as *AG* to *AF* or *AD* to *AC*, and *GF* is to *CD* as *AF* to *AC* or *AG* to *AD*.

---

[6] An English translation appeared in 1987 *[182a; pp. 69–143]*.

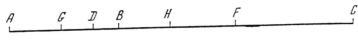

Figure 68

It follows that the branch AG is to its paired branch AC as the combined ratios of the shoot GD to the shoot CF and the shoot GF to the shoot CD, which is equal to the ratio of the rectangle on the shoots of the pair GD, GF to the rectangle on the shoots of the corresponding pair CD, CF.

It follows that the rectangle on the shoots GB, GH—the "twin" of the rectangle GD, FG—is to its corresponding rectangle CB, CH, the twin of the rectangle CD, CF, as the rectangle GD, GF—the "twin" of the rectangle GB, GH—is to its corresponding rectangle CD, CF, the "twin" of the rectangle CB, CH (*[142, pp. 116–118]*).

Thus beginning with $AG/AF = AD/AC$ and $AF/AC = AG/AD$, Desargues obtains $AG/AF = GD/CF$ and $AF/AC = GF/CD$, and from these, by composition of ratios (in his terminology, *combination of ratios*), the equalities $AG/AC = (GD/CF) \cdot (GF/CD) = (GD \cdot GF)/(CF \cdot CD)$. In the same way he obtains for analogous "rectangles" (in his terminology, *twins*)

$$AG/AC = (GB/CH) \cdot (GH/CB) = (GB \cdot GH)/(CH \cdot CB),$$

that is,

$$(GD \cdot GF)/(CF \cdot CD) = (GB \cdot GH)/(CH \cdot CB).$$

This can be rewritten as the equality of two cross ratios

$$(GD/CD)/(GB/CB) = (GF/CF)/(GH/CH)$$

that is,

$$(GC, DB) = (GC, FH).$$

Here any pair of corresponding "knots" can be replaced by any other. This means that Desargues proved that his correspondence between the "knots" of a "tree" is a projective correspondence on the points of the line. Also, in view of the complete equality of pairs of corresponding "knots," we can interchange these pairs in a cross ratio without altering its value. Since an interchange of the pairs of points in a cross ratio changes its value to its reciprocal, we conclude that all cross ratios in the Desargues correspondence are equal to 1 if we ignore orientations of segments, and to -1 if we take such orientations into consideration; that is, the corresponding point sets are harmonic. But then the fixed points of the Desargues correspondence divide each pair of corresponding points harmonically.

Desargues also considers the case when the "trunk" is between the points of each pair. Whereas in the first case the points of corresponding pairs did

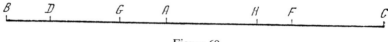

Figure 69

Figure 69

not separate each other (were "apart"), they do so in the second case (are "mixed"). The correspondence is again a projective transformation of the line.

Thus if there are given three pairs of points B, H; C, G; D, F such that the points of each pair are simultaneously mixed (Figure 69) or separated with respect to the points of every other pair, and if the corresponding rectangles made up of segments between these points are to each other as their twins taken in the same order, then this disposition of three pairs of points on a straight line is called here an involution *[142, p. 119]*.

The term *involution*, introduced at this point by Desargues, is also of botanical origin, and denotes, literally, the twisted state of young leaves. Since involution is a projective transformation of a line that coincides with its inverse, the term *involution* is today used to designate just such transformations. In other words, a transformation T is an involution if $T = T^{-1}$ or, equivalently, if T^2 is the identity transformation.

Incidentally, the analytic descriptions, relative to an affine coordinate system, of the involutions introduced by Desargues are

$$x' = a^2/x, \tag{3.8}$$

and

$$x' = -a^2/x. \tag{3.9}$$

In the first case, the involution has two real fixed points ($x = \pm a$) and is called *hyperbolic* (such involutions are exemplified by the mappings induced on the diameters of a circle of radius a by an inversion relative to this circle). In the second case, the involution has no real fixed points (the values of x such that $x' = x$ are $x = \pm ai$) and is called *elliptic* (the reason for these names is that these involutions are induced on the line at infinity by conjugate diameters of, respectively, a hyperbola and an ellipse).

Desargues was the first to consider "polar transformation" relative to a conic. This transformation associates with a point P the locus of points S such that P, S divide harmonically the points Q, R in which the line PS intersects the conic. This locus turns out to be a straight line. Today, this line is called the *polar* of the point, and the point is called its *pole*.[7]

---

[7] The word *pole* is derived from the Greek *polos* (axis) and originally denoted a point of intersection of a rotating sphere with the axis of rotation. The terms *pole of a line* and *polar of a point*—both with respect to a conic—were introduced by François Joseph Servois (1767–1847) and Joseph Diaz Gergonne (1771–1859), respectively.

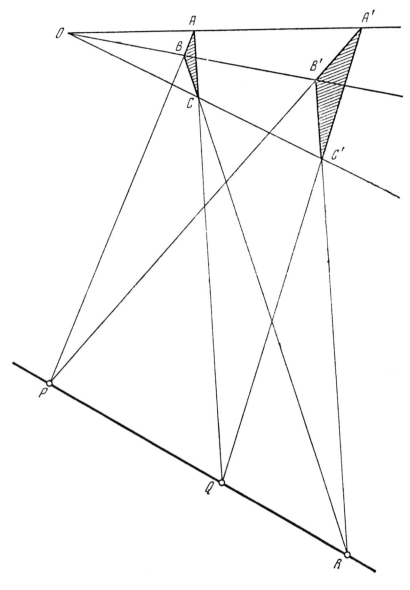

Figure 70

Desargues showed that a conic defines on each straight line in its plane an involution that associates with a point on such a line the point of intersection of that line with the polar of the point in question. If the line intersects the conic (in two points), then the involution is hyperbolic (and the points of intersection of the line and the conic are its fixed points). If the line and the conic have no common points, then the involution is elliptic.

Desargues showed that the pairs of points on an arbitrary straight line in a plane that are its points of intersection with the conics of a pencil of conics (for example, circles) passing through two points define an involution on that line. Desargues also showed that if one can draw two tangents from a point in the plane to a conic, then the line joining the points of tangency is the polar of the point; that the pole of the line at infinity relative to an ellipse or a hyperbola is its center; and that the polar of a point on the conic is the tangent to the conic at that point. Desargues used these theorems to solve construction problems—for example, the problem of finding the axes of the conic that is the image under projection of a circle.

Desargues also proved the theorem which is today known as *Desargues' theorem*. The theorem states that if two triangles ABC and $A'B'C'$ are such that the lines AA', BB', and CC' are concurrent (at some point O), then the points P, Q, R of intersection of the pairs of lines AB and $A'B'$, AC and $A'C'$, BC and $B'C'$ are collinear (Figure 70). In that case, there exists a special projective transformation—a *homology*[8] (with *center O* and *axis PQR*) that maps the triangle ABC onto the triangle $A'B'C'$. Therefore Desargues' theorem is sometimes referred to as the *theorem on homologous triangles*. Axial affinities and central dilatations (both discussed previously) are instances of homologies. An *axial affinity* is a homology whose center is a point at infinity not on its axis, and a *central dilatation* is a homology whose axis is the line at infinity and whose center is the center of the central dilatation. The center of a homology may lie on its axis. If that is the case, and if the axis is the line at infinity, then the homology is a *translation*. If the axis is not the line at infinity, then the homology is a *shear*.

Influenced by Desargues, Blaise Pascal (1623–1662), the great French philosopher, physicist, and mathematician, published his *Essay on conics* (Essay pour les coniques. Paris, 1604) *[407]*.[9] Pascal was then only 16, and his *Essay* was a one-page poster. In it Pascal proved *Pascal's theorem*, that the points of intersection of the pairs of opposite sides of a hexagon inscribed in a conic are collinear. If the conic reduces to a pair of lines, then Pascal's theorem reduces to Pappus' theorem discussed earlier.

Projective and Birational Transformations in the Works of Newton

The great English mathematician and physicist Sir Isaac Newton (1642–1727) used various geometric transformations extensively. In his *Enumeration of lines of the third order* (Enumeratio linearum tertii ordinis), written before 1670 but published together with his *Optics* only in 1704 *[388]*[10], Newton gives

---

[8] The term *homologia* (correspondence) was introduced in this sense by Desargues *[142]*.
[9] An English translation *[182a, pp. 180–184]* appeared in 1987.
[10] Newton's notes on this work, containing the proofs omitted from the printed text, have been published by Derek Thomas Whiteside *[631, vol. 2, pp. 10–89]*.

a classification of cubics (curves "of the second genus") based on the fact that all these curves can be obtained by central projection ("by casting shadows from a bright point") of five *diverging parabolas*

$$y^2 = ax^3 + bx^2 + cx + d$$

that differ by the nature of the roots of the polynomial $ax^3 + bx^2 + cx + d$. Newton shows that

If we cast on the infinite plane the shadows of figures from a bright point, then the shadows of conics will always be conics; the shadows of curves of second genus will always be curves of second genus; and the shadows of curves of third genus will always be curves of third genus, and so on to infinity. And just as, by casting shadows, the circle gene-rates all conics, so too five diverging parabolas generate and procure all other curves of the second genus *[388, pp. 28–29]*.

In this famous *Mathematical principles of natural philosophy* (Principia mathe-matica philosophiae naturalis) *[387]*, in which he formed classical mechanics and—in connection with the requirements of dynamics—presented the foundations of the differential and integral calculus, Newton uses projective transformations to solve concrete problems. In chapter 5 of book I, devoted to the definition of the orbits of moving bodies (they are conics), we find lemma 22:

To transform figures into other figures of the same kind *[387, p. 90]*.

Here Newton defines the following transformation of a curve *HGI* into a curve *hgi* (Figure 71). Project an arbitrary point *G* on the curve *HGI*, parallel to the line *AO*, onto the point *D* on the line *AB*. Then project *D* from *O* to *d* on *BL*. Next draw the line *dg* at some fixed angle α with the line *BL* and lay off on the line *dg* a segment *dg* such that $dg : Od = DG : OD$. If we refer the curve *HGI* to the skew coordinate system with axes *AB* and *OA*, then the co-ordinates of the point *G* are $X = AD$ and $Y = DG$. If we refer the curve *hgi* to

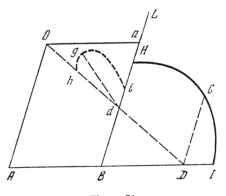

Figure 71

the skew coordinate system with axis of abscissas BL, origin at the point a of intersection of the axis BL, and the line Oa parallel to the axis AB, and coordinate angle α, then the coordinates of the point g on the curve are $x = ad$ and $y = dg$. Putting $AB = p$, $OA = q$, we can rewrite the preceding proportion as $y:p = Y:X$. On the other hand, the similarity of the triangles Oad and OAD implies that $ad/Oa = OA/AD$, that is, $x:p = q:X$. But then Newton's transformation can be written as $X = pq/x$, $Y = qy/x$.

This transformation is projective. Newton shows that if the point G traces a straight line, a conic, or, more generally, an nth-order algebraic curve, then the point g traces a corresponding straight line, conic, or nth-order algebraic curve, and that this transformation maps lines that intersect on the axis of abscissas into parallel lines. Newton's comment on this lemma is that it

> serves to solve difficult geometric problems by transforming the given figures into simpler ones.... After solving the problem for the transformed figure we need only transform it into the original figure and thus obtain the required solution for the latter *[387, p. 91]*.

In the *Enumeration of curves of the third order* Newton also considers more general transformations, known today as *birational transformations*, given by invertible functions $x' = f(x, y)$, $y' = g(x, y)$, $x = \phi(x', y')$, $y = \chi(x', y')$, such that f, g, ϕ, and χ are rational functions.[11] The simplest example of such a transformation is inversion in a circle (3.3), mapping lines into lines or circles, and inversions in an ellipse, hyperbola and parabola described by Apollonius (see above, p. 115). In his classification of cubics Newton lists nine curves, which he calls *hyperbolisms of conics*: four *hyperbolisms of a hyperbola* (types 57–60 of his classification), three *hyperbolisms of an ellipse* (types 61–63), and two *hyperbolisms of a parabola* (types 64–65). He writes:

> By a hyperbolism I mean a figure whose ordinate is obtained if one takes the product of the ordinate of that figure by the given straight line, divided by the common abscissa. In this way a straight line becomes a conic hyperbola, and every conic becomes one of the figures called here hyperbolisms of conics *[388, pp. 23–24]*.

The analytic description

$$x = \frac{pz + qv + r}{Az + Bv + C}, \qquad y = \frac{Pz + Qv + R}{Az + Bv + C},$$

of a general collineation of the plane first appeared in *Analytic exercises on algebraic equations and properties of curves* (Miscellanea analytica de aequationibus algebraicis et curvarum proprietatibus. Cambridge, 1762) by Edward Waring (1734–1798), a mathematician of Newton's school *[620, p. 82]*.

---

[11] In connection with birational transformations in Whiteside's edition of the works of Newton see the paper *[535]* of Galina Shkolenok.

Affine Transformations in the Works of Clairaut and Euler

General affine transformations first turned up in Europe in a paper by the 18-year-old A. C. Clairaut, entitled *On curves obtained by intersecting a curved surface by a plane in a given position* (Sur les courbes que l'on forme en coupant une surface courbe quelconque par un plan donné de position. Paris, 1733) *[116]*. In this paper Clairaut gives a proof of a result in Newton's *Enumeration* without relying on the assertion that all cubics can be obtained by central projection from five diverging parabolas. Clairaut proves that sections of a cubic cone $xy^2 = ax^3 + bx^2z + cxz^2 + dz^3$ by planes $x = $ const. are diverging parabolas, and the remaining species of cubics are sections by other planes. Clairaut defines affine transformations as follows:

> Here we regard two curves as being of the same species if they differ only in that their coordinates do not form the same angle, or the abscissas and ordinates of one of them are always the same parts of the abscissas and ordinates of the other, much as is the case with one ellipse relative to another if their axes are not in the same ratio *[116, p. 486]*.

Clairaut writes this transformation as

$$x = (c/d)u, \qquad y = (b/e)s,$$

where x, y and u, s denote coordinates in two systems whose axes and angles are, in general, different.

Affine transformations were also considered by Euler in the 18th chapter—*On similarity and affinity of curved lines*—of the second volume of his *Introduction to infinitesimal analysis* (Introductio in analysin infinitorum. Lausanne, 1748) *[176, vol. 8, 9]*. First Euler considers similar figures and similarity transformations given by

$$x = X/n, \qquad y = Y/n,$$

notes that all circles form a class of similar figures, and states:

> Whereas in the case of similar curves homologous abscissas and ordinates increase or decrease in the same ratio, in the case when the abscissas follow one ratio and the ordinates another the curves are no longer similar. Since the resulting curves are nevertheless related, we shall call them *affine* curves. Thus affinity includes similarity as a special case, for affine curves become similar if the two ratios, followed separately by the abscissas and the ordinates, become equal. Given some curved line *AMB* (Figure 72a) one can obtain from it an infinity of affine curves *amb* (Figure 72b) as follows: one must choose the abscissa *ap* so that $AP:ap = 1:m$, then draw the ordinate *pm* so that $PM:pm = 1:n$. In this way, by changing both ratios $1:m$ and $1:n$, or just one of them, one can obtain an infinity of curves that are affine with respect to the first curve *AMB* *[176, vol. 9, p. 323]*.

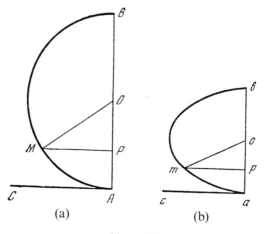

(a)   (b)

Figure 72

Then Euler represents affine transformations by means of the formulas

$$x = X/m, \qquad y = Y/n$$

and notes that they map circles onto ellipses, hyperbolas onto hyperbolas, and parabolas onto parabolas.

The terms *affine* and *affinity* (*affinitas*—literally, legal relationship, relationship by marriage) occur here for the first time. Undoubtedly, Euler's term reflects the terms *figures of the same kind* and *figures of the same species* used by Newton and Clairaut, respectively. By introducing this term Euler apparently wished to emphasize that "affine curves" are less closely related than similar curves and far less closely related than "similar and equal"—that is, congruent—curves.

Euler also gives general formulas for affine and similarity transformations obtained by combining the preceding affine and similarity transformations with rotations about a point.

We note that in the 15th chapter—*On curves with one or several diameters* —of the same work Euler actually undertakes to classify plane motions. By a *diameter of a curve*—more precisely, an *orthogonal diameter of a curve*—Euler means a straight line that halves all the chords perpendicular to it, that is, an axis of symmetry. The chapter's main concern is to clarify the conditions under which a curve has one or more axes of symmetry. Basically, however, the more fundamental problem here is to clarify the conditions under which a curve is "similar and equal"—that is, congruent—to itself. The fact that the curves under consideration are algebraic makes it possible to speak not of congruence of the curve in the large but rather of the curve as having two "similar and equal parts." While discussing different cases of the disposition of two "similar and equal parts" of a curve, Euler notes all types of plane motions (Figures 73a–73d): (a) translation, (b) rotation, (c) reflection in a line,

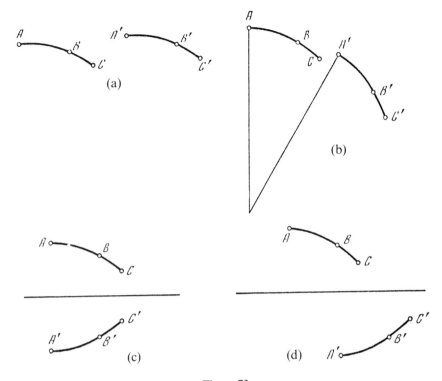

Figure 73

and (d) glide reflection (that is, reflection in a line followed by translation along that line).

Euler shows that an algebraic curve cannot be mapped onto itself by a translation and, except for a circle (which is mapped onto itself by a rotation through every angle), can be mapped onto itself only by a rotation through an angle commensurable with a right angle; if the curve has n axes of symmetry, then all of them are concurrent and successive axes form angles of π/n. Euler finds that for an algebraic curve $F(x, y) = 0$ to have n axes of symmetry the polynomial $F(x, y)$ must be

some rational function of the expressions $x^2 + y^2$ and $x^n - \dfrac{n(n-1)}{1.2}$

$x^{n-2}y^2 + \dfrac{n(n-1)(n-2)(n-3)}{1.2.3.4} \, x^{n-4}y^4$—and so on *[176, vol. 9, p. 194]*.

The reason is that any rotation about the origin transforms $x^2 + y^2$ into itself, and the only rotations that transform $(x + iy)^n$—the real part of which is the second of the expressions given by Euler—into itself are rotations through $2\pi/n$ and their multiples. This is so because a rotation through $2\pi/n$ multiplies $x + iy$ by $\cos(2\pi/n) + i \sin(2\pi/n)$ and thus $(x + iy)^n$ by $[\cos(2\pi/n) + i\sin(2\pi/n)]^n = 1$. At this point Euler makes no explicit use of complex numbers, even though in volume I of the *Introduction to analysis [176, vol. 8, p. 140]* he gives the de Moivre formula in the form we are familiar with today:

$$(\cos z \pm \sqrt{-1} \sin z)^n = \cos nz \pm \sqrt{-1} \sin nz.$$

The *Introduction to analysis* does not include the results in Euler's paper *On certain properties of conics shared by infinitely many curved lines* (Sur quelques propriétés des sections coniques qui convennent à une infinité d'autres lignes courbes. Berlin, 1746) *[176, vol. 27, pp. 51–73]*, in which he studied curves with arbitrary *diameters*, that is, with axes of symmetry associated with skew, as well as the usual right, affine reflections. All affine reflections—skew as well as right—are special axial affinities, and axial affinities are themselves special affine transformations. All diameters of conics are diameters with respect to suitable (skew or right) affine reflections. Unlike *orthogonal diameters*, such diameters can be parallel; this is so, for example, in the case of the diameters of a parabola. Euler shows that if an algebraic curve is mapped onto itself by skew reflections in two lines, then it is mapped onto itself by the affine transformation that is the product of these skew reflections. A detailed analysis is carried out in the case when a skew reflection in one diameter maps a second diameter onto a third. Some of the properties of affine transformations used by Euler in this paper are that they map lines onto lines, parallel lines onto parallel lines, and midpoints onto midpoints.

Finally, we mention Euler's paper of 1777 *On the center of similarity* (De centro similitudinis. Petersburg, 1795) *[176, vol. 26, pp. 276–285]*, which shows that for any two similar figures in the plane there exists a *center of similarity*—a point Γ such that if a, b and A, B are two pairs of corresponding points of these figures then the triangles Γab and ΓAB are similar; essentially Euler proves that every similarity that is not an isometry has a fixed point.

Conformal Transformations in the Works of Euler and Lagrange

In his *Discourse on orthogonal trajectories* (Considerationes de traiectoriis orthogonalibus. Petersburg, 1770) *[176, vol. 28, pp. 99–119]*, Euler investigated another class of transformations. He found that one way of obtaining orthogonal trajectories—that is, pairs of families of curves such that members of different families intersect each other at right angles—was to associate with each point of the plane with rectangular coordinates x, y the complex number $x + iy$ (in Euler's notation: $x + y\sqrt{-1}$) and to use functions which he wrote down in the form

$$x + y\sqrt{-1} = \text{funct}\,(T + V\sqrt{-1}),$$
$$x - y\sqrt{-1} = \text{funct}\,(T - V\sqrt{-1}).$$

Here Euler has in mind analytic functions, and his defining condition means that the functions can be expanded in power series with real coefficients. Such functions define a conformal mapping of the complex plane that maps the orthogonal families of lines $T = $ const. and $V = $ const. into orthogonal families of curves. Euler pays special attention to the case when the functions are polynomials and the curves in the orthogonal families are algebraic. One such case is that of quadratic polynomials, where the corresponding curves are confocal ellipses and hyperbolas. Euler lays special emphasis on functions of the form

$$x + y\sqrt{-1} = \frac{f + g(T + V\sqrt{-1})}{h + k(T + V\sqrt{-1})},$$

that is, fractional linear transformation. In geometric terms, these functions define the so-called circular transformations of the plane, generated by similarities and inversions in circles. Euler used conformal mappings in his geographic papers, where he constructed a conformal mapping of the sphere into the plane consisting of a stereographic projection followed by a certain conformal mapping (see p. 130). The same problem was treated by Joseph Louis Lagrange (1736–1813) in a paper entitled *On the construction of geographic charts* (Sur la construction des cartes geographiques. Berlin, 1781) *[298, vol. 4, pp. 639–692]*. Lagrange made use of conformal mappings effected by analytic functions

$$x + iy = f(u + it), \qquad x - iy = \varphi(u - it)$$

that can be expanded in power series with complex coefficients. Lagrange selected the functions f and φ so that they mapped the meridians and parallels on the sphere onto a prescribed orthogonal system of curves in the plane. The term *conformal projection* for an angle-preserving mapping of a surface onto the plane first appeared in the paper *On geographic projection of an elliptical spheroid* (De projectione sphaeroidis ellipticae geographica. Petersburg, 1789) *[573]* by Friedrich Theodor Schubert [Fedor Ivanovič Šubert] (1758–1825), one of Euler's students.

Projective Transformations in the Works of Monge and Carnot

Interest in synthetic projective geometry was revived at the end of the 18th century as a result of the appearance of *Descriptive geometry* (Géométrie descriptive. Paris, 1799) *[372]* by Gaspard Monge (1746–1818), an eminent French mathematician and revolutionary. The main part of this work is devoted to the exposition of *Monge's method*—widely used in technical drafting even in our own time and consisting in the orthogonal projection of 3-dimensional figures to two perpendicular planes that are subsequently superposed—and of central projection. In addition, it contains a number of theorems of projective geometry and their proofs. One such theorem is that the tangents from a point to a quadric (a quadratic surface) touch that surface at the points of a plane curve. The plane of this curve is called the *polar plane* of the point, and the point is called the *pole* of that plane. The polar transformation with respect to a quadric is the 3-dimensional analogue of the polar transformation with respect to a conic in the plane (discussed previously).

Monge influenced his student Lazare Carnot (1753–1823), also a revolutionary and the "Organizer of victory," to concern himself with problems of projective geometry. In the paper *On the correlation of figures in geometry* (De la corrélation des figures de géométrie. Paris, 1801) *[90]*, Carnot considers continuous, mostly projective, transformations of figures, which he calls *correlations* (*corrélations*). Carnot calls the *principle of correlation* the preservation of the properties of figures under these transformations and the passing of numerical magnitudes to their limiting values. He distinguishes *direct correlations*, in which the numerical magnitudes that characterize the system do not change sign; *indirect correlations*, in which some of these magnitudes vanish and change sign; and *complex correlations*, in which some of these magnitudes become imaginary, such as, for example, the "correlation" between the circle $x^2 + y^2 = a^2$ and the equilateral hyperbola $x^2 - y^2 = a^2$, and between the ellipse $x^2/a^2 + y^2/b^2 = 1$ and the hyperbola $x^2/a^2 - y^2/b^2 = 1$. When two figures are linked by a "correlation," then properties of one can be deduced from those of the other.

In his *Geometry of position* (Géométrie de position. Paris, 1803) *[91]*, Carnot defines a projective invariant of four points—their cross ratio (3.4). What is new in his definition is that the segments involved are oriented and their lengths are signed. Carnot shows that the cross ratio (3.4) is positive or negative according as the pairs of points A, B and C, D do not, or do, separate each other and proves the equality of the cross ratios of the quadruples of the points in which various lines—*transversals*—intersect four lines of a pencil. We saw earlier that—apart from consideration of signs—this theorem had been proved by Pappus. In his *Essay on the theory of transversals* (Essai sur la théorie des transversales. Paris, 1806) *[92, pp. 65–112]*, Carnot continues the study of cross ratios and proves the theorem—also proved by Pappus without consideration of signs—that two diagonals of a complete quadrilateral intersect the third in two points which, together with the two vertices on the latter diagonal, form a harmonic tetrad. In his study of the complete quadrilateral Carnot took as his starting point Menelaus' theorem about this figure.

Projective Transformations in the Works of Poncelet

Carnot's work was continued by another of Monge's students, the military engineer and participant in Napoleon's invasion of Russia, Jean Victor Poncelet (1788–1867). Poncelet formulated his ideas on projective geometry while a Russian prisoner of war in Saratov. Upon his return to France, he published these ideas in his *Treatise on projective properties of figures* (Traité des propriétés projectives des figures. Paris, 1822) *[434]*. After defining central projection and describing its properties, Poncelet writes:

> All these properties of central projection can be deduced purely geometrically from its own nature and from the most generally accepted principles, and there is no need to resort to algebraic analysis for their definition and proof: for example, in order to prove that an nth-order curve remains a curve of this order under projection, it suffices to note that the first curve can be intersected by a straight line in its plane in at most n points, and this must necessarily hold for the second curve, for the projection of a straight line is invariably a straight line which must pass through all the points that correspond to the points of the first.
>
> According to the generally accepted definition of Apollonius, a *conic section*, or simply a *conic*, is the curve of intersection of a plane and a circular cone, and thus nothing other than the projection of a circle. Since a line in the plane of a circle intersects the latter in at most two points, it follows from the above that a conic is also a curve of order two.
>
> In what follows, a figure, all of the graphical dependencies of whose parts are of the kind discussed above—that is, they are dependencies that are not destroyed by projection—will be called a *projective figure*.
>
> Similarly, we shall call these dependencies and, more generally, all relations and properties which hold at once for the given figure and its projection, *projective relations or properties* *[434, pp. 4–5]*.

Having defined *projective properties*, Poncelet shows that all conics are projective figures and—like Newton—suggests that in order to solve a difficult problem involving a conic one should project the conic onto a circle, solve the corresponding problem for the circle, and apply the inverse mapping. In developing Carnot's idea of complex correlations, Poncelet introduces the concept of imaginary points in the plane and uses them to prove certain theorems—for example, the theorem that all circles in the plane pass through two imaginary *cyclic points* at infinity, and the theorem that the foci of a conic are the points of intersection of the tangents to the conic from the cyclic points.

Geometric Transformations in the Works of Möbius

The projective geometry of Monge, Carnot, and Poncelet was synthetic. In just a few years after the publication of Poncelet's *Treatise on projective properties of figures* there appeared *The barycentric calculus* (Der barycentrische Calcul. Leipzig, 1827) *[369, vol. 1, pp. 5–388]*—an analytic treatment of affine and projective transformations by the German mathematician and astronomer August Ferdinand Möbius (1790–1868). The name of the book derives from Möbius' use of "barycentric coordinates" of points: if m_1, m_2, m_3 are masses located at the vertices of a fixed triangle $E_1 E_2 E_3$, then the center of gravity (*barycenter*) of these masses can be characterized—up to a constant multiple—by the numbers m_1, m_2, m_3. These numbers, called *barycentric coordinates*, are a special case of homogeneous point coordinates. If m_1, m_2, m_3 may take on negative values, then such coordinates can also be assigned to points of the plane outside the triangle. Barycentric coordinates can also be used to define points at infinity in the plane.

Möbius showed that projective transformations can be described by linear transformations of barycentric coordinates. Möbius introduces the general concept of a geometric transformation—a one-to-one correspondence between figures—and calls it a *relationship* (*Verwandtschaft*). This term is apparently a translation of Euler's "affinitas." Möbius calls an affine transformation an "affine relationship" or "affinity," and a projective transformation a "collinear relationship" or "collineation." Möbius was the first to consider general projective transformations of space mapping points into planes and collinear points into coaxial planes. He called such transformations *correlations*—a term borrowed from Carnot. An example of a correlation is a polarity with respect to a quadric. In this *Textbook of statics* (Lehrbuch der Statik. Leipzig, 1837) *[369, vol. 3, pp. 1–497]*, Möbius studies another type of correlation—a *null-system*—closely connected with dynamical screws determined by systems of forces. If we define a point by means of projective coordinates x^0, x^1, x^2, x^3, then a plane

$$\sum_i u_i x^i = u_0 x^0 + u_1 x^1 + u_2 x^2 + u_3 x^3 = 0 \qquad (3.10)$$

is defined by the "tangential coordinates" u_0, u_1, u_2, u_3, a collineation can be written as

$$x'^i = \sum_j a^i_j x^j, \qquad (3.11)$$

and a correlation can be written as

$$u_i = \sum_j a_{ij} x^j. \qquad (3.12)$$

In particular, a polarity with respect to a quadric

$$\sum_i \sum_j a_{ij} x^i x^j = 0 \qquad (3.13)$$

is a correlation (3.12) whose coefficients a_{ij} satisfy the symmetry condition $a_{ij} = a_{ji}$ and coincide with the coefficients in eq. (3.13), and a null-system is a correlation whose coefficients satisfy the skew-symmetry condition $a_{ij} = -a_{ji}$.

Circular and Conformal Transformations

We saw earlier that Euler and Lagrange considered circular and conformal transformations of the plane defined by fractional linear transformations of the complex plane and arbitrary complex analytic functions. The French mathematician Joseph Liouville (1809–1882), in an appendix to an edition of Monge's *Applications of analysis to geometry* (Applications de l'analyse à la géométrie. Paris, 1850) *[373, pp. 609–616]*, investigated conformal mappings in space and showed that, in contrast to the planar case, every conformal transformation of space maps spheres into spheres or planes and is the space analogue of a circular transformation.

Liouville's study prompted Möbius to investigate circular transformations without using complex numbers. He did this in the paper *The theory of circular relationships presented purely geometrically* (Die Theorie der Kreisverwandtschaften in rein geometrischer Darstellung. Leipzig, 1855) *[369, vol. 2, pp. 245–314]*.

Chapter 4
Geometric Algebra and the Prehistory of Multidimensional Geometry

Geometric Algebra

Our terms *square* and *cube* go back to the Pythagoreans, for whom *quadratic numbers* and *cubic numbers* were special cases of *figurate numbers*. These included *plane numbers* $m \cdot n$, *solid numbers* $l \cdot m \cdot n$, as well as the more complex *triangular numbers* $n(n + 1)/2$, *pentagonal numbers* $n(3n - 1)/2$, *pyramidal numbers* $n(n + 1)(n + 2)/2 \cdot 3$, and so on.[1] This terminology derives from the notion that points—which the Pythagoreans identified with units—are distributed in a discrete manner in figures in accordance with definite rules.

The Pythagoreans called a quadratic number *tetragonos* (quadratic, quadrangular) and *dynamis* (potency), and a cubic number *kybos* (cubic). The definitions of *square, cubic, plane*, and *solid numbers* in Euclid's *Elements* (definitions 17–20 in book VII) are taken over from the Pythagoreans *[173, vol. 2, p. 278]*. On the other hand, for geometric magnitudes Euclid used geometric algebra, in which rectangles played the role of products of segments and, in particular, squares played the role of products of segments by themselves. For example, in proposition 4 of book II Euclid formulates the algebraic identity $(a + b)^2 = a^2 + 2ab + b^2$ as follows:

> If a straight line be cut at random, the square on the whole is equal to the squares on the segments and twice the rectangle contained by the segments.

> For let the straight line AB be cut at random at C; I say that the square on AB is equal to the squares on AC, CB and twice the rectangle contained by AC, CB (Figure 74) *[173, vol. 1, p. 373]*.

---

[1] Pentagonal numbers turn up unexpectedly in geometry. Thus the family of $(n - 1)$-dimensional plane generators of an $(N - 1)$-dimensional quadric surface depends on $Nn - [n(3n - 1]/2$ parameters *[464, p. 280]*.

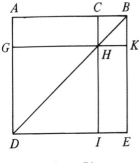

Figure 74

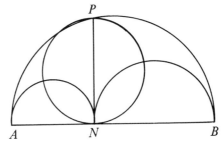

Figure 75

Proposition 4 of Archimedes' *Book of lemmas* is equivalent to this identity:

If AB be the diameter of a semicircle and N any point on AB (Figure 75) and if semicircles be described within the first semicircle and having AN, BN as diameters respectively, the figure included between the circumferences of the three semicircles is "what Archimedes called an *arbēlos*"; and its area is equal to the circle on PN as diameter, where PN is perpendicular to AB and meets the original semicircle in P *[25, p. 304]*.

In fact, the areas of the semicircles with diameters $AN = a$, $BN = b$, and $AB = a + b$ are $\frac{1}{8}\pi a^2$, $\frac{1}{8}\pi b^2$, and $\frac{1}{8}\pi(a + b)^2$, respectively, and the area of the *arbēlos* is $\frac{1}{4}\pi ab = \frac{1}{4}\pi(PN)^2$.

Geometric Names of Powers in the Works of Heron and Diophantus and in the Medieval East

It seems that it was precisely Euclid's geometric algebra that blocked consideration of powers higher than the third. The fourth power first appeared in the first century A.D., in the *Metrica* (Metrika) of Heron, for whom algebra was a purely computational rather than geometric subject. Heron called the fourth power *dynamodynamis* (square-square) *[224, v. 3, p. 48]*.

In the part of Diophantus' *Arithmetica* (Arithmétika) (third century A.D.) that has been preserved in Greek we already find six powers. For the fourth power Diophantus uses Heron's term. For the fifth and sixth powers he uses the analogous terms *dynamokybos* (square-cube) and *kybokybos* (cubo-cube) *[151, v. 2, pp. 449–514]*. In a recently discovered part of the *Arithmetica* that has come down to use in an Arabic translation by Qusṭā ibn Lūqā (d. ab. 912) *[524, pp. 318, 393]* we find also the eighth power (*square-square-square-square* or *cubo-cubo-square*) and the ninth power (cubo-cubo-cube). In the Arabic translation of Diophantus and in other works the mathematicians of the Near and Middle East called x^2 *māl* (property), x^3 *kaʿb* (cube), x^4 *māl*

māl, x^5 *māl ka'b*, x^6 *ka'b ka'b*, x^7 *māl māl ka'b*, x^8 *māl ka'b ka'b*, x^9 *ka'b ka'b ka'b*. In the 11th and 12th centuries these mathematicians went beyond Diophantus' powers.

In his *Brilliant [book] on the science of arithmetic* (Al-bāhir fī 'ilm al-ḥisāb) *[502]* (see also *[469]*) Samaw'al al-Maghribī (d. 1175) notes that the Iranian mathematician Abū Bakr Muḥammad al-Karajī (d. 1016) stated, in one of his algebraic treatises, the binomial formula for integer coefficients,

$$(a + b)^n = a^n + na^{n-1}b + \binom{n}{2}a^{n-2}b^2 + \cdots + \binom{n}{m}a^{n-m}b^m + \cdots$$

$$+ \binom{n}{2}a^2b^{n-2} + nab^{n-1} + b^n,$$

and gave a table of binomial coefficients as well as the rule of their formation,

$$\binom{n}{m} = \binom{n-1}{m-1} + \binom{n-1}{m}.$$

We find these rules in the *Collection on arithmetic with slate and dust* (Jāmi' al-ḥisāb bi-l-takht wa-l-turāb) *[592, 594]* of Naṣīr-al-Dīn al-Ṭūsī and in the *Key of arithmetic* (Miftāḥ al-ḥisāb) *[269, pp. 41–44]* of the Samarkand mathematician Ghiyāth al-Dīn Jamshīd al-Kāshī (d. c. 1430). The latter lists the six powers above and goes on:

Then square-square-cube, then square-cubo-cube, then cubo-cubo-cube; then [every time] the word 'cube' is replaced by 'square-square,' the second 'square' is replaced by 'cubo,'and so on ad infinitum *[269, p. 30]*.

We also note that the cited works of al-Ṭūsī and al-Kāshī present what is now known as the *Ruffini-Horner method* for extracting roots of integer degree of natural numbers. Apparently this method was taken over from the Chinese. It seems that it was first used in the Islamic world by 'Umar Khayyām in the lost *Problems of arithmetic* (Mushkilāt al-ḥisāb) which he mentions in his algebraic treatise *[272, pp. 74–75]*. All three writers use for roots of arbitrary degree the word *ḍil'*—side, edge—which signifies side of a square, edge of a cube, and base in the case of higher degrees. But in spite of their use of geometric terminology the three writers did not give geometric interpretations of powers higher than the third. Such an interpretation is attempted in the *Book of ingenious spiritual methods and natural mysteries about subtleties of geometric figures* (Kitāb al ḥiyal al-rūḥāniyya wa-l-asrār al-ṭabī'iyya fī daqā'iq al-ashkāl al-handasiyya) of the famous philosopher Abū Naṣr al-Fārābī (c. 870–950) *[178, pp. 91–231]*, and in the *Book about geometric constructions indispensable for craftsmen* (Kitāb fīmā yaḥtāju ilayhi al-ṣāni' min a'māl al-handasiyya) *[5]* by the mathematician and astronomer Abū-l-Wafā' al-Būzjānī (940–998).

After constructing the side of a square whose area is the sum of the areas of three equal squares, in the form of the diagonal of a cube on one of these squares as base, both authors go on to say that

things are exactly the same if we wish to [contruct] a square of more than three or less than three squares *[5, p. 118; 178, p. 200]*.

It is very likely that these words mean that "in the case when the number of squares exceeds three" the side of the required square is equal to the diagonal of the "square-square" and "square-cube" regarded as multidimensional generalizations of the cube. It is possible that this problem was discussed in the lost *Book of introduction into imaginary geometry* (Kitāb al-madkhal ilā al-handasa al-wahmīyya) of al-Fārābī *[236, vol. 2, p. 136]*.

A four-dimensional sphere was first considered by Abū Saʿīd al-Sijzī in his *Book on measuring of spheres by spheres* (Kitāb fī misāḥa al-ukar bi al-ukar) *[537, p. 331]*.[2]

Whereas Diophantus and the mathematicians of the Near and Middle East used the additive principle in forming names of powers ($x^5 = x^{2+3}$, $x^6 = x^{3+3}$, ...), the Indian mathematicians used in its place the multiplicative principle, that is, by *square-square* (*varga-varga*) they meant $x^4 = x^{2 \cdot 2}$, by *cubo-square* (*ghāna-varga*) $x^6 = x^{3 \cdot 2}$, and by cubo-square-square (*ghāna-varga-varga*) $x^{12} = x^{3 \cdot 2 \cdot 2}$. The Indians called x^5 the *product of a square and a cube* (*varga-ghāna-ghata*, $x^5 = x^2 \cdot x^3$), and x^7 the *product of a square, square and cube* (*varga-varga-ghāna-ghata*, $x^7 = x^2 \cdot x^2 \cdot x^3$).

We know of just one Arabic work whose author used the additive principle in forming names of powers. The work in question, *Book on cube, square and proportion numbers* (Kitāb al kaʿb wa al-māl wa-al-aʿdād al-mutanāsiba) by the Ṣābian mathematician Sinān ibn al-Fatḥ (10th century), recently discovered by R. Rashed *[446, p. 21]*.

Hypergeometric Names of Powers in the Works of Byzantine and Italian Mathematicians

In the works of European mathematicians, who were under the influence of Eastern mathematicians, we encounter the additive as well as the multiplicative principles of naming powers.

In a paper on Diophantus, the Byzantine mathematician Michael Psellus (1018–1078) writes that a third-century Alexandrian mathematician named Anatolius, whom Psellus calls a *computer* (logist), used for x^5 the term *alogos prōtos*—the "first inexpressible"—and for x^7 the term *alogos deuteros*—the "second inexpressible" *[574, vol. 2, pp. 430–432]*.

In his *Book of the abacus* (Liber abaci) *[318, p. 446]*, which was strongly influenced by Arabic writings, Leonardo Pisano called x^2 *census*, x^4 *census*

---

[2] This was pointed out to the author by E. I. Slavutin.

census, x^6 *census census census,* and x^8 *census census census census.* The additive principle is also found in an Italian manuscript of the 15th century, whose author calls x^2 both *quadrato* and *censo,* x^3 *cubo,* x^4 *censo di censo,* and x^5 *censo di cubo* *[324, p. 288]*.

But most Italian mathematicians of the 15th century preferred the multiplicative principle. In his *Summary of arithmetic, geometry, ratios and proportionality* (Summa de arithmetica, geometria, proportioni et proportionalita. Venice, 1494) *[403]*, Luca Pacioli calls x^2 *censo* (in abbreviated form, *ce.*), x^3—*cubo* (*cu.*), x^4—*censo de censo* (*ce. ce.*), x^5—*primo relato* ($p^0 r^0$), x^6—*censo de cubo* (*ce. cu.*), x^7—*secondo relato* ($2^0 r^0$), x^8 *censo de censo de censo* (*ce. ce. ce.*), x^9—*cubo de cubo* (*cu. cu.*), x^{10}—*censo de primo relato* (*ce.* $p^0 r^0$), x^{11}—*tertio relato* ($3^0 r^0$), and so on, up to x^{29}. The adjectives *primo, secondo,* and *tertio* in the names of x^5, x^7, and x^{11} stand for first, second, and third. Similarly, x^{13}, x^{17}, x^{19}, and x^{23} are called fouth, fifth, sixth, and seventh relato *[403, f. 67 v.]*. The names used by Pacioli for x^5 and x^7 remind one of the corresponding names used by Psellus. Pacioli's *relato* for *alogos* is a distorted translation or the translation of a distortion: The usual Latin translation of *alogos*—"inexpressible"—is *irrationalis,* where the Latin negation *ir* corresponds to the Greek negation *a*. It is possible that the translator confused *alogos* with *ho logos*—ratio—and rendered it as *relatum*—ratio. Another possibility is that *alogos* was interpreted as nonratio, was rendered as *irrelatum,* and this was later shortened to *relatum.* Pacioli's *relato* is the Italian form of the Latin *relatum.*

Three-Dimensional Geometric Algebra

In the previously mentioned treatise of al-Sijzī, *Book of measuring of spheres by spheres,* we find three-dimensional analogues of the propositions of Euclid and Archimedes that interpret the identity $(a + b)^2 = a^2 + 2ab + b^2$. Thus in proposition 2 of this treatise al-Sijzī proves that every sphere on whose diameter are constructed two tangent spheres tangent to the large sphere is such that the surplus of the large sphere over the two spheres is equal to the sphere whose diameter is the edge of a cube equal to three equal solids each of which is limited by the diameter of the large sphere and by the diameters of the two spheres *[537, p. 326]*.

Al-Sijzī considers the "sphere *AB*" (i.e., the sphere with diameter *AB*) and the two spheres *AC* and *BC* (Figure 76) and says that the surplus *DD* is equal to the spheres whose diameters are equal to the edge of the cube which is equal to three parallelepipeds with edges *AB, AC,* and *CB.* The solid *DD* of al-Sijzī is the solid of revolution of Archimedes' *arbēlos.* If *AC* = *a* and *CB* = *b,* then the three parallelepipeds of al-Sijzī form the surplus of the cube $(a + b)^3$ over the cubes a^3 and b^3. Al-Sijzī represent this cube in Figure 77, where the decomposition of the cube interprets the identity

$$(a + b)^3 = a^3 + 3ab(a + b) + b^3, \tag{4.1}$$

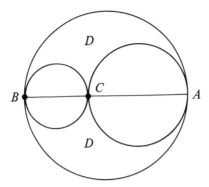

Figure 76

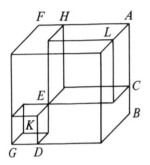

Figure 77

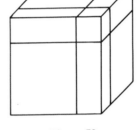

Figure 78

equivalent to the identity

$$(a + b)^3 = a^3 + 3a^2b + 3ab^2 + b^3. \tag{4.2}$$

In the proposition 3 al-Sijzī also gives the interpretation of the identity (4.2). The decomposition of the cube into two cubes and six parallelepipeds interpreting this identity was proposed in the *Algebra* (Coss) of Christoff Rudolff (ab. 1500–ab. 1545) for whom it was named *Christoff's cube*) *[558, f. 173 r.]* (Figure 78).

The rule of decomposition of a cube (probably Christoff's) was used by the Italian algebraists of the 16th century Niccolo Tartaglia (c. 1500–1557) and Girolamo Cardano (1501–1576) (who also used the Italian *relato* and the Latin *relatum*) in their solutions of cubic equations by radicals. Tartaglia discovered a method for solving equations by radicals, and Cardano published Tartaglia's solution in his *Great art, or on algebraic rules* (Ars magna sive de regulis algebraicis. Nürnberg, 1545) *[88, 89]* and supplied his own proof. Cardano's proof (see *[195]*) was based on the rule of decomposition of $(a + b)^3$. To solve an equation $x^3 + ax = b$ one put $x = U - V$. Substitution in the equation yielded the equality $U^3 - 3U^2V + 3UV^2 - V^3 + aU - aV = b$, where $U^3 - V^3 = b$ and $3\ UV = a$. Similarly, to solve an equation $x^3 = ax + b$ one put $x = U + V$; here $U^3 + V^3 = b$, $3UV = a$.

Hypergeometric Names of Powers in the Works of German Cossists

The additive principle of naming powers occurs also in the early algebra handbooks by the German *cossists*. The German name *Coss* for algebra derives from the Italian *cosa*—thing—itself a translation of the Arabic *shay'* and the Latin *res*. Following the Eastern mathematicians, the Italian mathematicans used this term to denote the unknown in algebraic equations. For example, in the Dresden manuscript C. 80 (ab. 1480) we find x—*res* (r); x^2—*zensus* (z); x^3—*cubus* (c); $x^4 = zz$; $x^5 = rzz$; $x^6 = zzz$; $x^7 = czz$; $x^8 = zzzz$, $x^9 = rzzzz$; $x^{10} = zzzzz$. But already in the Vienna manuscript 3277 (ab. 1500) x^5 is denoted by *alt* and is called *quadrangularis* (a distortion of *alogos?*) and x^6 is denoted by $z + c$ (*quadratus et cubus*) *[591, vol. 2, p. 148]*. All subsequent cossist algebras use the multiplicative principle. In a manuscript of 1525, whose author refers to himself as *Initius Algebras* (founder of algebra) *[244]*, as well as in its abbreviated version entitled *Algebra* (Die Coss) and written a year earlier by Adam Riese (1492–1559), x^2 is called *Zensus* or *Quadrat* and is denoted by ʒ, and the names and abbreviations of x^3 though x^9 are x^3—*Cubus* (ᴄ); x^4—*Zensus de Zensu* (ʒʒ); x^5—*Sursolidum* (ß); x^6—*Zensicubus* (ʒᴄ); x^7—*Bissursolidum* (bíß); x^8—*Zensus Zensui de Zensu* (ʒʒʒ); x^9—*Cubus de Cubo* (ᴄᴄ). Higher powers are denoted in the manuscript of Initius Algebras as follows: $x^{10} - $ ʒß; $x^{11} = $ terß; $x^{12} = $ ʒʒᴄ; $x^{13} = $ quadrß; $x^{14} = $ ʒbíß; $x^{15} = $ ᴄß; $x^{16} = $ ʒʒʒʒ; $x^{17} = $ quintß; $x^{18} = $ ʒᴄᴄ *[455, p. 5; 244, p. 474]*.

Initius Algebras makes the following comment about the names of x^5 and x^7:

> Note that these two symbols are called *sursolida*, that is, *surda solida*, for they are obtained from solids and surfaces, multiplied by one another in surdic and irrational section, arising at the fifth and seventh steps of multiplication" *[244, p. 477]*.

The Latin *surdus*—that gives rise to *surda solida* and *surdisch* (literally: deaf)—is a translation of the Arabic *aṣamm* (dumb, deaf), the Arabic term for the Greek *alogos*—inexpressible. The Greek, Islamic, and Western European scholars used (respectively) the terms *alogos*, *aṣamm*, and *surdus* for irrational roots. Later, *surdus* came to denote irrational numbers. This explains the term *surdic and irrational section*. It means that it is not possible to obtain square and cube roots of x^5 and x^7, that is, such roots of these powers of integers are irrational. Also, we are explicitly told that *sursolidum* is short for *surdum solidum*.

Riese's definition points to the same origin of the word *sursolidum*:

> sursolidum is a deaf number (ist eine taube zal) that has nothing in common with either square or cube *[455, p. 35]*.

The German *taub* is the equivalent of the Latin *surdus*. It seems that the

Cossist terminology derived from some lost source used by Psellus and Pacioli. What is new here is the use of *solidum*—solid—in the names of prime powers. This indicates that the Cossists (or the sources they relied on) viewed these powers as generalizations of cubes. Similar names of powers up to x^9 occur in Rudolff's *Algebra*:

x^2—Zensus (ʒ); x^3—Cubus (ᴄ); x^4—Zensdezens (ʒʒ); x^5—Sursolidum (ß); x^6—Zensicubus (ʒᴄ); x^7—Bsursolidum (*Bß*); x^8—Zenszensdezens (ʒʒʒ); x^9—Cubus de Cubo (ᴄᴄ) *[558, f. 63].*

Rudolff defines sursolidum as

an awkward number (ungeschickte zal) without square or cube root *[558, f. 63 v.].*

In his supplements (Königsberg, 1553) to Rudolff's *Algebra* Michael Stifel, the greatest of the cossist scholars, introduced the following symbols for higher powers: $x^{10} = $ ʒß; $x^{11} = C$ß; $x^{12} = $ ʒʒᴄ; $x^{13} = D$ß; $x^{14} = $ ʒ*B*ß; $x^{15} = $ ᴄß; $x^{16} = $ ʒʒʒʒ; $x^{17} = E$ß; $x^{18} = $ ʒᴄᴄ; $x^{19} = F$ß; $x^{20} = $ ʒʒß; $x^{21} = $ ᴄ*B*ß; $x^{22} = $ ᴄ*C*ß *[558, f. 160 v.].* In his *Complete arithmetic* (Arithmetica integra. Nürnberg, 1544) *[559]* Stifel called x^5 *surdesolidum* rather than *sursolidum* and used the symbols $x^7 = b$ß; $x^{11} = c$ß; $x^{13} = d$ß; $x^{17} = e$ß, and so on *[559, f. 32 r.].*

Later mathematicians, however, viewed *sursolidum* as an abbreviation of *supersolidum*—"supersolid" (*sur*—"above" in French—is an abbreviation of the Latin *super*). This name is found, for example, in *On the occult part [of the study] of numbers called algebra* (De occulta parte numerorum quam algebram vocant. Paris, 1560) *[414]* of Jacques Pelletier (1517–1582).

The view of the fifth power as a "supersolid" is found in the 17th century in the works of René Descartes (1596–1650), the creator of symbolic algebra. In his *Geometry* (La géométrie. Paris, 1637) *[143, vol. 3, pp. 307–485; 144]*, he gives the names of the powers x^2 through x^6 as

square or cube, or square of square, or supersolid or square of cube (le quarré, ou le cube, ou le quarré de quarré, ou le sursolide, ou le quarré de cube) *[143, vol. 3, p. 373].*

Descartes also used the term *sursolide* on other occasions. In one paper *[143, vol. 3, p. 188]* he used Stifel's term *B-sursolide* for x^7. He called the third book of his *Geometry*, which was devoted to the solution of problems of degree 3 or higher, *On the construction of solid or supersolid problems* (De la construction des problèmes solides ou plus que solides) *[143, vol. 3, p. 442].* Franz van Schooten (1615–1660) gave the following Latin translation of this quotation from Descartes:

quadratum, sive cubus, sive quadrato-quadratum, sive surde-solidum, sive quadrato-cubus *[146, p. 5].*

Further on von Schooten translated *sursolide* as *surde-solidum [146, pp. 108–109].* Descartes himself used in one of his Latin letters *[143, vol. 3, p. 265]* the term *supersolidum.*

Multidimensional Generalizations of the Cube
in the Work of Stifel

The "hypergeometric" names of powers and the interpretation of *sursolidum* as *supersolidum* rather than as the earlier *surdesolidum* suggested to Stifel the idea of a multidimensional generalization of a cube. Stifel's idea was quite precise. In his reworked version of Rudolff's *Algebra*, in the supplement to book 1, Stifel wrote:

> Such [geometric] progressions are named after true geometric progressions in the proper sense of the word, in which the first to be represented is a point as the element of lines; the second to be listed is a line (long or short); the third [to be constructed] is a plane square figure named after the measure of the drawn line by length and width; the fourth is the cube, for which the drawn line is a cubic root or the measure by all of its three dimensions—length, width and breadth. The geometric progression does not go further to other, larger dimensions. That is why every geometric progression is translated into arithmetic: unity for a point, the first number for a line, the second number for a square plane figure, and the third number for a cubic solid. Whereas in arithmetic we are allowed to invent many things, even if they are completely devoid of form, in geometry we must not assume corporeal lines and surfaces (cörperliche linien und superficies) and go beyond the cube as if there were more than three dimensions (über den cubum hinauss faren gleych als weren mehr denn drey dimensiones), for this would be unnatural. In that case, the geometric progression would go further and further, without any purpose and conclusion, the cube would be regarded as a corporeal point, after which one would set the corporeal line, then the corporeal surface, then the cube, to be followed by others, as just indicated, without stopping. But then one would have to make a considerable allowance for a beautiful and remarkable application of algebra *[558, f. 9r.–9v.]*.

Regarding a line as the trace of a moving point, Stifel called a cube a *corporeal point*, and by a *corporeal line* and *corporeal surface* he meant the result of the motion of a cube in one or two directions perpendicular to all dimensions of the cube. In another place, generalizing the notion of a solid line, Stifel defined a *cossic line* as the trace of a moving magnitude of arbitrary dimension, which he illustrated by the trace of a moving line *ac*: if this line moves along a line *ab*, then the rectangle abcd is a *cossic line [558, f. 173 r.]*. Having presented the binomial formula $(a + b)^n$ in his *Complete arithmetic*, Stifel, in his reworking of Rudolff's *Algebra*, used his ideas on multidimensional cubes to illustrate this formula. After citing the geometric interpretation of the binomial formula for $n = 3$ by means of Christoff's cube, Stifel writes:

Just as square binomials decompose into 4 parts, and cubic ones into 8 parts, square-square binomials (die binomia zensizensica) decompose into 16 parts, and the supersolids (die sursolida) into 32 parts, and this continues in a similar manner in accordance with the double progression *[558, f. 482 v.]*.

The idea of a space of more than three dimensions appeared at the end of the 16th century in the commentaries on Aristotle's *Physics* by the Portuguese Jesuit Manuel de Gois [read: *Goish*], who worked in Coimbra University. These commentaries are famous because Cantor took from them the term *transfinite* to denote the cardinality of infinite sets—in de Gois' words

actually (actu) consisting of infinitely many parts *[363, p. 258]*.

Following Bradwardine—of whom more later—de Gois introduced the notion of *imaginary space* and conjectured that it was the habitat of God *[210, p. 561]*. He went on to say that

this imaginary space is not a real magnitude possessing three dimensions *[210, p. 562]*.

This property of *imaginary space* supports the conjecture that this notion goes back (through a great many intermediate links) to the *imaginary geometry* of al-Fārābī.

First Attempts at a Geometric Interpretation of Functions of Many Variables

Another road leading to multidimensional geometry is connected with attempts at geometric interpretation of functions of many variables. The first such attempt goes back to the French mathematician Nicole Oresme (ab. 1323–1382) and his treatise *On the configurations of qualities and motions* (De configurationibus qualitatum et motum) *[397]*. About a *linear quality* (a function of one variable) Oresme wrote:

Every linear quality is figured [that is, represented by figures] by means of surfaces perpendicularly situated on the surface informed with quality at its base *[397, p. 177]*,

that is, by a graph in the plane. About a *surface quality* (a function of two variables) he wrote:

A surface quality is imaginable as a corporeal figure perpendicularly situated on the surface, furnished with quality at its base *[397, p. 208]*.

Further Oresme defines a *corporeal quality*, that is, a function of three variables:

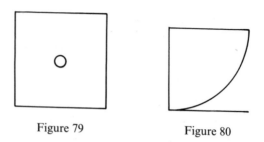

Figure 79   Figure 80

By following the present imagery with respect to [a quality being represented as] outside of the subject in its every part, a corporeal quality is
figured according to the figuration of [all of the] surface qualities of the
same body. It is clear from the statements made earlier that some
corporeal quality can be completely imagined or figured by every kind of
solid figure, so long as a perpendicular line can be drawn to the base of
that figure from any point of the figure by which the quality of this kind
is designated. And therefore, no quality is designated by a perforated
figure of this sort (Figure 79):

and by a subconcave figure, that is, one that is concave opposite the
base, as is this figure (Figure 80):

or by some such figure *[397, p. 210]*.

Thus the graph of a linear quality is situated in the plane perpendicular to the
linear base of this quality, the graph of a surface quality is also perpendicular
to the plane base of this quality, and the graph of a corporeal quality consists
of lines perpendicular to all dimensions of the corporeal base of this quality,
that is, this graph is situated in the fourth dimension.

The same idea occurs in the 17th century in the work of one of the founders
of analytic geometry, Pierre Fermat (1601–1665), namely in his *New analytic
treatment of unknowns of second and higher order* (Novus secundarum et
ulterioris ordinis radicum in Analyticis usus) *[182, vol. 3, pp. 157–163]*. After
investigating an equation in two unknowns and showing that such an equation defines a curve, Fermat writes:

If the problem involves three unknowns, then to satisfy the equation one
has to find not only a point or a curve but an entire surface. In this way
surface loci arise, and so on *[182, vol. 3, p. 161]*.

It is clear that by *and so on*, that is, by further geometric loci, Fermat meant
multidimensional geometric interpretations of equations with more than three
unknowns. Thus by the 17th century the need for geometric interpretations
of, first, algebraic powers beyond the third, and then functions of more than
two unknowns, led mathematicians close to the idea of multidimensional
space.

The Geometric Algebra of Viète

Analytic geometry was created almost simultaneously by Fermat in his *Intro-duction to plane and solid loci* (Ad locos planos et solidos isagoge) *[182, vol. 1, pp. 91–103]*, and by Descartes in his *Geometry* (1637). The former was written somewhat earlier and, though not published until after Fermat's death, was read by Paris mathematicians in manuscript.

It seems that the two creators of analytic geometry were inspired by a small work by Viète entitled *First notes on the logistic of types* (Ad logisticem speciosam notae priores. Paris, 1631) *[603, pp. 13–41]*.[3] This was published after his death and shortly before the appearance of the works of Fermat and Descartes.

In proposition 46 of *First notes* Viète solves the following problem:

Given two right triangles to find a third right triangle. Let two right triangles be given and let the hypotenuse of the third be similar to what is obtained by multiplying the hypotenuse of the first by the hypotenuse of the second, namely Z and X *[603, p. 34]*.

A comment is called for. It is required to find a triangle whose hypotenuse is equal to the product of the hypotenuses of the first two triangles. Viète says that the hypotenuse of the third triangle is "similar" to the product of the hypotenuses of the first two triangles because he adheres to the homogeneity principle adopted in antiquity and so regards the product of the two hypote-nuses as a rectangle and the third hypotenuse as the base of a rectangle of equal area whose height is equal to a unit segment. Viète considers two right triangles with hypotenuses Z and X, horizontal legs D and G, and vertical legs B and F, respectively, and points out that the problem has two solutions:

In the first case the first side is B by $G + D$ by F and the second side is B by $F=D$ by G; in the second case the first side is B by $G=D$ by F and the second side is B by $F + D$ by G (Figure 81) *[603, p. 34]*.

Viète's sign "$=$" denotes subtraction of the smaller of two magnitudes from the larger, that is, "$A=B$" stands for $|A - B|$. Viète calls his composition of triangles *generation of triangles* (genesis triangulorum).

In proposition 48 Viète considers the composition of two equal triangles and calls the result a *double angle triangle*. In proposition 49 he composes the original triangle with the *double angle triangle* obtaining a *triple angle triangle*; repetition of the process in propositions 50 and 51 produces a *quadruple angle triangle* and *quintuple angle triangle*, respectively. Viète calls his operation *parting* (diductio) of right triangles and formulates its general rule:

---

[3] Although this is one of the most interesting of Viète's works, it is little studied by historians of science. Attention to it was called by Isabella Grigor'evna Bašmakova and Evgeniĭ Iosifovič Slavutin, who have investigated in *[38]* various algebraic aspects of this work. See also the book *[39]* by Bašmakova and Slavutin.

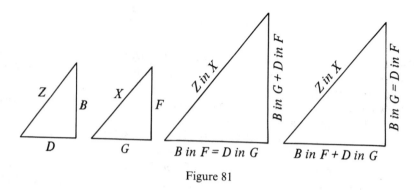

Figure 81

if one forms an arbitrary power of a binomial [made up of two] roots and separates the resulting individual homogeneous terms into two parts, with positives (adfirmata) followed by negatives (negata), then the first of these parts will be similar to the base [that is, the horizontal leg—B. R.] of another triangle, the second to [its] perpendicular [that is, the vertical leg—B. R.], and the hypotenuse [of that triangle] is similar to the power itself. As for the triangle from which the similar one has been obtained, its base is equal to one of the combined roots, its perpendicular to the other, and its name derives from the angle subtended by its perpendicular. Now the triangle obtained by extending the [triangle constructed] on the roots is named for the angle that is a multiple of that angle, whatever the order of the corresponding power, according to the property of that power, namely, double if the power is a square, triple if a cube, quadruple if a square-square, quintuple if a square-cube, and so on to infinity *[603, p. 37]*.

In propositions 48–51 of the *First remarks* there are drawings of double, triple, quadruple, and quintuple triangles, and next to their legs are given their lengths expressed in terms of the lengths B and D of the legs of the initial triangle:

$$Dq - Bq, \qquad D \text{ in } B2,$$

$$Dc - D \text{ in } Bq3, \qquad Dq \text{ in } B3 - Bc,$$

$$Dqq - Dq \text{ in } Bq6 + Bqq, \qquad B \text{ in } Dc4 - Bc \text{ in } D4,$$

$$Dqc - Dc \text{ in } Bq10 + D \text{ in } Bqq5, \qquad B \text{ in } Dqq5 - Bc \text{ in } Dq10 + Bqc.$$

In modern terms, these are

$$D^2 - B^2, \quad 2DB,$$

$$D^3 - 3DB^2, \quad 3D^2B - B^3,$$

$$D^4 - 6D^2B^2 + B^4, \quad 4BD^3 - 4B^3D,$$

$$D^5 - 10D^3B^2 + 5DB^4, \quad 5BD^4 - 10B^3D^5 + B^5$$

and thus particular cases of the expressions

$$D^n - \binom{n}{2} D^{n-2} B^2 + \binom{n}{4} D^{n-4} B^4 - \cdots + (-1)^m \binom{n}{2m} D_n^{n-2m} B^{2m} + \cdots,$$

and

$$nBD^{n-1} - \binom{n}{3} B^3 D^{n-3} + \binom{n}{5} B^5 D^{n-5} - \cdots$$

$$+ (-1)^m \binom{n}{2m+1} B^{2m+1} D^{n-2m-1} + \cdots.$$

The coefficients $\binom{n}{m}$ denote the number of combinations of n objects taken m at a time and are thus the binomial coefficients.

The formulations of these propositions shows that Viète realized that his composition of triangles resulted in the addition of the angles between hypotenuse and base and, in the special case of the composition that he called *parting*, in the multiplication of the angle by a natural number. Viète used this fact to solve problems on the division of an angle as well as on the division of a circle into an equal number of parts; the solutions appear in the papers *Theorems on the division of angles* (Ad angulares sectiones theoremata) and *An answer to the problem Adrian van Roomen asked the mathematicians of the whole world to solve* (Ad problema, quod ombibus mathematicis totius orbis construendum proposuit Adrianus Romanus, responsum) *[603, pp. 287–324]*. (The fact in question follows from formulas Viète was familiar with, that express the cosine and sine of the sum and difference of two angles in terms of their sines and cosines. The expressions for the legs of the triangle constructed in proposition 46 reduce to these formulas if the hypotenuses of the first two triangles are 1.)

In these papers, in which Viète dealt only with angles of triangles, he used an analogous composition but only up to similarity. The following two theorems are formulated in the two papers (the first of which contains the proofs given by Viète's student A. Anderson):

Theorem I. If there are three right triangles and the acute angle of the first differs from the acute angle of the second by the acute [angle] of the third, with the excess on the side of the first, then the sides of the third are obtained by means of the following similarities: the hypotenuse is similar to the rectangle on the hypotenuses of the first and the second, the perpendicular is similar to the rectangle on the perpendicular of the first and the base of the second minus the rectangle on the perpendicular of the second and the base of the first, the base [is similar] to the rectangle on the bases of the first and second plus the rectangle on their perpendiculars. . . .

Theorem II. If there are three right triangles and the acute angle of the first added to the acute angle of the second equals the acute [angle] of the third, then the sides of the third are obtained by means of the following similarities: the hypotenuse is similar to the rectangle on the hypotenuses of the first and second, the perpendicular is similar to the rectangle on the perpendicular of the first and the base of the second plus the rectangle on the perpendicular of the second and the base of the first, the base [is similar] to the rectangle on the bases of the first and second minus the rectangle on their perpendiculars *[603, pp. 287–289, 314–315]*.

In the reply to van Roomen each of these theorems is followed by examples: for theorem I three triangles whose *perpendiculars* are 1 and whose *bases* are, respectively, 2, 3, and 7; for theorem II three triangles whose *perpendiculars* are again 1 and whose *bases* are, respectively, 7, 3, and 2. Viète points out that

these two theorems form the foundation of the whole doctrine of the division of angles *[603, p. 315]*.

Since the expressions for the legs of the triangle obtained by composing two triangles coincide with the real and imaginary parts of the product of the complex numbers $D + Bi$ and $G + Fi$ or $G - Fi$, the composition of Viète's triangles essentially coincides with the geometric interpretation of the multiplication of complex numbers: the "similarity" between the hypotenuse of the third triangle and the product of the hypotenuses of the first two triangles corresponds to the fact that the modulus of the product of two complex numbers equals the product of the moduli of the factors, and the additive property of the angles of these triangles corresponds to the fact that the argument of the product of two complex numbers equals the sum of the arguments of the factors. In particular, the legs of the triangles constructed in propositions 48–51 are equal, respectively, to the real and imaginary parts of the powers

$$(D + Bi)^2 = D^2 + 2DBi - B^2,$$

$$(D + Bi)^3 = D^3 + 3D^2 Bi - 3DB^2 - B^3 i,$$

$$(D + Bi)^4 = D^4 + 4D^3 Bi - 6D^2 B^2 - 4DB^3 i + B^4,$$

$$(D + Bi)^5 = D^5 + 5D^4 Bi - 10D^3 B^2 - 10D^2 B^3 i + 5DB^4 + B^5 i,$$

and the legs of the triangle obtained in a similar way in the general case are equal to the real and imaginary parts of the power

$$(D + Bi)^n = D^n + nD^{n-1} Bi - \binom{n}{2} D^{n-2} B^2 - \binom{n}{3} D^{n-3} B^3 i +$$

$$+ \binom{n}{4} D^{n-4} B^4 + \binom{n}{5} D^{n-5} B^5 i - \cdots + (-1)^m \binom{n}{2m} D^{n-2m} B^{2m} +$$

$$+ (-1)^m \binom{n}{2m + 1} D^{n-2m-1} B^{2m+1} i + \cdots .$$

Since neither the *First remarks* nor other works of Viète refer to complex numbers, it is appropriate to regard his composition of triangles as the geometric interpretation not of the multiplication of complex numbers but of their equivalents, the Hamilton number pairs.

Thus in Viète's *First remarks*, we encounter a correspondence between the points in the upper right quarter of the plane and the upper vertices of Viète's triangles. On the one hand this correspondence associates with every point in question a pair of segments, the *base* and *perpendicular*, coinciding with the rectangular coordinates of that point, and on the other hand the hypotenuse and the angle between the hypotenuse and base, coinciding with the polar coordinates of that point—its radius vector and polar angle.

If we think of the hypotenuses of the triangles constructed by Viète in the *First remarks* as *vectors* representing the radius vectors of points, then the hypotenuses of the triangles constructed in the *Theorem on the division of angles* and in the *Reply to van Roomen* represent vectors, defined up to multiples, known in modern geometry as *pseudovectors*.

We note that Viète does not define addition of his vectors and pseudovectors but only their multiplication, analogous to the multiplication of complex numbers.

It seems that Viète came to his composition of triangles by starting with the rules for the multiplication of the cosines and sines of sums and differences of angles. But another possibility is suggested by Isabella Grigor'evna Bašmakova. On the basis of her analysis *[151a, p. 218]* of problem 19 in book III of Diophantus' *Arithmetica* she thinks it likely that Diophantus was familiar with the identity

$$(a^2 + b^2)(c^2 + d^2) = (ac + bd)^2 + (ad - bc)^2$$
$$= (ad + bc)^2 + (ac - bd)^2,$$

equivalent to the law of multiplication of Hamilton's number pairs. This identity was applied in the *Treatise on the construction of rectangular triangles with rational sides* (Risāla fī inshā' al-muthallathāt al-qāima al-zawāya al-muntaqa al-aḍlāʿ) *[275, p. 172]* (see also *[446, p. 217]*) of the 10th-century mathematician Abū Ja'far Muhammad Ibn al-Husain al-Khāzin, and in the *Book of Squares* (Liber quadratorum) of Leonardo Pisano *[317]*, who had a good knowledge of the Arabic mathematical literature. It is therefore very likely that Viète borrowed the rule of composition of triangles from Pisano's *Book of squares* and gave it a geometric form. It is altogether possible that Pisano borrowed this rule from works of Diophantus that have not come down to us, or from Arabic works inspired by Diophantus.

It is natural to ask for motivating sources for the rectangular and polar coordinates introduced by Viète in the *First remarks*. Although in the *Conics* Apollonius systematically used oblique and rectangular coordinates—admittedly linked closely to particular curves—and in *On spirals* Archimedes used polar coordinates—also closely linked to particular spirals—the only known scholar before Viète who simultaneously used rectangular and polar coordinates and established the rules of transition from one to the other was Thābit ibn Qurra. In the first chapter of his *Book on time instruments called sundials [241; 492, p. 254]* Ibn Qurra determined the position of the end of the shadow of a gnomon by means of the *length of the shadow* and the *azimuth of the shadow* as well as *parts of the length* and *parts of the width*. If was therefore likely that in the matter of coordinates Viète was influenced by Ibn Qurra rather than by Apollonius or Archimedes. Viète may have known some of Ibn Qurra's results through al-Battānī. It is difficult to tell which of the three roads led Viète to his doctrine of composition of triangles; in fact, it is quite reasonable to assume that his teaching was the result of their simultaneous influence.

In discussions of the emergence of analytic geometry at the beginning of the 17th century it is customary to point out the role of the Latin translations by Commandino and Maurolico of Apollonius' *Conics*. In particular, our terms *abscissa, ordinate,* and *applicate* (the last two, derived from Commandino's *ordinatim applicata*—applied in succession—meant the same thing) are Latin equivalents of Apollonius' terms. There is no doubt about the influence of these translations on Fermat and Descartes: Fermat used the term *applicata* and Descartes the term *appliquée par ordre*—the French translation of Commandino's term. On the other hand, there is a marked similarity between the triangles Fermat used in his *Introduction* to introduce rectangular coordinates and Viète's triangles. Specifically, Fermat associates with every point *I* of the upper right quarter of the plane its rectangular coordinates *A* and *E* (Figure 82). These coordinates are denoted by vowels, in accordance with Viète's principle of denoting his "required quantities." Actually Fermat, like Viète, characterizes each point *I* not only by means of its rectangular coordinates but also by its radius vector and polar angle, except that he does not introduce notations for the latter and considers exclusively equations stated in terms of the coordinates *A* and *E*. The fact that the *Introduction* was written soon after the publication of Viète's *First remarks* strongly suggests that Fermat arrived at the idea of analytic geometry not only under the influence of Apollonius' *Conics* but also Viète's *First remarks*. Apparently as a result of reading Viète's work, Fermat thought of "disconnecting" Apollonius' abscissas and ordinates of points from the curves under consideration, which is why he drew the coordinates of points together with their radius vectors.

It is also possible that Descartes too came to his calculus of segments, which

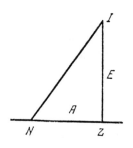

Figure 82

is the basis of his analytic geometry, after pondering Viète's *calculus of triangles*. Although it is true that Descartes's point coordinates are very close to Apollonius', and that he represented coordinates but not radius vectors, it is nevertheless likely that his rejection of the homogeneity principle used by Viète and Fermat was linked to the representation of the hypotenuse of one triangle as the product of the hypotenuses of two other triangles (the unit segment indispensable for the definition of such a product also plays an essential role in Descartes's calculus of segments).

After the publication of Descartes's *Geometry*, symbolic algebra and analytic geometry developed in forms imparted to them by Descartes, and Viète's works were all but ignored. This explains why John Wallis failed to use the ideas in Viète's *First remarks* when attempting to construct a geometric interpretation of complex numbers in his *A treatise of Algebra* (London, 1685) *[618]*. Had Wallis combined these ideas with his own idea that the imaginary magnitude $\sqrt{-bc}$ was the mean proportional between $-b$ and c or b and $-c$, that is, a segment perpendicular to the segments b and c laid off on one line on each side of the perpendicular *[618, p. 56]*, he would have been able to construct an entirely satisfactory geometric interpretation of complex numbers of the kind actually obtained only at the end of the 18th and the beginning of the 19th century.

Descartes's Geometric Algebra

Descartes's *Geometry* has played an exceptional role in the history of science in that it marked the beginning of a new period in the history of mathematics. It should be borne in mind that the *Geometry* was intended as an illustration of the application of Descartes's general philosophical machinery to a concrete problem. Already in his *Rules for the direction of the mind* (Regulae ad directionem ingenii), written in the twenties of the 17th century but published posthumously in Amsterdam in 1701 *[145, vol. 1, pp. 7–78]*, Descartes wrote:

> This made me realize that there must be a general science which explains all the points that can be raised concerning order and measure irrespective of the subject-matter, and that this science should be termed *mathesis universalis*—a venerable term with a well-established meaning ...

This means that by *universal mathematics* Descartes meant algebra, in the form of the symbolic algebra which he subsequently created in the *Geometry*. The universal method of solution of the most varied problems consists in the following:

> *If we perfectly understand a problem we must abstract it from every superfluous conception, reduce it to its simplest terms and, by means of an enumeration, divide it up into the smallest possible parts ... We should make a direct survey of the problem to be solved, disregarding the fact*

that some of its terms are known and others unknown ... For this purpose only four operations are required: addition, subtraction, multiplication and division ... Once we have found that equations, we must carry out the operations which we have left aside ... If there are many equations of this sort, they should all be reduced to a single one ... [145, v. 1, pp. 51, 70, 71, 76].

In other words, the method is to reduce every problem to a mathematical problem, formulate the mathematical problem in terms of algebra, and reduce the resulting algebraic problem to the solution of some algebraic equation.

Descartes's idea of reducing a large variety of problems to mathematical problems was generally recognized only in the 20th century, when the rise of new branches of mathematics made possible the genuine mathematization of many disciplines to which—it was thought earlier—mathematics could not be applied. Descartes's historical limitation was that he restricted "universal mathematics" to algebra. In his subsequent *Discourse on the method of rightly conducting one's reasoning well and seeking the truth in the sciences* (Discours de la Méthode pour bien conduire sa raison et chercher la vérité dans les sciences. Paris, 1637) *[145, v. 1, pp. 109–175] (vol. 6 of [143])*, in which the *Geometry [144]* was the first of three supplements, Descartes formulated the general rules of the scientific method and went on to say:

I observed that in order to know these proportions I would need sometimes to consider them separately, and sometimes merely to keep them in mind or understand many together. And I thought that in order the better to consider them separately I should suppose them to hold between lines, because I did not find anything simpler, nor anything that I could represent more distinctly to my imagination and senses. But in order to keep them in mind or understand several together, I thought it necessary to designate them by the briefest possible symbols *[145, v. 1, p. 121]*.

The reduction of all magnitudes to lines was realized in the *Geometry*. Whereas Fermat formulated the equations of "geometric loci" by using the terminology of the geometric algebra of antiquity in which the product of two linear magnitudes is a rectangle, the product of three linear magnitudes a solid, and so on (the example of Viète's formulation of the spherical cosine theorem illustrates the clumsiness of this terminology), in Descartes's new geometric algebra the product of lines is again a line. At the very beginning of the *Geometry* Descartes writes:

All problems in geometry can be reduced to such terms that a knowledge of the lengths of certain straight lines is sufficient for their construction. Just as arithmetic consists of only four or five operations, addition, subtraction, multiplication, division, and the extraction of roots, which may be considered a kind of division, so in geometry, to find required lines it is merely necessary to add or subtract other lines; or else, taking one line which I shall call the unit in order to relate it as closely as possible to numbers, and which can in general be chosen arbitrarily, and having given two other lines, to find a fourth line which shall be to one of the given lines as the other is to the unit (which is the same as multiplication); or, again, to find a fourth line which is to one of the given lines as the unit is to the other (which is equivalent to division); or, finally, to find one, two, or several mean proportionals between the unit and some other line (which is the same as extracting the square root, cube root, etc., of the given line) *[144, pp. 2, 5]*.

Then Descartes constructs the segment *ab* as the fourth proportional to the segments 1, *a*, *b*, the segment *a/b* given the same segments, and the segment $\sqrt{a}$ given the segments 1 and *a*.

Leibniz's Idea of the "Geometry of Position"

Descartes's idea of the need to mathematize natural science was further developed by the German philosopher and mathematician Gottfried Wilhelm Leibniz (1646–1716). But, unlike Descartes, Leibniz no longer tried to reduce all mathematics to algebra. Rather, he regarded the differential and integral calculus as universal mathematics and advanced a very general *continuity principle [312, vol. 6, pp. 129–135]*. One of Leibniz's ideas that played an extremely important role in the history of geometry is connected with "geometric algebra" understood in a new sense. Leibniz stated this idea in a letter to Christian Huygens (1629–1695) dated September 8, 1679 *[312, vol. 2, pp. 17–25, 315, pp. 248–258]*, in which he said

I am still not satisfied with algebra because it does not give the shortest methods or the most beautiful constructions in geometry. This is why I believe that, so far as geometry is concerned, we need still another analysis which is distinctly geometric or linear and which will express *situation [situs]* directly as algebra expresses *magnitude* directly. And I believe that I have found the way and that we can represent figures and even machines and movements by characters, as algebra represents numbers or magnitudes. I am sending you an essay which seems to me to be important....

I have discovered certain elements of a new characteristic, which is entirely different from algebra and which will have great advantages for representing to the mind exactly, and in a way faithful to its nature, even without figures, everything which depends on sense perception. Algebra is a characteristic for undetermined numbers and magnitudes only, but it does not express situation, angles, and motions directly. Hence it is often difficult to analyze the properties of a figure by calculation and still more difficult to find very convenient geometric demonstrations. But this new characteristic, which follows the visual figures, cannot fail to give the solution, the construction and the geometric demonstration, all at the same time, and in a natural way and in one analysis, that is through determined procedure. Algebra is compelled to presuppose the elements of geometry, this characteristic instead carries the analysis through to its end. If it were completed in the way in which I think of it, one could write down the description of a machine, no matter how complicated, in characters which would be merely the letters of the alphabet, and provide the mind with a method of knowing the machine and all its parts, their motion and use without use of any figures or models and the need of imagination. Yet the figure would inevitably be present to the mind to interpret the characters. One could also give exact descriptions of natural things, for example, the structure of plants and animals. With its aid people who find it hard to draw could explain a matter perfectly, provided they have it before them or in their minds, and could transmit their thoughts or experiences to posterity—a thing which cannot be done today because the words of our languages are not sufficiently fixed or well enough fitted for good explanation without figures.

This is the least useful aspect of this characteristic, however, for if only description were involved, it would be better—assuming that we can and are willing to bear the expense—to have figures and even models or, better still, the original things themselves. But its chief value lies in the reasonings which can be done and the conclusions which can be drawn by operations with its characters, which could not be expressed in figures and still less in models without multiplying these too greatly, or without confusing them with too many points and lines in the course of the many futile attempts one is forced to make. This method, by contrast, will guide us surely and without effort. I believe that by this method one could treat mechanics almost like geometry, and one could even test the qualities of materials, because this ordinarily depends on certain figures in their sensible parts. Finally, I have no hope that we can get very far in physics until we have found some such method of abridgment to lighten its burden of imagination. For example, we see what a series of geometrical reasoning is necessary merely to explain the rainbow, one of the simplest effects of nature; so we can infer what a chain of conclusions would be necessary to penetrate into the inner nature of complex effects whose structure is so subtle that the microscope which can reveal more than the hundred-thousandth part does not explain it enough to help us much. Yet there would be some hope of achieving this goal, at least in part, if this truly geometrical analysis were established *[315, pp. 248–250]*.

Leibniz denotes given points by the first letters of the alphabet $A, B, \ldots$ and the unknowns by the last letters $X, Y, \ldots$ He introduces the congruence sign ୪ (Descartes's equality sign ∞ rotated through 90°) and the symbol (Y) that stands for "for all Y" (in modern mathematical logic such a symbol is called a *universal quantifier* and is denoted by $\forall Y$). He writes the equation of a sphere as $AB \, 8 \, BX$ ("the segment BX with fixed endpoint is congruent to the fixed segment AB"), the equation of a plane as $AX \, 8 \, BX$, the equation of a circle in space as $ABC \, 8 \, ABX$ ("the triangle ABX with fixed side AB is congruent to the fixed triangle ABC"), and the equation of a line in space as $AY \, 8 \, BY \, 8 \, CY$. By means of these equations Leibniz shows that the intersection of two spheres is a circle, and the intersection of two planes is a line.

Leibniz's letter was published in his collected works shortly after his death and was well known to mathematicians of the 18th and 19th centuries. The term *geometry of position* first appeared in Euler's paper *Solution of a problem pertaining to the geometry of position* (Solutio problematis ad Geometriam situs continens. Petersburg, 1736) *[176, vol. 7, pp. 1–10]* devoted to the proof of the topological problem of the impossibility of the successive crossing of the seven bridges joining the banks of the river Pregel in Königsberg with two islands in it. Euler understood the term *geometry of position* in the sense of topology. This was also the sense given to the term by J. B. Listing, who coined the term *topology*, and by Riemann and Poincaré, the creators of combinatorial topology, who called topology by Leibniz's term *Analysis situs*. We note that in his reply to the letter of November 1689 by the Italian geometer Vitale Giordano, in which the latter criticized Leibniz's definition of a line, Leibniz wrote:

Does he sin who takes as the foundation the concepts of plane and solid, and is he not, rather, deserving of praise?

and praised

the view that the concept of a solid precedes the concepts of surface and line *[312, vol. 1, p. 19]*.

Thus Leibniz shared the opinion of those who felt that one should first define a solid, then a surface, and then a line. We saw earlier that this view—a consequence of Aristotle's teaching on mathematical concepts—was advanced in the Middle Ages by al-Fārābī, and was applied in the geometric part of the mathematical and astronomical handbook of al-Bīrūnī.[4]

In *A dialogue for the introduction to arithmetic and algebra* (Ein Dialog zur Einführung in die Arithmetik und Algebra) *[314]*, published only in 1976, Leibniz introduces a "real square-square" and, in reply to the question: "Can you proceed to higher powers?" says

I can do it endlessly *[314, pp. 100–103]*.

This idea of multidimensional space is a development of the previously mentioned idea of Descartes, since Leibniz gives as an example of a five-dimensional object the impulse of a falling heavy mass.

---

[4] Al-Fārābī stated this view in commentaries to Euclid's *Elements [178, pp. 233–276; see pp. 238–239]*. Al-Bīrūnī applied this idea in the geometric part of *The book of instruction in the elements of the art of astrology [59, p. 1; 53, vol. 6, pp. 22–23]*.

Prehistory of Vector Calculus Linked to Geometry

Another interpretation of the term *geometry of position* is found in the work of Lazare Carnot, who used this term, *géometrie de position*, as the title of a book published in 1802 *[90]*. Earlier we mentioned the theorems of projective geometry considered in this work. But another very important characteristic of this book is that it concerns itself with oriented segments and angles (we owe to Carnot the symbol $\overline{AB}$ for an oriented segment with beginning A and endpoint B); these oriented segments represent *vectors* in geometric form. Two more interpretations of the term *geometry of position* derive from Carnot. Thus the 19th-century German geometers Theodor Reye *[451]* and Christian von Staudt *[553]* meant by this term (*Geometric der Lage*) projective geometry, and Hermann Grassmann *[211, vol. 1]*, inspired by Leibniz's idea, which he interpreted as the idea of vector calculus, was one of the creators of just such a calculus. We also note the geometric calculus of oriented segments presented by Bellavitis (1803–1880) in his *Calculus of equipollence* (Calcolo delle equipolenze. Padova, lenza. Venice, 1835) *[41]*. Bellavitis called oriented segments equipollent if they had the same length and direction. Thus a class of segments equipollent in Bellavitis' sense is what we now call a *free vector*.

Prehistory of Vector Calculus Linked to Mechanics

In addition to this purely geometric source, vector calculus had two other sources, namely mechanics and algebra.

Concrete vector magnitudes—velocities and forces—first turned up in mechanics. Already in *Mechanical problems* (Problēmata mēchanika), written in the school of Aristotle, we come across composition of *motions*, that is, velocities, according to the parallelogram law:

Let the ratio of the two motions (i.e. speeds) be that which AB has to AC (Figure 83). Let AC be moved (parallel to itself) towards B, and let AB be moved down (also parallel to itself) towards GC. Let A have reached D and B have reached E. Then, if during the two motions the speeds have been in the ratio of AB to AC, AD must bear to AE the same

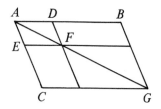

Figure 83

ratio. Therefore, the small quadrilateral (parallelogram) [ADFE] must be similar to the greater one so that they have the same diameter [AFG] and A (at the moment in question) will be at F. The same proof will apply at whatever point the motion is interrupted. The moving object will always be on the diameter [219, p. 230].

Since Aristotle thought that force was proportional to "motion," this composition rule referred, essentially, to forces. Composition of "motions" was also used by Archimedes in his treatise On spirals:

> If a straight line of which one extremity remains fixed be made to revolve at a uniform rate in a plane until it returns to the position from which it started, and if, at the same time as the straight line revolves, a point move at a uniform rate along the straight line, starting from the fixed extremity, the point will describe a spiral in the plane. [25, p. 154]

and then by Ptolemy in his Almagest [441, p. 442ff.] in the determination of the motion of planets by means of deferents and epicycles. Composition of "motions" was also used by the astronomers of the medieval East in their expositions and modifications of Ptolemy's theory (see [270]). We quote from al-Bīrūnī's exposition of Ptolemy's theory in his Canon of Masʿūd:

> The motion of the center of the epicycle of each of the two lower [planets] is equal to the motion of the body of the Sun ... similarly, the motion of each of the three upper planets on the circle of its epicycle is equal to the sum of the motion of the center of its epicycle and the motion of the Sun [57, p. 1165; 53, vol. 5, part 2, p. 351].

We note that in his Book of optics (Kitāb al-manāẓir) [19] [5] Ibn al-Haytham uses a mechanical model of the reflection of light and describes experiments involving the throwing of a metallic sphere on a surface of a metallic mirror and the rebounding of the sphere from the mirror. The "force of motion" of the rebounding sphere is regarded as the sum of "forces" directed perpendicularly to the surface of the mirror and in the direction of this surface. Ibn al-Haytham writes:

> The motion of a body moving at an angle to an obstacle will be composed of a motion in the direction of the perpendicular erected at the point of contact with the surface of the obstructing body, and in a direction perpendicular to it drawn in the surface of the obstructing body (see [562, pp. 90–92]).

European scholars continued the investigation of parallelograms of motions and forces. The Flemish scholar Simon Stevin (1548–1620) wrote

---

[5] The edition [19] reproduces a medieval Latin translation of the Book of optics. In 1983 there appeared an Arabic edition with an English translation by Abdelhamid Ibrahim Sabra (b. 1924) [238].

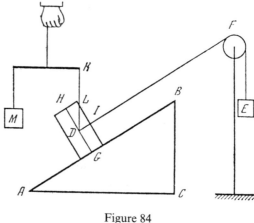

Figure 84

in his *Elements of statics* (De Beghinselen der Weeghconst) *[557, vol. 1, pp. 35–285]:*

Let us draw the vertical from the center of prism D as DK (Figure 84), meeting the side of the prism in L. This being so, the triangle LDI is similar to the triangle ABC, for the angles ACB and LID are right angles, and LD is parallel to BC, and DI to AB. Therefore, as AB is to BC so is LD to DI. But as AB is to BC so is the prism to the weight E.... Therefore, as LD is to DI, so is the prism to the weight E. Let us now attach at the line KD the vertical lifting weight M of equal apparent weight to the prism. The weight M will be of equal weight to the prism.... Therefore, as LD is to DI, so is M to E *[557, vol. 1, pp. 182–183]*.

We see from this that Stevin assumes that the force E directed along one leg of a right triangle, together with the force of reaction of the inclined plane directed along the other leg, are in equilibrium with the force M directed along the hypotenuse; that is, three forces applied to a single point are in equilibrium if they are respectively parallel and proportional to three sides of a right triangle; later *[557, vol. 1, pp. 182–183]* Stevin also proved this assertion in the case when the forces are parallel and proportional to the sides of an arbitrary triangle. Stevin's assertion is equivalent to the parallelogram law of forces.

In his treatise *Mechanics, or On motion, a geometric treatise* (Mechanica, sive De motu, tractatus geometricus, 1669) *[617, vol. 1, pp. 578–1063]* John Wallis formulated the rules of the parallelogram of forces and parallelepiped of forces and the rules of composition of directed segments used by him to denote forces, displacements, velocities, and accelerations. Wallis writes that, in particular,

If a motion is composed of three simple motions, whose directions and velocities are represented by the straight lines $A\alpha$, AB, AC (Figure 85),

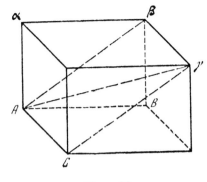

Figure 85

then it is equivalent to the simple motion along $A\gamma$ with velocity $A\gamma$ *[617, vol. 1, p. 999].*

In the 19th century, using mechanics as the starting point, Ademar Jean Claude Barré de Saint Venant (1797–1886) constructed a calculus of vectors in the paper *On geometric sums and differences and their application to the simplification of the exposition of mechanics* (Sur les sommes et differences géométriques et leur application pour la simplification de la exposition de la mécanique. Paris, 1845) *[498].*

Prehistory of Vector Calculus Linked to Algebra

The algebraic source of vector calculus was the complex numbers. Complex numbers first turned up in Cardano's *Great Art [88, 89]* in connection with the solution of the problem of dividing 10 into two parts whose product is 40. Cardano wrote down the solution in the form $5\tilde{p}$: $R\tilde{m}$: 15 and $5\tilde{m}$: $R\tilde{m}$: 15, that is, using modern notations, $5 \pm i\sqrt{15}$. Complex numbers also occur in the solution of cubic equations. Thus in the so-called irreducible case the real root of a cubic equation $x^3 = px + q$ with $(q/2)^2 - (p/3)^3 < 0$ is represented as the sum of two conjugate complex numbers. Although already Raphael Bombelli (ab. 1526–1573) had set down the rules for operating with complex numbers in his *Algebra* (L'Algebra. Bologna, 1572) *[70]*, even Leibniz, who used complex numbers for integrating certain real functions, called them in his *New example of analysis for the science of the infinite in connection with sums and quadratures* (Specimen novum Analyseos pro Scientia infini circa Summas et Quadraturas. Leipzig, 1702) *[312, vol. 5, pp. 350–360]*

a miracle of analysis, a monster in the world of ideas, an amphibian between existence and nonexistence *[312, vol. 5, p. 357].*

The mystical aura surrounding complex numbers dispersed only in the 18th century, when Jean Le Rond d'Alembert (1717–1783) showed in his *Essay of a new theory of resistance of liquids* (Essai d'une nouvelle théorie sur la resistance des fluides. Paris, 1752) *[134]* that the coordinates P, Q of the velocity of a moving fluid at a point with coordinates x, y are proportional to expressions that d'Alembert wrote as

$$\Delta\left(x + \frac{y}{\sqrt{-1}}\right) + \Delta\left(x - \frac{y}{\sqrt{-1}}\right),$$

$$\frac{1}{\sqrt{-1}}\left[\Delta\left(x + \frac{y}{\sqrt{-1}}\right) - \Delta\left(x - \frac{y}{\sqrt{-1}}\right)\right],$$

that is, to the real and imaginary parts of a function $\Delta(z)$ of the complex variable $z = x + (y/\sqrt{-1})$ that could be expanded in a power series with real coefficients; here we encounter for the first time the *Cauchy-Riemann conditions*

$$\frac{\partial u}{\partial x} = \frac{\partial v}{\partial y}, \qquad \frac{\partial u}{\partial y} = -\frac{\partial v}{\partial x}$$

of analyticity of a function $w = f(z)$ of a complex variable ($w = u + iv$, $z = x + iy$).[6]

We find similar arguments in Euler's *Continuation of the investigations into the theory of motion of fluids* (Continuation des recherches sur la théorie du mouvement des fluides. Berlin, 1757) *[177, vol. 12, pp. 92–132]*. We see that even the algebraic road to vectors led, to a significant extent, through problems of mechanics.

Incidentally, implicit geometric interpretations of complex numbers occur in Viète's *First remarks on logistica speciosa* and in Euler's *Introduction to Analysis*. Specifically, in Viète's case the cosine and sine of an angle $n\alpha$ turn out to be equal, respectively, to the real and imaginary parts of the complex number $(\cos\alpha + i\sin\alpha)^n$, and in Euler's case a (plane) rotation through an angle $2\pi/n$ is connected with the real part of the complex number $(x + iy)^n$ *[176, vol. 9, p. 194]*. We also saw that Euler used conformal mappings of the plane effected by analytic functions of a complex variable (that admit expansions in power series with real coefficients) in his papers on orthogonal trajectories and cartography *[176, vol. 28, pp. 99–119, 248–297]*.

The interpretation of complex numbers as vectors in the plane with explicit geometric meanings for their operations is found for the first time in the paper *Attempt at an analytic representation of direction and an attempt to apply it, mainly for the solution of plane and spherical polygons* (Om directiones analytiske betegning et forsög anwendt fornemellig til plane og sphaeriske polygoners oplösning. Copenhagen, 1799) *[624]* by Caspar Wessel (1745–1818).

---

[6] The Cauchy-Riemann equations that occur in the work of d'Alembert, Euler, and Cauchy were deduced and thoroughly investigated by Riemann in his doctoral dissertation *Foundations of a new general theory of functions of a variable complex magnitude* (Grundlagen für eine allgemeine Theorie der Functionen einer veränderlichen complexen Grösse) *[454, pp. 3–48]*.

In addition to dealing with plane vectors Wessel advanced the idea of vectors in space and attempted to define a multiplication for such vectors *[624, pp. 23–28]*. Mathematicians became aware of Wessel's paper only after the publication of a French translation (in 1897).

The idea of a geometric interpretation of the operations on complex numbers was also advanced by Jean Robert Argand (1768–1822) in his *Attempt to represent complex numbers in a certain way by means of geometric constructions* (Essai sur une manière de représenter les quantités dans les constructions géométriques. Paris, 1806) *[27]*, and became widely known after the publication of the *Course of algebraic analysis* (Cours d'analyse algébrique. Paris, 1821) *[100, vol. 3]* of Augustin-Louis Cauchy (1789–1867), and the *Theory of biquadratic residues II* (Theoria residorum biquadratorum. Commentatio secunda. Göttingen, 1832) of Gauss *[196, vol. 2, pp. 95–178]*. That is why 19th-century mathematicians frequently referred to the complex plane as the *Cauchy plane* or the *Gauss plane*. In the first half of the 19th century Hamilton and other English algebraists tried to generalize complex numbers to three- and then four-dimensional space. These attempts led them first to various *triplets* and then to *quaternions*, whose calculus contains the algebra of vectors in three-dimensional space.

The Idea of Multidimensional Space in the 18th Century and at the Beginning of the 19th Century

The idea of multidimensional space was first advanced in a very clear form by the German philosopher Immanuel Kant (1724–1804) in his early paper *Thoughts on the correct assessment of living forces* (Gedanken von der wahren Schätzung der lebendigen Kräfte. Königsberg, 1876) *[260, pp. 1–181]*. The young Kant tried to give physical reasons for the three-dimensionality of space, writing that

> It is likely that three-dimensionality is the result of the fact that in the world around us substances interact so that the forces involved are inversely proportional to the squares of distances . . . another law would imply a space form with other properties and dimensions. A science of all such space forms would undoubtedly be the most sublime geometry (*höchste Geometrie*) which finite reason could pursue. . . . If the existence of space forms with other dimensions is possible, then it is very likely that God has realized them somewhere *[260, p. 23]*.

Also in the 18th century, the concept of the fourth dimension was first linked with time. In the paper *Dimension*, in the *Encyclopedia, or rational dictionary of the sciences, arts and handicrafts* (Encyclopédie, ou Dictionnaire raisonné des sciences, des arts et des métiers. Paris, 1764) *[135]*, published by d'Alembert and Diderot, d'Alembert wrote:

A clever acquaintance of mine believes that it is possible to think of time as a fourth dimension, so that the product of time and solidity would in some sense be the product of four dimensions; it seems to me that this idea, while debatable, has certain merits—at the very least the merit of novelty *[135, p. 1010]*.

The development of multidimensional spaces was strongly influenced by Lagrange's introduction—in his *Analytical Mechanics* (Mécanique analytique. 1788) *[299, vol. 1, p. 36]*—of generalized coordinates of mechanical systems, which he called "variables ξ, ψ, φ, . . . ," and characterized as independent variables, since they determine all change in the position of the system *[299, vol. 1, p. 36]*. Although Lagrange himself emphasized in the introduction that his

methods require neither constructions nor geometric or mechanical considerations; they require just algebraic operations *[299, vol. 1, p. 1]*,

there can be little doubt that his readers, forced to make their own drawings, must have associated with his generalized coordinates some geometric interpretation—visual for one and two degrees of freedom and mental for a larger number.

Four-dimensional space, with the reservation that it "cannot be imagined," was introduced by Möbius in 1827 in his *Barycentric calculus*. In connection with congruent sets of points that cannot be brought into coincidence by means of continuous transformations but can be obtained from one another by means of reflection in a plane, Möbius writes:

Suppose that A, B, C, D, . . . and A', B', C', D', . . . are two equal and similar systems in three-dimensional space, and that the points D, E, . . . and D', E', . . . lie on differently named sides of the planes ABC and $A'B'C'$. One might conclude by analogy that in order for the systems to coincide we must carry out a half turn in the space of four dimensions. But since such a space cannot be imagined, the coincidence is, in this case, impossible *[369, vol. 1, p. 172]*.

Möbius's idea was later used by H. G. Wells (1866–1946) in his science fiction *Story of Platner*, whose hero visited the fourth dimension and returned with his heart on the right-hand side.

Chapter 5
Philosophy of Space

The evolution of philosophical notions of space played an important role in the preparation of the discovery of non-Euclidean geometry and the subsequent generalizations of the idea of space. That is why we give a brief survey of this evolution.

The Idea of the Infinity of Space and of the Finiteness of the World

The idea of the infinity of space seems to have arisen in ancient Greece, notwithstanding the fact that the ancient Greek scholars viewed the world as finite. The idea of the finiteness of the world was spelled out in its most obvious form by Plato. In the *Timaeus* (Timaios), Plato (425–347 B.C.) says of God:

> And for shape he gave it that which is fitting and akin to its nature. For the living creature that was to embrace all living creatures within itself, the fitting shape would be the figure that comprehends in itself all the figures there are; accordingly, he turned its shape rounded and spherical, equidistant every way from centre to extremity—a figure the most perfect and uniform of all; for he judged uniformity to be immeasurably better than its opposite *[425, p. 54]*.

Elsewhere in the same dialogue Plato assigns to the atoms of the four elements—fire, air, water, and earth—the forms of four regular polyhedra ("platonic solids")—the tetrahedron, the octahedron, the icosahedron, and the cube, and says that

> There still remained one construction, the fifth; and the god used it for the whole, making a pattern of animal figures thereon *[425, p. 218]*.

That is, Plato connected the fifth regular polyhedron—the dodecahedron—with the Universe. Here "animal figures" are zodiacal signs, one for each face of the dodecahedron.

Next we describe the position of Democritus (460?–370? B.C.). The writings of Democritus have not come down to us, and we know his views only from quotations by later philosophers. Thus Joannes Philoponus (sixth century A.D.), in his commentaries on Aristotle's *Physics*, described Democritus' position as follows:

> Democritus assumed the existence of infinite worlds, supposing that the vacuum is infinite. For on the basis of what principle of distribution would one part of the vacuum be filled with some world and another not? Thus, if the world exists in a certain part of the vacuum then, obviously, [it exists] in all of the vacuum. But since the vacuum is infinite, the worlds are also infinite (see *[344, p. 207]*).

The Epicurean Lucretius describes the doctrine of the ancient atomists on the infinity of space in his poem *On the nature of things* as follows:

> space is without bound or limit, and I have shown in many words, and it has been proved by true reasoning, that it spreads out immeasurable towards every quarter everywhere *[342, p. 241]*.

In the *Timaeus*, Plato enters into a polemic with the doctrine of Democritus on the multiplicity of worlds:

> Accordingly, to the end that this world may be like the complete Living Creature in respect of its uniqueness, for that reason its maker did not make two worlds nor yet an indefinite number; but this Heaven has come to be and is and shall be hereafter one and unique. *[425, p. 49]*,

and thus rejects Democritus' doctrine of the infinity of worlds.

Aristotle considers the question of the infinite in his *Physics*. He admits only the potential infinite:

> For generally the infinite has this mode of existence: one thing is always being taken after another, and each thing that is taken is always finite, but always different *[29, vol. 2, p. 206ᵃ]*.

Aristotle thinks that time, motion, and thought are infinite but spatial magnitudes are not:

> Time indeed and movement are infinite, and also thinking, in the sense that each part that is taken passes in succession out of existence.
> Magnitude is not infinite either in the way of reduction or of magnification in thought *[29, vol. 2, p. 208ᵃ]*.

Aristotle thought that mathematicians require just the notion of the potential infinite:

Our account does not rob the mathematicians of their science, by disproving the actual existence of the infinite in the direction of increase, in the sense of the untraversable. In point of fact they do not need the infinite and do not use it. They postulate only that the finite straight line may be produced as far as they wish. It is possible to have divided in the same ratio as the largest quantity another magnitude of any size you like. Hence, for the purposes of proof, it will make no difference to them to have such an infinite instead, while its existence will be in the sphere of real magnitudes *[29, vol. 2, p. 207^b]*.

But, at the same time, the previously mentioned doctrine of Aristotle that mathematical concepts are obtained by abstracting from objects of the real world enables one to disengage oneself from the finiteness of physical magnitudes. That is why Ibn Rushd (Averroes, 1126–1198), a follower of Aristotle, wrote in his commentaries on Aristotle's *Physics [28]* that a geometer "can admit" an arbitrarily large magnitude—something a physicist cannot do, and "having thought of an arbitrarily large magnitude he can take an even larger one." In postulate II of his *Elements [173, vol. 1, p. 154]* Euclid required the possibility of indefinite extension of every straight line ("To produce a finite straight line continuously in a straight line").

The questions of the vacuum and of space, infinite in Aristotle's sense, were extensively debated by theologians and philosophers in the 14th century — by Henry de Gondavo, Richard of Middleton, Walter Burleigh, and Thomas Bradwardine (see *[291; 292, pp. 33–84]*); the latter, in the treatise *Of the cause of God* (De causa Dei), called empty space beyond the world the *imaginary vacuum* (vacuum imaginarium) (see *[292, p. 78]*; also, see p. 155 of this book. Since, in chronological terms, the *imaginary vacuum* of Bradwardine occurs between the *imaginary geometry* of al-Fārābī and the *imaginary space* of de Gois (see p. 161), it is possible that Bradwardine represented one of the intermediate links between al-Fārābī and de Gois.

The Doctrine of Independence of Space from Matter and Its Critique

The ancient scholars had also created the doctrine of the independence of space from matter. There is no doubt that Democritus, who referred to space as the "vacuum," supported this view. The explicit formulation of this doctrine is due to Aristotle, who referred to space as "locus"—place. In his *Physics* Aristotle wrote:

For place is supposed to the something like a vessel—the vessel being a transportable place. But the vessel is no part of the thing.

In so far then as it is separable from the thing, it is not the form: *qua* containing, it is different from the matter. *[29, vol. 2, p. 209^b]*.

The doctrine of independence of space from matter was criticized by Descartes, who wrote in his *Principles of Philosophy* (Principia Philosophiae. Paris, 1644):

Space, or interior place, differs from the physical substance contained in it only in our thought. Actually, extension in length, width, and depth that makes up space also makes up a body. The only difference between them is that we ascribe to a body a definite extension and think that one, together with the other, changes place when it moves; whereas to space we ascribe extension so general and indefinite that whenever we remove from some space the body that fills it we do not assume that we have also moved the extension of this space which, in our view, exists unchanged as long as it has the same magnitude and figure and does not change position relative to outside bodies by means of which we define this space *[143, vol. 8, p. 45]*.

In the same work Descartes said this of "empty space":

As for empty space, in the sense in which philosophers understand this word, that is, space devoid of all substance, it is clear that there is no space in the world that is such, for the extension of space as inner place does not differ from the extension of a body. And just as from the single fact that a body extends in length, width and depth we correctly conclude that it is substance (for it is impossible that "nothing" has extent) so too with respect to space, supposed empty, it must be concluded that as soon as there is in it extension then, necessarily, there must as well be in it substance *[143, vol. 8, p. 49]*.

Descartes's view of the impossibility of empty space was further developed by Leibniz in his *New essays on human understanding* (Nouveaux Essais sur Entendement humain. Paris. 1765), written in 1704. Leibniz wrote:

It is necessary rather to conceive space as full of a matter originally fluid, susceptible of all the divisions, and even actually subject to divisions and subdivisions to infinity, but with this difference, however, that it is divisible unequally in different parts on account of the motions which more or less concur there. This it is which causes matter to have everywhere a degree of rigidity as well of fluidity, and no body to be hard or fluid in the highest degree, i.e., no atom to be found of an insurmountable hardness nor any mass entirely indifferent to division *[312a, p. 54]*.

Unlike Descartes and Leibniz, Newton favored the doctrine of independence of space from matter. In the *Principia* he states that

absolute, true and mathematical time, of itself, and from its own nature flows equally without relation to anything external and by another name is called duration *[387, p. 6]*,

defines "relative, apparent or ordinary time," and then gives the following definition of absolute space:

Absolute space, in its own nature, without relation to anything external, remains always similar and immovable *[387, p. 6]*.

Then follows the definition of "relative" space.

It is clear that the epithets *true* and *mathematical* that Newton applied to "absolute time" apply equally to his "absolute space." For Newton, the basic characteristics of *mathematical space* were its uniformity, immovableness, and independence "of everything external." Newton's "absolute space" became one of the cornerstones of the theoretical mechanics founded in the *Principia* and developed by the great 18th-century mathematicians and mechanists Euler, d'Alembert, and Lagrange.

We note that the ancient doctrine of independence of the "vacuum" or "place" from the objects in it was also directed against the view of the Pythagoreans, who identified the points of space with the souls of the dead or of the unborn, and that Newton's doctrine of "absolute space" was directed against Leibniz's doctrine of monads—simple substances that make up the multiformity of the world and that Leibniz identified with the points of space.

Descartes's viewpoint was supported by Mihail Vasil'evič Lomonosov (1711–1765) in the same *Discourse on the solidity and fluidity of bodies* (Rassuždenie o tverdosti i židkosti tel. Petersburg, 1760) in which he formulated the "universal natural law" of conservation of matter and motion *[336, pp. 340–352]*. Here Lomonosov criticizes Newton's law of universal gravitation:

At this point, might not someone ask that I show the cause, or the matter, or the manner by which are held together the very indivisible particles of particles, compressed by the liquid matter that bathes them. Must I not here admit, someone will say, the existence of an attractive force? Absolutely not.

Then Lomonosov tries to explain the attraction of bodies by means of properties of space:

Whoever knows the difference between absolutely necessary properties of bodies and their variable characteristics can see clearly that when it comes to all that is indispensable for things and their existence one can neither show the cause nor should one ask for it; for example, why does a triangle have three sides; why is a body extended, and similar questions; for one should look for the cause of conjunction where we see that insensitive particles are in a state of conjunction at one moment and forfeit it at another, that its force now increases now decreases. Here one might ask why it is so and not otherwise. And in the conjunction of insensitive particles that make up bodies change is not declared; for this reason one must not ask for a cause. The philosophical *justification* called *sufficient reason* does not extend to the indispensable properties of bodies. From this improper usage has arisen the debate, famous in the scholarly world, about single substances, that is, about particles with no extention whatever. Since extension is an indispensable property of a body without which it cannot be a body, and since virtually all of the force of the definition of a body consists in extension, the question and argument about inextensive particles of an extensive body is futile; for in this case one must look for proofs of the definition instead of, as is customary, deriving consequences, in good order, from the definition *[336, pp. 342–243]*.

We see that Lomonosov criticizes Newton's doctrine as well as Leibniz's doctrine, in many respects its opposite, of the principle of "sufficient reason" and of "simple substances"—Leibniz's monads. His own position corresponds to the tradition of the ancient atomists. Lomonosov sees space as "a conjunction of particles insensitive" as well as "inextensive." What is most important for us, however, is that he connects gravitational properties of matter with properties of space.

The Doctrine of the a Priori Nature of Space

Aristotle's doctrine that mathematical concepts are obtained by abstraction from objects of the real world was directed against the Pythagoreans, who explained all regularities in the world by numerical regularities, and against Plato, who explained them by geometric regularities. The reason for Plato's pushing geometry into the foreground (the inscription over the door of Plato's academy was *Medeis ageōmetrētos eisitō*—let no one ignorant of geometry enter here) was the collapse of the Pythagoreans' worldview after their discovery of incommensurable magnitudes, that is, magnitudes whose ratios can not be expressed as ratios of natural numbers. Earlier we mentioned the role in the structure of the world that Plato assigned to regular polyhedra. Aristotle points out that a mathematician "effects an abstraction, for in thought it is possible to separate figures from motion" and adds that

> without realizing it, the philosophers who teach about ideas do the same thing: they abstract physical properties, less separable than mathematical ones" *[29, vol. 2, pp. 193ᵇ–194ᵃ]*.

"Philosophers who teach about ideas" are Plato and his students. For Plato mathematical concepts are a special case of the "ideas" that Aristotle also regards as abstractions from "things," that is, objects of the real world. Plato thinks that mathematical, and, in particular, geometric notions are inborn and for proof describes in the dialogue *Meno* (Mēnōn) an experiment in which a bright slave boy, prompted by a number of leading questions, proves that a square constructed on the hypotenuse of a right isosceles triangle is twice as large as the square on one of its legs. Socrates, the principal figure of the dialogue, concludes:

Then he who does not know may still have true notions of that which he does not know? . . . And at present these notions have just been stirred up in him as in a dream; but if he were frequently asked the same questions in different forms, he would know as accurately as anyone at last *[426, p. 284]*.

The doctrine of the a priori nature of geometric views in its most distinct form was expressed by Immanuel Kant in his fundamental philosophical work *The critique of pure reason* (Kritik der reinen Vernunft. Königsberg, 1781) *[261]*. In this connection Kant writes:

1. Space is not a conception which has been derived from outward experiences. For, in order that certain sensations may relate to something without me, (that is, to something which occupies a different part of space from that in which I am); in like manner, in order that I may represent them not merely as without of and near to each other, but also in separate places, the representation of space must already exist as a foundation. Consequently, the representation of space cannot be borrowed from the relations of external phenomena through experience; but, on the contrary, this external experience is itself only possible through the said antecedent representation. . . .

2. Space then is a necessary representation *à priori*, which serves for the foundation of all external intuitions. We never can imagine or make a representation to ourselves of the nonexistence of space, though we may easily enough think that no objects are found in it. It must, therefore, be considered as the condition of the possibility of phenomena, and by no means as a determination dependent on them, and is a representation *à priori*, which necessarily supplies the basis for external phenomena *[261, pp. 23–24]*.

However, later Kant writes:

It is, moreover, not necessary that we should limit the mode of intuition in space and time to the sensuous faculty of man. It may well be, that all finite thinking beings must necessarily in this respect agree with man (though as to this we cannot decide), but sensibility does not on account of this universality cease to be sensibility *[261, p. 43]*.

These words of Kant show that he did not entirely rule out the possibility of generalizing the notion of space; we saw earlier (p. 179) that the young Kant suggested the idea of one such generalization.

The most striking proof of the falsity of the doctrine of the a priori nature of geometric notions is the origin of universally accepted names of geometric objects *sphere* from *sphaira*—ball; *cylinder* from *kylindros*—roller; *cone* from *konos*—pine cone; *prism* from *prisma*—something sawn off; *trapezium* from *trapezion*—small table; *rhombus* from *rhombos*—top, and so on. These names show that, originally, bodies of the form of a sphere, cylinder, cone, and so on, were named for concrete objects having these respective forms.

The doctrine of the a priori nature of geometric notions was criticized by many philosophers and mathematicians. In chapter VI we will consider its critique by Lobačevskiĭ that helped him make his remarkable discovery. At this point we quote the critique of Kant's doctrine by Timofeĭ Fedorovič Osipovskiĭ (1765–1832), rector of Har'kov University, in his work *On space and time* (0 prostranstve i vremeni. Har'kov, 1807):

I agree with the founder of critical philosophy [that is, Kant] that it is impossible to deduce irrefutable synthetic conclusions from notions acquired from experience when this acquisition is understood precisely in the sense in which he assumes it, that is, when one acquires ideas about certain special cases belonging to a single whole but not marked with the stamp of universality. But the notion of space is acquired in a very different manner: it begins with the whole and the parts are already contained in it; for everybody knows, and Mr. Kant himself says, that the notion of space precedes the notion of all things that borrow parts of this whole. The possibility of obtaining an idea of the whole space together with its parts is implicit, in the first place, in the very manner in which we acquire this notion, that is, in our sense of vision that is so constructed that the whole is imprinted on it together with all its parts; and in the second place, in that the whole is uniform in its entirety and continuous *[398, pp. 14–15]*.

And as for Kant's thought that if space were the condition of the existence of things and therefore "were in them and not in us" then we could not be sure that this property of space that our senses impart to a certain object actually belongs to it, Osipovskiĭ replies:

No one will take it upon himself to prove that the space that we perceive in things is completely the same in them as we perceive it; it is enough that there is in them something that corresponds to what we observe, and that it corresponds in accordance with a constant law of dependence between what is in them and that which it imprints on our sensations. If, on the contrary, there is nothing in the thing that corresponds to the notion, related to space, that is born in us when we sense it, that is, if there is no mutual dependence whatever between this thing and our notion, then why will the notion relate to the thing? For example, if nothing corresponds in a sphere to the roundness that we perceive when we look at a sphere why then can we link the notion of roundness to the notion of sphere; for then these ideas will be totally unrelated to one another. In that case, all synthetic chains of ideas proposed and proved in Mathematics in relation to space would be pure chimeras, that arise just in our heads in an involuntary but incoherent manner, have no relation whatever to things and are therefore incapable of any application to them; but it is well known that no one ever said anything more true than Euclid in his elements [that is, the *Elements*] and nowhere is there a more precise correspondence than the one between the truths proposed in these elements and what is actually observed in things *[398, pp. 15–16]*.

Finally, Osipovskiĭ arrives at the following conclusion:

All that has been said above makes one think that space and time are conditions for the existence of things that exist in nature and in themselves and not only in our form of sensation. As regards space, my view is this: the notion about it arises from impressions that originate in it with the aid of [the action of] our outer senses on our inner senses *[398, p. 16]*.

One of the founders of Marxist philosophy, Friedrich Engels (1820–1895), engaged in a polemic with the well-known botanist Karl Wilhelm Nägeli (1817–1891). In the paper *The limits of natural scientific knowledge* (Die Schranken der naturwissenschaftlichen Erkenntniss. München, 1877) Nägeli wrote:

We know exactly the meaning of an hour, a metre, a kilogram, but we do not know what time, space, force and matter, motion and rest, cause and effect are.

Engels's response, contained in his *Dialectic of nature* (Dialektik der Natur. 1925), was:

It is the old story. First of all one makes sensuous things into abstractions and then one wants to know them through the senses, to see time and smell space. The empiricist becomes so steeped in the habit of empirical experience, that he believes that he is still in the field of sensuous experience when he is operating with abstractions. We know what an hour is, or a metre, but not what time and space are! As if time was anything other than just hours, and space anything but just cubic metres! The two forms of existence of matter are naturally nothing without matter, empty concepts, abstractions which exist only in our minds *[169, p. 235]*.

We see that Engels calls space and time *two forms of existence of matter*. In his *Anti-Dühring, Herr Eugen Dühring's revolution in science* (Herrn Eugen Dührings Umwälzung der Wissenschaft. London, 1878) Engels wrote:

But it is not at all true that in pure mathematics the mind deals only with its own creations and imaginations. The concepts of number and figure have not been derived from any source other than the world of reality. The ten fingers on which men learnt to count, that is, to perform the first arithmetical operation, are anything but a free creation of the mind. Counting requires not only objects that can be counted, but also the ability to exclude all properties of the objects considered except their number—and this ability is the product of a long historical evolution based on experience. Like the idea of number, so the idea of figure is borrowed exclusively from the external world, and does not arise in the mind out of pure thought. There must have been things which had shape and whose shapes were compared before anyone could arrive at the idea of figure. Pure mathematics deals with the space forms and quantity relations of the real world—that is, with material which is very real indeed. The fact that this material appears in an extremely abstract form can only superficially conceal its origin from the external world. But in order to make it possible to investigate these forms and relations in their pure state, it is necessary to separate them entirely from their content, to put the content aside as irrelevant; thus we get points without dimensions, lines without breadth and thickness, a and b and x and y, constants and variables; and only at the very end do we reach the free creations and imaginations of the mind itself, that is to say, imaginary magnitudes. Even the apparent derivation of mathematical magnitudes from each other does not prove their *a priori* origin, but only their rational connection. Before one came upon the idea of deducing the *form* of a cylinder from the rotation of a rectangle about one of its sides, a number of real rectangles and cylinders, however imperfect in form, must have been examined. Like all other sciences, mathematics arose out of the *needs* of men: from the measurement of land and the content of vessels, from the computation of time and from mechanics. But, as in every department of thought, at a certain stage of development the laws, which were abstracted from the real world, become divorced from the real world, and are set up against it as something independent, as laws coming from outside, to which the world has to conform. That is how things happened in society and in the state, and in this way, and not otherwise, *pure* mathematics was subsequently *applied* to the world, although it is borrowed from this same world and represents only one part of its forms of interconnection—and it is only *just because of this* that it can be applied at all *[170, pp. 58–59].*

Continuity and Discreteness of Space in Antiquity and in the Middle Ages

The question of whether space is discrete or continuous was debated already by the ancient philosophers. In his commentaries on Aristotle's *Physics*, Simplicius tells us that Anaxagoras (fifth century B.C.), one of the ancient philosophers, maintained that

For in small there is no least but only a lesser *[148, p. 83]*,

that is, he subscribed to the principle of infinite divisibility, apparently applying it to matter as well as to space. His older contemporary Leucippus and younger contemporary Democritus favored atomistic notions of matter as well as of space. In connection with these views Aristotle wrote in *De caelo* that

There is, further, another view—that of Leucippus and Democritus of Abdera—the implications of which are also unacceptable. The primary masses, according to them are infinite in number and indivisible in mass *[29, vol. 2, p. 303ᵃ]*.

In his *Letter to Eratosthenes on the method of mechanical theorems* (Peri tōn mēchanikōn theōrematon pros Eratosthenē ephodos)[1] Archimedes wrote:

in the case of the theorems the proof of which Eudoxus was the first to discover, namely that the cone is a third part of the cylinder, and the pyramid of the prism, having the same base and equal height, we should give no small share of the credit to Democritus who was the first to make the assertion with regard to the said figure though he did not prove it *[25, App., p. 13]*.

Archimedes did not consider Democritus' reasoning as a proof, for in his time proofs of theorems about areas and volumes were thought rigorous only if they employed the so-called method of exhaustion. Since the theorems mentioned by Archimedes are based on the theorem about the equality of volumes of two pyramids with the same base and height, there is no doubt that Democritus' arguments were based on the idea of two such pyramids as made up of layers of indivisible "primary magnitudes," that is, geometric atoms. All this indicates that Democritus viewed space as consisting of atoms of finite size and every finite body as consisting of a "pretersensually" large number of such atoms.

Another variant of atomistic notions of space is found among the Pythagoreans. It is not clear whether these notions arose in the original school of Pythagoras or later, under the influence of Democritus. The fact remains that the later Pythagoreans thought of solids as made up of discrete points. This notion is at the basis of the Pythagorean doctrine of "figurate numbers" that we mentioned in chapter 4. It is set forth in the *Treatises of the Brethren of Purity* (Rasā'īl Ikhwān al-Ṣafā'), which is strongly influenced by the Pythagoreans of the philosophical school that functioned in Basra and other towns in the 10th century. The *Treatises*, published under the collective pseudonym of the school, represent an encyclopedia of a number of sciences as well as—in accordance with Pythagorean traditions—a number of mystical doctrines.

In the geometric treatise the Brethren of Purity wrote:

The shortest line—of two points, thus . . , then of three points, thus . . . , then of four points, thus . . . . , then of five, thus . . . . . , then they increase by one, like the numbers in the natural sequence. The tiniest trigonal

---

[1] In English works this is usually referred to as the *Method* and will be so referred to in the sequel. (Translator.)

figure—of three points, thus $\therefore$, then of six points, thus $\therefore\cdot$, then of ten points, thus $\cdot\cdot\cdot\cdot$, and they increase according to this sample *[243, vol. 1, p. 83]* (see also *[586]*).

It is clear that the Pythagoreans viewed geometric figures as made up of discrete points separated by finite, although very small, distances.

Aristotle supported the opposite point of view. He wrote in *Physics*:

> Nothing that is continuous can be composed of indivisible parts: e.g., a line cannot be composed of points, the line being continuous and the point indivisible *[29, vol. 2, p. 231ᵃ]*.

As we have already pointed out, Aristotle held that atomistic views lead to a contradiction. He found contradictory the notion that a continuous magnitude, for example a line, consists of points; for on this view a line of finite length, a plane figure of finite area and a body of finite volume consist of points whose length, area, and volume are each equal to zero. Actually, atomistic notions included attempts to eliminate this contradiction by ascribing to "points" small but finite dimensions, as was done by Democritus, or by linking the length of a line to the distances between points and by representing plane figures and solids as, respectively, plane or spatial lattices or other plane and spatial configurations, as was done by the Pythagoreans. Aristotle solved this problem differently. He regarded lines, planes, and solids as infinitely divisible but did not view them as collections of points but rather as loci where points can be located. This notion of Aristotle was taken over by Euclid who in proposition 10 of book I of the *Elements* solved the problem:

> To bisect a given finite straight line *[173, vol. 1, p. 267]*

for every segment, including arbitrarily small ones. It was this view that gave rise to the term *locus* that is now thought of as meaning a "set of points." At the same time, some of the definitions of Euclid's *Elements* bear traces of atomistic traditions. For example, consider the definitions 1, 2, and 5 of book I:

> A *point* is that which has no part;
> A *line* is breadthless length;
> A *surface* is that which has only length and breadth *[173, vol. 1, p. 153]*.

Here "that, which has no part" is an indivisible "primary magnitude," an atom of space; "breadthless length" is a chain of atoms; and "only length and breadth" is a layer of atoms of the type considered by Democritus in his theorem on the volume of a pyramid.

We have already mentioned Archimedes' remarks about the atomistic reasoning of Democritus. In the *Method* he wrote:

> for certain things first became clear to me by a mechanical method, although they had to be demonstrated by geometry afterwards because their investigation by the said method did not furnish an actual demon-

stration. But it is of course easier, when we have previously acquired, by the method, some knowledge of the questions, to supply the proof than it is to find it without any previous knowledge *[25, App., p. 13]*.

In *The Sand-reckoner* (Psammitēs) Archimedes mentions an assertion of Aristarchus of Samos (310?–230? B.C.):

His hypotheses are that the fixed stars and the sun remain unmoved, that the earth revolves about the sun in the circumference of a circle, the sun lying in the middle of the orbit, and that the sphere of the fixed stars, situated about the same centre as the sun, is so great that the circle in which he supposes the earth to revolve bears such a proportion to the distance of the fixed stars as the centre of the sphere bears to its surface *[25, p. 222]*

and continues:

Now it is easy to see that this is impossible; for, since the centre of the sphere has no magnitude, we cannot conceive it to bear any ratio whatever to the surface of the sphere *[25, p. 222]*.

It is clear that Euclid's younger contemporary Aristarchus thought of the center of a sphere as a geometric atom of finite dimensions.

Although the majority of mathematicians of the medieval East shared the viewpoint of Aristotle and Euclid, atomistic views were not entirely unknown. We have already mentioned the atomistic views of the Brethren of Purity. Similar views were held by Muslim theologians, the *muʿtazila* and the *mutakallimūn*. Among the *muʿtazila* there were adherents of both types of mathematical atomism, the one represented by Democritus and the one espoused by the Pythagoreans. Abū-l-Hāshim al-Jubbāʾī (820–933) belonged to the first school and Abū-l-Qāsim al-Kaʿbī (d. 932) to the second. Al-Kaʿbī's nickname ("cubical") shows that in his doctrine a key role was played by the arrangement of atoms as knots of a cubical lattice. The book *The contentious questions between the Basrians and the Baghdadians* (al-Masāʾil fīʾl-khalaf bayna al-Baṣriyyīn wa l-Baghdādiyyīn) of Abū Rāshid al-Nayshābūrī *[424, p. 2]* deals with the issues dividing the two doctrines. Al-Nayshābūrī states that the main argument was about whether or not "an atom partakes of extension." Al-Jubbāʾī's answer is in the affirmative since, in his view, it is difficult to imagine a body made up of atoms without extension. Al-Kaʿbī argued that the extension of bodies is due not to the atoms but to the distances between them *[424, p. 7]*.

The doctrine of the *mutakallimūn* was set forth by the famous philosopher Moses Maimonides (1135–1204) in *The guide of the perplexed* in which he says about the *mutakallimūn* that

They thought that the whole world—I mean to say every body in it—is composed of very small particles that, because of their subtlety, are not subject to division. The individual particle does not possess quantity in any respect. However, when several are aggregated, their aggregate possesses quantity and has thus becomes a body *[349, p. 195]*.

The *mutakallimūn* held similar views of time and, since "time is made up of 'now,'" they concluded that God creates the world anew every instant and claimed on this basis that all events in the world "come from God."

We have already mentioned that the Brethren of Purity developed a mathematical atomism of the Pythagorean type. Among the followers of mathematical atomism in· the medieval East there were also philosophers who developed the materialistic traditions of ancient philosophy. One such scholar was Abū Bakr Muḥammad ibn Zakariyā al-Rāzī (865–952), who, in many respects, followed the "line of Democritus." In his *Book on matter* (al-Kitāb al-hayūlā) al-Rāzī wrote that

> the structure of all bodies is the result of the mixing of particles of primary matter with particles of the vacuum, that is, absolute space *[424, p. 104]*.

These "particles of the vacuum" are mathematical atoms. Mathematical atomism explains the title of al-Rāzī's (lost) *Treatise that [the assertion that] the diagonal of a square is incommensurable with a side is not geometric* (Risāla fī anna quṭr al-murabbaʿ lā yushāriku al ḍilʿ min ghayr handasa) *[236, vol. 1, p. 309]*: in atomistic geometry all segments, including the diagonal of a square and a side, are commensurable. In the treatise *Concerning what was between him and Abū-l-Qāsim al-Kaʿbī on the question of time* (Mā jāra baynahi wabayna Abū'l-Qāsim al-Kaʿbī fī l-zamān) *[52, p. 11]* al-Rāzī debated with al-Kaʿbī questions about the atomistic structure of space. In his philosophical correspondence with ibn Sīnā, al-Bīrūnī wrote:

> If for each of these things, that is, boundaries of a body, there are two sides and one middle then division is indefinite, and this is impossible *[527, p. 40]*.

Al-Bīrūnī himself had a great deal of sympathy for atomists of the al-Rāzī variety and in his second letter to ibn Sīnā wrote:

> Why does Aristotle regard as fallacious the doctrine of indivisibility of a particle if the assertion of the indefinite divisibility is even more fallacious.... Atomists are characterized by quite a few [debatable] assertions well known among geometers but the words of those who oppose the atomists are even less acceptable *[54, pp. 13–14]*.

In his reply to al-Bīrūni, Ibn Sīnā defended the viewpoint of Aristotle. Al-Bīrūnī also mentions atomists on other occasions. Thus in one of the astronomical books of the *Canon of Masʿūd* he writes:

> When we speak of moving [bodies] there is no limit that one can apply to oneself in the matter of rigor.... To approximate the truth it is necessary to repeat the process of making things accurate. Finally, the solution of

this is possible only after the resolution of the contention between the "followers of the particle" [that is, atomists] and the followers of its rejection *[57, p. 937; 53, vol. 5, part 2, p. 214]*.

'Umar Khayyām also admits the possibility of the triumph of mathematical atomism. In a chapter of his *Commentaries on the difficulties in the premises of Euclid's book* devoted to the theory of ratios Khayyām considers first the ratios of pairs of magnitudes "in which the smaller is a fraction or fractions of the larger." After mentioning the third possibility—"they may have no numerical ratio, which characterizes only geometric magnitudes"—Khayyām writes:

If they will say that there is no third case altogether and there are only two numerical cases, we will reply that consideration of the rules of ratios and proportions of magnitudes in these three cases does not confound us, and if this case is refuted they will not reproach us for anything, but, since it has not been refuted, we consider it and complete the two indicated cases *[272, p. 129]*.

The mathematician Quṭb al-Dīn al-Shīrāzī (1236–1311) wrote in his commentaries to Naṣīr al-Dīn al-Ṭūsī's *Treatise on the motion of rolling and the relation between the straight and the curved* (Rīsāla fī ḥarakat al-daḥraja wa'l-nisba bayn al-mustawī wa'l munḥanī):

As to those who mention that [points and lines] follow one after the other and recognize formation [of lines from points and of surfaces from points and lines]. . . since for them the equality and the inequality [of figures] is [established] only by the number of points, for them is possible equality and inequality between line and body, between line and surface and between surface and body *[529, p. 197]*.

Thus al-Shīrāzī recognizes the existence of mathematical atomism in his time.

Mathematical atomism existed also in ancient and medieval India. In his *Description of three worlds* (Tiloyapaññati) the Jainist scientist Yativrisabha (fifth century) defined an atom of space called *pradeśa*. Yativrisabha considers *pradeśa* to be the smallest unit of length and gives the names of the units of length equal to $8, 8^2, \ldots, 8^{11}$ *pradeśas*. The last of these units is *angula*—finger (width of a man's finger) *[250, pp. 22–23]*. He also gives the least even and odd numbers of *pradeśas* for various geometric figures: for lines they are 2 and 3, for squares 4 and 9, for cubes 8 and 27, for rectangles $6 = 2 \cdot 3$ and $15 = 3 \cdot 5$, for parallelepipeds $12 = 2 \cdot 2 \cdot 3$ and $45 = 3 \cdot 3 \cdot 5$, for triangles 6 and 3, for tetrahedra 4 and 35, and for "circles" (squares without angle points) 12 and 5 *[250, p. 48]*.

Atomistic ideas took a different form in Buddhist doctrines. For Buddhists the world is a set of physiopsychological elements called *dharmas* (see the studies of the eminent Soviet indologist Fedor Ippolitovič Ščerbatskoĭ [Stcherbatsky] *[534; 554]*) and space is a form of ordering these elements.

Atomistic views of space also developed in Western Europe. In his *Treatise on the continuum* (De continuo tractatus) Bradwardine (ab. 1290–1349) wrote:

there are five famous opinions concerning the composition of the continua among ancient and modern philosophers. For certain [philosophers], like Aristotle, Averroes and most of the moderns, hold that a continuum is not composed of atoms but rather of parts divisible without end. Others, however, hold with its composition out of indivisibles. But there are two variants [of this position]. For Democritus maintains that a continuum is composed of indivisible bodies. Others claim it is composed of points, and fall into two groups. For Pythagoras, the father of this sect, Plato and Walter the modern contend that a continuum is composed of a finite number of indivisibles. Others, however, [believe] in its composition out of an infinite number [of indivisibles], and these [indivisibilists] are [again] twofold. For certain of them, like Henry the modern, say a continuum is composed of an infinite number of indivisibles immediately joined to one another; others still, like [the Bishop of] Lincoln side with an infinity of indivisibles which are mediate to one another *[210, p. 314; 652, pp. 402–403]*.

Here *Walter* is Walter Catton (d. 1342), who worked in Oxford; *Henry* is Henry Harley (ab. 1270–1317); *Robert of Lincoln* is the archbishop Robert Grosseteste (ab. 1175–1255). Mathematical atomism was also developed in France by Gerald Odonis (d. 1349) and his follower Nicola Bonetus (see *[652, p. 102]*).[2]

Continuity and Discreteness of Space in Newer Times

In the 17th century these questions were taken up by mathematicians. Thus Kepler in his *New stereometry of wine barrels* (Stereometria nova doliorum vinariorum. Linz, 1615) *[271, vol. 9, pp. 5–133]* and in his *Archimedes' Geometry* (Messekunst Archimedis) *[271, vol. 9, pp. 135–274]* restores Archimedes' atomistic arguments employed by him to compute areas and volumes. Kepler was not aware of Archimedes' the *Method* and used these methods to determine the volumes of many new solids.

At the same time the Italian mathematician Bonaventura Cavalieri advanced what is now known as the *principle of Cavalieri* in his *Geometry of indivisibles of the continuous developed by a new method* (Geometria indivisiblibus continuorum nova quadam ratione promota. Bologna, 1635) *[102]*. According to this principle, comparing areas of plane figures reduces to comparing "all lines" of these figures, and comparing volumes of bodies reduces to comparing "all planes" of these bodies; by "all lines" and "all planes" Cavalieri has in mind parallel sections of the various figures. Cavalieri

---

[2] Concerning the history of mathematical atomism see also the book of Anatoliĭ Nikolaevič Vyal'cev *[616]*.

formulated his principle as follows:

Regardless of whether the continuous consists of indivisibles or not, totalities of indivisibles can be compared with one another and their magnitudes are in a definite ratio to one another *[102, p. 18]*.

These words of Cavalieri show that in spite of the obvious atomistic origin of his principle he formulates it so that it should be true in both cases—in the case of the atomistic structure of space as well as in the case of its continuity and indefinite divisibility.

Of great interest are the views on the structure of space held by the creators of the differential and integral calculus, Newton and Leibniz. When he defines a derivative as the "ultimate ratio of evanescent quantities" Newton does not identify these "evanescent quantities" in the last moment before their vanishing with mathematical atoms. He writes in the *Principia*:

It may also be objected, that if the ultimate ratios of evanescent quantities are given, their ultimate magnitudes will also be given: and so all quantities will consist of indivisibles, which is contrary to what *Euclid* has demonstrated concerning incommensurables in the tenth Book of his *Elements*. But this objection is founded on a false supposition *[387, p. 39]*.

At the same time Newton introduces the notion of *moments*—magnitudes that are not zeros but at the same time are not finite variables—and views the derivative as the ratio of the *moment* of the function and the *moment* of its argument.

These quantities I here consider as variable and indetermined, and increasing or decreasing, as it were, by a continual motion or flux; and I understand their momentary increments or decrements by the name of moments; so that the increments may be esteemed as added or affirmative moments; and the decrements as subtracted or negative ones. But take care not to look upon finite particles as such. Finite particles are not moments but the very quantities generated by the moments. We are to conceive them as the just nascent principles of finite magnitudes *[387, p. 249]*.

In the differential calculus of Leibniz the counterpart of Newton's "moment" is Leibniz's *differential*. Unlike Newton, Leibniz admitted the atomistic structure of lines, and in his paper *A new method for maxima and minima as well as tangents which is neither impeded by fractional nor irrational quantities, and a remarkable type of calculus for them* (Nova methodus pro maximis et minimis, itemque tangentibus, quae nec fractas, nec irrationales quantitates moratur et singulare pro illis calculi genus. Leipzig, 1684) *[312, vol. 1, pp. 220–226]* he wrote:

To find a tangent is to pass a straight line that joins two points of a curve the distance between which is infinitesimal, or to produce a side of an infinite-angled polygon which is for us equivalent to a curve *[312, vol. 1, pp. 223]*.

In spite of their different attitudes with respect to *indivisibles* the views of Leibniz and Newton on the structure of space were very close. What was indispensable for the differential calculus was the continuity of lines, surfaces, and all space. At the same time, the study of functions required transition from point to point. Thus neither Newton nor Leibniz could be satisfied with the views of the ancients concerning the mutual relation of points and space. They somehow had to synthesize these views, and this could not be achieved without the introduction of *moments* and *differentials*—neither zeros nor finite quantities but the "just nascent principles of finite magnitudes"— actual infinitesimals, that could not, in the 17th century, be defined without contradictions.

In his work *The historical course of development* (Der historische Entwick-lungsgang) Karl Marx (1818–1883), the founder of Marxist philosophy who took a lively interest in the philosophical questions of mathematics, character-ized the differential calculus of Newton and Leibniz as a *mystical differential calculus [355, p. 165]* and summarized a survey of it in the following manner:

> They themselves believed in the secret character of the newly discovered calculus that gave correct (and in the case of geometric applications truly astounding) results by means of an approach that was mathematically absolutely incorrect. Thus they mystified themselves and all the more valued the new discovery *[355, p. 169]*.

Continuity and Discreteness of Space in the 19th Century and in the Beginning of the 20th Century

We conclude our survey of the philosophy of space up to the beginning of the 19th century with a look at the reflections on the continuity and discreteness of space by the greatest philosopher of the beginning of the 19th century, one of the founders of the dialectic method and the creator of a famous idealistic system Georg Wilhelm Friedrich Hegel (1770–1831). These reflections are contained in his *Philosophy of nature* (Philosophie der Natur) *[221]*. Hegel defines space as follows:

> The first or immediate determination of Nature is *Space*: the abstract *universality of Nature's self-externality*, self-externality's mediationless indifference. It is a wholly ideal *side-by-sideness* because it is self-externality; and it is absolutely *continuous*, because this asunderness is still quite *abstract*, and contains no specific difference within itself. *[221, p. 28]*.

Then Hegel considers the points of space:

It is not permissible to speak of *points of space*, as if they constituted the positive element of space, since space, on account of its lack of difference, is only the possibility and not the actual *positedness* of being-outside-of-one-another and of the negative, and is therefore absolutely continuous; the point, the being-for-self, is consequently rather the *negation* of space, a negation which is posited in space *[221, p. 29]*.

Thus Hegel contrasts *continuous space*, which he calls "self-externality" or "being-outside-of-one-another" with *points*, which he calls "being-for-self." He regards continuity as the fundamental property of space and discreteness as the fundamental property of points and arrives at the following conclusion.

The unity of these two moments, discreteness and continuity, is the objectively determined Notion of space. This Notion, however, is only the abstraction of space, which is often regarded as absolute space *[221, p. 30]*.

These words of Hegel mean that he regards mathematical space, which he, like Newton, calls *absolute space*, as an abstraction from physical space, and considers as the essence of the concept of this mathematical space the *unity of discreteness and continuity*—two properties of space that are, in the terminology of Hegel's philosophy, dialectical opposites.

In his synopsis of Hegel's *Lectures on the history of philosophy* the great Marxist philosopher, Vladimir Il'ič Lenin (1870–1924) wrote:

Motion is the essence of time and space. Two fundamental concepts express this essence: continuity (Kontinuität) and "punctuality" (= denial of continuity, *discontinuity*). Motion is a unity of (infinite) continuity (of time and space) and discontinuity (of time and space) *[313,.p. 258]*.

Here *punctuality* is the fact that that space is a set of points. As in Hegel, this property of space and time is opposed to continuity, but in contradistinction to Hegel, V. I. Lenin emphasizes the connection of space and time with motion, called by him *the essence of time and space*.

Positivists on Space

In the middle of the 19th century the French philosopher Auguste Comte (1798–1857) advanced his so-called positivist philosophy, whose followers became known as positivists. Comte tried to rely exclusively on experience and regarded any theory of matter and all forms of cognition that cannot be tested as "metaphysics." On the whole, Comte's own position was closer to materialism than to idealism, but later positivists interpreted the principles of "positivist philosophy" in the spirit of subjective idealism. Comte's philosophical system assigned great importance to mathematics and, in particular, to geometry.

Comte's doctrine of the origin of geometric notions is very close to Aristotle's materialist doctrine of the derivation of these notions by way of abstraction from objects of the real world. Therefore Comte, like al-Fārābī in the 10th century and Leibniz in the 17th century, criticizes Euclid's exposition of the foundations of geometry; Euclid first defines point, then line, then surface, and then body, and Comte demands that the order of exposition of these notions be reversed. Thus Comte's view of space is intimately tied to matter and differs radically from the Newtonian notion of absolute space that was universally accepted at the time. In his *Course of positive philosophy* (Cours de philosophie positive. Paris, 1830–1842) *[124]* Comte writes:

I put first the question of *space*, that has served the metaphysicians as the subject of so many sophistic debates and empty and childish arguments. If this notion is reduced to its positive sense then it will turn out that it is simply a matter of considering extension not in bodies but in some indefinite medium of which we assume that it holds within it all bodies in the universe. This notion arises in a natural way from observation, namely as the idea of an *imprint* which a body placed in a fluid leaves in it. Indeed, it is clear that from the geometric point of view such an *imprint* can be substituted for the body without the slightest change in our discourse.

As for the physical nature of this indeterminate *space* we should, for greater simplicity, think of it as similar to the real medium in which we live; indeed, if this medium were not gaslike but liquid then we would think of space as fluid. It is obvious that this matter is of secondary significance and that the main purpose of such a representation is to enable us to consider extension independently of body. Its a priori importance derives from the fact that it enables us to study geometric notions in and of themselves by eliminating all other phenomena that permanently accompany the latter in the case of physical bodies but have no effect whatsoever on them....

If geometric considerations were endowed with an abstract character in the indicated manner then they would become not only simpler but also more general. As long as extension was considered in connection with bodies one could take as a subject of investigation only those forms that actually exist in nature, and this restricted the scope of geometric investigations to an extraordinary degree. If we adopt the opposite viewpoint and think of extension as belonging to *space* then the human spirit is free to consider all imaginable forms; this generalization is indispensable before geometry can be given an entirely rational character. The sole purpose of the concepts of surface and line, considered in and of themselves, is to enable us to discuss with greater ease these two forms of extension while setting aside all that need not be taken into consideration. To this end it suffices to imagine that the dimension we wish to exclude decreases ever more while other dimensions remain the same and reaches such limits of smallness that it is no longer capable of attracting our attention. It is in this way that we naturally learn the true doctrine of a *surface*, and by repeating this operation, that is, by eliminating width just as one earlier eliminated depth, also the notion of a *line*. Finally, if we repeat this process once more, then we arrive at the notion of *point* or of extension, considered exclusively relative to place, completely independent of its magnitude and intended exclusively for the precise designation of position. . . .

From the above it is obvious how devoid of any sound sense are the arguments of metaphysicians concerning the foundations of geometry. It must also be noted that the geometers do not present these primary ideas in a sufficiently philosophical manner for, to give an example, they set forth the notions on different forms of extension in an order that is the very opposite of their natural connection, and this gives rise to very serious difficulties in elementary teaching *[124, vol. 1, part 2, pp. 144–146]*.

The Austrian physicist, mechanist and philosopher Ernst Mach (1836–1916) advanced his philosophical doctrine at the end of the 19th century. Mach himself called his teaching the *newest positivism* and thereby stressed his indebtedness to Comte. But Mach brought the efforts of the positivists to base themselves on experience alone to the subjective-idealist teaching according to which the "elements of the world" are identical with our sensations.

In *The science of mechanics presented in a historical-critical manner* (Die Mechanik in ihrer Entwicklung historisch-kritisch dargestellt. Prague, 1883), Mach, proceeding from his subjective-idealistic teaching, defined space and time as "well-ordered systems of sensations" *[345, p. 484]* and in *Knowledge and error* (Erkenntnis und Irrtum. Vienna, 1905) he wrote:

As regards physiology, time and space are systems of sensations of orientation that determine the release of sensations proper and of biologically appropriate reactions of adaptation. As regards physics, they are special dependences of physical elements on each other *[347, p. 339]*.

We must not forget that, in line with the "principle of economy of thought," Mach meant by "physical elements" sensations.

Mach's aspiration to proceed solely from experience resulted in his critique of Newton's teaching about absolute space and time, a view that has played an important role in the history of physics. In *The science of mechanics* Mach wrote:

In the quoted argument Newton betrays his intention to investigate the *factual* alone. No one can say anything about absolute space and absolute motion; these are purely abstract things that cannot be experimentally observed *[345, pp. 222–223]*.

Mach continues:

But if we do not wish to leave the realm of facts then we know only of *relative* spaces and motions *[345, p. 226]*.

In *Knowledge and error* Mach returns to this issue and says:

Considering that Newton's gravitational mechanics could no longer regard the fixed stars as an absolutely unchanging, stationary and rigid system, his daring attempt to relate the whole of dynamics to an absolute space, and correspondingly to absolute time, appears in some measure intelligible. In practice, this seemingly senseless assumption did not alter the use of the fixed stars as space-time coordinates, so that it remained harmless and long escaped serious criticism *[347, p. 345]*.

Mach's critique of the Newtonian doctrine of inertia, linked by him to mass, is closely related to his critique of the doctrine of absolute space.

Proceeding from experience Mach, like Comte, held to the Aristotelian view of the origin of geometric notions and wrote in *Knowledge and error*:

A point, by its motion, generates a one-dimensional line, a line a two-dimensional surface, and a surface a three-dimensional solid space. No difficulties are presented by this concept to minds at all skilled in abstraction. If suffers, however, from the drawback that it does not exhibit, but on the contrary artificially conceals, the natural and actual way in which the abstractions have been reached. A certain discomfort is therefore felt when the attempt is made from this point of view to define the measure of surface or unit of area after the measurement of lengths has been discussed.

A more homogeneous conception is reached if every measurement be regarded as a counting of space by means of immediately adjacent, spatially identical, or at least hypothetically identical, bodies, whether we be concerned with volumes, with surfaces, or with lines. Surfaces may be regarded as corporeal sheets, having everywhere the same constant thickness which we may make small at will, vanishingly small; lines, as strings or threads of constant, vanishingly small thickness. A point then becomes a small corporeal space from the extension of which we purposely abstract, whether it be part of another space, of a surface, or of a line. The bodies employed in the enumeration may be of any smallness or any form which conforms to our needs. Nothing prevents our idealizing in the usual manner these images, reached in the natural way indicated, by simply leaving out of account the thickness of the sheets and the threads. The usual and somewhat timid mode of presenting the fundamental notions of geometry is doubtless due to the fact that the infinitesimal method which freed mathematics from the historical and accidental shackles of its early elementary form, did not begin to influence geometry until a later period of development, and that the frank and natural alliance of geometry with the physical sciences was not restored until still later, through Gauss. But why the elements shall not now partake of the advantages of our better insight, is not to be clearly seen. Even Leibniz adverted to the fact that it would be more rational to begin with the solid in our geometrical definitions *[347, p. 270]*.

We see that in this matter Mach simply refers to Leibniz. Elsewhere in the same book Mach wrote:

The fundamental truths of geometry have thus, unquestionably, been derived from physical experience, if only for the reason that our visualizations and sensations of space are absolutely inaccesssible to measurement and cannot possibly be made the subject of metrical experience. But it is no less indubitable that when the relations connecting our visualizations of space with the simplest metrical experiences have been made familiar, then geometrical facts can be reproduced with great facility and certainty in the imagination alone—that is by purely mental experiment *[347, p. 291]*.

Since he was well acquainted with the progress of contemporary mathematics, Mach mentions in the same book many achievements of 19th-century geometry:

Analogues of the geometry we are familiar with, are constructed on broader and more general assumptions for any number of dimensions, with no pretension to being regarded as more than intellectual scientific experiments and with no idea of being applied to reality. In support of my remark it will be sufficient to advert to the advances made in mathematics by Clifford, Klein, Lie, and others. Seldom have thinkers become so steeped in reverie, or so far estranged from reality, as to imagine for our space a number of dimensions exceeding the three of the given space of sense, or to conceive of representing that space by any geometry that appreciably departs from the Euclidean. Gauss, Lobachevsky, Bolyai, and Riemann were perfectly clear on this point, and certainly cannot be held responsible for the grotesque fictions subsequently stated in this field *[347, pp. 322–323]*.

Mach presents the work of Saccheri, Lambert, Lobačevskiĭ, Riemann, and other geometers *[347, pp. 309–322]* and engages in a polemic with the followers of Kant's doctrine of the a priori nature of geometric knowledge:

If further we may assume without scruple that physiological space is innate, it shows too slight an agreement with geometrical space to be considered as an adequate basis for an a priori development of geometry in Kant's sense *[347, p. 256]*.

Here Mach refers to Listing's book *Preliminary studies on topology*, which we will discuss in chapter 8.

In *The science of mechanics* Mach wrote:

It is well known that as a result of efforts by Lobačevskiĭ, Bolyai, Gauss, and Riemann the view has established itself that what we call *space* is a *special real* case of a *more general conceivable* case of a manifold of a greater number of dimensions. The space of our vision and our feelings is a *triple* manifold, it has three dimensions, and each place in it can be determined by three independent indices. What is *conceivable* is a manifold of four and possibly more folds. Even the genus of a manifold is *conceivable* differently than in given space. The credit for this clarification, which we regard as very important, goes to Riemann. The properties of given space immediately present themselves to us as objects of *experience*, and all geometric pseudotheories that aim to establish them by philosophizing alone fall away *[345, p. 467]*.

Then Mach considers the "grotesque fictions" that arose as a result of the rise of multidimensional geometry:

The fourth dimension came opportunely for some theologians, intent on the complete destruction of hell, as well as for spiritualists. For spiritualists the utility of the fourth dimension consists in the following. From a bounded line one can pass into a second dimension without passing through its end points; from a surface bounded by a curved line one can pass into a third dimension and from closed space into a fourth without penetrating boundaries. Thanks to the fourth dimension even the innocent things performed by conjurers in three dimensions take on some new halo. All tricks of the spiritualists, such as the forming of knots in closed threads or the removal of objects from closed spaces succeed only whenever this is quite beside the point. All reduces to endless tricks. The obstetrician who could deliver a baby through the fourth dimension is, as yet, unborn *[345, p. 468]*.

Here Mach has in mind the "experiments" of Zöllner, mentioned by Klein and Engels, that we will speak of in chapter 7.

At the time atoms were not objects of "experience." Considering them as purely conceivable schemes Mach admitted the possibility of their interpretations in multidimensional spaces. In *The history and roots of the law of conservation of work* (Die Geschichte und die Wurzel des Satzes der Erhaltung der Arbeit. Prague, 1872) Mach wrote:

The larger the number of atoms in a molecule the larger must be the number of dimensions of the space if all conceivable connections among them are to be actually realized. We give just an example that shows in how narrowminded a manner we proceed if we imagine chemical elements to be set out one after the other in space (of three dimensions) and how very many of the relations among elements can escape our notice if we express them by means of a formula that cannot fully encompass them *[346, p. 29]*.

Mach notes that he arrived at this idea already in 1865 after familiarizing himself with multidimensional geometry which was being developed at that time *[346, p. 55]*.

We mention two further interesting thoughts of Mach on space and time. With regard to the chapter "Time and space from a physical viewpoint" of his book *Knowledge and error* he makes the following observation:

> The considerations of this chapter show that space and time cannot well be severed during investigation *[347, p. 350]*.

At the end of the same chapter Mach writes:

> That physiologically time and space represent only an apparent continuum and are probably composed of discontinuous though imprecisely discriminable elements may bè mentioned in passing. How far in physics we can uphold the assumption of spatial and temporal continuity is merely a question of what is appropriate and what agrees with experience *[347, p. 349]*.

These thoughts of Mach, as well as his critique of Newtonian dogmas, have played a positive role in the history of physics, for the physicists interpreted his term *experience* in a materialist sense, notwithstanding the fact, mentioned previously, that Mach himself interpreted it in the sense of subjective idealism and viewed space as one of the ways of ordering our sensations. A similar view of space was espoused by Henri Bergson (1859–1941), founder of the "philosophy of life." Like the positivists, Bergson tries to stand "above" materialism and idealism by assuming that "life" is neither matter nor spirit, both of which are products of its disintegration.

We note that Mach's *physical elements* are very close to Buddhist *dharmas*. F. I. Ščerbatskoĭ *[534; 554]* pointed out the strong similarity between positivist philosophy, and especially the philosophy of Bergson, and Buddhist philosophy.

The philosophical views of the positivists and, specifically, their views of space and time, were criticized by Lenin in his philosophical work *Materialism and empiriocriticism* (Materializm i empiriokriticizm. Petersburg, 1909) *[316; 313, vol. 18]*. (The reference to the paragraph *Space and time* is *[316, pp. 181–195]*.)

Chapter 6
Lobačevskian Geometry

N. I. Lobačevskiĭ's Discovery

Centuries of attempts to prove the parallel postulate led to the discovery of non-Euclidean geometry made at the beginning of the 19th century. This discovery was first published by the great Russian mathematician and professor at Kazan University Nikolaĭ Ivanovič Lobačevskiĭ in the paper *On the principles of geometry* (O načalah geometrii. Kazan, 1829) *[333, vol. 1, pp. 185–261]*. The first public announcement about this discovery was made during a meeting of the division of the physicomathematical sciences of Kazan University and took the form of a lecture entitled *A brief exposition of the principles of geometry including a rigorous proof of the theorem on parallels* (Exposition succincte des principes de la Géométrie avec une démonstration rigoureuse du théorème des parallèles). Lobačevskiĭ notes that he drew on this lecture for the first part of the memoir "On the principles of geometry." In the beginning of this part he writes:

> Who would not agree that a Mathematical discipline must not start out with concepts as vague as those with which we, in imitation of Euclid, begin Geometry, and that nowhere in Mathematics should one tolerate the kind of insufficiency of rigor that one was forced to allow in the theory of parallel lines.... The initial concepts with which any discipline begins must be clear and reduced to the smallest possible number. It is only then that they can provide a firm and adequate foundation for the discipline. Such concepts must be learned by the senses—the inborn ones must not be trusted *[333, vol. 1, pp. 185–186]*.

After introducing the basic concepts of geometry that do not depend on the parallel postulate Lobačevskiĭ writes:

> The sum of the angles in a rectilinear triangle cannot be $> \pi$; the sum of the angles in a spherical triangle is, on the contrary, always $> \pi$ *[333, vol. 1, p. 192]*.

And further:

We saw that the sum of the angles in a rectilinear triangle cannot be $> \pi$. It remains to assume that this sum is $= \pi$ or $< \pi$. Both can be assumed without any subsequent contradiction, and this gives rise to two Geometries: one, *in common use* to this day owing to its simplicity, actually agrees with all measurements; the other, an *imaginary* one, more general and therefore more difficult in its calculations, admits the possibility of dependence of lines on angles.

If one assumes that the angle sum is π in one rectilinear triangle then it will be that in all. If, on the contrary, we allow it to be less than π in one then it is easy to show that it decreases as the sides of the triangle increase.

Thus two lines cannot meet in the plane if they form with a third angles whose sum is π. They need not intersect in the case when this sum is $< \pi$ provided that we make the additional assumption that the angle sum in a triangle is $< \pi$.

Thus, relative to a line, all lines in the plane can be divided into those that meet it and those that do not. The latter will be called parallel (to that line) if they represent a limit, or, to put it differently, mark the transition from those in one category to those in the other among all lines issuing from one point. . . .

A consequence of the assumption that the angle sum in a triangle is $< \pi$ is that, with increasing radius, a circle tends not to a straight line but to a special kind of curve that we will call a *limit circle*. In this case, a sphere will tend to a curved surface which we will similarly call a *limit sphere*. The intersection of this surface with a plane is either a circle or a limit circle.

Geometry on the limit sphere is exactly the same as we know it in the plane. The limit circle takes the place of the straight line in the latter and the angles between the planes in which the limit (circles) lie replace the angles between straight lines. The shorter their arcs, the closer the limit circles are to straight lines, so that the difference in ratio to the length of an arc can be made arbitrarily small. Therefore, whatever applies to the first applies to the second, provided that we suppose the first and the second extremely small.

Thus if the Geometry of nature is such that two parallel lines must be inclined to a third at angles whose sum is $< \pi$, then the Geometry in common use is a Geometry of extremely short lines in comparison with those (of the geometry) where the angle sum in a triangle is perceptibly different from π *[333, vol. 1, pp. 194–196]*.

We note that the name *imaginary geometry* that Lobačevskiĭ gave to the geometry he discovered echoes the name *imaginary numbers* that he used for complex numbers. This name emphasizes that the relation of the geometry he discovered to the commonly used Euclidean geometry is the same as the relation of the complex numbers to the real numbers. Lobačevskiĭ's statement that "concepts must be learned by the senses—the inborn ones must not be trusted" shows that since the rejection of the parallel postulate had not led to contradictory consequences, he repudiated the notion that Euclidean geometry is the only conceivable consistent geometry and concluded that different consistent geometric systems are in fact, conceivable. Having inferred the consistency of "imaginary geometry," Lobačevskiĭ refuted the idealistic doctrine of the inbornness of our notions of space—an idea that originated with Plato and was subsequently redeveloped by Kant. In this connection Lobačevskiĭ posed the question of measuring the angle sum of a triangle with very large sides. Using the data supplied by the latest astronomical calendar Lobačevskiĭ computes the angle sum in a triangle whose vertices are the star Sirius and two diametrically opposed positions of the earth and finds that this angle sum differs from π by less than 0.000372 seconds. So small a difference could not be measured with contemporary angle-measuring intruments (actually, at this point, Lobačevskiĭ had made a mistake in his computations: the difference in question is 100 times smaller than his figure).

Having carried out these computations, Lobačevskiĭ writes:

Thus, the smaller the triangle, the less its angle sum differs from two right angles. After this, one can imagine to what extent this difference, on which our theory of parallels is based, supports the accuracy of all calculations of ordinary Geometry and lends support to the attitude of regarding the principles of the latter to have been, presumably, rigorously established [333, vol. 1, p. 209].

This passage explains the sense of the words "including a rigorous proof of the theorem on parallels" in the title of Lobačevskiĭ's lecture of 1826: by "a rigorous proof" of "the principles of ordinary geometry" Lobačevskiĭ means the impossibility of experimental determination of which of the two geometries—the "imaginary" or the one in common use—obtains in the real world. This implies the complete suitability of the "geometry in common use" for practical applications.

Lobačevskiĭ's Struggle for the Recognition of His Discovery

Lobačevskiĭ's memoir *On the principles of geometry* met with incomprehension. In 1832, Lobačevskiĭ, then rector of Kazan University, asked the university council to send his paper for review to the Petersburg Academy. Academician Mihail Vasil'evič Ostrogradskiĭ (1801–1862) was assigned to review the memoir. Ostrogradskiĭ ignored the geometric questions in the memoir and concentrated on two definite integrals computed by Lobačevskiĭ by geometric reasoning. Ostrogradskiĭ's review, contained in the transactions of the Academy, states:

Having pointed out that of the two definite integrals Mr. Lobačevskiĭ claims to have computed by means of his new method one is already known and the other is false, Mr. Ostrogradskiĭ notes that, in addition, the work has been carried out with so little care that most of it is incomprehensible. He therefore is of the opinion that the paper of Mr. Lobačevskiĭ does not merit the attention of the Academy *[254, p. 252]*.

In 1834 there appeared in the Petersburg literary journals *Syn otečestva* (Son of the fatherland) and *Severnyĭ arhiv* (Northern archive), published by the noted reactionaries N. Greč and F. Bulgarin, a review by a certain "S.S." entitled *On the principles of geometry, a work of Mr. Lobačevskiĭ* that was an insulting lampoon. Having failed to understand the essence of the new geometry and having concluded that one of the integrals considered by Lobačevskiĭ is "now $\pi/4$, now ∞," the author of the review says ironically:

Glory to Mr. Lobačevskiĭ who took upon himself the labor of revealing, on the one hand, the insolence and shamelessness of false new inventions, and on the other the simpleminded ignorance of those who worship their new inventions.

However, while I realize the full value of Mr. Lobačevskiĭ's work, I cannot but hold it against him that, having failed to give his book an appropriate title, he forced us to think for a long time in vain. For instance, why not write, instead of *On the principles of geometry: A satire on geometry* or *A caricature of geometry* or a similar thing *[254, p. 247]*.

In the view of A. P. Kotel'nikov *[254, pp. 250–251]*, whose geometric papers we will discuss in chapter 10, this review was written by one, and possibly two, of Ostrogradskiĭ's students, S. A. Buraček and S. I. Zelenyĭ, the first of whom was on the staff of *Syn otečestva*. The integral mentioned in both reviews was an integral depending on a parameter; it took on different values for different values of that parameter.

It was thus an irony of fate that Ostrogradskiĭ, a former student of T. F. Osipovskiĭ, whose views on Kant's doctrine of space were very close to those of Lobačevskiĭ, and who, as we will see in Chapter 7, played a nontrivial part in the rise of another generalization of ordinary geometry, namely n-dimensional geometry, should have failed to understand Lobačevskiĭ's discovery and, by the weight of his authority, delayed its recognition.

Another Petersburg academician who failed to understand Lobačevskiĭ's discovery was Victor Yakovlevič Bunyakovskiĭ (1804–1889). Bunyakovskiĭ himself had studied the theory of parallel lines and in 1853, after the publication of the basic papers of Lobačevskiĭ, published the book *Parallel lines* (Parallel'nye linii. Petersburg, 1853) *[82]* containing a survey and classification of the "proofs" of the parallel postulate and his own "proof" based on the definition of a straight line as a curve all of whose points have the same properties. The book made no mention of Lobačevskiĭ. Later Bunyakovskiĭ devoted a special paper—*Reflections on certain singularities in the construc-tions of non-Euclidean geometry* (Considérations sur quelques singularités qui se présentent dans la constructions de la géométrie non euclidienne. Petersburg, 1872) *[77]*—to a discussion of Lobačevskiĭ's geometry. In this paper he tried to show that there is a contradiction between the geometry of Lobačevskiĭ and the visual notions of space, but, unlike Ostrogradskiĭ and his students, he spoke respectfully of Lobačevskiĭ's talents.

Lobačevskiĭ's Further Papers on Non-Euclidean Geometry

Lack of recognition by the Academy of Sciences and the (vicious) review in the literary journals, clearly inspired by Ostrogradskiĭ, failed to break Lobačevskiĭ. His first memoir was followed by other papers in which he developed his discovery: *Imaginary geometry* (Voobražaemaya geometriya. Kazan, 1835; French translation: Berlin, 1836) *[333, vol. 3, pp. 16–70]*; *Applications of imaginary geometry to certain integrals* (Primenenie voo-bražaemoĭ geometrii k nekotorym integralam. Kazan, 1836) *[333, vol. 3, pp. 181–294]*; *New principles of geometry with a complete theory of parallels* (Novye načala geometrii s polnoĭ teorieĭ parallel'nyh. Kazan', 1835–1838) *[333, vol. 2, pp. 147–457]*; *Geometrical researches on the theory of parallel lines* (Geometrische Untersuchungen zur theorie der Parallellinien, Berlin, 1840) *[333, vol. 1, pp. 79–127; 332]*; *Pangeometry* (Pangeometriya. Kazan, 1855; French translation: Kazan, 1856) *[333, vol. 3, pp. 435–524]*. The word *Pangeometry* (Universal geometry) used by Lobačevskiĭ in the last of these papers shows that he viewed his geometry as the general case of which Euclidean geometry is a special (more precisely, a limiting) case. What played an essential role in convincing Lobačevskiĭ of the consistency of his geometry was that he could use it to compute by means of geometric considerations certain definite integrals that he thought of as expressions for areas of surfaces and volumes of solids in *imaginary geometry*. (We recall that some of these applications were already present in his first memoir and provoked Ostro-gradskiĭ's attacks.) In some cases Lobačevskiĭ found new ways of evaluating known integrals and in other cases both the integral and the method of evaluation were new.

Lobačevskiĭ's Philosophy of Space

We saw that Lobačevskiĭ's discovery was closely linked to his philosophical views on space and to his critique of Kant's philosophical views. Already in his *Geometry* (Geometriya) written in 1823, but published only in the 20th century (Kazan, 1909), Lobačevskiĭ writes in the introduction:

A geometric body takes from natural bodies just the property of *extension*. Extension is the property of bodies such that when they dilate they touch *[333, vol. 2, p. 43]*.

A systematic exposition of the basic notions of geometry was given by Lobačevskiĭ in his *New principles of geometry with a complete theory of parallels* (Kazan, 1835–1838). This exposition begins as follows:

> *Touching* is a distinctive state of bodies and gives them the name *geometric* when we retain in them this property and disregard all others, whether essential or accidental.... When they *touch*, two bodies *A* and *B* form a single geometric body *C*, where the component parts *A* and *B* remain distinct and are not lost in the whole of *C*. Conversely, an arbitrary *section S* divides a body *C* into two parts *A* and *B*.... In this way, one can imagine all bodies in nature to be parts of one whole that we call *space [333, vol. 2, p. 168]*.

Then Lobačevskiĭ defines congruence ("sameness") and equality of size ("equality") of two bodies. After defining a body *A* that *fills* a location *B*, Lobačevskiĭ states that

> all other bodies that, without any interchange with them, also fill location *B* will be geometrically *the same* in all respects. Two bodies are merely *equal* if the parts of one must be rearranged before it can fill the location of the other *[333, vol. 2, p. 189]*.

Having defined a body and three "principal sections" that divide it into eight parts Lobačevskiĭ defines surface, line, and point:

> After three sections have been made in a body and eight mutually touching parts have resulted, then, relative to the first section two parts touch *along a surface*, relative to two sections two parts touch *along a line* and relative to all three sections two parts on opposite sides touch *at a point [333, vol. 2, p. 173]*.

Next distance is defined:

> The relative location of two points is called their *distance* and it is determined by the touching of two bodies in which one allows all transformations that do not transform the points themselves *[333, vol. 2, p. 178]*.

Using the notion of distance, Lobačevskiĭ defines a *sphere* as the locus of points equidistant from a point, a *plane* as the locus of points equidistant from two points, and a *line* as the locus of points in a plane equidistant from two of its points.

We see that these definitions are very close to the definitions given by Leibniz. Like Leibniz in his letter to Vitale Giordano (and like al-Fārābī and al-Bīrūnī)[1] Lobačevskiĭ starts out with the definition of a geometric body and then defines surface, line, and point.

Like Leibniz in his letter to Huygens, Lobačevskiĭ defines a plane by means

---

[1] See footnote 4 in Chapter 4 (p. 173).

of the distances of its points from two points. Lobačevskiǐ's definition of a straight line is somewhat different from Leibniz's definition.

In *Geometry* and in *New principles* Lobačevskiǐ stresses that a geometric body "retains" only the geometric properties of physical bodies, that is, is a geometric abstraction from such bodies.

What is very interesting is Lobačevskiǐ's attempt, sketched in *Geometry* and consistently implemented in *New principles*, of basing geometry on purely topological properties of contact and section.

The following thought, likewise expressed by Lobačevskiǐ in *New principles*, after the exposition of the geometry he discovered, is especially noteworthy:

> Strictly speaking, all we know in nature is motion, without which sense impressions are impossible. Thus all other notions, for example, Geometric ones, are artificially made by our minds, for they are abstracted from properties of motion: therefore space in itself, separately, does not exist for us. This being so, no contradiction can arise in our minds if we allow that certain forces in nature follow one, and others their own particular Geometry *[333, vol. 2, p. 147]*.

Lobačevskiǐ's idea that the geometric properties of space may be different in different parts of space and may depend on "forces," that is, on matter, is a distant anticipation of an idea of Einstein's general theory of relativity.

We have adduced a number of Lobačevskiǐ's thoughts. They show that he reflected very broadly on the fundamental notions of geometry and that his discovery of non-Euclidean geometry was the result of his reflections on only one of the questions he contemplated.

The Work of János Bolyai

Simultaneously with Lobačevskiǐ, the same discovery was made by the remarkable Hungarian mathematician János Bolyai (1802–1860), son of Farkas Bolyai whom we mentioned earlier, and by Carl Friedrich Gauss (1777–1855), the greatest German mathematician of the end of the 18th and the first half of the 19th century. The young Bolyai became interested in the theory of parallel lines under the influence of his father and continued to study it in spite of his father's opposition. J. Bolyai published his discovery in the form of a supplement to F. Bolyai's book published in Maros-Vasarhely in 1832. The full title of this work is *Supplement containing the absolutely true science of space, independent of the truth or falsity of Euclid's axiom XI*[2] (*that can never be decided a priori*) (Appendix scientiam spatii absolute veram exhibens: a veritate aut falsitate Axiomatis XI Euclidis (a priori haud umquam deciden-

---

[2] In some editions of Euclid's *Elements* "Axiom XI" is the same as the parallel postulate.

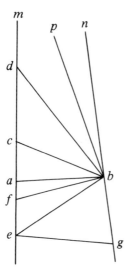

Figure 86

da) independentem *[65; 66]*. Hence the usual references to this work as the *Appendix*.

The *Appendix* is written in an extremely condensed manner, using a great many symbolic notations. For example, *ab* is the infinite straight line passing through the points *a* and *b*; *ab̃* is the ray with vertex at *a* passing through *b*; *abc* is the angle with sides *bã* and *bc̃*; *R* stands for a right angle. The *Appendix* opens with the words:

If the ray *am̃* is not cut by the ray *bñ*, situated in the same plane, but is cut by every ray *bp̃* comprised in the angle *abn*, we will call ray *bñ parallel* to ray *am̃*; this is designated by *bñ‖am̃* (Figure 86).

It is evident that *there is one such ray bñ, and only one*, passing through any point *b* (taken outside of the straight *am̃*), and that the sum of the angles *bam*, *abn* cannot exceed 2*R*; for in moving *bc̃* around *b* until *bam* + *abc* = 2*R*, somewhere ray *bc̃ first* does not cut ray *am̃*, and it is then *bc̃‖am̃*. It is clear that *bñ‖em̃*, wherever the point *e* be taken on the straight *am̃* *[65, p. 41; 66, p. 5]*.

If the parallel postulate (Bolyai's "axiom XI") holds, then the straight lines *bc* and *am* are parallel. If the parallel postulate does not hold, then these straight lines are parallel in the sense of Lobačevskiĭ (*bn* does not intersect *am* but can be obtained by passage to the limit from those *bc* that intersect *am*). Bolyai calls "the system of geometry that takes the hypothesis of Euclid's axiom XI to be true" the *system* $\sum$, and the system based on the "opposite hypothesis" the *system S*. What holds in both systems Bolyai calls *absolute* and this is what he means by "the absolutely true science of space independent of the truth or falsity of Euclid's axiom XI"; this branch of geometry is now

called *absolute geometry*. Bolyai tries to set forth as many facts of absolute
geometry as he can. To illustrate Bolyai's notation we write down his version
of the law of sines,

$$\frac{\sin A}{\bigcirc a} = \frac{\sin B}{\bigcirc b} = \frac{\sin C}{\bigcirc c},$$

where $\bigcirc r$ is the circumference of a circle of radius r, a result that is true in
Euclidean as well as in Lobačevskian geometry.

Gauss's Notes and Letters

Gauss, who independently made the same discovery as Lobačevskiĭ and
Bolyai, stated his views on this question only in rough notes and in letters. In
1799 he wrote to F. Bolyai about his study of the theory of parallel lines:

> It is true that I have come upon much which by most people would
> be held to constitute a proof [of the parallel postulate]: but in my eyes
> it proves as good as *nothing*. For example, if one could show that a
> rectilinear triangle is possible, whose area would be greater than any
> given area, then I would be ready to prove the whole of geometry
> absolutely rigorously. Most people would certainly let this stand as an
> Axiom; but I, no! It would indeed be possible that the area might always
> remain below a certain limit, however, far apart the three angular points
> of the triangle were taken. I have many such statements but I find none
> of them satisfactory *[196, vol. 8, pp. 159–160; 71, pp. 65–66]*.

One recognizes in Gauss's statement a well-known fact of Lobačevskian
geometry, which is that in this geometry the area of a triangle is proportional
to its *defect*, defined as the difference between π and the angle sum of the
triangle. If the defect of a triangle is δ and the area is $k\delta$ then, since $\delta < \pi$, we
see that $k\delta$ cannot exceed $k\pi$.

In 1804 Gauss wrote to F. Bolyai in connection with his attempts to prove
the parallel postulate in his *Theory of parallels*:

> Your method does *not* yet satisfy me. I will try to make the critical point
> (which belongs to the same kind of obstacles that made my own efforts
> so futile) as clear as I can. I still hope that these cliffs will be navigated
> eventually, and this before I die *[196, vol. 8, p. 160]*.*

---

* This is not a literal translation. The original German text is:

> Dein Verfahren mir noch *nicht* Genüge leistet. Ich will versuchen, den Stein des
> Anstosses, den ich noch darin finde (und der auch wieder zu derselben *Gruppe* von Klippen
> gehört, voran meine Versuche bisher scheiterten) mit so vieler Klarheit als mir möglich
> ist, ans Licht zu ziehen. Ich habe zwar noch immer die Hoffnung, dass jene Klippen einst,
> und noch vor meinem Ende, eine Durchfahrt erlauben werden.

(*Translator*).

These words show that at that time Gauss had not yet given up trying to prove the parallel postulate.

By finding ever more consequences of the denial of the parallel postulate Gauss penetrated ever deeper into what we call Lobačevskian geometry. In 1816, in a letter to his former student the Königsberg astronomer Christian Ludwig Gerling (1788–1864), Gauss wrote that the denial of the parallel postulate would imply the existence of an absolute measure of length:

It seems paradoxical that there could be a constant straight line given as if a priori but I do not find in this any contradiction. In fact, it would be desirable that Euclidean geometry were not true, for we would then have a universal measure a priori. One could use the side of an equilateral triangle with angle = $59°59'59''$, 9999 as a unit of length *[196, vol. 8, p. 169]*.

Here too the words "it would be desirable that Euclidean geometry were not true" show that Gauss still regards it as true.

But in 1817, in a letter to his old friend Heinrich Wilhelm Olbers (1758–1840), Gauss writes:

I am ever more convinced that the necessity of our geometry cannot be proved—at least not by *human* reason for human reason. It is possible that in another lifetime we will arrive at other conclusions on the nature of space that we now have no access to. In the meantime we must not put geometry on a par with arithmetic that exists purely a priori but rather with mechanics *[196, vol. 8, p. 177]*.

These words of Gauss point to the source of his doubts concerning the existence of a geometry other than Euclidean: Gauss initially adhered to Kant's doctrine of the a priori nature of mathematical concepts, but, as a result of reflecting on the theory of parallel lines, he arrived at the conclusion that, whereas the concepts of arithmetic are a priori, the concepts of geometry, like those of mechanics, are abstracted from the material world. It is possible that this is why, after having concluded that *non-Euclidean geometry* (which is what Gauss called the new geometry in an 1831 letter to Heinrich Christian Schumacher (1780–1850) *[196, vol. 8, p. 216]* is noncontradictory Gauss did not publish his results. In 1818, in a letter to Gerling, he wrote:

I am glad that you have the courage to express yourself as if you acknowledged the falsity of our theory of parallels and with it of all our geometry. But the wasps whose nest you stir up will fly at your head *[196, vol. 8, p. 179]*.

A. P. Norden's *[396]* view that by "stirred-up wasps" Gauss meant the proponents of the a priori nature of mathematical concepts is very plausible. In 1832 Gauss read J. Bolyai's *Appendix* and wrote to his father:

If I commenced by saying that I *must not praise this work* you would certainly be surprised for a moment. But I cannot say otherwise. To praise it, would be to praise myself. Indeed the whole contents of the work, the path taken by your son, the results to which he is led, coincide almost entirely with my meditations, which have occupied my mind partly for the last thirty or thirty-five years. So I remained quite stupefied. So far as my own work is concerned, of which up till now I have put little on paper, my intention was not to let it be published during my lifetime. Indeed the majority of people have not clear ideas upon the questions of which we are speaking, and I have found very few people who could regard with any special interest what I communicated to them on this subject. To be able to take such an interest it is first of all necessary to have devoted careful thought to the real nature of what is wanted and upon this matter almost all are most uncertain. On the other hand, it was my idea to write down all this later so that at least it should not perish with me. It is therefore a pleasant surprise for me that I am spared this trouble, and I am very glad that it is just the son of my old

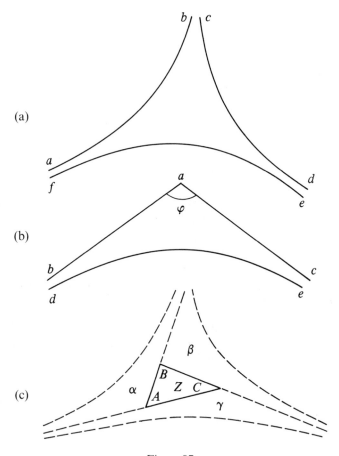

Figure 87

friend who takes the precedence of me in such a remarkable manner
[71, p. 100].

Then Gauss gives some advice on terminology. He suggests the name
parasphere for J. Bolyai's "surface *F*" (Lobačevskiĭ's *limit sphere*), *paracycle*
for "line *F*" (Lobačevskiĭ's *limit circle*), *hypercycle* for the locus of points of
the plane equidistant from a line, and *hypersphere* for the locus of points in
space equidistant from a plane. Also, he gives an original

proof of the proposition that the difference between the angle sum in a
triangle and 180° is proportional to the area of the triangle *[196, vol. 8,
p. 221]*.

Gauss denotes by *t* the least upper bound of the area of a triangle. This means
that *t* stands for the area of the part of the plane between three pairwise
parallel lines *ab*, *cd* and *fe* (Figure 87a). The part of the plane between the
angle *bac*, equal to *φ*, and the line *de* parallel to its sides (Figure 87b) is to *t*
as 180° − *φ* is to 180°. It follows that the areas *α*, *β*, *γ* in Figure 87c are equal
to *At*/180°, *Bt*/180°, *Ct*/180°, respectively. If *Z* is the area of the triangle *ABC*
in Figure 87c then we have $t = \dfrac{At + Bt + Ct}{180°} + Z$, that is,

$$Z = \frac{180° - (A + B + C)}{180°} t.$$

Gauss had a very high opinion of the papers of Lobačevskiĭ; he read a
German version of *Geometrical investigations* and studied Russian in order
to be able to read the other papers in the original.

The Papers of Wachter, Schweikart, and Taurinus

In addition to Gerling there were other correspondents who discussed
with Gauss the issue of the parallel postulate. They were Friedrich Ludwig
Wachter (1792–1818), Ferdinand Carl Schweikart (1780–1859), and Franz
Adolf Taurinus (1794–1874).

Wachter was Gauss's student at Göttingen. Later he became a teacher
of mathematics in a gymnasium in Danzig (now Gdańsk) and published a
brochure *Proof of Euclid's eleventh geometric axiom* (Demonstratio axiomatis
geometrici in Euclidis undecimi. Danzig, 1817) *[607]* containing a "proof"
of the parallel postulate based on the premise that four arbitrary points
determine a "surface of four points" and two such surfaces intersect along a
single line determined by three points. Wachter tried to prove that the "surface
of four points" is a sphere and the "line of three points" is a circle but this is
impossible without rigorous definitions of these surfaces and lines. More
important was Wachter's letter to Gauss, written in 1816. Here Wachter
considered the limit of a sphere as its radius tends to infinity and claimed that,

even if the parallel postulate does not hold, the geometry of this surface is Euclidean. We know that if the parallel postulate holds then the surface in question is a plane, and if it does not hold then this surface is Lobačevskiĭ's "limit sphere," for which Wachter's claim is true.

Schweikart was a lawyer by profession. He began his study of geometry with the publication of the book *The theory of parallel lines including a proposal that it be banned from geometry* (Die Theorie der Parallellinien nebst dem Vorschlage ihrer Verbannung aus der Geometrie. Leipzig, 1808) *[518]*. Contrary to the promise of the title, the book contained a false "proof" of the parallel postulate. But after 1812–1817, during which time Schweikart was a professor of jurisprudence at Har'kov University, he changed his point of view. It is possible that this was due to the influence of Osipovskiĭ (who was rector of Har'kov University in 1813–1820), author of the previously mentioned book *On time and space [398]* that was a critique of Kant's views on the a priori nature of the concepts of space and time. Be that as it may, in 1818 Schweikart gave Gerling a note for Gauss which stated, among other points, that

There are two kinds of geometry—a geometry in the strict sense—the *Euclidean*; and an astral science of magnitudes.

Triangles in the latter have the property that the sum of their three angles is not equal to two right angles. *This* being assumed we can prove rigorously:

(a) That the sum of the three angles of a triangle is *less* than two right angles;

(b) that the sum becomes ever less, the greater the area of the triangle;

(c) that the altitude of an isosceles right-angled triangle continually grows as the sides increase but it can never become greater than a certain length which I call the *Constant*.... *[196, vol. 8, pp. 180–181; 71, p. 76]*.

In calling non-Euclidean geometry the "astral science of magnitudes" Schweikart was making the assumption that it held somewhere in the universe.

Upon receipt of Schweikart's note Gauss wrote to Gerling:

Professor Schweikart's note has given me a great deal of pleasure and I ask to convey to him for this my very best wishes. Almost all of this is copied from my soul *[196, vol. 8, p. 181]*.

Under the influence of his uncle Schweikart, Taurinus got interested in the problem of the parallel postulate and published two brochures entitled, respectively, *Theory of parallel lines* (Theorie der Parallellinien. Köln, 1825) *[168, pp. 255–266]* and *First elements of geometry* (Geometriae prima elementa. Köln, 1826) *[168, pp. 267–283]*. In the first of these Taurinus refuted the obtuse-angle hypothesis and showed that in case of the acute-angle hypothesis there must be a Schweikart "constant," which he called a *param-*

eter. He rejected the acute-angle hypothesis because it led to the possibility of many values of the "parameter." In the second brochure he developed the consequences of the acute-angle hypothesis even further. He found the trigonometric formulas of this geometry and showed that they can be obtained from the formulas of spherical trigonometry by replacement of the radius of the sphere by a pure imaginary number. He expressed his "parameter" in terms of this imaginary magnitude and computed the circumference of a circle, the areas of a circle and of a sphere, and the volume of a sphere. Taurinus expressed trigonometric formulas of this geometry in terms of the hyperbolic functions

$$\cosh x = \frac{e^x + e^{-x}}{2} = \cos ix, \qquad \sinh x = \frac{e^x - e^{-x}}{2} = \frac{1}{i} \sin ix. \quad (6.1)$$

Gauss, who corresponded with Taurinus, terminated the correspondence when Taurinus, in the preface to his brochure, asked him to state his views on the subject. Gauss's reaction reduced Taurinus to despair, and he burned all copies of the brochure in his possession.

The Struggle for Recognition of Lobačevskian Geometry

During Lobačevskiĭ's lifetime there was only one mathematician who publicly accorded high praise to his work in geometry. He was Petr Ivanovič Kotel'nikov (1809–1879), professor of mathematics at Kazan University and father of A. P. Kotel'nikov, whom we mentioned earlier. In an address entitled *On bias against mathematics* (O predubeždenii protiv matematiki), delivered in the university *aula* on May 31, 1842, Kotel'nikov said, among other things:

> In this connection I cannot pass over in silence that the futile millennial attempts to prove with all mathematical rigor one of the fundamental theorems of geometry, to the effect that the angle sum in a rectilinear triangle is equal to two right angles, inspired Mr. Lobačevskiĭ, a revered and meritorious professor of our university, to undertake the prodigious task of building a whole science, a geometry based on the new assumption that the angle sum in a triangle is less than two right angles—a task that is bound to gain recognition sooner or later *[444, pp. 9–10]*

It is possible that P. I. Kotel'nikov understood Lobačevskiĭ's ideas after reading the excellent exposition of his discovery contained in *Geometrical researches on the theory of parallel lines..* This work, sent by Lobačevskiĭ to Gauss, made the latter study Russian and familiarize himself with Lobačevskiĭ's Russian memoirs. In February of 1841 Gauss wrote to J. F. Encke:

> I am making reasonable progress in learning to read Russian and this gives me a great deal of pleasure. Mr. Knorre sent me a small memoir of Lobačevskiĭ (in Kazan), written in Russian, and this memoir, as well as

his small German book on parallel lines (an absurd note about it has appeared in Gersdorff's *Repertorium*) have awakened in me the desire to find out more about this clever mathematician. As Knorre told me, many of his papers are in the Russian *Proceedings of Kazan University* [*196, vol. 8, p. 232*].

On September 28, 1846, Gauss wrote to Schumacher:

Lately I had reason to reread the small work of Lobačevskiĭ (*Geometrische Untersuchungen zur Theorie der Parallellinien,* at G. Fincke, 4 signatures). This work contains the foundations of the geometry that would obtain, and form a coherent whole, if Euclidean geometry were not true. A certain Schweikart called this geometry *Astralgeometrie.* Lobačevskiĭ calls it *imaginary geometry.* You know that for 54 years (since 1792) I have shared the same views with some additional development of them that I do not wish to go into here; thus I have found nothing actually new for myself in Lobačevskiĭ's work. But in developing the subject the author followed a road different from the one I took; Lobačevskiĭ carried out the task in a masterly fashion and in a truly geometric spirit. I see it as my duty to call your attention to this work that is bound to give you truly exceptional pleasure [*196, vol. 8, pp. 238–239*].

Although Gauss, for reasons indicated previously, published nothing about non-Euclidean geometry, it was as a result of his suggestion that Lobačevskiĭ was elected a corresponding member of the Göttingen Scientific Society.

Gauss's letter to Schumacher, quoted previously, was published shortly after his death in the book *The correspondence between C. G. Gauss and H. C. Schumacher* (Briefwechsel zwischen C. F. Gauss und H. C. Schumacher. Altona, 1860–1865); the remaining letters of Gauss were published in the eighth volume of his collected works [*196, vol. 8. Göttingen, 1900*].

The publication of Gauss's letter to Schumacher about Lobačevskiĭ made a strong impression on European mathematicians. The activities of a number of advocates of the new geometry date to the sixties of the last century.

At that time there appeared the *Notes on Lobatschewsky's Imaginary Geometry* (London, 1865) by the English algebraist and geometer Arthur Cayley (1821–1895) [*103, vol. 5, pp. 471–472*], in which he compared the trigonometric formulas of Lobačevskiĭ and the formulas of spherical trigonometry. And although this note shows that the author of the theory of projective metrics (which we will discuss later) failed to understand the essence of Lobačevskiĭ's discovery, it helped to make it known.

The French mathematician Jules Hoüel (1823–1866) published a French translation of Lobačevskiĭ's *Geometrical researches* together with excerpts from the correspondence between Gauss and Schumacher (Etudes géométriques sur la théorie des parallèles, suivie d'un extrait de la correspondance de Gauss et de Schumacher. Bordeaux, 1866; Paris, 1866) [*331*] and a book—A *critical essay on the fundamental principles of geometry* (Essai critique sur les principes fondamentaux de la géométrie, Paris, 1867) *[234]*— containing an exposition of the basic ideas of Lobačevskiĭ's geometry. The German Richard Baltzer (1818–1887) set forth the foundations of Lobačevskian geometry in the second edition of his *Elements of mathematics* (Die Elemente der Mathematik. Dresden, 1867) *[35]*. The Italian mathematician Giuseppe Battaglini published a paper *On the imaginary geometry of Lobačevskiĭ* (Sulla geometria immaginaria di Lobatschewsky. Naples, 1867) *[40]* as well as an Italian translation of Lobačevskiĭ's *Pangeometry* (Naples, 1867) *[335]* and J. Bolyai's *Appendix* (Naples, 1868). The Moscow mathematician Alekseĭ Vasil'evič Letnikov (1837–1888) published in an early volume of the journal *Matematičeskiĭ sbornik* (The mathematical collection) a Russian translation of Lobačevskiĭ's *Geometrical researches* (Moscow, 1868) *[334]* with an introduction in which he described his geometric works as "remarkable but not well known" and said that these works were likely to contribute to the improvement of teaching methods and to destroy all hope of proving the parallel postulate. The Kazan mathematician Erast Petrovič Yaniševskiĭ (1829–1906) published a *Historical note on the life and work of N. I. Lobačevskiĭ* (Istoričeskaya zapiska o žizni i deyatel'nosti N. I. Lobačevskogo. Kazan, 1868) *[643]* that was soon translated into French and Italian. Also in 1868 there appeared a paper by E. Beltrami, to be discussed in greater detail later, devoted to an interpretation of the geometry of Lobačevskiĭ. Through these publications the geometry of Lobačevskiĭ became known by about 1870 to the geometers of the most important countries in Europe.

Lobačevskiĭ's Trigonometry

After setting forth the basis of *imaginary geometry* in *The principles of geometry* Lobačevskiĭ defines the *angle of parallelism*. He drops a perpendicular of length *a* from a point *B* to a straight line and draws through *B* a parallel to that line (Figure 88). The *angle of parallelism* is the angle between the perpendicular and the parallel. In Euclidean geometry this angle is always

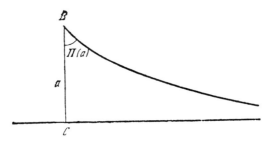

Figure 88

equal to $\pi/2$, whereas in Lobačevskian geometry it is acute and is a function of a. Lobačevskiĭ denotes this function here by $F(a)$ but in later papers by $\Pi(a)$. It is clear that

$$\lim_{a \to 0} \Pi(a) = \frac{\pi}{2}, \qquad \lim_{a \to \infty} \Pi(a) = 0.$$

Lobačevskiĭ extends this function to all real values of a by putting $\Pi(0) = \pi/2$ and $\Pi(-a) = \pi - \Pi(a)$, and shows that for every angle A, acute or obtuse, there is a value a ($a > 0$ if A is acute and $a < 0$ if A is obtuse) such that $A = \Pi(a)$.

Then Lobačevskiĭ finds the trigonometric formulas for rectilinear and spherical triangles in his space. In the first case he expresses these formulas in terms of the function $F(a)$. In the second case his formulas coincide with the formulas of spherical trigonometry in Euclidean space:

Thus in a right triangle with legs a and b, opposite angles A and B, hypotenuse C, we have, in the rectilinear case,

$$\sin F(c) = \sin F(a) \sin F(b),$$

$$\tan F(c) = \tan F(a) \sin A,$$

$$\cos F(b) = \cos F(c) \cos A,$$

$$\sin F(c) = \tan A \tan B, \qquad (14)$$

$$\tan A = \cos F(a) \tan F(b),$$

$$\sin B = \sin F(a) \cos A;$$

and in the spherical case

$$\cos c = \cos a \cos b,$$

$$\sin a = \sin A \sin c,$$

$$\tan a = \tan c \cos B,$$

$$\cos c = \cot A \cot B, \qquad (15)$$

$$\tan a = \tan A \sin b,$$

$$\cos A = \cos a \sin B$$

—well-known formulas of spherical geometry by means of which it is easy to show that in every spherical triangle with sides a, b, c, and opposite angles A, B, C we have

$$\cos A \sin b \sin c + \cos b \cos c = \cos a,$$

$$\sin a \sin B = \sin b \sin A,$$

$$\cot A \sin C + \cos C \cos b - \cot a \sin b = 0, \qquad (16)$$

$$\cos a \sin B \sin C - \cos B \cos C = \cos A.$$

This shows that measurement of spherical triangles does not depend on the assumption concerning parallels. This is not so for measurement of rectilinear triangles. Just as equations (15) imply (16), so too with the aid of equations (14) we can show that in every rectilinear triangle with sides a, b, c and opposite angles A, B, C we have

$$\tan F(a)\sin A = \tan F(b)\sin B,$$

$$\cos A \cos F(b)\cos F(c) + \frac{\sin F(b)\sin F(c)}{\sin F(a)} - 1 = 0,$$

$$\cot A \sin B \sin F(c) + \cos B - \frac{\cos F(c)}{\cos F(a)} = 0, \qquad (17)$$

$$\cos C + \cos A \cos B - \frac{\sin A \sin B}{\sin F(c)} = 0$$

[333, vol. 1. pp. 205–206].

The first of the formulas in (16) is the spherical cosine theorem (1.7); the second formula in (16) is the spherical sine theorem (1.5); the fourth formula in (16) is the dual spherical cosine theorem (1.13); and the third formula in (16) is an algebraic consequence of these formulas. The first, second, third, and sixth formulas in (15) are obtained from the formulas in (16) for $C = \pi/2$, and the fourth and fifth formulas in (15) are algebraic consequences of the others; the first of these formulas is the *spherical Pythagorean theorem* and the fifth coincides, essentially, with the spherical tangent theorem.

At the end of the paper Lobačevskiĭ writes:

After we have found equations (17) which represent the dependence of the angles and sides of a triangle; when, finally, we have given general expressions for elements of lines, areas and volumes of solids, all else in the Geometry is a matter of analytics, where calculations must necessarily agree with each other, and we can not discover anything new that is not included in these first equations from which must be taken all relations of geometric magnitudes, one to another. Thus if one now needs to assume that some contradiction will force us subsequently to refute the principles that we accepted in this geometry, then such contradiction can only hide in the very equations (17). We note, however, that these equations become equations (16) of spherical Trigonometry as soon as, instead of the sides a, b, c, we put $a\sqrt{-1}$, $b\sqrt{-1}$, $c\sqrt{-1}$; but in ordinary Geometry and in spherical Trigonometry there enter everywhere only ratios of lines: therefore ordinary Geometry, Trigonometry and the new Geometry will always agree among themselves *[333, vol. 1, p. 261].*

If the sides a, b, c of a spherical triangle are measured in radians and if we denote the radius of the sphere by r then formulas (1.5), (1.7), and (1.13) of the spherical sine, cosine, and dual cosine theorems can be written as

$$\frac{\sin\dfrac{a}{r}}{\sin A} = \frac{\sin\dfrac{b}{r}}{\sin B} = \frac{\sin\dfrac{c}{r}}{\sin C}, \tag{6.2}$$

$$\cos\frac{a}{r} = \cos\frac{b}{r}\cos\frac{c}{r} + \sin\frac{b}{r}\sin\frac{c}{r}\cos A, \tag{6.3}$$

$$\cos A = -\cos b\cos C + \sin b\sin C\cos\frac{a}{r}. \tag{6.4}$$

Multiplication of the sides of the triangle by i is equivalent to multiplying the radius of the sphere by i. Putting $r = qi$ and making use of formulas (6.1) we can rewrite these formulas as

$$\frac{\sinh a/q}{\sin A} = \frac{\sinh b/q}{\sin B} = \frac{\sinh c/q}{\sin C}, \tag{6.5}$$

$$\cosh a/q = \cosh b/q\cosh c/q - \sinh b/q\sinh c/q\cos A, \tag{6.6}$$

$$\cos A = -\cos B\cos C + \sin B\sin C\cosh a/q. \tag{6.7}$$

Formulas (6.5)–(6.7) are the formulas of Lobačevskian trigonometry in the form in which they were written down by Taurinus.

The *asymptotic triangle* of the Lobačevskian plane shown in Figure 88 can be obtained by passage to the limit from a right spherical triangle with right angle C by movement of its vertex A to infinity, in which case the angle A tends to zero. If we apply this passage to the limit to formula (6.7) and put in it $A = 0$, $B = \Pi(a)$, and $C = \pi/2$ then we obtained the relation

$$1 = \sin\Pi(a)\cosh a/q,$$

which we write as

$$\sin\Pi(a) = \frac{1}{\cosh a/q}.$$

The latter implies the relations

$$\cos\Pi(a) = \frac{\sinh a/q}{\cosh a/q} = \tanh a/q$$

and

$$\tan^2\frac{\Pi(a)}{2} = \frac{1 - \cos\Pi(a)}{1 + \cos\Pi(a)} = \frac{1 - \tanh a/q}{1 + \tanh a/q} = \frac{\cosh a/q - \sinh a/q}{\cosh a/q + \sinh a/q}$$

$$= \frac{e^{-a/q}}{e^{a/q}} = e^{-2(a/q)},$$

so that

$$\tan\frac{\Pi(a)}{2} = e^{-a/q}, \tag{6.8}$$

that is,

$$\Pi(a) = 2\arctan(e^{-a/q}). \tag{6.9}$$

It is easy to check that $\Pi(0) = 2\arctan 1 = \pi/2$, $\lim_{a\to\infty}\Pi(a) = 2\arctan 0 = 0$. Lobačevskiĭ notes that

if a, b, c, are supposed very small, so that it is permissible to ignore powers and products whose dimensions are higher *[333, vol. 1, p. 206]*.

Then the formulas of his trigonometry reduce to those of ordinary trigonometry. In fact

$$\sin x = x - \frac{x^3}{3!} + \cdots, \qquad \cos x = 1 - \frac{x^2}{2!} + \frac{x^4}{4!} - \cdots,$$

$$\sinh x = x + \frac{x^3}{3!} + \cdots, \qquad \cosh x = 1 + \frac{x^2}{2} + \frac{x^4}{4!} + \cdots$$

Also, if in formulas (6.2)–(6.3) and (6.5)–(6.6) we replace $\sin x$ and $\sinh x$ by x, $\cos x$ by $1 - x^2/2$, and $\cosh x$ by $1 + x^2/2$, and in formulas (6.4) and (6.7) we replace $\cos x$ and $\cosh x$ by 1, then we obtain the formulas of plane trigonometry

$$\frac{a}{\sin A} = \frac{b}{\sin B} = \frac{c}{\sin C}, \tag{6.10}$$

$$a^2 = b^2 + c^2 - 2bc\cos A, \tag{6.11}$$

$$\cos A = -\cos B\cos C + \sin B\sin C = -\cos(B + C). \tag{6.12}$$

The last of these formulas is equivalent to $\cos A = \cos(\pi - B - C)$, that is, to $A + B + C = \pi$. We note that $\cos(a/r) < 1$ and formula (6.4) imply that $\cos A < -\cos B\cos C + \sin B\sin C = -\cos(B + C)$; that is, $\cos A < \cos(\pi - B - C)$. This means that in spherical trigonometry $A > \pi - B - C$, that is, $A + B + C > \pi$. In much the same way $\cosh a/q < 1$ and formula (6.7) imply that $\cos A > -\cos B\cos C + \sin B\sin C = -\cos(B + C)$; that is, $\cos A > \cos(\pi - B - C)$. This means that in Lobačevskian geometry $A < \pi - B - C$; that is, $A + B + C < \pi$.

It is not difficult to see that if the angle sum in a triangle is greater than or less than π then the *angular excess* $\varepsilon = A + B + C - \pi$ or *angular defect* $\delta = \pi - A - B - C$ of a triangle made up of two triangles is equal to the sum of the excesses or defects of the component triangles. In fact, if the triangle ABC consists of the triangles ACD and CBD (Figure 89) and $A + C_1 + D_1 = \pi + \varepsilon_1$, $B + C_2 + D_2 = \pi + \varepsilon_2$ then $A + B + C_1 + C_2 + D_1 + D_2 = A + B + C + \pi = 2\pi + \varepsilon_1 + \varepsilon_2$, that is, $A + B + C = \pi + \varepsilon_1 + \varepsilon_2$, so that $\varepsilon = \varepsilon_1 + \varepsilon_2$; putting $\varepsilon_1 = -\delta_1$, $\varepsilon_2 = -\delta_2$, we obtain $\delta = \delta_1 + \delta_2$. Since the measure of area of a plane region is an additive function of the region that is invariant under motions and positive (for a region containing an arbitrarily small triangle), it follows that the angular excess ε or the angular defect δ is a measure of the areas of triangles; that is, the measure of areas in terms of

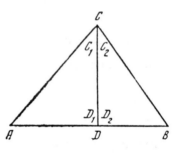

Figure 89

squares of the measures of length is proportional to ε or δ. To determine each of the proportionality coefficients we use the fact that, in the small, spherical geometry and Lobačevskian geometry are close to Euclidean geometry and compute the area of a small isosceles right triangle with legs a. In view of formulas (6.4) and (6.7), putting $C = \pi/2$ and $A = B$, we have, respectively, $\cos A = \sin A \cos \dfrac{a}{r}$, that is, $\cot A = \cos a/r$, and $\cos A = \sin A \cosh a/q$, that is, $\cot A = \cosh a/q$. Since in the first case $A = \pi/4 + \varepsilon/2$ and in the second case $A = \pi/4 - \delta/2$, expansion of $\cot x$ in a Taylor series

$$\cot x = \cot(x_0 + h) = \cot x_0 - \frac{h}{\sin^2 x_0} + \cdots$$

about the point $x_0 = \pi/4$ yields the formulas

$$\cot x = 1 - \frac{\varepsilon}{2} \frac{1}{\sin^2 \dfrac{\pi}{4}} + \cdots = 1 - \varepsilon + \cdots,$$

$$\cot x = 1 + \frac{\delta}{2} \frac{1}{\sin^2 \dfrac{\pi}{4}} + \cdots = 1 + \delta + \cdots,$$

and comparison of these with the expansions

$$\cos \frac{a}{r} = 1 - \frac{a^2}{2r^2} + \cdots, \qquad \cosh \frac{a}{q} = 1 + \frac{a^2}{2q^2} + \cdots,$$

yields the relations $a^2/2r^2 = \varepsilon$, $a^2/2q^2 = \delta$. On the other hand, the Euclidean formula for the area S of the small triangle yields $S = a^2/2$. Replacing in the last relations $a^2/2$ by S we obtain expressions for S in terms of ε and δ, respectively. Since S is proportional to ε and δ, these expressions hold, respectively, for arbitrary spherical triangles and for triangles in the Lobačevskian plane.

For spherical geometry the expression in question is

$$S = r^2\varepsilon, \tag{6.13}$$

and for Lobačevskian geometry it is

$$S = q^2 \delta. \tag{6.14}$$

We know that already Lambert knew that in the case of the "acute-angle hypothesis" S and δ are proportional.

Consistency of Lobačevskian Geometry

We saw that, in Lobačevskiĭ's view, what attested to the consistency of the geometry he discovered was that its trigonometric formulas are obtained from the corresponding formulas of spherical geometry by multiplying the sides of a triangle by an imaginary unit.

The inadequacy of these arguments is that here the consistency of plane Lobačevskian geometry is based on the consistency of its trigonometric formulas, but all that is proved in the paper is that these formulas are consequences of the assumptions of Lobačevskian geometry. To deduce the consistency of Lobačevskian geometry from the consistency of its trigonometric formulas one must show the opposite. Specifically, one must show that all propositions of Lobačevskian geometry are consequences of its trigonometric formulas and the propositions of absolute geometry, that is, the propositions that are independent of the parallel postulate. Lobačevskiĭ gives a relevant proof in his *Imaginary geometry*, where he writes at the beginning:

> Now, putting aside geometric constructions and choosing a short reverse road, I intend to show that the principal equations I found [in the paper quoted previously] for the connections between the sides and angles of a triangle in imaginary geometry can be profitably adopted in Analytics and will never lead to conclusions that are false in any manner whatever *[333, vol. 3, p. 17]*.

Then Lobačevskiĭ defines the function $\Pi(a)$ by means of the relation (6.8) and adjoins to the propositions of absolute geometry the trigonometric relations in a right triangle that are equivalent to the preceding relations (14) of his previous paper. From these Lobačevskiĭ first derives the preceding trigonometric relations (17) ((13) in that paper) in an arbitrary triangle and then the assertion that the angle sum in a triangle is less than two right angles. Since the latter assertion is equivalent to the parallel postulate of Lobačevskian geometry, he thereby showed that all propositions of his geometry that he derived in the previous paper from the propositions of absolute geometry and the parallel postulate of his geometry can also be derived from the propositions of absolute geometry and the indicated trigonometric formulas.

By comparing the mentioned trigonometric formulas and the formulas (16) of spherical geometry in his previous paper ((15) in that paper) Lobačevskiĭ again arrives at the conclusion that

nothing in the theory prevents us from taking the angle sum of a right
triangle to be less than two right angles,

and that

with this assumption equations (13) replace equations (15) and cannot
lead to false conclusions *[333, vol. 3, p. 26]*.

Lobačevskiĭ's arguments do not represent a finished proof of the consistency of his plane geometry. It seems that what Lobačevskiĭ meant by consistency is the consistency of the totality of these formulas and the axioms of
absolute geometry. But the formulas of spherical trigonometry, which imply
that the angle sum in a triangle is greater than π, contradict the axioms of
absolute geometry when viewed as axioms of plane trigonometry. Thus
Lobačevskiĭ's reasoning proves only the internal consistency of the trigonometric formulas, and this alone is no proof of the consistency of his geometry.

Nevertheless one can begin with Lobačevskiĭ's arguments and prove the
consistency of his geometry rather than of its trigonometric formulas. To do
this it is necessary to make use of the idea of a *complex space*. We saw that
Poncelet introduced the notion of *imaginary points* of space in his *Treatise on
projective properties of figures. Complex Euclidean space* is defined as the
totality of imaginary points of Euclidean space together with its real points;
É. Cartan suggested that the latter be called a *spatial chain [94, p. 126]*. Every
algebraic or analytic curve and surface in real space can be regarded as part
of a curve or surface in complex space defined by the same equations; the
points of the given complex curve or surface that do not belong to the *spatial
chain* are the imaginary points of the corresponding curve or surface. The
distance between two points of complex space is expressed in terms of their
coordinates by the same formulas as in real space, and a corresponding
statement holds for angles between straight lines. The consistency of complex
Euclidean space is proved with the help of its complex arithmetic model just
as the consistency of real Euclidean space is proved with the help of its real
arithmetic model. Just as the real dimension of a complex line interpreted in
the complex plane is 2, so the real dimension of three-dimensional complex
space is 6.

Since distances and angles in complex space are defined by the same
formulas as in real space, it follows that the formulas of plane and spherical
trigonometry in complex space also coincide with the corresponding formulas
in real space. In particular the formulas of spherical trigonometry on a sphere
of complex radius r are formulas (6.2)–(6.4), where the lengths of the sides a,
b, c and the angles A, B, C are complex. If one writes the complex radius r in
the form qi then these formulas can also be written in the form (6.5)–(6.7),
where a, b, c and A, B, C are also complex. This shows that Lobačevskian
geometry is realized in a certain set of points of a sphere with pure imaginary
radius qi (q real) of complex Euclidean space. The points with real rectangular
coordinates x, y ($x = \bar{x}, y = \bar{y}$) and pure imaginary coordinate z ($z = -\bar{z}$)

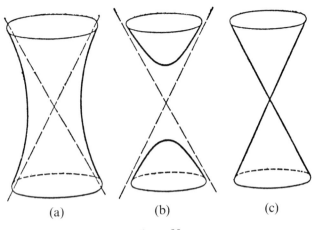

(a)   (b)   (c)

Figure 90

form such a set. Such a set of points in complex Euclidean space was first considered by the French mathematician Henri Poincaré and the German mathematician Hermann Minkowski in connection with an interpretation of the special theory of relativity. Such a set of points can also be characterized by means of three real coordinates x, y, z, but then the distance between two points with respective coordinates x_1, y_1, z_1 and x_2, y_2, z_2 is defined not by means of the usual formula of Euclidean space

$$d^2 = (x_2 - x_1)^2 + (y_2 - y_1)^2 + (z_2 - z_1)^2, \qquad (6.15)$$

but by means of the formula

$$d^2 = (x_2 - x_1)^2 + (y_2 - y_1)^2 - (z_2 - z_1)^2. \qquad (6.16)$$

The space of points with the distance formula (6.16) is now called *pseudo-Euclidean space*.[3] In this space there are point pairs with $d^2 > 0$, $d^2 < 0$, and $d^2 = 0$. The corresponding straight lines are called, respectively, *lines with real length*, *lines with imaginary length*, and *isotropic lines*. In this space there are three types of planes: *Euclidean planes*, all of whose straight lines are lines of real length; *pseudo-Euclidean planes*, with straight lines of all three types; and *isotropic planes*, with straight lines of the first and third types only. Similarly, this space contains three types of spheres (Figures 90a, 90b, and 90c), namely spheres with real, pure imaginary, and null radii, respectively. All spheres are given by equations of the form

$$x^2 + y^2 - z^2 = r^2 \qquad (6.17)$$

but their radii satisfy the respective relations $r^2 > 0$, $r^2 < 0$, and $r = 0$, and the spheres themselves are represented by a one-sheeted hyperboloid (Fig-

---

[3] Concerning pseudo-Euclidean space see p. 264.

ure 90a), a two-sheeted hyperboloid (Figure 90b), and a cone (Figure 90c), respectively.

A model of the Lobačevskian plane is one sheet of a sphere with pure imaginary radius in pseudo-Euclidean space or a sphere of pure imaginary radius with identified antipodal points.[4] This explains why the trigonometric formulas (6.5)–(6.7) of the Lobačevskian plane are obtained from formulas (6.2)–(6.4) of spherical trigonometry by replacement of r by qi. This sphere of pure imaginary radius is the "imaginary sphere" anticipated by Lambert (see p. 101). The existence of this model of the Lobačevskian plane proves its consistency.

Two straight lines in the Lobačevskian plane intersect if the diametric planes of the sphere with pure imaginary radius that correspond to these lines intersect along a straight line of imaginary length and do not intersect if the corresponding diametric planes intersect along a straight line of real length or along an isotropic straight line. In the second of the two latter cases the nonintersecting straight lines of the Lobačevskian plane can be obtained from intersecting ones by passage to the limit and are called *parallel straight lines* in the sense of Lobačevskiĭ. In the first of these two cases the nonintersecting lines cannot be obtained from intersecting ones by passage to the limit and are called *diverging straight lines*. Two diverging straight lines have a common perpendicular that corresponds to the diametric plane of the sphere of imaginary radius that is perpendicular to the two diametric planes corresponding to the given straight lines; these straight lines move apart on each side of the common perpendicular. In the case of the acute-angle hypothesis, such a common perpendicular for the sides of a Khayyām-Saccheri quadrilateral is its lower base. Parallel lines behave in a radically different way in the sense that they approximate each other asymptotically in the *direction of parallelism*. If we intersect a sphere of pure imaginary radius by means of a nondiametric pseudo-Euclidean plane then we obtain a curve that is equidistant from the straight line that corresponds to the diametric plane parallel to the given nondiametric one. If we intersect the sphere by means of an isotropic plane we obtain Lobačevskiĭ's "limit circle." The two latter curves are now known as the *equidistant curve* and the *horocycle*, respectively. Whereas circles are orthogonal trajectories of pencils of intersecting straight lines, equidistants are orthogonal trajectories of *pencils of diverging straight lines* that correspond to pencils of diametric planes of the sphere with an axis of real length; such straight lines are perpendiculars to a single straight line. Similarly, horocycles are orthogonal trajectories of "pencils of parallel straight lines" that correspond to pencils of diametric planes of the sphere with an isotropic axis. By rotating an equidistant curve about one of the straight lines of its associated pencil we obtain an *equidistant of a plane*—a surface equidistant from (the corresponding) plane. By rotating a horocycle

---

[4] This interpretation is the basis of the exposition of Lobačevskian geometry in the author's books *[465, see p. 151 ff, 466, p. 119 ff]*.

about one of the straight lines of its associated pencil we obtain a *horosphere*—Lobačevskiĭ's limit sphere. By a similar interpretation of Lobačevskian space in four-dimensional pseudo-Euclidean space it is possible to show in a very intuitive manner that the geometry of an equidistant surface of a plane is Lobačevskian (with a different constant q) and the geometry of a horosphere is Euclidean. Lobačevskiĭ used this fact to prove the trigonometric formulas of the geometry he discovered.

On a sphere of imaginary radius it is possible to introduce coordinates that are similar to latitude and longitude on an ordinary sphere. If we denote by ρ the spherical distance of a point on the sphere from its point of intersection with the Oz-axis and by φ the angle between the plane passing through the point and the Oz-axis and the plane xOz, then the expressions for the coordinates x, y, z in terms of ρ and φ are:

$$x = q \cosh\frac{\rho}{q}\cos\varphi, \qquad y = q\cosh\frac{\rho}{q}\sin\varphi, \qquad z = q\sinh\frac{\rho}{q}. \qquad (6.18)$$

These coordinates were first introduced by Carl Weierstrass (1815–1897) in a seminar on Lobačevskian geometry conducted by him in the late sixties at Berlin University, which is why these coordinates are now called *Weierstrass coordinates* (see *[283, p. 189]*).

The Beltrami Model

The first model of the Lobačevskian plane was given by Eugenio Beltrami (1835–1900) in his *Attempt at an interpretation of non-Euclidean geometry* (Saggio di interpetrazione della geometria non euclidea. Naples, 1868) *[42, pp. 374–405]*. In this paper Beltrami showed that Lobačevskian geometry is realized on surfaces of constant curvature that he called *pseudospherical surfaces*. We will consider this model in chapter 8 in connection with the discussion of the intrinsic geometry of a surface. Beltrami used coordinates related to the Weierstrass coordinates by means of the equations

$$u = ax/z, \qquad v = ay/z, \qquad (6.19)$$

that is,

$$u = a\tanh\frac{\rho}{q}\cos\varphi, \qquad v = a\tanh\frac{\rho}{q}\sin\varphi \qquad (6.20)$$

These coordinates are now called *Beltrami coordinates*.

It is easy to see that

$$u^2 + v^2 = a^2\tanh^2\frac{\rho}{q} < a^2, \qquad (6.21)$$

whence

$$\rho = q \operatorname{arctanh} \frac{\sqrt{u^2 + v^2}}{a},$$

or

$$\rho = \frac{q}{2} \ln \frac{a + \sqrt{u^2 + v^2}}{a - \sqrt{u^2 + v^2}}. \tag{6.22}$$

Formulas (6.19) show that the coordinates u, v can be regarded as the rectangular coordinates of the projection of a sphere of imaginary radius from its center to the plane $z = a$. Here the whole sphere of imaginary radius is represented by the interior of the circle $u^2 + v^2 = a^2$. The diametric sections of the sphere that correspond to the straight lines of the Lobačevskian plane intersect this plane along straight lines, so that the straight lines of the Lobačevskian plane are represented by chords of the indicated circle (Figure 91). Figure 92 shows a straight line a as well as straight lines through a point A variously related to a. Thus the straight lines b and b' are parallel to a, c and c' intersect a, and d and d' diverge from a.

Beltrami considers u and v as rectangular coordinates of an auxiliary plane. In Beltrami's words,

If we denote by the letters x and y the rectangular coordinates of points of an auxiliary plane, then the questions

$$x = u, \qquad y = v$$

determine a representation of the region under investigation in which to every point of the region there corresponds a uniquely determined point of the plane and conversely; and the whole region turns out to be represented in the interior of a circle of radius a with center at the origin

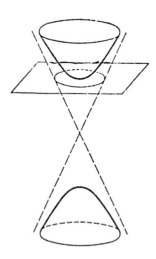

Figure 91

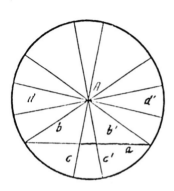

Figure 92

that we will call the *limit circle*. In this representation the chords of the limit circle correspond to the geodesics of the surface and, in particular, the parallels to the coordinate axes correspond to the coordinate geodesic lines *[42, p. 379]*.

Beltrami gave no formula for the distance between two arbitrary points and did not explain how the motions of the Lobačevskian plane are represented in his model, but this model provided the first proof of the consistency of Lobačevskian plane geometry for it represented the entire Lobačevskian plane in the Euclidean plane.

Cayley's Projective Metrics

A paper by Arthur Cayley, published a few years before the appearance of Beltrami's paper, essentially answered the questions left open by Beltrami. Cayley was then a practicing lawyer but later became a professor of mathematics at Cambridge University. Cayley was one of the creators of the theory of invariants of algebraic forms. In his algebraic papers Cayley made extensive use of geometric interpretations. In *A sixth memoir upon quantics* (London, 1859) *[103, vol. 2, pp. 561–592]* he introduces the notion of a projective metric in the plane. Given a conic (conic section) in the projective plane, whose equation Cayley writes in the form

$$(a, b, c, f, g, h \; \mathfrak{X} \; x, y, z)^2 = 0$$

and which he calls *Absolute*, it is possible to associate to two points with respective coordinates x, y, z and x', y', z' a distance, denoted by Cayley as

$$\cos^{-1} \frac{(a, \ldots \mathfrak{X} \; x, y, z \; \mathfrak{X} \; x', y', z')}{\sqrt{(a, \ldots \mathfrak{X} \; x, y, z)^2}\sqrt{(a, \ldots \mathfrak{X} \; x', y', z')^2}}$$

($\cos^{-1}$ = arc cos), such that, in view of what he had proved earlier,

if P, P', P'' be points on the same line, then we have, as we ought to have,

$$\text{Dist.}(P, P') + \text{Dist.}(P', P'') = \text{Dist.}(P, P'').$$

[103, vol. 2, p. 587].

The left side of the equation of the *Absolute* is a quadratic form. If we denote the coordinates x, y, z by x_0, x_1, x_2 and the coefficients a, b, c, d, e, f of the quadratic form by $a_{00}, a_{11}, a_{22}, a_{01} = a_{10}, a_{12} = a_{21}, a_{20} = a_{02}$, then the quadratic form can be written as $\sum_i \sum_j a_{ij} x_i x_j$. The numerator of the argument of $\cos^{-1}$ is a bilinear form. If we write y_0, y_1, y_2 instead of x', y', z', then we can write this bilinear form as $\sum_i \sum_j a_{ij} x_i y_j$. Thus, the equation of the absolute takes the form

$$\sum_i \sum_j a_{ij} x_i x_j = 0, \tag{6.23}$$

and the distance ρ between points with respective ordinates x_i and y_j takes the form

$$\rho = \cos^{-1} \frac{\sum_i \sum_j a_{ij} x_i y_j}{\sqrt{\sum_i \sum_j a_{ij} x_i x_j} \sqrt{\sum_i \sum_j a_{ij} y_i y_j}}. \tag{6.24}$$

Cayley notes that

The general formulæ suffer no *essential* modification, but they are greatly simplified in form by taking for the point-equation of the Absolute

$$x^2 + y^2 + z^2 = 0,$$

or, what is the same, for the line-equation

$$\xi^2 + \eta^2 + \zeta^2 = 0.$$

In fact, we then have for the expression of the distance of the points (x, y, z), (x', y', z'),

$$\cos^{-1} \frac{xx' + yy' + zz'}{\sqrt{x^2 + y^2 + z^2} \sqrt{x'^2 + y'^2 + z'^2}};$$

for that of the lines (ξ, η, ζ), (ξ', η', ζ'),

$$\cos^{-1} \frac{\xi\xi' + \eta\eta' + \zeta\zeta'}{\sqrt{\xi^2 + \eta^2 + \zeta^2} \sqrt{\xi'^2 + \eta'^2 + \zeta'^2}};$$

and for that of the point (x, y, z) and the line (ξ', η', ζ'),

$$\sin^{-1} \frac{\xi'x + \eta'y + \zeta'z}{\sqrt{x^2 + y^2 + z^2} \sqrt{\xi'^2 + \eta'^2 + \zeta'^2}}.$$

Suppose (x, y, z) are ordinary rectangular coordinates in space satisfying the condition $x^2 + y^2 + z^2 = 1$; then the point having (x, y, z) for its coordinates will be a point on the surface of the sphere, and (the last-mentioned equation always subsisting) the equation $\xi x + \eta y + \zeta z = 0$ will be a great circle of the sphere; and since we are only concerned with the ratios of ξ, η, ζ, we may also assume $\xi^2 + \eta^2 + \zeta^2 = 1$. We may of course retain in the formulæ the expressions $x^2 + y^2 + z^2$ and $\xi^2 + \eta^2 + \zeta^2$, without substituting for these the values unity, and it is in fact convenient thus to preserve all the formulæ in their original forms. We have thus a system of spherical geometry; and it appears that the Absolute in such system is the (spherical) conic, which is the intersection of the sphere with the concentric cone or evanescent sphere $x^2 + y^2 + z^2 = 0$. The circumstance that the Absolute is a proper conic, and not a mere point-pair, is the real ground of the distinction between spherical geometry and ordinary plane geometry, and the cause of the complete duality of the theorems of spherical geometry *[103, vol. 2, pp. 590–591]*.

By a *spherical conic* is meant the imaginary circle of intersection of all spheres in space with the plane at infinity; this circle is also the intersection of the sphere $x^2 + y^2 + z^2 = 1$ and the imaginary *concentric cone* $x^2 + y^2 + z^2 = 0$ that Cayley also calls the *evanescent sphere*. In this case Cayley has in mind the realization of his projective metric in the plane at infinity of Euclidean space.

Cayley also notes that

In ordinary plane geometry, the Absolute degenerates into a pair of points, viz. the points of intersection of the line infinity with any evanescent circle, or what is the same thing, the Absolute is the two circular points at infinity. The general theory is consequently modified, viz. there is not, as regards points, a distance such as the quadrant, and the distance of two lines cannot be in any way compared with the distance of two points; the distance of a point from a line can be only represented as a distance of two points *[103, vol. 2, p. 292]*.

By *distance of two lines* Cayley means the metric invariant of two straight lines, that is, the angle between two intersecting lines and the distance between two parallel lines. When he speaks of the *Absolute* degenerating into a pair of points Cayley has in mind not the usual conic but the *tangential conic*

$$\sum_i \sum_j a_{ij} u_i u_j = 0, \qquad (6.25)$$

where u_i are the tangential coordinates of the straight line $\sum_i u_i x_i = 0$, that is, the totality of tangents to the conic. If an ordinary conic is singular (that is, if the determinant of its matrix is zero), then it splits into a pair of real or imaginary straight lines, and if a tangential conic is singular then it splits into a pair of real or imaginary pencils of straight lines, that is, a pair of real or imaginary points. In that case, formula (6.24) defines an angle between two lines, but the distance between two points cannot be so expressed and for them there is no "distance similar to a quadrant," that is, equal to $\pi/2$.

We note that the collineations of the projective plane that preserve a "spherical conic" represent the rotations of a sphere, and the collineations of the projective plane that preserve the line at infinity of a Euclidean plane and its pair of circular ("cyclic") points are not motions but similarities of the Euclidean plane.

If the absolute of a projective plane with a Cayley metric is an imaginary conic then this plane is isometric to a sphere of radius 1 with identified antipodal points. Such a plane is now called an *elliptic* or *Riemannian non-Euclidean plane*. Cayley did not study the case when the absolute is a real conic and the case when this conic splits into a pair of real points. The corresponding planes are now called the *hyperbolic* or *Lobačevskian non-*

Euclidean plane and the *pseudo-Euclidean* plane, respectively. Nevertheless, he was aware of the exceptional importance of the projective metrics he introduced, and at the end of the memoir he wrote:

Metrical geometry is thus a part of descriptive geometry, and descriptive geometry is *all* geometry, and reciprocally *[103, vol. 2, p. 592]*.

Cayley's *descriptive geometry* is projective geometry.

Cayley failed to see the connection between his metrics and Lobačevskian geometry. His previously mentioned *Notes on Lobatschewsky's Imaginary Geometry [103, vol. 5, pp. 471–472]* shows that although he was familiar with Lobačevskiĭ's papers he did not at all understand the paper of the Russian geometer he quoted.

Klein's Model

The connection between the results of Cayley and Lobačevskian geometry was established by Felix Klein (1849–1925), whose paper, published shortly after Beltrami's, provided the essential development of the Beltrami model discussed previously. Klein, a student of Plücker, was a professor at the universities of Erlangen and Göttingen. In his paper *On the so-called non-Euclidean geometry* (Über die sogenannte Nicht-Euklidische Geometrie. Leipzig, 1871–1872) *[282, vol. 1, pp. 254–305]* Klein showed that if the Cayley absolute is a real curve then the part of the projective plane in its interior is isometric to the hyperbolic plane, and he constructed a similar model for space. Klein writes:

My definition of a projective metric generalizes somewhat the definition given by Cayley. In order to define the distance between two points I represent them as joined by means of a straight line. It intersects the fundamental curve in two other points that are in a definite cross ratio with the two given points. *I call the distance between the two points the logarithm of this cross ratio multiplied by an arbitrary, but permanently fixed, constant c.* In order to define the angle between two straight lines I pass through their point of intersection the two tangent lines to the fundamental curve. Together with the two given straight lines they determine a certain cross ratio. *I call the angle between the two straight lines the logarithm of this cross ratio multiplied by another arbitrary, but permanently fixed, constant c'.* These geometric definitions coincide with Cayley's analytic definitions as soon as we assign particular values to c and c', putting both equal to $\sqrt{-1}/2$ *[282, vol. 1. p. 255]*.

To see that this is indeed the case we note that the right side of formula (6.24), which is virtually the same as Cayley's distance formula in the metric defined by him, is the cross ratio (xy, wz) of the points x, y with coordinates x_i and y_i and the points z, w of intersection of the straight line they define with the polar planes of these points relative to the "absolute." Also, the points i, j of intersection of this straight line and the absolute divide harmonically the pair of points x, z and the pair of points y, w.

It is easy to check that the cross ratios (xy, wz) and (ij, xy) are connected by the relation

$$(xy, wz) = \frac{((ij, xy) + 1)^2}{4(ij, xy)} \tag{6.26}$$

(for proof it suffices to map the points x, y, i, j to the respective points $1, a, 0, \infty$ by means of a fractional linear transformation that necessarily preserves cross ratios. Then $(ij, xy) = a$, $z = -1$, $w = -a$, and $(xy, wz) = \dfrac{(a+1)^2}{4a}$).

If we denote the distance between the points x, y by ρ, then Klein's definition can be written as

$$\rho = c \ln(ij, xy). \tag{6.27}$$

Hence, in view of (6.26) with $c = i/2$ we have

$$(xy, wz) = \frac{(e^{\rho/c} + 1)^2}{4e^{\rho/c}} = \left(\frac{e^{\rho/2c} + e^{-\rho/2c}}{2}\right)^2 = \left(\frac{e^{\rho/i} + e^{-\rho/i}}{2}\right)^2 = \cos^2 \rho.$$

On the other hand, for a real $c = 1/2$ we have

$$(xy, wz) = \frac{(e^{\rho/c} + 1)^2}{4e^{\rho/c}} = \left(\frac{e^{\rho/2c} + e^{-\rho/2c}}{2}\right)^2 = \left(\frac{e^\rho + e^{-\rho}}{2}\right)^2 = \cosh^2 \rho.$$

Klein notes that the latter case obtains for a real conic and, most importantly, that in this case—the case of so-called hyperbolic geometry—*the geometry realized in the interior of the conic is Lobačevskian geometry.*

To see this we consider in greater detail some propositions of hyperbolic geometry (they will be enclosed in quotation marks):

"Through a point in the plane one can pass two parallels to a given straight line, that is, there are two straight lines that intersect a given straight line at points at infinity." These are straight lines that join the point to the two points of intersection of the given line and the fundamental conic section... *[282, vol. 1, p. 289].*

For angles with vertices in the interior of the conic Klein takes $c' = i/2$.

It is easy to see that the Beltrami model in a circle is a special case of the Klein model in the case when the conic is a circle. Also, if the point x is the center of the circle and the point y in the interior of the circle has rectangular coordinates u, v and is at a distance $\sqrt{u^2 + v^2}$ from the center, then the cross ratio $(ij, xy) = (-a, a; 0, y)$ has the value $(a + \sqrt{u^2 + v^2})/(a - \sqrt{u^2 + v^2})$ and formula (6.22) is a special case of formula (6.27) for $c = q/2$. Thus, apart from the element of generality, formula (6.27) determines the distance between two arbitrary points in the Beltrami model.

Klein also showed that the motions of the Lobačevskian plane, that is, the one-to-one distance-preserving transformations of this plane, are represented in his model by the collineations that map the fundamental conic onto itself. Collineations preserve cross ratios, that is, if a collineation maps four points i, j, x, y of one straight line of which i, j are on the conic onto four points i', j', x', y' of another straight line of which i', j' are on the conic, then $(ij, xy) = (i'j', x'y')$. In view of (6.27), this implies that the distance ρ of x and y is equal to the distance ρ' of x' and y'. This implies that in the Beltrami model the motions of the Lobačevskian plane are given by the collineations that map the "limit circle" onto itself.

If in the case of the elliptic plane we put the constant c in formula (6.22) equal to $ri/2$ rather than $i/2$ then we get the elliptic plane obtained by identifying antipodal points of a sphere with arbitrary radius r. Similarly, if in the case of the hyperbolic plane we put the same constant equal to $q/2$ rather than $1/2$ then we obtain a hyperbolic plane with arbitrary constant q. In all three geometries we put $c' = i/2$ so as to ensure the usual measure of angles.

In the case of a pair of conjugate complex points Klein restricts himself to considering the interior of the conic; in the exterior of the conic, that is, in the so-called ideal region of the Lobačevskian plane, and in the case of a pair of real points, that is, in pseudo-Euclidean geometry, the measure of angles requires that the constant c' be real.

Klein also defines projective metrics in space:

The basis of a general projective metric in space is provided by an arbitrary *fundamental surface of the second order.*

To define the distance between two points one joins them by a straight line. It intersects the fundamental surface in two new points that are in a definite cross ratio with the two given points. *The logarithm of this cross ratio multiplied by an arbitrary constant c yields what one should call the distance between the two given points.*

The angle between two planes is defined in a similar manner. One passes through their line of intersection the two tangent planes to the fundamental surface. Together with the given planes they determine a certain cross ratio. The angle between the planes is the logarithm of this cross ratio multiplied by an arbitrarily chosen constant c' *[282, vol. 1, p. 301].*

Here Klein also considers three cases: the case of an imaginary fundamental surface is the case of *elliptic* geometry; when that surface is "*real, not ruled and surrounds us*" we have the case of *hyperbolic*, that is, Lobačevskian, geometry; and in the "limiting case," when the fundamental surface degenerates into an imaginary conic in a certain plane, we have the *parabolic* case of Euclidean geometry (the imaginary conic is the imaginary *spherical circle* at infinity). For the first two geometries the distance ρ is given by the same formula (6.27) as in the plane, but in the case of elliptic geometry the constant

c is pure imaginary, and in the case of Lobačevskian geometry it is real. In all three cases angular measure is determined by the same formula with $c' = i/2$. This ensures the usual measure of angles.

Klein considers neither the exterior of an oval quadric, that is, the ideal region of Lobačevskian space, nor a ruled fundamental surface, nor the case when the fundamental surface degenerates into a real conic—the case of pseudo-Euclidean space; that is, he does not consider the cases for which c' is real.

The Poincaré Model in a Half-Plane and in a Circle

The great French mathematician and physicist Henri Poincaré (1854–1912), who worked at the Sorbonne and at the École Polytechnique, proposed two versions of a model of Lobačevskian geometry. In 1882 he published a paper entitled *The theory of Fuchsian groups* (Théorie des groupes fuchsiens. Stockholm, 1882) *[431, vol. 2, pp. 108–168]* devoted to fractional linear transformations of a complex variable (the groups are named for the German mathematician Lazarus Fuchs (1833–1902)). When he considered complex fractional linear transformations with real coefficients ("real substitutions") he noticed that they preserve the complex upper half-plane and that this half-plane provides a model for the Lobačevskian plane. In this paper Poincaré writes:

We will say that two figures are *congruent* if one is the image of the other under some *real* substitution. Since real substitutions form a group it is clear that two figures congruent to a third are congruent to each other.

First of all one can formulate the following theorems.

Homologous angles of two congruent figures are equal.

If in two congruent figures a point γ is homologous to α and a point δ is homologous to β then

$$(\alpha, \beta) = (\gamma, \delta)$$

[431, vol. 2, pp. 112–113].

By (α, β) Poincaré means the cross ratio

$$(\alpha\beta, \bar{\alpha}\bar{\beta}) = \frac{\alpha - \bar{\alpha}}{\alpha - \bar{\beta}} \bigg/ \frac{\beta - \bar{\alpha}}{\beta - \bar{\beta}}.$$

After noticing that points α, β, $\bar{\alpha}$, $\bar{\beta}$ lie on a circle with center on the real X-axis and denoting the points of intersection of this circle with the X-axis by h (on the arc $\beta\bar{\beta}$) and k (on the arc $\alpha\bar{\alpha}$) Poincaré introduces another cross ratio

$$[\alpha, \beta] = \frac{\alpha - h}{\alpha - k} \bigg/ \frac{\beta - h}{\beta - k}$$

and shows that

$$(\alpha, \beta) = \frac{4[\alpha, \beta]}{([\alpha, \beta] + 1)^2},$$

where $[\alpha, \beta]$ is real and greater than 1, and that

if γ is a point of the circle $\alpha\beta$ then

$$[\alpha\gamma][\gamma\beta] = [\alpha\beta]$$

[431, vol. 2, p. 114].

If the points α, β lie on a perpendicular to the X-axis, then the points $\bar{\alpha}$, $\bar{\beta}$ lie on the same perpendicular. Thus these perpendiculars are considered as special cases of circles orthogonal to the X-axis.

By ignoring higher-order infinitesimals Poincaré shows that

$$[z, z + dz] = 1 + \frac{|dz|}{y}$$

and that

$$\ln[z, z + dz] = \frac{|dz|}{y}.$$

Poincaré calls the integral

$$\int \frac{|dz|}{y},$$

"taken along an arc of some curve, the length L of this arc," and the double integral

$$\int\int \frac{dx\, dy}{y^2},$$

"taken over some plane figure, the area S of this figure." He continues:

From what was said earlier it follows that two congruent arcs of a curve have the same L and two congruent plane figures have the same S. The arc $\alpha\beta$ of a circle with center on the X-axis has L equal to $[\alpha, \beta]$.

I cannot ignore the connection that exists between the notions just introduced and the non-Euclidean geometry of Lobačevskiĭ.

We will agree not to assign to the words *straight line, length, distance, area* their usual meanings. We will call a straight line any circle with center on the X-axis, the length of a given curve, the magnitude that we denoted by L, the distance between two points the magnitude L of the arc that joins them and that belongs to a circle with center on the X-axis and, finally, the area of a plane figure that which we called earlier the magnitude S.

Furthermore, we will consider that the words *angle* and *circle* have their usual meanings but that the center of a circle is the point that is at the same distance (we have in mind "distance" in the new sense of the word) from all points of the circle, and that the distance itself is the radius of the circle.

The theorems of Lobačevskian geometry hold for the magnitudes understood in the new sense, that is, all theorems of ordinary geometry other than those that are consequences of Euclid's postulate (on parallels) can be applied to them.

Such geometric terminology has turned out to be tremendously useful to me in the course of my investigations *[431, vol. 2, p. 114]*.

The application of Lobačevskian geometry to the theory of Fuchsian groups was that the Poincaré model made it possible to represent these groups as discrete groups of motions of the Lobačevskian plane.

If we denote the cross ratio $[\alpha, \beta]$ of two complex numbers α, β and two real numbers h, k by $(hk, \alpha\beta)$ then we can write Poincaré's definition of the distance ρ between α and β as

$$\rho = \ln(hk, \alpha\beta). \tag{6.28}$$

By mapping the complex upper half-plane onto the unit disk $|z| \leq 1$ by means of a fractional linear transformation we obtain a form of the Poincaré model in which, as in the Beltrami model, the Lobačevskian plane is represented by the interior of a circle. In contrast to the Beltrami model, straight lines are represented in this model by arcs of circles orthogonal to the limit circle and by diameters of this circle, and the distance between two points represented by complex numbers α and β is given by (6.28), where h and k are the points of intersection of the limit circle with the circle through α and β orthogonal to the limit circle (Figure 93). In one form of the Poincaré model parallel straight lines are represented by semicircles or rays with common endpoint on the X-axis and in the other by circular arcs or diameters with a common endpoint on the limit circle. If in this description of parallel straight lines we stipulate that the various pairs of curves have no points in common then we obtain a description of diverging straight lines. Poincaré's statement that "the words *angle* and *circle* have the usual meaning" indicates that in

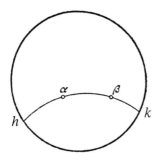

Figure 93

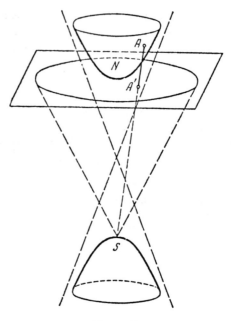

Figure 94

both forms of his model angles between straight lines in the Lobačevskian plane are the same as angles in the complex plane with the usual Euclidean metric, and circles in the Lobačevskian plane are represented by circles that have no points in common with the X-axis and the limit circle, respectively.

We will prove these two remarkable properties of the Poincaré model by employing a stereographic projection of a sphere of imaginary radius from one of its poles to a tangent plane to that sphere that is a Euclidean plane (Figure 94). Just as in the case of the stereographic projection of a sphere to a plane in Euclidean space considered in chapter 3, one can show that the stereographic projection of a sphere of imaginary radius has three properties analogous to those of the former projection: (a) projections to the plane of plane sections of the sphere of imaginary radius are circles or, if the sections of the sphere pass through a pole, straight lines; (b) angles between curves on the sphere are the same as the angles between their projections in the plane; (c) a rotation of the sphere about the diameter passing through its pole induces a rotation in the plane through the same angle about its point of tangency with the sphere. In this projection one of the planes of the sphere of imaginary radius is represented by the interior of the "limit circle," that is, the intersection of the plane with the isotropic cone, and the second plane is represented by the exterior of this circle; here it is easy to check that pairs of antipodal points of the sphere are represented by pairs of points of the plane that are mutually inverse with respect to the "limit circle." It follows that diametric sections of the sphere of imaginary radius are represented by circles

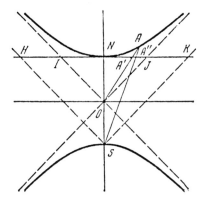

Figure 95

that are mapped onto themselves under this inversion, that is, by circles orthogonal to the "limit circle." The central angle NOA subtended by the arc NA of a great circle of the sphere of imaginary radius (Figure 95) is proportional to the length ρ of this arc, where $\rho = \frac{q}{2}\ln(IJ, NA')$ and can be expressed in terms of the cross ratio, equal to the cross ratio (IJ, NA') of the straight lines OI, OJ, ON, OA that project the points I, J, N, A from the center of the sphere, in the form $\angle NOA = c'\ln(OI, OJ: ON, OA)$. The inscribed angle NSA, subtended by the same arc, is equal to half the angle NOA, that is,

$$\angle NSA = \frac{c'}{2}\ln(OI, OJ; ON, OA) = \frac{c'}{2}\ln(IJ, NA').$$

On the other hand, this angle can be expressed in terms of the cross ratio of the straight lines SH, SK, SN, SA that project the points H, K, N, A from the pole S, as well as in terms of the equal cross ratio (HK, NA''), by the formula

$$\angle NSA = c'\ln(SH, SK; SN, SA) = c'\ln(HK, NA'').$$

Therefore

$$\ln(HK, NA'') = \frac{1}{2}\ln(IJ, NA'),$$

and the distance ρ is expressed in terms of the cross ratio (HK, NA'') by the formula

$$\rho = q\ln(HK, NA'').$$

If we regard the points H, K, N, A'' as points of the complex plane and replace them with the corresponding complex numbers h, k, α, β, whose cross ratio $(hk, \alpha\beta)$ is equal to the cross ratio (HK, NA''), and if we apply an arbitrary fractional linear transformation of the complex plane that preserves the limit circle and denote the images of these points by the same letters, then

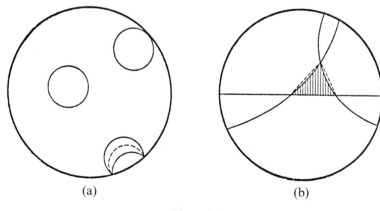

(a)　　　　　　　　　　　　　　　(b)

Figure 96

we obtain an expression for the distance ρ between two points of the Lobačevskian plane in terms of the corresponding complex numbers α, β and the points h, k of intersection of the limit circle and the circle $\alpha\beta$ orthogonal to it in the form

$$\rho = q \ln(hk, \alpha\beta). \tag{6.29}$$

If we compare formulas (6.29) and (6.28) then we see that we have obtained the Poincaré model in a circle that differs from the earlier exposition of this model only in that all distances have been multiplied by a factor q.

That the circles of the Lobačevskian plane are represented by circles that have no points in common with the limit circle follows from property (a) of the stereographic projection. The same property implies that the horocycles are represented by circles that touch the limit circle, and the two branches of an equidistant curve by two circular arcs that intersect the limit circle in two points (under inversion in the limit circle each of these arcs is mapped onto the supplement of the other). Figure 96a shows a circle, a horocycle, and an equidistant curve in the Poincaré model in a circle.

That angles between curves in the Poincaré plane are measured the Euclidean way is a consequence of property (b) of stereographic projection. Figure 96b shows a triangle of the Lobačevskian plane in the Poincaré model of this plane in a circle (it is visually obvious that the angle sum of this triangle is smaller than π). The motions of the Lobačevskian plane are represented in the two versions of the Poincaré model by fractional linear transformations that map onto themselves the real axis and the limit circle, respectively; these transformations can also be viewed as circular transformations of the plane.

In the paper *Non-Euclidean geometries* (Les géométries non-euclidiennes. Paris, 1891) *[432]* Poincaré extended the two versions of his model to space. In one version of this model Lobačevskian space is represented by a half-space and in the other by the interior of a sphere; straight lines of Lobačevskian space are represented in one version by circular arcs (and rays) orthogonal to the boundary plane of the half-space and in the other by circular arcs orthogonal to the limit sphere and by its diameters; the distance between two points α and β is given by the same formula, (6.28) or (6.29), as in the case of the plane (h and k are the endpoints of the arc or segment $\alpha\beta$). Angles are measured the Euclidean way; spheres of Lobačevskian space are represented in both versions by spheres without common points with the boundary plane of the half-space and with the limit sphere, respectively; horospheres by spheres that touch the boundary plane and the limit sphere, respectively; the two sheets of the equidistant surface of a plane are given in the spherical version of the model by segments of a sphere that intersect the limit sphere along a circle (under inversion in the limit sphere each segment is mapped onto the supplement of the other).

In his popular account *Science and hypothesis* (Le science et l'hypothèse. Paris, 1902) *[433, pp. 1–197]* Poincaré returns to the first version of his model. He writes:

Let us consider a certain plane, which I shall call the fundamental plane, and let us construct a kind of dictionary by making a double series of terms written in two columns, and corresponding each to each, just as in ordinary dictionaries the words in two languages which have the same signification correspond to one another:—

Space ⋯ ⋯ ⋯ The portion of space situated above the fundamental plane.
Plane ⋯ ⋯ ⋯ Sphere cutting orthogonally the fundamental plane.
Line ⋯ ⋯ ⋯ Circle cutting orthogonally the fundamental plane.
Sphere ⋯ ⋯ ⋯ Sphere.
Circle ⋯ ⋯ ⋯ Circle.
Angle ⋯ ⋯ ⋯ Angle.
Distance between
two points ⋯ ⋯ ⋯ Logarithm of the anharmonic ratio of these two points and of the intersection of the fundamental plane with the circle passing through these two points and cutting it orthogonally.
Etc.   Etc.

Let us now take Lobatschewsky's theorems and translate them by the aid of this dictionary, as we would translate a German text with the aid of a German-French dictionary. *We shall then obtain the theorems of ordinary geometry [433, pp. 59–60].*

Poincaré's Model on a Hyperboloid

Poincaré proposed another model of the Lobačevskian plane in the paper "On the fundamental hypotheses of geometry" (Sur les hypothèses fondamentales de la géométrie. Paris, 1887) *[431, vol. 11, pp. 79–91]*. Here he wrote:

If we restrict ourselves to two dimensions then Riemann's geometry admits of a very simple interpretation: as is well known, there is no difference between it and spherical geometry provided that we agree to call the great circles on the sphere straight lines.

First, I will generalize this interpretation so that it can also be extended to Lobačevskian geometry. We consider any quadric surface. We agree to call its plane diametric sections *straight lines* and its plane nondiametric sections *circles*.

It remains to determine what is to be meant by the angle between two intersecting lines and by the length of a segment of a straight line.

Through a point on the surface we pass two plane diametric sections (that we agreed to call *straight lines*). Now we consider tangent lines to these two sections and two rectilinear generators of the surface passing through the selected point. These four straight lines (in the ordinary sense of the word) have a certain cross ratio. The angle that we wish to define will be equal to the logarithm of this cross ratio if the two generators are real, that is, if the surface is a one-sheeted hyperboloid, and to the quotient of this logarithm by $\sqrt{-1}$ otherwise.

Consider an arc of a conic section—part of a plane diametric section (what we agreed to call a *segment* of a straight line). Like any four points on a conic, the two endpoints of the arc and the two points at infinity of the conic section have a certain cross ratio. We agree to call the logarithm of this cross ratio the *length* of the *segment* under consideration if the conic section is an hyperbola and the quotient of this logarithm by $\sqrt{-1}$ if the conic section is an ellipse.

The angles and lengths defined in this way are connected by a number of relations that form a body of theorems analogous to the theorems of plane geometry.

This body of theorems may be called a *quadric geometry*, for our starting point was the study of a basic quadric surface (quadrique).

Since there are several types of quadric surfaces there are several types of quadric geometries.

If the basic surface is an ellipsoid then the quadric geometry does not differ from the geometry of Riemann.

If the basic surface is a two-sheeted hyperboloid then the quadric geometry does not differ from the geometry of Lobačevskiǐ.

If this surface is an elliptic paraboloid then the quadric geometry reduces to Euclidean geometry; it is a limiting case of the two previous ones *[431, vol. 11, pp. 80–81]*.

It is easy to see that if we introduce in space pseudo-Euclidean geometry, in which the square of the distance from the center of the hyperboloid to its surface is given by the left side of its equation, then the two-sheeted hyperboloid, on which, according to the preceding interpretation, is realized plane Lobačevskian geometry, is the very sphere of imaginary radius that we discussed previously (p. 246).

Chapter 7
Multidimensional Spaces

Multiple Integrals

Earlier we saw that the idea of a multidimensional space arose in connection with the geometric interpretation of, first, algebraic equations of degree higher than the third, and, later, of functions of three and more variables.

A genuine geometry of multidimensional spaces arose when the development of the theory of algebraic forms of n variables and of n-tuple integrals called for geometric interpretation of functions of many variables.

The theory of n-tuple integrals was effectively developed in the first half of the 19th century by many mathematicians. One of them was the Russian mathematician Mihail Vasil'evič Ostrogradskiĭ, who, in the *Note on the theory of heat* (Note sur la théorie de la chaleur. Petersburg, 1831) *[400; 399, pp. 131–141]*, discovered his famous formula of transformation of a triple integral over a volume into a double integral over its surface and extended it in the *Memoir on the calculus of variations of multiple integrals* (Mémoire sur le calcul des variations des intégrales multiples. Petersburg, 1835) *[401; 399, pp. 9–37]* by expressing an n-tuple integral over an n-dimensional region in terms of an $(n-1)$-tuple integral over its surface. Ostrogradskiĭ wrote his formula as *[399, p. 27]*

$$\int \left(\frac{dP}{dx} + \frac{dQ}{dy} + \frac{dR}{dz} + \cdots \right) dx\, dy\, dz \ldots = \int \frac{\left(P\frac{dL}{dx} + Q\frac{dL}{dy} + R\frac{dL}{dz} + \cdots \right) dS}{\sqrt{\frac{dL^2}{dx^2} + \frac{dL^2}{dy^2} + \frac{dL^2}{dx^2} + \cdots}},$$

where $L = 0$ is the equation of the surface that bounds the region,

$$\frac{\frac{dL}{dx}}{\sqrt{\frac{dL^2}{dx^2} + \frac{dL^2}{dy^2} + \frac{dL^2}{dz^2} + \cdots}}, \quad \frac{\frac{dL}{dy}}{\sqrt{\frac{dL^2}{dx^2} + \frac{dL^2}{dy^2} + \frac{dL^2}{dz^2} + \cdots}},$$

$$\frac{\dfrac{dL}{dz}}{\sqrt{\dfrac{dL^2}{dx^2} + \dfrac{dL^2}{dy^2} + \dfrac{dL^2}{dz^2} + \cdots}}, \ldots$$

are the components of the normal to this surface, and dS is its surface element (Ostrogradskiĭ denoted partial derivatives by an ordinary d).

In the same year there appeared the paper *On the transformation of two arbitrary homogeneous functions of the second order by means of linear substitutions into two others containing only squares of the variables*; together with many theorems on the transformation of multiple integrals (De binis quibuslibet functionibus homogeneis secundi ordinis per substitutiones lineares in alias binas transformandis, quae solis quadratis variabilium constant; una cum variis theorematis de transformatione integralium. Berlin, 1834) *[246, vol. 3, pp. 191–268]* by the German mathematician Carl Gustav Jacob Jacobi (1804–1851) devoted to both of the preceding problems, that is, the theory of quadratic forms in n variables and multiple integrals. First Jacobi poses the problem:

To find linear substitutions

$$y_1 = \alpha'_1 x_1 + \alpha'_2 x_2 + \cdots + \alpha'_n x_n,$$

$$y_2 = \alpha''_1 x_1 + \alpha''_2 x_2 + \cdots + \alpha''_n x_n,$$

$$\ldots\ldots\ldots\ldots\ldots\ldots\ldots\ldots\ldots\ldots\ldots\ldots$$

$$y_n = \alpha^{(n)}_1 x_1 + \alpha^{(n)}_2 x_2 + \cdots + \alpha^{(n)}_n x_n,$$

such that

$$y_1 y_1 + y_2 y_2 + \cdots + y_n y_n = x_1 x_1 + x_2 x_2 + \cdots + x_n x_n,$$

[246, vol. 3, p. 199]

and shows that the required substitution must satisfy the conditions

$$\alpha'_\kappa \alpha'_\lambda + \alpha''_\kappa \alpha''_\lambda + \cdots + \alpha^{(n)}_\kappa \alpha^{(n)}_\lambda = 0,$$

$$\alpha'_\kappa \alpha'_\kappa + \alpha''_\kappa \alpha''_\kappa + \cdots + \alpha^{(n)}_\kappa \alpha^{(n)}_\kappa = 1.$$

Jacobi's problem is the problem of finding the matrices of rotations of n-dimensional Euclidean space; his conditions are the **orthogonality** conditions for the matrix of the linear substitution. Jacobi solves the problem indicated in the title of his paper, that is, the problem of finding a linear substitution $x_\kappa = \sum_\lambda \beta^{(\lambda)}_\kappa y_\lambda$ that simultaneously reduces two quadratic forms $V = \sum_{\kappa, \lambda} a_{\kappa, \lambda} x_\kappa x_\lambda$, $W = \sum_{\kappa, \lambda} b_{\kappa, \lambda} x_\kappa x_\lambda$ to the respective forms

$$V = G_1 y_1 y_1 + G_2 y_2 y_2 + \cdots + G_n y_n y_n,$$

$$W = H_1 y_1 y_1 + H_2 y_2 y_2 + \cdots + H_n y_n y_n,$$

[246, vol. 3, p. 247],

or, briefly, the problem of finding a linear substitution that simultaneously reduces two quadratic surfaces to canonical form. At the end of the paper Jacobi solves a number of problems dealing with the computation of multiple integrals, including the problem of computing an $(n - 1)$-tuple integral over all (positive) values of the real variables $x_1, x_2, \ldots, x_n$ satisfying the equation

$$x_1 x_1 + x_2 x_2 + \cdots + x_n x_n = 1.$$

Jacobi writes down the answer in the following form:

If n is even then

$$S = \frac{\left(\dfrac{\pi}{2}\right)^{n/2}}{(n - 2)(n - 4)\ldots 2},$$

and if n is odd then

$$S = \frac{(\pi/2)^{(n-1)/2}}{(n - 2)(n - 4)\ldots 3}$$

[246, vol. 3, p. 267].

The integrals give the surface of the segment of the unit sphere in n-dimensional Euclidean space in the region $x_i \geq 0$. This means that in order to compute the surface of the whole sphere we must multiply the values of these integrals by 2^n. Thus if n is even then

$$S = \frac{2^{n/2}\,\pi^{n/2}}{(n - 2)(n - 4)\ldots 2},$$

and if n is odd then

$$S = \frac{2^{(n+1)/2}\,\pi^{(n-1)/2}}{(n - 2)(n - 4)\ldots 3};$$

For a sphere of radius r these expressions must be multiplied by r^{n-1}. It should be noted that neither Ostrogradskiĭ nor Jacobi employed geometric terminology.

Cayley's Analytical Geometry of n Dimensions

The term *geometry of* n *dimensions* first appeared in an early paper of Arthur Cayley, *Chapters in the analytical geometry of* (n) *dimensions* (Cambridge, 1843) *[103, vol. 1, pp. 55–62]*. The term *geometry of* n *dimensions* appears only in the title, and the paper is purely algebraic in nature. Cayley considers systems of several homogeneous linear equations in n variables of the form

$$A_1 x_1 + A_2 x_2 + \cdots + A_n x_n = 0,$$

$$\dots\dots\dots\dots\dots\dots\dots\dots\dots\dots$$

$$K_1 x_1 + K_2 x_2 + \cdots + K_n x_n = 0$$

and *reciprocal equations* of such systems relative to a homogeneous quadratic function U that Cayley writes in the form

$$2U = \sum (\alpha^2) x_\alpha^2 + 2 \sum (\alpha\beta) x_\alpha x_\beta.$$

One obtains these equations by equating to zero determinants made up of partial derivatives of the function U with respect to the x_α and the coefficients of the equations of the given system. Cayley shows that if the system has r linearly independent equations then the reciprocal system has $n - r$ linearly independent equations.

At the end of the paper Cayley writes that

in the case of four variables the investigations set forth above prove the following properties of quadric surfaces:

I. If a cone intersects a second-order surface then one can pass through their curve of intersection three different cones whose vertices lie in a plane that is the polar conjugate of the vertex of the intersecting cone.

II. If two planes intersect a surface of second order then it is possible to pass through their curves of intersection two cones whose vertices lie on a straight line which is the polar conjugate of the line of intersection of the two curves *[103, vol. 1, p. 62]*.

This example shows that Cayley treats the variables x_1, x_2, ..., x_n as projective coordinates not of n-dimensional but of $(n - 1)$-dimensional projective space and views systems of linear equations as equations of $(n - r - 1)$-dimensional planes. He views the equation $U = 0$, where U is a quadratic function, as the equation of a quadric surface. Then the "reciprocal equations" determine conjugate polar $(n - r - 1)$-dimensional and $(r - 1)$-dimensional planes. The general theorem, of which the example given by Cayley at the end of the paper is but a special case, can be formulated as follows: if a degenerate quadric surface with $(r - 1)$-dimensional "vertex" intersects a nondegenerate quadric surface then one can pass through the surface of their intersection $r + 1$ cones whose vertices lie on the $(n - r - 1)$-dimensional polar of the "vertex" of the degenerate quadric surface.

It is clear that in naming his paper *Chapters of analytical geometry of* (n) *dimensions* Cayley had in mind this theorem of multidimensional projective geometry, but in the absence of multidimensional geometric terminology he restricted himself to a "multidimensional" geometric title.

It is very likely that Cayley arrived at the notion of multidimensional space under the influence of Hamilton's discovery of the quaternions interpreted as vectors in a four-dimensional space. Be that as it may, two years later, in the paper *On Jacobi's elliptic functions and on quaternions* (London, 1845) *[103, vol. 1, p. 127]*, Cayley not only considers quaternions but also extends them to so-called Cayley numbers, or octaves, interpreted as vectors of an eight-dimensional space.

Grassmann's Science of Linear Extension

In 1844 there appeared *The science of linear extension* (Die lineale Ausdehnungslehre. Leipzig, 1844) *[211, vol. 1, part 1, pp. 1–319]*, Hermann Grassmann's (1809–1877) fundamental work on multidimensional geometry. Grassmann defines an *extended manifold of the first degree* as

> the totality of elements into which a generating element passes under continuous motion *[211, vol. 1, part 1, p. 48]*

and, in particular, he defines a

> simple extended manifold—one obtained as a result of continuous extension of one and the same basic variation *[211, vol. 1, part 1, p. 48]*

that is, he defines an oriented arc of a continuous line and, in particular, an oriented rectilinear segment. He regards segments as equal if they are generated by "one and the same variation" and associates to each class of equal oriented segments

> an extended magnitude or an extension of the first degree, or a stretch *[211, vol. 1, part 1, p. 49]*

that is, a *free vector*. He also defines a

> system of the first degree—the totality of elements that can be obtained by the extension of one and the same or of the opposite variation, *[211, vol. 1, part 1, p. 49]*

that is, an abstract straight line. Then Grassmann defines a *system of second degree*, that is, an abstract two-dimensional plane, as follows:

> To begin with, I take two basic variations of different kinds and subject an element of the first basic variation (or its opposite) to an arbitrary extension and then subject the thus modified element to the second mode of variation, likewise arbitrarily extended; I call the totality of elements formed in this manner a system of second degree *[211, vol. 1, part 1, p. 52]*.

Then, in an entirely analogous manner, Grassmann defines *systems of third and higher degrees*, that is, three-dimensional and multidimensional spaces:

> Further, if I take the third basic variation that does not transform that same initial element into an element of this system of second degree—

which I will therefore call independent of the first two—and subject an
arbitrary element of this system of second degree to this third variation
(or to its opposite) arbitrarily extended; then the totality of elements
formed in this way is a system of third degree, and since this method of
formation, according to the idea, is applicable without any restriction,
I can in this way define systems of arbitrarily high degree *[211, vol. 1,
part 1, p. 52]*.

After pointing our that a plane in ordinary space can be regarded as a
system of second degree and "all infinite space" as a system of third degree,
Grassmann notes that

> geometry goes no further but abstract science knows no bounds
> *[211, vol. 1, part, p. 53]*.

These words show that Grassmann meant by geometry only the geometry of
three-dimensional space, of plane and straight line, and regarded what we call
geometry of multidimensional space not as geometry but as an abstract
science of extension.

We note that Grassmann considers curves, surfaces, "all infinite space,"
and multidimensional manifolds as "totalities of elements."

To any two elements α and β Grassmann associates the "segment" $[\alpha\beta]$
and formulates the following theorem:

> If $[\alpha\beta]$ and $[\beta\gamma]$ represent arbitrary variations then $[\alpha\gamma] = [\alpha\beta] + [\beta\gamma]$
> *[211, vol. 1, part 1, p. 56]*.

It is clear that Grassmann's "segments" are *bound vectors* and his
"variations" are *free vectors*. Grassmann goes on to apply his concepts to
geometry (that is, to geometry of three-dimensional space, where to each pair
of points X, Y one can associate the "segment" $[XY]$), and to mechanics,
where "segments" represent velocities, accelerations, and forces.

Then Grassmann defines an *exterior product* of vectors:

> By the exterior product of n segments is meant an extended magnitude
> of n-th order that is obtained if every element of the first order gives rise
> to the second, every element formed in this way gives rise to the third,
> and so on *[211, vol. 1, part 1, pp. 89–90]*

—that is, the exterior, or outer, product of two segments is a parallelogram,
the outer product of three segments is a parallelepiped, and the outer product
of m segments is an m-dimensional parallelepiped. Grassmann uses outer
products of two and three segments to define the area of a parallelogram, the
volume of a parallelepiped, the static moment of a force, and the condition of
equilibrium of forces in mechanics.

Subsequently Grassmann reworked his book into *The science of extension*
(Die Ausdehungslehre. Leipzig, 1862) *[211, vol. 1, part 2, pp. 1–506]* where
he introduced the notion of linear dependence of magnitudes

$$a = \beta b + \gamma c + \cdots$$

("a is numerically derived from the magnitudes b, c, ... by means of the numbers β, γ, ..."); *units*—linearly independent basis elements; *extensive magnitudes*, "numerically derived from the system of units," which Grassmann wrote as $\alpha_1 e_1 + \alpha_2 e_2 + \cdots$ or, in abbreviated form, as $\sum \alpha e$; the sum and difference of extensive magnitudes $\sum \alpha e + \sum \beta e = \sum (\alpha + \beta)e$, $\sum \alpha e - \sum \beta e = \sum (\alpha - \beta)e$; the product of an extensive magnitude by a number $\sum \alpha e \cdot \beta = \sum (\alpha \beta)e$; the *inner product* (a, b) of two extensive magnitudes; and the *exterior products* $[a \ b]$, $[a \ b \ c]$, ... of two or more extensive magnitudes. Grassman's *extensive magnitudes* are, essentially, *vectors of an abstract vector space*. Grassmann also associated with them concrete representations in the form of directed segments that he called *Stab* (literally, stick). The inner product of extensive magnitudes coincides with the inner product of vectors, and the outer products of two and three segments in three-dimensional space coincide with the cross and mixed products, respectively. Outer products are represented as linear combinations of outer products of the basis vectors. Since the coordinates of these linear combinations, called *Grassmann coordinates* of m-dimensional planes, continue to be the fundamental characteristic of m-dimensional planes in n-dimensional space, manifolds of such planes are now called *Grassmann manifolds*.

Grassmann was well acquainted with the previously mentioned letter of Leibniz to Huygens, as witness his paper *Geometric analysis linked to the geometric characteristic found by Leibniz* (Geometrische Analyse geknüpft an die von Leibniz erfundene Geometrische Charakteristik. Leipzig, 1847) *[211, vol. 1, part 1, pp. 321–398]*. Undoubtedly, Grassmann viewed his geometric calculus as a realization of Leibniz's idea.

In *New geometry of space based on regarding straight lines as space elements* (Neue Geometrie des Raumes gegründet auf die Betrachtung der geraden Linien als Raumelement. Leipzig, 1868) *[430]*, the German geometer Julius Plücker (1801–1868) investigated the four-dimensional manifold of straight lines of three-dimensional space and characterized straight lines by means of *Plücker coordinates* that are *Grassmann coordinates* in the special case $n = 3$, $m = 1$.

Schläfli's Theory of Multiple Continuality

In 1851 the Swiss mathematician Ludwig Schläfli (1814–1895) finished a large monograph devoted to multidimensional Euclidean geometry called *Theory of multiple continuality* (Theorie der vielfachen Kontinuität. Basel, 1901) *[507, vol. 1, pp. 169–387]*, which he submitted to the Vienna Academy of Science. In the beginning of the book Schläfli writes:

The treatise that I have the honor of presenting to the Imperial Academy of Science contains an attempt to found and elaborate a new branch of analysis which is, in a way, a kind of analytic geometry of n dimensions and includes that for plane and space as special cases for $n = 2, 3$. I call it a theory of multiple continuality, generally speaking, in the sense in which, for example, the geometry of space can be called a theory of triple continuality. Just as in this theory a group of values of three coordinates determines a point, so too in that theory a group of values of n coordinates $x, y, \ldots$ determines a *solution* (Lösung). The reason I use this term is that this is precisely the name of any group of values satisfying one or more equations in many variables. The only thing that is unusual about the name is that I also use it in the case where the variables are not connected by even a single equation. In this case I call the totality of all solutions an n-tuple totality. If, on the contrary, 1, 2, 3, ... equations are given then I call the totality of their solutions an $(n - 1)$-tuple, $(n - 2)$-tuple, $(n - 3)$-tuple ... *continuum* [507, vol. 1, p. 171].

We see that Schläfli, like Grassmann, does not extend the terminology of geometry of three-dimensional space to multidimensional geometry and calls an *n-tuple totality* what we call n-dimensional space, a "solution" what we call a point, and an *m-tuple continuum* what we call an m-dimensional surface. Also, Schläfli assumes that it is possible to choose a system of variables such that the "the distance between two given solutions $(x, y, \ldots)$, $(x', y', \ldots)$" is equal to

$$\sqrt{(x' - x)^2 + (y' - y)^2 + \cdots}.$$

In this case he calls the system of variables *orthogonal*, as against *skew*. In the latter case the distance between two solutions is of the form

$$\sqrt{(x' - x)^2 + (y' - y)^2 + \cdots + 2k(x' - x)(y' - y) + \cdots}$$

[507, vol. 1, p. 172]. Then Schläfli takes two linear homogeneous polynomials $p = ax + by + cz + \cdots + hw$ and $p' = a'x + b'y + \cdots + h'w$ in the "orthogonal variables" $x, y, \ldots, w$ and considers the totality of solutions for which p and p' are simultaneously > 0. He thinks of this totality of solutions as a fraction of the whole *unbounded totality*, and if the denominator of this fraction is taken as 2π then he calls its numerator the "*angle* between the polynomials p and p'" and denotes it by $\sphericalangle(p, p')$. Schläfli states that the "*angle* between the polynomials," that is, in modern terminology, the angle between two $(n - 1)$-dimensional planes, is given by the formula—see [507, vol. 1, p. 172]—

$$\cos \sphericalangle(pp') = \frac{aa' + bb' + cc' + \cdots + hh'}{\sqrt{a^2 + b^2 + \cdots + h^2}\sqrt{a'^2 + b'^2 + \cdots + h'^2}},$$

where both roots in the denominator are supposed to be positive.

Schläfli calls a plane a *linear continuum*, and curved surfaces he calls *higher continua*; "one-tuple continua" (lines) he calls *paths*, and linear "one-tuple continua" he calls *rays*. His book consists of three parts: *The science of linear continua*, *The science of spherical continua*, and *Various applications of the theory of multiple continuality* that go beyond the "linear" and "spherical" (that is, deal with "quadratic" and "higher continua"). Although the instances of Schläfli's use of the term *distance* quoted earlier suggest that he thought of it as a real number, the fact is that he often used this term in the sense of a segment joining two points. For example, at one point Schläfli states:

Let $x, y, z, \ldots$ be the projections of the distance r and $x_1, y_1, z_1, \ldots$ the projections of another distance $r_1 \ldots$ We put

$$xx_1 + yy_1 + zz_1 + \cdots = rr_1 \cos w$$

and call w the *angle* of the directions of the two distances r and r_1 *[507, vol. 1, p. 172]*.

Then Schläfli defines the totality of *solutions* for which

x is contained between two constants whose difference is a, y between two linear functions of x whose difference is b, z between two linear functions of x, y whose difference is c, and so on. Then the totality is contained between n parallel continua; it is called a *parallelescheme* *[507, vol. 1, p. 181]*.

Schläfli's *parallelescheme* is a multidimensional parallelepiped. Schläfli shows that "the measure of a parallelescheme is equal to the determinant of the orthogonal projections of its edges" *[507, vol. 1, p. 182]*; that is, if a multidimensional parallelepiped is constructed out of "distances" with "projections" $x_0^i, x_1^i, \ldots, x_n^i$ then its volume is equal to the determinant (x_j^i). Then Schläfli defines multidimensional polyhedra, which he calls *polyschemes* and computes the volumes of multidimensional pyramids and other polyhedra. He also determines the volume of an m-dimensional plane region as follows: he finds the projections of this region to all $\binom{n}{m}$ m-dimensional coordinate planes and shows that

the measure of an arbitrary closed m-tuple continuum is equal to the square root of the sum of the squares of its projections *[507, vol. 1, p. 183]*.

In the last chapter of the first part of his book Schläfli proves the generalized Euler theorem which asserts that if a polyhedron in n-dimensional space is homeomorphic to a sphere then the numbers N_p of p-dimensional faces (for $p = 0$, vertices; for $p = 1$, edges) are connected by the relation

$$1 - N_0 + N_1 - N_2 + \cdots + (-1)^p N_p + \cdots + (-1)^n N_{n-1} - (-1)^n = 0$$

(for $n = 2$ a special case of this formula is the equality $N_0 = N_1$, and for $n = 3$ we have Euler's theorem $N_0 - N_1 + N_2 = 2$) *[507, vol. 1, p. 193]*, and constructs the theory of *regular polyhedra* in n-dimensional space. Here

Schläfli introduced for regular polyhedra in n-dimensional space the *Schläfli symbol* $\{p_1, p_2, \ldots, p_{n-1}\}$. The faces of such a polyhedron are regular polyhedra $\{p_1, p_2, \ldots, p_{n-2}\}$, and the *vertex figures*, being regular polyhedra with vertices in the midpoints of the edges with common ends in one vertex of the given polyhedron, are regular polyhedra $\{p_2, p_3, \ldots, p_{n-1}\}$. For a regular polygon with p vertices the Schläfli symbol is $\{p\}$. For regular polyhedra in three-dimensional space the Schläfli symbols are $\{3, 3\}$ for a tetrahedron, $\{4, 3\}$ for a cube, $\{3, 4\}$ for an octahedron, $\{5, 3\}$ for a dodecahedron, and $\{3, 5\}$ for an icosahedron. Schläfli shows that for $n \geq 5$ there are only three types of regular polyhedra, namely a *simplex* $\{3, 3, \ldots, 3, 3\}$, a *cube* $\{4, 3, \ldots, 3, 3\}$ and a *cross polyhedron* $\{3, 3, \ldots, 3, 4\}$ with $n + 1$, $2n$, and 2^n faces, respectively, and for $n = 4$ there are six types of regular polyhedra $\{3, 3, 3\}$, $\{4, 3, 3\}$, $\{3, 3, 4\}$, $\{3, 4, 3\}$, $\{5, 3, 3\}$ and $\{3, 3, 5\}$, with 5, 8, 16, 24, 120 and 300 three-dimensional faces, respectively *[507, vol. 1, pp. 212–226]*. The vertices of the polyhedron $\{3, 4, 3\}$ are 16 vertices of a cube and 8 points resulting from the reflection of the center of symmetry of this cube in its faces.

In the second part of the book Schläfli finds, among other things, the surface of an n-dimensional sphere that reduces to the integrals (7.1) and (7.2) above. In the third part Schläfli finds the center and principal axes of a *quadratic continuum*—a multidimensional quadric surface.

Schläfli's monograph was published in its entirety only in 1901, after the author's death. However, its most important results (including the results given above) were published in his papers *Reduction of a multiple integral containing as special cases the arc of a circle and the area of a spherical triangle* (Reduction d'une intégrale multiple qui comprend l'arc du cercle et l'aire du triangle spherique comme cas particuliers. Paris, 1855) and *On the multiple integral* $\int^n dx\, dy \ldots dz$ whose limits are $p_1 = a_1 x + b_1 y + \cdots + h_1 z > 0$, $p_2 > 0, \ldots, p_n > 0$ *and* $x^2 + y^2 + z^2 < 1$ (London, 1858–1860) *[507, vol. 2, pp. 164–190; 219–270]*. Schläfli's papers gained little fame in his lifetime, and regular polyhedra in n-dimensional space were rediscovered in the eighties of the 19th century by scholars such as Washington Irving Stringham (1847–1909) *[563]* and Reinhold Hoppe (1816–1900) *[233]*. Schläfli's work was continued by Peter Hendrik Schoute (1846–1913) in his *Multidimensional geometry* (Mehrdimensionale Geometrie. Leipzig, 1902–1905) *[509]*. In addition to developing Schläfli's analytic and synthetic geometry Schoute solved problems of multidimensional descriptive geometry. In particular he proposed the following method of representing a point in $2n$-dimensional Euclidean space in the plane, now known as the *Schoute diagram*: a point with coordinates $x_1, x_2, \ldots, x_{2n}$ is represented by n points with coordinates $x_1, x_2; x_3, x_4; \ldots; x_{2n-1}, x_{2n}$ *[509, vol. 1, p. 200]*.

In this famous paper of 1854 *On the hypotheses which lie at the foundations of geometry* (Über die Hypothesen welche der Geometrie zu Grunde liegen) *[454, pp. 272–287; 122, pp. 55–75 (English translation by W. K. Clifford*; the reader may wish to consult the more modern translation of M. Spivak *[552a, pp. 135–153])]* Bernhard Riemann (1826–1866) introduced a notion of an *n-ply extended manifold* far more general than those of Grassmann and Schläfli. We note that as early as 1872, after describing various groups of transformations on three-dimensional space, Klein wrote in the final paragraph of his *Erlangen program* (1872):

> It is obvious how to realize the transfer of the above space to the notion of a pure manifold *[282, vol. 1, pp. 486–487]*.

In the seventies Klein wrote a number of papers on multidimensional geometry. That of 1876, in which he showed that a closed curve with corners in three-dimensional space can be freed of corners in four-dimensional space, prompted the German astronomer and physicist F. Zöllner to make his notorious "attempt to enlist the aid of spirits for an experimental verification of this fact." [1]

The Terminology of Multidimensional Geometry

Modern geometric terminology first appeared in a paper by the Italian geometer Enrico Betti (1823–1892), *On spaces of an arbitrary number of dimensions* (Sopra gli spazi di un numero qualunque di dimensioni. Milan, 1871) *[55, vol. 2, pp. 273–290]*, that played an important role in the history of topology. Betti's paper opens with the following words:

> Let $z_1, z_2, \ldots, z_n$ be n variables that can take on all real values from $-\infty$ to $+\infty$. We will call the *n*-ply infinite field of systems of values of these variables a *space* of n dimensions and denote it by S_n. A system $(z_1^0, z_2^0, \ldots, z_n^0)$ will define a *point* L_0 of this space; we will call $z_1^0, z_2^0, \ldots, z_n^0$ the *coordinates* of this point. A system of m equations will determine a field of systems of values of $n - m$ independent variables that will be a space S_{n-m} of that number of dimensions contained in S_n. We will call a space of just one dimension, forming a simple infinity, a *line [55, vol. 2, p. 273]*.

One year later this terminology was taken over by the French mathematician Camille Jordan (1838–1922) in the note *Essay on the geometry of n dimensions* (Essai sur la géométrie à *n* dimensions. Paris, 1872) *[251, vol. 3, pp. 3–5]* soon to appear in enlarged form as a paper with the same title *[251, vol. 3, pp. 76–149]*. At the beginning of both papers it is stated that

> We will consider a point of a space of *n* dimensions that which is determined by *n* coordinates $x_1, x_2, \ldots, x_n$ *[251, vol. 3, pp. 3, 79]*.

Unlike Betti, for whom the space S_n and its subspaces S_{n-m} were arbitrary

---

[1] This attempt is described by Klein in *Lectures on the development of mathematics in the 19th century [283, pp. 169–170]* and by Engels in the paper *Natural science in the world of spirits* in his *Dialectic of nature [169, pp. 50–61]*.

manifolds, Jordan defines multidimensional Euclidean space and its planes:

> A linear equation between the coordinates determines a *plane*, k
> consistent linear equations, a *k-plane*; $n - 1$ equations, a *straight line.*
> The *distance* between two points will be $\sqrt{(x_1 - x'_1)^2 + \cdots}$ *[251, vol. 3,
> pp. 3, 79].*

This terminology coincides with the modern terminology except for the term *k-plane*, which now stands for a k-dimensional rather than an $(n - k)$-dimensional plane. We note that after analyzing the conditions of parallelism and perpendicularity of planes and transformations of coordinates, Jordan finds in this paper the metric invariants of a *k-plane* and an *l-plane*—stationary angles and smallest distances between planes. He defines stationary angles as follows:

> A system consisting of a k-plane P_k and an l-plane P_l passing through a point of the space has ρ different invariants, where ρ is the smallest of the numbers k, l, $n - k$, $n - l$. These invariants can be regarded as angles between the multidimensional planes (multiplans) *[251, vol. 3, pp. 4, 110].*

Jordan defines the squares of the cosines of these angles as the eigenvalues of a certain matrix of order ρ. Here he also finds the canonical form of a rotation of n-dimensional space *[251, vol. 3, p. 128].*

Algebraic Manifolds

In the seventies of the last century there arose a special branch of multidimensional geometry, namely the geometry of algebraic manifolds (surfaces) in multidimensional spaces. This theory was a generalization of the theory of algebraic curves in the complex plane that arose in the first half of the 19th century. Whereas Newton, in *Enumeration of curves of the third order [388]*, and Euler, in volume II of *Introduction to infinitesimal analysis [176, vol. 9, pp. 122–155]*, classified curves of the third and fourth orders in the real plane, Plücker, in *Theory of algebraic curves* (Theorie der algebraischen Curven. Bonn, 1839) *[430]*, constructed a theory of algebraic curves of all orders in the complex projective plane. In addition to the usual "point" equations of curves Plücker introduced *tangential equations* of curves: since the equation of a straight line in the projective plane is

$$u_0 x_0 + u_1 x_1 + u_2 x_2 = 0,$$

the numbers u_0, u_1, u_2 can be viewed as coordinates of a straight line u in the projective plane, and an equation

$$\varphi(u_0, u_1, u_2) = 0,$$

which determines a one-parameter family of straight lines, can be viewed as the equation of the envelope of this family of straight lines. The equation is called a *tangential* equation for it defines a curve in terms of its tangents rather than in terms of its points. The degree of the point equation of a curve is called its *order*, and the degree of its tangential equation is called its *class*. Plücker showed that in the complex projective plane the order n of a curve, its class m, the number δ of its double points, the number κ of its recurrence points, the number ι of its double tangents, and the number τ of its inflection points—in the absence of other singularities—are connected by the "Plücker formulas"

$$m = n(n-1) - 2\delta - 3\kappa, \quad n = m(m-1) - 2\iota - 3\tau,$$

$$\tau = 3n(n-2) - 6\delta - 8\kappa, \quad \kappa = 3m(m-2) - 6\iota - 8\tau.$$

By comparing the number of parameters the curve depends on with the number of constants in its equation Plücker discovered a number of errors Euler had made in his classification of curves of the fourth order.

In *The theory of Abelian functions* (Theorie der Abelschen Functionen. Göttingen, 1857) *[454, pp. 88–142]* Riemann introduced an important characteristic of plane algebraic curves that he denoted by the letter p. Alfred Clebsch (1833–1872), in the paper *On plane curves whose coordinates are rational functions of one parameter* (Über diejenigen ebenen Curven deren Coordinaten rationale Functionen eines Parameters sind. Berlin, 1865) *[118]* called p the *genus (Geschlecht)* of a curve. Riemann showed that for $p = 0$ the coordinates of a curve can be expressed by means of rational functions of one parameter; for $p = 1$ they can be expressed by means of elliptic integrals, and for $p > 1$ they can be expressed by means of hyperelliptic abelian integrals. For $p = 0$ the curves are called *unicursal*, for they can be drawn in the projective plane with one stroke of the pen, and for $p = 1$ they are called *elliptic* or *bicursal*.

In the paper *On singularities of algebraic curves* (Über die Singularitäten algebraischer Curven. Berlin, 1965) *[119]* Clebsch showed that in the case of curves for which the Plücker formulas hold, the genus p and the numbers n, m, δ, κ, ι, and τ are connected by the relation

$$p = \frac{(n-1)(n-2)}{2} - \delta - \kappa = \frac{(m-1)(m-2)}{2} - \tau - \iota.$$

Of other papers on the theory of plane curves we mention that of Gustav Roch (1839–1866) *On the number of arbitrary constants in algebraic functions* (Über die Anzahl der willkürlichen Constanten in algebraischen Functionen. Berlin, 1864) *[458]* in which Roch, in the course of developing the ideas of Riemann's *Theory of Abelian functions*, proved what is known as the *Riemann-Roch theorem*.

Significant results in the theory of algebraic surfaces of three-dimensional complex projective space were obtained by Max Noether (1844–1921) in *On the theory of univalent correspondence of algebraic configurations* (Zur Theorie des eindeutigen Entsprechens algebraischer Gebilde. Leipzig, 1870–1875) *[390]* and in other papers, of which we mention *Extension of the Riemann-Roch theorem to algebraic surfaces* (Extension du théorème de Riemann-Roch aux surfaces algébriques. Paris, 1886) *[391]*, by Federigo Enriques (1871–1946) in *Introduction to geometry on algebraic surfaces* (Introduzione alla geometria sopra le superficie algebraiche. Rome, 1896) *[171]*, and by Francesco Severi (1879–1961) in many papers generalized in the *Treatise on algebraic geometry* (Trattato di geometria algebraica. Bologna, 1926) *[525]*. Of Severi's papers we mention the article *On the Riemann-Roch theorem and on continuous families that belong to algebraic surfaces* (Sul teorema de Riemann-Roch e sulle serie continue appartenenti ad una superficie algebraica. Turin, 1905) *[526]* devoted to further generalization of Riemann-Roch theorem. Of the many papers on multidimensional algebraic manifolds we mention the article of the famous chess master Emmanuel Lasker (1868–1941) *On the theory of modules and ideals* (Zur Theorie der Moduln und Ideale. Leipzig, 1905) *[306]* in which he gave a criterion for when a given algebraic equation is one of the equations of an algebraic manifold.

Enumerative geometry is a branch of mathematics that goes back to Plücker's method of counting the number of parameters of algebraic curves and their equations. It was founded by Hermann Schubert (1848–1911) in *Calculus of enumerative geometry* (Kalkül der abzählenden Geometrie. Leipzig, 1879) *[514]*, Schubert also extended these methods to multidimensional geometry in the paper n-*dimensional generalizations of the fundamental numerical characteristics of our space* (Die n-dimensionalen Verallgemeinerungen der fundamentalen Abzahlen unseres Raumes. Leipzig, 1886) *[515]*. In this paper Schubert showed, among other things, that the dimension of a *Grassmann manifold*, that is, the manifold of all m-dimensional planes of n-dimensional space, is $(m + 1)(n - m)$, and enumerated the dimensions of the so-called Schubert manifolds—manifolds of planes whose intersections with a nested system of fixed planes have prescribed dimensions (such a system of planes is now called a *flag*). Schubert's work was continued by the geometer and historian of mathematics Hieronymus Georg Zeuthen (1839–1920) in his *Textbook of enumerative methods of geometry* (Lehrbuch der abzählenden Methoden der Geometrie. Leipzig, 1914) *[647]*. The principles of enumeration of parameters, not adequately justified by Schubert himself, were justified by topological methods by Bartel Lendert van der Waerden (b. 1903) in the *Topological foundatians of enumerative geometry* (Topologische Begründung der abzählenden Geometrie. Leipzig, 1929) *[609]*.

By the end of the 19th century the idea of multidimensional space had become an integral part of mathematics. In this connection we quote the opening words of Henri Poincaré's famous memoir *Analysis situs* (Paris, 1895), devoted to combinatorial topology:

> Geometry of n dimensions is concerned with the investigation of reality; no one doubts this. Bodies in hyperspace are carefully defined

just as bodies in ordinary space and while we cannot represent them we can imagine and study them. And, whereas, say, mechanics of more than three dimensions must be considered pointless, the position of hypergeometry is completely different.

Indeed, the aim of geometry is not only the direct description of bodies apprehended by our sense organs: above all, it is the analytic investigation of a certain group, and, consequently, nothing prevents us from studying other groups that are analogous and more general.

But immediately there arises the question: should we replace the language of analytic investigation by the language of geometry which loses all its advantages as soon as the possibility of using the senses has vanished? It turns out that this new language is more accurate; also, the analogy with ordinary geometry can give rise to fruitful associations of ideas and suggest useful generalizations *[431, vol. 6, p. 193].*[2]

Axiomatics of Euclidean Space

The broad view of space that marked the end of the 19th century called for the elaboration of a firm logical basis for Euclidean space as well as its various generalizations. The first such attempt was made by Moritz Pasch (1843–1930) in *Lectures on modern geometry* (Vorlesungen über neuere Geometrie. Leipzig, 1882) *[408].* Of other relevant works one should mention two books by Giuseppe Peano (1858–1932), *Geometric calculus according to Grassmann's Ausdehnungslehre preceded by an exposition of the operations of deductive logic* (Calcolo geometrico secondo l'Ausdehnungslehre di Grassmann preceduto dalle operazioni della logica deduttiva. Turin, 1888) *[410]* and *Logical exposition of the foundations of geometry* (I principii di geometria logicamente espositi. Turin, 1889) *[411]*, and the book of Peano's student Mario Pieri (1860–1913) entitled *On elementary geometry as a hypothetical deductive system* (Della geometria elementare come sistema ipotetico deduttivo. Turin, 1899) *[420].* Pasch's book, the second of the books by Peano just mentioned, and Pieri's book are devoted to the axiomatics of three-dimensional Euclidean space, whereas the first of the two Peano books, based on Grassmann's *Ausdehnungslehre*, contains an axiomatization of n-dimensional linear space.

David Hilbert's (1862–1943) *The foundations of geometry* (Grundlagen der Geometrie. Leipzig, 1899) *[229; 230]* was the most popular work on problems of axiomatics to appear at the end of the 19th century. Hilbert defines three-dimensional Euclidean space simply as a set of elements of arbitrary nature subdivided into three systems. Hilbert writes:

Let us consider three distinct systems of things. The things composing the first system, we will call *points* and designate them by the letters *A*,

---

[2] What Poincaré meant by "reality" is discussed on pp. 268–270.

$B, C, \ldots$; those of the second, we will call *straight lines* and designate them by the letters $a, b, c, \ldots$; and those of the third system, we will call *planes* and designate them by the Greek letters $\alpha, \beta, \gamma, \ldots$ The points are called the *elements of linear geometry*; the points and straight lines, the *elements of plane geometry*; and the points, lines, and planes, the *elements of the geometry of space* or the *elements of space [229, p. 3]*.

Then Hilbert states that points, straight lines, and planes are thought of as subject to certain relations called *lying on, between, congruent, parallel*, and *continuous*. A precise, and for mathematical purposes complete, description of these relations is obtained by means of axioms of geometry that Hilbert divides into five groups:

  I, 1–8. Axioms of *incidence*.
  II, 1–4. Axioms of *order*.
  III, 1–5. Axioms of *congruence*.
  IV.   Axiom of *parallels*.
  V, 1–2. Axioms of *continuity [229, p. 3; 230, p. 2]*.

The *incidence axioms* define the relation "a point lies on a straight line," "a point lies on a plane," and so on. The *order axioms* define the relation "a point lies on a straight line between two points" by means of which one defines a segment of a straight line with given endpoints. Here is also included *Pasch's axiom*: if A, B, C are three points that do not lie on the same straight line and a is a straight line in the plane ABC that does not pass through any of the points A, B, C, then, if a passes through one of the points of the segment AB, it must also pass through one of the points of the segment AC or one of the points of the segment BC.[3] The congruence axioms define the relation of congruence of segments and angles. The parallel axiom is equivalent to Euclid's postulate V (= Euclid's parallel postulate).[4] One of the two continuity axioms is Archimedes' axiom, which states that for any two segments there is a natural number n such that if we lay off the smaller segment n times then we obtain a segment larger than the larger segment; the second is Cantor's axiom, which states that every nested sequence of intervals has a common point.[5]

Weyl's Axiomatics of n-Dimensional Euclidean Space

A well-known axiomatics of n-dimensional Euclidean space was given by Hermann Weyl (1885–1955), who worked in Switzerland, Germany, and

---

[3] This axiom was first stated explicitly by Pasch in the book *[408]*.
[4] See p. 36.
[5] Archimedes' axiom was formulated by him in the treatise *On the sphere and cylinder [24, p. 5]*, but it appears earlier (in a slightly different form) as definition 4 of book V of Euclid's *Elements [173, vol. 2, p. 120]*. Since book V is a reworked version of one of the works of Eudoxus, Archimedes' axiom is sometimes called the *Eudoxus-Archimedes axiom*. Cantor's axiom was formulated by him in the paper *[86]*.

the United States of America, contained in his book *Space-Time-Matter* (Raum-Zeit-Materie. Berlin, 1918) *[626]*. Weyl added to Peano's axioms of *n*-dimensional linear space axioms that deal with the connection between points and vectors and axioms of an inner product. His undefined terms are *vector* and *point*. Vectors are subject to the following axioms:

1. Vectors

Two vectors **a** and **b** uniquely determine a vector **a** + **b** as their sum. A number λ and a vector **a** uniquely define a vector λ**a**, which is "λ times **a**" (multiplication). These operations are subject to the following laws:—

(α) Addition—
   (1) **a** + **b** = **b** + **a** (Commutative Law).
   (2) (**a** + **b**) + **c** = **a** + (**b** + **c**) (Associative Law).
   (3) If **a** and **c** are any two vectors, then there is one and only one value of **x** for which the equation **a** + **x** = **c** holds. It is called the difference between **c** and **a** and signifies **c** − **a** (Possibility of Subtraction).
(β) Multiplication—
   (1) $(\lambda + \mu)$**a** = $(\lambda$**a**$)$ + $(\mu$**a**$)$ (First Distributive Law).
   (2) $\lambda(\mu$**a**$)$ = $(\lambda\mu)$**a** (Associative Law).
   (3) 1**a** = **a**.
   (4) λ(**a** + **b**) = $(\lambda$**a**$)$ + $(\lambda$**b**$)$ (Second Distributive Law) *[626, p. 17]*.

Weyl observes that for rational multipliers λ and μ the laws (β) follow from the axioms of addition provided that multiplication by such multipliers is defined by means of addition. Then one can use the principle of continuity to extend multiplication by rational multipliers to multiplication by arbitrary real multipliers. However, Weyl prefers to introduce separate multiplication axioms

because they cannot be derived in the general form from the axioms of addition by logical reasoning alone *[626, p. 17]*.

Weyl points out that

By refraining from reducing multiplication to additon we are enabled through these axioms to banish continuity, which is so difficult to fix precisely, from the logical structure of geometry *[626, p. 17–18]*.

After defining linear dependence and independence of vectors Weyl formulates the *Axiom of Dimensionality* (γ) in the form:

There are *n* linearly independent vectors, but every *n* + 1 are linearly dependent on one another *[626, p. 19]*.

Further Weyl formulates axioms about the connection between points and vectors:

2. Points and Vectors

1. Every pair of points A and B determines a vector $\mathbf{a}$; expressed symbolically $\overrightarrow{AB} = \mathbf{a}$. If A is any point and $\mathbf{a}$ any vector, there is one and only one point B for which $\overrightarrow{AB} = \mathbf{a}$.

2. If $\overrightarrow{AB} = \mathbf{a}$, $\overrightarrow{BC} = \mathbf{b}$, then $\overrightarrow{AC} = \mathbf{a} + \mathbf{b}$ *[626, p. 18]*.

The set of points and vectors satisfying Weyl's axioms I and II form an n-*dimensional affine space* E_n. The points B for which the vector $\overrightarrow{AB}$ is a linear combination of m linearly independent vectors form an m-dimensional plane; for $m = 1$ they form a straight line.

To obtain n-*dimensional Euclidean* space R_n Weyl adds the following axiom:

METRICAL AXIOM: *If a unit vector* $\mathbf{e}$, *differing from zero, be chosen, every two vectors* $\mathbf{x}$ *and* $\mathbf{y}$ *uniquely determine a number* $(\mathbf{x}, \mathbf{y}) = Q(\mathbf{x}, \mathbf{y})$; *the latter, being dependent on the two vectors, is a symmetrical bilinear form.* The quadratic form $(\mathbf{x}, \mathbf{x}) = Q(\mathbf{x})$ which arises from it is positive definite. $Q(\mathbf{e}) = 1$.

We shall call Q the **metrical groundform**. We then have that *an affine transformation which, in general, transforms the vector* $\mathbf{x}$ *into* $\mathbf{x}'$ *is a congruent one if it leaves the metrical groundform unchanged:—*

$$Q(\mathbf{x}') = Q(\mathbf{x}) \quad . \quad . \quad . \quad .$$

Two geometrical figures which can be transformed into one another by a congruent transformation are congruent [626, p. 28].

The fact that the inner product is a bilinear form implies that

$$(\lambda \mathbf{a}, \mathbf{b}) = \lambda(\mathbf{a}, \mathbf{b}), \qquad (\mathbf{a} + \mathbf{a}', \mathbf{b}) = (\mathbf{a}, \mathbf{b}) + (\mathbf{a}', \mathbf{b})$$

(and similarly for the second factor) and that one has the commutative law $(\mathbf{a}, \mathbf{b}) = (\mathbf{b}, \mathbf{a})$.

The form $Q(\mathbf{x}, \mathbf{y})$ can be used to define the *length* of a vector $\mathbf{a}$ as $|\mathbf{a}| = \sqrt{Q(\mathbf{a})}$, the *distance* between points A and B as the length of the vector AB, and the angle φ between vectors $\mathbf{a}$ and $\mathbf{b}$ as the number

$$\cos \varphi = \frac{(\mathbf{a}, \mathbf{b})}{|\mathbf{a}| \, |\mathbf{b}|}.$$

If in Weyl's axiomatics we require only that the form $Q(\mathbf{x})$ is a nonsingular form of index l (instead of requiring it to be positive definite), that is, if we stipulate that among n basis vectors $\mathbf{e}_i$ satisfying the orthogonality condition $(\mathbf{e}_i, \mathbf{e}_j) = 0$ for $i \neq j$ there are l vectors $\mathbf{e}_\alpha$ with $Q(\mathbf{e}_\alpha) > 0$ and $n - l$ vectors $\mathbf{e}_w$ with $Q(\mathbf{e}_w) < 0$, then we obtain an axiomatic definition of n-*dimensional pseudo-Euclidean space* $^l R_n$ *of index* l. In the previous chapter we considered the special case $n = 3$, $l = 2$, of such a space (we saw that one model of the Lobačevskian plane is a sphere of imaginary radius in this space with identified antipodal points).

Spacetime of the Special Theory of Relativity as a Pseudo-Euclidean Space

In Chapter 6 we mentioned that pseudo-Euclidean space appeared for the first time in connection with the special theory of relativity in the papers of Poincaré and Minkowski.

The special theory of relativity was founded by Einstein in the paper *On the electrodynamics of moving bodies* (Zur Electrodynamik der bewegter Körper. Leipzig, 1905) *[340, pp. 37–65]*. After analyzing the electromagnetic phenomena associated with a moving conductor and a stationary magnet as well as a stationary conductor and a moving magnet, and the famous Michelson experiment that aimed to determine the absolute motion of the earth, Einstein arrived at the following conclusion:

Examples of this sort, together with the unsuccessful attempts to discover any motion of the earth relatively to the "light medium," suggest that the phenomena of electrodynamics as well as of mechanics possess no properties corresponding to the idea of absolute rest. They suggest rather that, as has already been shown to the first order of small quantities, the same laws of electrodynamics and optics will be valid for all frames of reference for which the equations of mechanics hold good. We will raise this conjecture (the purport of which will hereafter be called the "Principle of Relativity") to be status of a postulate, and also introduce another postulate, which is only apparently irreconcilable with the former, namely, that light is always propagated in empty space with a definite velocity c which is independent of the state of motion of the emitting body *[340, pp. 37–38]*.

From this Einstein concluded that upon transition from one inertial coordinate system to another the space coordinates x, y, z and the time coordinate t transform linearly, and that in the case of motion along the Ox-axis with velocity v this transformation can be written as

$$x' = \frac{x - vt}{\sqrt{1 - \dfrac{v^2}{c^2}}}, \qquad y' = y, \qquad z' = z, \qquad t' = \frac{t - \dfrac{v}{c^2}x}{\sqrt{1 - \dfrac{v^2}{c^2}}}. \qquad (7.1)$$

(Here c stands for the velocity of light; although Einstein denoted the velocity of light by V in the previously mentioned paper he replaced V by c in the subsequent papers.)

These transformations are called *Lorentz transformations* for the Dutch physicist Hendrik Antoon Lorentz (1853–1928), who first introduced such a transformation in the paper *Electromagnetic phenomena in a system moving with any velocity smaller than that of light* (Amsterdam, 1904) *[340, pp. 1–36]*.

Almost simultaneously Poincaré advanced similar ideas in the paper *On the dynamics of the electron* (Sur la dynamique de l'electron. Palermo, 1906) *[431, vol. 9, pp. 494–550]*. Referring to the experiments of Fresnel and Michelson, Poincaré, like Einstein, arrived at the following conclusion:

It appears that the impossibility of demonstrating experimentally absolute motion of the earth is a law of nature; we essentially are led to accept this law, which we will call the relativity postulate, and to accept it unconditionally *[431, vol. 9, p. 494]*.

Although he did not penetrate as deeply as Einstein into the laws of physics, Poincaré in this paper formulated a number of important mathematical results: he showed that

the Lorentz transformations form a group *[431, vol. 9, p. 514]*

and gave it the now universally accepted name *the Lorentz group*, and he then defined what we now call pseudo-Euclidean space:

We will regard

$$x, y, z, t\sqrt{-1},$$
$$\delta x, \delta y, \delta z, \delta t\sqrt{-1},$$
$$\delta_1 x, \delta_1 y, \delta_1 z, \delta_1 t\sqrt{-1}$$

as the coordinates of three points P, P', P'' in four-dimensional space.

It is easy to see that the Lorentz transformations represent just rotations in this space about the origin regarded as fixed *[431, vol. 9, p. 542]*.

In chapter 6 we saw that, essentially, when he considered Lobačevskian geometry on a hyperboloid in the paper *On the fundamental hypotheses of geometry* (1887) *[431, vol. 11, pp. 79–91]*, Poincaré considered a pseudo-Euclidean space in which this hyperboloid played the role of a sphere of imaginary radius. Be that as it may, in *Science and hypothesis* (1906) Poincaré pointed out that in addition to Euclidean, Lobačevskian, and elliptic geometries there is a "fourth geometry," which is readily seen to be pseudo-Euclidean geometry. While mentioning "hidden axioms" Poincaré wrote:

Among these implicit axioms there is one which seems to merit some attention, because when it is abandoned a fourth geometry can be reconstructed as coherent as those of Euclid, Lobačevskiĭ and Riemann.

To prove that a perpendicular may always be erected at the point A to a straight [line] AB we consider a straight [line] coincident with the fixed straight [line] AB, and we make it turn about the point A until it comes into the prolongation of AB.

Thus two propositions are presupposed: First that such a rotation is possible, and next that it may be continued until the two straight [lines] come into the prolongation one of the other.

If the first point is admitted and the second rejected, we are led to a

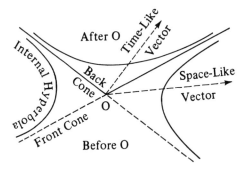

Figure 97

series of theorems even stranger that those of Lobačevskiĭ and Riemann, but equally exempt from contradiction.

I shall cite only one of these theorems and that not the most singular: *A real straight [line] may be perpendicular to itself [433, p. 62].*

In Chapter 9 we shall see that the notion of pseudo-Euclidean space, ruled out by his hypothesis IV, is already found in the work of Helmholtz (see pp. 336–337).

Another mathematician who arrived at the idea of the spacetime of the special theory of relativity was Hermann Minkowski. In his paper *Time and space* (Zeit und Raum. Leipzig, 1909) *[340, pp. 73–91]* Minkowski wrote:

The world-postulate permits identical treatment of the four co-ordinates x, y, z, t. By this means, as I shall now show, the forms in which the laws of physics are displayed gain in intelligibility. In particular the idea of acceleration acquires a clear-cut character.

I will use a geometrical manner of expression, which suggests itself at once if we tacitly disregard z in the triplex x, y, z. I take any world-point O as the zero-point of spacetime. The cone $c^2 t^2 - x^2 - y^2 - z^2 = 0$ with apex O (Figure 97) consists of two parts, one with values $t < 0$, the other with values $t > 0$. The former, the front cone of O, consists, let us say, of all the world-points which "send light to O," the latter, the back cone of O, of all the world-points which "receive light from O." The territory bounded by the front cone alone, we may call "before" O, that which is bounded by the back cone alone, "after" O. The hyperboloidal sheet already discussed

$$F = c^2 t^2 - x^2 - y^2 - z^2 = 1, \quad t > 0$$

lies after O. The territory between the cones is filled by the one-sheeted hyperboloidal figures

$$-F = x^2 + y^2 + z^2 - c^2 t^2 = k^2$$

for all constant positive values of k. We are specially interested in the hyperbolas with O as centre, lying on the latter figures. The single branches of these hyperbolas may be called briefly the internal hyperbolas with centre O. One of these branches, regarded as a world-line, would represent a motion which, for $t = -\infty$ and $t = +\infty$, rises asymptotically to the velocity of light, c.

If we now, on the analogy of vectors in space, call a directed length in the manifold of x, y, z, t a vector, we have to distinguish between the time-like vectors with directions from O to the sheet $+ F = 1, t > 0$, and the space-like vectors with directions from O to $-F = 1$. The time axis may run parallel to any vector of the former kind. Any world-point between the front and back cones of O can be arranged by means of the system of reference so as to be simultaneous with O, but also just as well so as to be earlier than O or later than O. Any world-point within the front cone of O is necessarily always before O; any world-point within the back cone of O necessarily always after O *[340, pp. 83–84]*.

Poincaré's paper *On the dynamics of the electron* appeared in a specialized mathematical journal, and for a long time physicists had no knowledge of it. This explains why the four-dimensional pseudo-Euclidean space which models the spacetime of the special theory of relativity is often referred to as *Minkowskian space* rather than, more appriopriately, *Poincaré space*.

Poincaré's Philosophy of Space

In this book we come across the work of Poincaré on three main occasions. In chapter 6 he appears as the author of two models of Lobačevskian geometry, in chapter 8 he is mentioned in connection with the part he played in the creation of topology, and in this chapter he appears in connection with his part in the evolution of the theory of relativity. Poincaré was one of the greatest mathematicians, mechanists, and astronomers of the end of the 19th and the beginning of the 20th century. In his books *Science and hypothesis* (La science et l'hypothèse. Paris, 1906) *[433, pp. 1–197]* and *The value of science* (La valeur de la science. Paris, 1913) *[433, pp. 199–355]* Poincaré also considered a number of philosophical questions pertaining to space. Poincaré's philosophical views are very close to those of Mach.

We mentioned that in the paper *On the dynamics of an electron* Poincaré arrived at a "relativity principle" very close to the relativity principle of Einstein. After this, in *Science and hypothesis* and in *The value of science*, Poincaré began to treat the relativity of space and time in a philosophical sense. In particular, in *Science and hypothesis* he wrote:

> Another frame that we impose on the world is space. Whence come the first principles of geometry? Are they imposed on us by logic? Lobačevskiĭ has proved not, by creating non-Euclidean geometry. Is

space revealed to us by our senses? Still no, for the space our senses could show us differs absolutely from that of the geometer. Is experience the source of geometry? A deeper discussion will show us it is not. We therefore conclude that the first principles of geometry are only conventions, but they are not arbitrary, and if we transferred to another world (that I call the non-Euclidean world and seek to imagine) then we should have been led to adopt others.

In mechanics we should be led to analogous conclusions and should see that the principles of this science, though more directly based on experiment, still partake of the conventional character of the geometrical postulates *[433, p. 29]*.

At the end of his paper *On the fundamental hypotheses of geometry* quoted previously Poincaré wrote:

The fundamental hypotheses of geometry are not facts based on experiment. Rather, observation of certain physical phenomena makes us choose these very hypotheses out of all possible ones.

On the other hand, the collection we choose is simply more convenient than others and we cannot claim that Euclidean geometry is true and Lobačevskian geometry is false any more than we can say that Cartesian coordinates are true and polar coordinates are false *[431, vol. 11, p. 91]*.

In *The value of science* Poincaré wrote:

Next must be examined the frames in which nature seems enclosed and which are called time and space. In "Science and Hypothesis" I have already shown how relative their value is; it is not nature which imposes them upon us; it is we who impose them upon nature because we find them convenient. But I have spoken of scarcely more than space, and particularly qualitative space, so to say, that is of the mathematical relations whose aggregate constitutes geometry. I should have shown that it is the same with time as with space and still the same with "qualitative space"; in particular I should have investigated why we attribute three dimensions to space *[433, p. 207]*.

We note that, as he explains later, by "qualitative space" Poincaré means topology. Then Poincaré defines space and time as a peculiar "language" of science:

... without this language most of the ultimate analogies of things would have remained forever unknown to us; and we should forever have been ignorant of the internal harmony of the world which is, we shall see, the only true objective reality *[433, p. 207]*.

However, by "objective reality" Poincaré means not what exists outside human cognition but, on the contrary, human sensations:

Does the harmony the human intelligence thinks it discovers in nature exist outside of this intelligence? No, beyond doubt a reality completely independent of the mind which conceives it, sees or feels it, is an impossibility. A world as exterior as that, even if it existed, would for us be forever inaccessible. But what we call objective reality is, in the last analysis, what is common to many thinking beings, and could be common to all, this common part, we shall see, can only be the harmony expressed by mathematical laws. It is this harmony then which is the sole objective reality, the only truth we can attain *[433, p. 209]*.

We see that Poincaré considers space and time as well as all laws of nature as mere symbols created by men for their convenience. This explains the sharp critique of the philosophical works of Poincaré in V. I. Lenin's *Materialism and empiriocricitism [316, p. 148]*.

We see that at the beginning of the 20th century Poincaré came close to the greatest discoveries in physics. Undoubtedly, what prevented him from making these discoveries was his philosophical bias.

The Special Theory of Relativity and Lobačevskian Geometry

The application of Lobačevskian geometry to problems of the special theory of relativity is based on the fact that, as we saw, the spacetime of special relativity is a four-dimensional pseudo-Euclidean space, and three-dimensional Lobačevskian space can be regarded as a sphere of imaginary radius in that space with identified antipodal points. As noted already by Poincaré, the Lorentz transformations of spacetime can be interpreted as rotations of pseudo-Euclidean space, so that the group of Lorentz transformations is locally isomorphic to the group of motions of three-dimensional Lobačevskian space.

The direct connection between Lobačevskian geometry and special relativity was established by the German physicist Arnold Sommerfeld (1868–1951) in the paper *On the composition of velocities in relativity theory* (Über die Zusammensetzung der Geschwindigkeiten in der Relativtheorie. Leipzig, 1909) *[548]*.

This connection can be obtained as follows. Let a material particle move in the direction Ox_1 of some coordinate system moving rectilinearly and uniformly. This motion can be described by a graph (Figure 98) in the plane $x_1 O x_4$ which can be thought of as the pseudo-Euclidean plane $^1 R_2 (x_4 = ct)$. The velocity of this particle is $v = \dfrac{dx_1}{dt} = c \dfrac{dx_1}{dx_4}$. The differential dx_1 is on a spacelike line and the differential dx_4 is on a timelike line. Therefore the length of the differential dx_1 is real and the length of the differential dx_4 is imaginary and equal to idx_4. The tangent of the angle between the tangent to our curve and the Ox_1-axis is imaginary and equal to idx_4/dx_1. This angle is imaginary

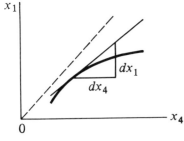

Figure 98

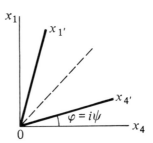

Figure 99

and we denote it by $\varphi = i\psi$. Therefore

$$\tan\varphi = \frac{\sin\varphi}{\cos\varphi} = \frac{i\sinh\psi}{\cosh\psi} = i\tanh\psi = i\frac{dx_1}{dx_4},$$

and the velocity of the particle is

$$v = c\frac{dx_1}{dx_4} = c\tanh\psi \qquad (7.2)$$

The relation (7.2) is invariant under a transition from one coordinate system to another which moves with respect to the first system with constant velocity v along the Ox_1-axis provided that the origins of the two systems coincide at $t = 0$ (Figure 99). In that case the Lorentz transformation (7.1) is of the form

$$x_1' = x_1\cosh\psi - x_4\sinh\psi, \qquad x_2' = x_2, x_3' = x_3,$$

$$x_4' = -x_1\sinh\psi + x_4\cosh\psi.$$

The formula (7.2) shows that the velocity of a particle in a coordinate system is determined by the angle between the tangent to the world line described by this particle and the time axis of the system. Therefore the velocity of the motion of a coordinate system moving rectilinearly and uniformly relative to another such system is determined by the angle $\varphi = i\psi$ between the time axes of these systems. If we imagine a hemisphere of radius i in the space $^1R_4$, then these axes define two points on this hemisphere whose spherical distance is ψ. But this hemisphere can be considered as the Lobačevskian space $^1S_3$ with curvature -1.

Let us consider two coordinate systems with the time axes represented by the points A and B of the Lobačevskian space $^1S_3$ and a moving particle with a tangent to its world line represented by the point M of that space (Figure 100). Then the velocities of the particles in these systems are connected with the distances $\psi_1 = AM$, $\psi_2 = BM$ by the relations

$$v_1 = c\tanh\psi_1, \qquad v_2 = c\tanh\psi_2 \qquad (7.3)$$

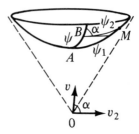

Figure 100

and the velocity of the second system relative to the first one is connected with the distance $\psi : AB$ by the relation (7.1).

Let us consider the triangle ABM in the space $^1S_3$. Then the angle between the velocities v and v_2 is equal to the angle adjacent to the angle B of this triangle. The law of cosines for the triangle ABM has the form

$$\cosh \psi_1 = \cosh \psi \cosh \psi_2 - \sinh \psi \sinh \psi_2 \cos(\pi - \alpha)$$

$$= \cosh \psi \cosh \psi_2 + \sinh \psi \sinh \psi_2 \cos \alpha.$$

In view of (7.1) and (7.2) we have

$$\frac{1}{\sqrt{1 - \dfrac{v_1^2}{c^2}}} = \frac{1 + \dfrac{v v_2}{c^2} \cos \alpha}{\sqrt{1 - \dfrac{v^2}{c^2}}\sqrt{1 - \dfrac{v_2^2}{c^2}}}.$$

This yields the following formula for addition of velocities in the special theory of relativity:

$$\frac{v_1}{c} = \frac{\sqrt{\dfrac{v^2}{c^2} + \dfrac{v_2^2}{c^2} + 2\dfrac{v v_2}{c^2} \cos \alpha - \left(\dfrac{v v_2}{c^2} \sin \alpha\right)^2}}{1 + \dfrac{v v_2}{c^2} \cos \alpha}.$$

Although Sommerfeld established the connections between the formula for the addition of velocities in the theory of relativity and the trigonometric formulas for hyperbolic functions he was not aware that these formulas are formulas of Lobačevskian geometry. This was shown by the Yugoslav geometer Vladimir Varičak (1865–1942) in the paper *On the non-Euclidean interpretation of the theory of relativity* (Über die nichteuklidische Interpetation der Relativtheorie. Leipzig, 1912) *[601]* (see also his book *Representation of the theory of relativity in three-dimensional Lobačevskian space* (Darstellung der Relativitätstheorie im dreidimensionalen Lobatschefski-jschen Raume. Zagreb, 1924) *[600]*. The geometry of Lobačevskian three-space has been applied, respectively, to problems of relativistic physics and kinematics by N. A. Černikov in *Lectures on Lobačevskian geometry and the theory of relativity* (Lekcii po geometrii Lobačevskogo i teorii otnositel'nosti. Novosibirsk, 1965) *[110]* and by Yakov Abramovič Smorodinskiĭ (b. 1917) in *Kinematics of collisions presented geometrically* (Kinematika stolknoveniĭ v geometričeskom izloženii. Yerevan, 1963) *[543]*.

Infinite-Dimensional Spaces

At the end of the 19th century mathematicians had become so used to the notion of a multidimensional space that they posed the question of defining an infinite-dimensional space. In the paper *Remarks on the geometry of a function space* (Cenno sulla geometria dello spazio funzionale. Palermo, 1896–1897) *[423, vol. 1, pp. 368–377]*, in which he generalized his recent work, the Italian mathematician Salvatore Pincherle (1853–1936) wrote as follows:

> In some of my recently published papers I introduced a notion that can be conveniently regarded in many investigations of analysis as the totality of analytic functions of one variable or, for greater definiteness, as the totality of series of positive integer powers of x—a totality, or space, of which every series is an element. Such a manifold, which obviously has an infinite number of dimensions, may be called a function space. Every power series in x is a point of this space and the coefficients of the series can be regarded as the coordinates of this point *[423, vol. 1, p. 368]* (see also *[360, p. 75]*).

Pincherle also considered linear operators on his "functional space" and, among other things, showed that in that space, the role played in finite-dimensional spaces by the discrete set of eigenvalues of a linear operator is played by a continuous set of values now known as the *spectrum* of a linear operator.

An important class of infinite-dimensional spaces that can be regarded as a direct multidimensional analogue of Euclidean space was introduced into mathematics by Hilbert and is now known as the class of *Hilbert spaces*. In a series of papers entitled *Foundations of a general theory of linear integral equations* (Grundzüge einer allgemeinen Theorie der linearen Integralgleichungen. Göttingen, 1904–1910) *[227]* Hilbert proposed that an integral equation

$$f(s) = \varphi(s) + \int_a^b K(s, t)\varphi(t)\, dt$$

be regarded as the limiting case of systems of linear equations

$$a_p = x_p + \sum_q a_{pq} x_q \qquad (7.4)$$

as the number of variables x_p tends to infinity. Hilbert generalized his own and his students' work in a monograph of the same title (Leipzig, 1912) *[228]*. In connection with the investigation of infinite systems of the form (7.4) Hilbert also considered linear forms $L(x) = \sum_p l_p x_p$ in infinitely many variables x_p such that the series $\sum_p l_p^2$ converges and called the convergence condition the condition of *boundedness* of the form $L(x)$. Hilbert regarded the coefficients l_p, as well as the variables x_p, subject to the convergence of the series $\sum_p x_p^2$, as coordinates of "infinite-dimensional vectors" for which he introduced an inner product $(u, v) = \sum_p u_p v_p$, as well as orthogonal transformations $x'_p = \sum_q o_{pq} x_q$.

Hilbert linked "infinite-dimensional vectors" to continuous functions by looking at the coordinates a_p as Fourier coefficients of a continuous function $f(s)$ with respect to functions $\Phi_1(s)$, $\Phi_2(s)$, ... that form, as he put it, a "complete orthonormal system" on an interval $a \leq s \leq b$. By the orthogonality condition Hilbert meant the conditions

$$\int_a^b \Phi_p(s)\Phi_q(s)\,ds = \begin{cases} 0 & (p \neq q), \\ 1 & (p = q), \end{cases}$$

and by the completeness condition he meant the condition

$$\sum_p \left(\int_a^b u(s)\Phi_p(s)\,ds \right)^2 = \int_a^b [u(s)]^2\,ds. \tag{7.5}$$

Hilbert wrote the Fourier coefficients of a function $f(s)$ with respect to the "orthonormal system of functions" in the form

$$a_p = \int_a^b f(s)\Phi_p(s)\,ds. \tag{7.6}$$

Thus Hilbert actually considered two models of *denumerably infinite-dimensional linear space*, now denoted by l^2 and L^2, respectively. The space l^2 consists of sequences of numbers u_p such that $\sum_p u_p^2 < \infty$ with the inner product (u, v) given above, and the space L^2 consists of functions on a real interval $[a, b]$ such that the squares of their absolute values are Lebesgue integrable on $[a, b]$, with the inner product

$$(f, g) = \int_a^b f(s)g(s)\,ds, \tag{7.7}$$

where the integral is a Lebesgue integral. Hilbert's completeness condition (7.5) means that the inner product of the function $u(s)$ with itself is equal to the sum of the squares of its "coordinates" (the so-called Parseval equality that is an infinite-dimensional generalization of Pythagoras' theorem; the French mathematician Marc Antoine Parseval (1775–1836) found this condition in 1805 for trigonometric series).

The spaces l^2 and L^2, as well the more general spaces l^p and L^p obtained from l^2 and L^2 by replacing the squares and square roots in the "vector moduli" $\sqrt{\sum_p x_p^2}$ and $\sqrt{\int_a^b [f(s)]^2\,ds}$ by p-th powers and p-th roots, were introduced by the Hungarian mathematician Frigyes Riesz (1880–1956) in *Investigations on systems of integrable functions* (Untersuchungen über Systeme integrierbarer Funktionen. Leipzig, 1910) *[456, vol. 1, pp. 441–497]*.

The *orthogonal systems of functions* that Hilbert referred to had appeared already in the 19th century. The most important system of this kind is the system of functions 1, $\cos \omega t$, $\cos 2\omega t$, $\cos 3\omega t$, ..., $\sin \omega t$, $\sin 2\omega t$, $\sin 3\omega t$, ..., that form a complete orthonormal system on the interval $[0, T]$ ($\omega = 2\pi/T$).

Representations of functions as linear combinations

$$f(t) = a_0 + \sum_p a_p \cos p\omega t + \sum_p b_p \sin p\omega t \tag{7.8}$$

of these functions were widely used already in the 18th century. Today, series of the form (7.8) are called *Fourier series* after Jean Baptiste Fourier (1768–1830), who constructed a theory of such series in his famous *Analytic theory of heat* (Théorie analytique de chaleur. Paris, 1822), and, in particular, computed the *Fourier coefficients* a_p by means of formula (7.6).

More general orthogonal systems of functions are defined by replacing the inner product (7.7) with the more general inner product

$$(f, g) = \int_a^b w(s) f(s) g(s) \, ds, \tag{7.9}$$

where $w(s)$ is a so-called weight function assumed to be nonnegative.

If $a = 0$, $b = 1$, $w(s) = s$ then the orthogonal functions are the *Bessel functions* (*cylindrical functions*); if $a = -1$, $b = 1$, and $w(s) = 1$ then the orthogonal functions are the *Legendre polynomials* (*spherical functions*) obtained by orthogonalizing the monomials 1, s, s^2, s^3, ...; if $a = -1$, $b = 1$, and $w(s) = (1 - s)^\alpha (1 - s)^\beta$ then the orthogonal functions are the *Jacobi polynomials* (*hypergeometric functions*); if $a = -\infty$, $b = \infty$, and $w(s) = e^{-s}$ then the orthogonal functions are the *Čebyšev-Hermite functions* obtained by orthogonalizing the system of functions $e^{-t^2/2}$, $te^{-t^2/2}$, $t^2 e^{-t^2/2}$, ...; and if $a = 0$, $b = \infty$, and $w(s) = s^\alpha e^{-s}$ then the orthogonal functions are the *Čebyšev-Laguerre functions*.

We see that orthogonal systems of functions were used by such eminent mathematicians of the 18th and 19th centuries as Euler, Jacobi, Hermite, Laguerre, and Pafnutiĭ L'vovič Čebyšev (1821–1894). In connection with the investigation of orthogonal systems of functions at the beginning of the 20th century we mention the works of Vladimir Andreevič Steklov (1864–1926), of which the most important is the memoir *On the theory of closedness of systems of orthogonal functions that depend on an arbitrary number of variables* (Sur la théorie de fermature des systèmes de fonctions orthogonales dépendent d'un nombre quelconque de variables. Petersburg, 1911) *[556]*; by *closedness* Steklov meant completeness of a system of functions. Using Hilbert's term, Steklov showed that the functions he considered "by the presently accepted terminology ... form an orthogonal sequence." In this paper Steklov obtained the completeness condition for orthogonality with respect to an arbitrary weight function,

$$\sum_p a_p^2 = \int_a^b w(s) f^2(s)\, ds,$$

that generalizes Hilbert's condition (7.5), as well as an analogous condition for the orthogonality of a system of functions of many variables. We note that, without realizing it, Steklov worked at the time in the area of infinite-dimensional geometry. Nevertheless, following the traditions of the Petersburg mathematical school and its preference for concrete results in mathematical physics, he rejected not only infinite-dimensional but also multidimensional geometry and shared the viewpoint of his friend and teacher Aleksandr Mihaĭlovič Lyapunov (1857–1918), who, in the article *The life and works of P. L. Čebyšev* (Žizn' i trudy P. L. Čebyševa, 1985) *[109, pp. 7–26]*, wrote:

> At a time when worshippers of the very abstract ideas of Riemann become ever more absorbed in function-theoretic investigations and pseudogeometric researches in spaces of four and more dimensions and go so far in these researches that it is impossible to see their significance with respect to any applications not only in the present but in the future, P. L. Čebyšev and his followers always stay on solid ground and are guided by the viewpoint that the only valuable researches are those that are inspired by applications (scientific or practical) *[109, pp. 19–20]*.

We note that after mentioning "pseudogeometric researches" Lyapunov remarks:

> These researches have recently been linked to the deep geometric investigations of N. I. Lobačevskiĭ with which, however, they have nothing in common. Like P. L. Čebyšev, this great geometer always remained on real ground and would hardly see in these researches of a transcendental nature the development of his ideas *[109, p. 20]*.

Steklov's works, with their important applications to problems of mathematical physics, have made it obvious that "pseudogeometric investigations" in spaces of not just "four and more" but even infinitely many dimensions turned out to be extremely useful for the solution of very concrete problems.

The requirements of quantum mechanics have resulted in extensive use of complex Hilbert spaces, in which a point is a sequence of complex numbers (coordinates) x_p such that the series $\sum_p \bar{x}_p x_p = \sum_p |x_p|^2$ converges, with the inner product

$$(u, v) = \sum_p \bar{u}_p v_p,$$

or else a function of a complex variable on an interval $[a, b]$ such that the square of its absolute value is Lebesgue integrable on $[a, b]$, with the inner product

$$(f, g) = \int_a^b \overline{f(s)} g(s) \, ds, \tag{7.10}$$

where the integral is also a Lebesgue integral. This space was first axiomatically defined by John von Neumann (1903–1957), who worked in Hungary, Germany, and the United States of America, in his *Mathematical foundation of quantum mechanics* (Mathematische Begründung der Quantenmechanik. Göttingen, 1927) *[385, vol. 1, pp. 151–207]*. This axiomatic definition, together with a finite-dimensional analogue of the space, was reproduced in von Neumann's book *Mathematical foundations of quantum mechanics* (Mathematische Grundlagen der Quantenmechanik. Berlin, 1932) *[386, pp. 36–46]*.

Von Neumann developed the theory of self-adjoint (Hermitian-symmetric) operators and, in particular, showed that just as Hermitian-symmetric matrices in finite-dimensional spaces can be represented relative to a basis of eigenvectors as sums $\sum_i \lambda_i E_{ii}$, where E_{ii} is a matrix whose element a_{ii} is 1 and the other elements are 0, so too self-adjoint operators can be expressed as Lebesgue-Stieltjes integrals $\int_{-\infty}^{\infty} \lambda \, dE_\lambda$, where λ varies over the "spectrum" of the operator, and dE_λ is the differential operator ($\int_{-\infty}^{\infty} dE_\lambda$ is the unit operator).

Other models of real and complex Hilbert spaces are furnished by spaces of, respectively, real- and complex-valued functions of several variables defined on various manifolds.

The basis for the application of complex Hilbert spaces to quantum mechanics is that the elementary particles of quantum mechanics (electrons, photons, and so on) are characterized by *wave functions* $\psi^\alpha(x, y, z, t)$ defined on a certain region D of spacetime. If one defines the inner product of two such functions φ and ψ by means of the integral

$$(\varphi, \psi) = \int_D \overline{\varphi} \psi \, dV \tag{7.11}$$

then these functions form a space that satisfies von Neumann's axioms.

To physical magnitudes there correspond self-adjoint operators A on this space. The probability that a particle, determinable by a given wave function ψ, is characterized by the values of a given physical magnitude within given bounds $[a, b]$ and is in a given region of spacetime is expressed by means of the integral over this region of the expression $(\psi, (\int_a^b \lambda \, dE_\lambda) \psi)$. In particular, to the space coordinates x^i and to the time coordinate t of a particle there correspond operators that determine multiplication of the wave functions by, respectively, x^i and t, and to the coordinates p_i of the impulse and to the energy H of the particle there correspond the operators $\dfrac{h}{2\pi i} \dfrac{\partial}{\partial x^i}$ and $\dfrac{ih}{2\pi} \dfrac{\partial}{\partial t}$, where h is the so-called Planck constant.

In the space of wave functions there also act unitary operators U that determine unitary linear representations of noncompact Lie groups and, in particular, the Lorentz group. These representations turn out to be very useful for the solutions of many problems of quantum physics.

Infinite-Dimensional Analogs of Pseudo-Euclidean and Non-Euclidean Spaces

The problems of quantum mechanics not only provided an impulse for the development of Hilbert space, which is an infinite-dimensional analogue of Euclidean space, but also for the development of an infinite-dimensional analog of pseudo-Euclidean space. In the paper *The physical interpretation of quantum mechanics* (London, 1942) *[153]* Paul Adrian Dirac (1902–1984) posed the problem of modifying the formalism of quantum mechanics so as to allow for probabilities of the appearance of particles with positive as well as negative energies (that could be viewed as probabilities of their emission as well as absorption). In the paper *On Dirac's new method of field quantization* (London, 1943) *[409]*, Wolfgang Pauli (1900–1958) analyzed Dirac's method and wrote that "in Dirac's formalism of field quantization one generalizes the usual metric in the Hilbert space of states of a system" and that the resulting modification amounts to replacing the integral (7.11) with the integral

$$\int_D \bar{\varphi}\eta\psi \, dV, \tag{7.12}$$

where the operator η can be reduced "to a normal form that is diagonal, where, however, each diagonal element is 1 or -1" by a suitable coordinate transformation. Pauli points out that the usual theory is obtained in the case when η is the identity operator and notes that

> We obtain something essentially new if we use indefinite bilinear forms to define the length of vectors in Hilbert space. They have as a consequence that operators with exclusively positive eigenvalues can have negative values of mathematical expectations. One can also express this by saying that one introduces negative probabilities of the realization of certain positive eigenvalues *[409, p. 177]*.

It is possible to define an inner product analogous to the inner product (7.12) by means of the equality (7.9) by admitting weight functions $w(s)$ that take on both positive and negative values.

Shortly thereafter, infinite-dimensional analogs of pseudo-Euclidean spaces were considered by Soviet mathematician Lev Semenovič Pontryagin (b. 1908) in the paper *Hermitian operators in a space with indefinite metric* (Érmitovy operatory s indefinitnoĭ metrikoĭ. Moscow, 1944) *[438]*. Such spaces were also studied in a number of papers by Mark Grigor'evič Kreĭn (b. 1907) and his students. One of these papers is *Twist lines in Lobačevskian space of an infinite number of dimensions and Lorentz transformations* (Vintovye linii v prostranstve Lobačevskogo beskonečnogo čisla izmereniĭ i lorencevy preobrazovaniya. Moscow, 1948) *[294]*. In this paper Kreĭn investigates an infinite dimensional analog of pseudo-Euclidean space of index *l*; the term *Lobačevskian space of an infinite number of dimensions* refers to one sheet of a sphere in this space. So far, Kreĭn's paper is the only piece of research dealing with infinite-dimensional non-Euclidean geometry, although it is not difficult to define infinite-dimensional analogues of elliptic space and of hyperbolic spaces of arbitrary index.

Another branch of infinite-dimensional geometry was introduced by Abram Mironovič Lopšic (1891–1984) in *Certain problems of tensor algebra and linear dimensionless spaces* (Nekotorye zadači tenzornoĭ algebry v lineĭnyh bezrazmernyh prostranstveh. Moscow, 1948) *[338]* and Detlef Laugwitz (b. 1932) in *Differential geometry without the axiom of dimension* (Differentialgeometrie ohne Dimensionaxiom. Berlin, 1954) *[307]*; Lopšic's *dimensionless space* is a linear space without the dimension axiom. This incomplete system of axioms is satisfied not only by *n*-dimensional space but also by its infinite-dimensional analogue. In the work *Certain questions of projective, affine and descriptive geometry in dimensionless space* (Nekotorye voprosy proektivnoĭ, affinnoĭ i načertatel'noy geometrii v bezrazmernom prostranstve. Moscow, 1956) *[339]* Lopšic studied analogous spaces obtained by not including the dimension axiom in the axioms for projective and affine spaces. The resulting incomplete axiom systems are satisfied by finite-dimensional affine and projective spaces as well as by their infinite-dimensional analogs. Undoubtedly, unification of the approaches of Kreĭn and Lopšic will lead to the study of a large class of infinite-dimensional analogs of non-Euclidean spaces.

We note recent works in this area by Tat'yana Borisovna Tapero (b. 1942): *Metric invariants of pairs of planes in infinite-dimensional space* (Metričeskie invarianty par ploskosteĭ v beskonečnomernom prostranstve. Leningrad, 1978) *[576]*, *Common perpendiculars of pairs of planes in infinite-dimensional space* (Obščie perpendikulary ploskosteĭ v beskonečnomernom prostranstve. Ul'yanovsk, 1979) *[577]*, and her survey of the history of infinite-dimensional spaces *[575]*, and the book of Viktor Egorovič Fomin *Differential geometry of Banach manifolds* (Differencial'naya geometriya banahovyh mnogoobraziĭ. Kazan, 1983) *[183]*.

Chapter 8
The Curvature of Space

Curvature and Intrinsic Geometry of a Surface in the Works of Euler

By the *curvature* of a curve at a point we mean the limit of the ratio of the angle $\Delta\alpha$ between the tangents at the endpoints of an arc to the length Δs of that arc as the latter contracts to the point; that is, the limit

$$k = \frac{d\alpha}{ds} = \lim_{\Delta s \to 0} \frac{\Delta\alpha}{\Delta s}. \tag{8.1}$$

k is also the reciprocal of the *radius of curvature* at the point in question, that is, the radius of the osculating circle at that point. (The osculating circle at a point P of a curve is defined as the limit of circles determined by three points on the curve as they tend to P.) The concepts of the curvature of a curve and of the osculating (literally "kissing," from the Latin *osculans*) circle were already known to Leibniz.[1] Leibniz also suggested the possibility of characterizing the curvature of a surface by means of an osculating sphere.[2]

Euler characterized the curvature of a surface in his *Investigations of the curvature of surfaces* (Recherches sur la courbure des surfaces. Berlin, 1767) *[176, vol. 28, pp. 1–22]*. Euler called a normal section of a surface $z = f(x, y)$ perpendicular to the xOy plane a *principal section*. By varying the angle φ between a normal section and the principal section Euler found that at each point of the surface there is a maximal radius of curvature f and a minimal radius of curvature g, that their planes are perpendicular to one another, and

---

[1] Leibniz first considered osculation of curves—in particular, osculation of curves and circles—in his *New reflections on the nature of the angles of tangency and osculation and their use in mathematical practice to replace complex figures with simpler ones* (Meditatio nova de natura Anguli contactus et osculi, horumque usu in practica Mathesi ad figuras faciliores succedaneas difficilioribus substituendas. Leipzig 1686) *[312, v. 7, pp. 326–329]*.

[2] In a letter to Johann Bernoulli, dated July 29, 1698.

Figure 101

that the connection between an arbitrary radius of curvature $r = r(\varphi)$ and the radii of curvature f and g is given by

$$r = \frac{2fg}{f + g - (f - g)\cos 2\varphi}.$$

Euler writes:

For a simple derivation of this formula we join the maximal and minimal osculating radii by putting $Of = f$ and $Og = g$ and describe on fg a semi-ellipse with one focus at O (Figure 101). Then for the section MN one must take the angle fOr, doubled with respect to the angle EZM, and the line Oz will be equal to the osculating radius for the section MN. Thus drawing conclusions about the curvature of surfaces, at first glance a matter of great complexity, reduces for each element to the determination of two osculating radii one of which is maximal and the other minimal for that element; these two things entirely determine the nature of the curvature, and we obtain the curvature of all possible sections perpendicular to the given element [176, vol. 28, p. 21].

In 1837 Charles Dupin (1784–1873) [157, p. 109] modified Euler's formula to read

$$\frac{1}{r} = \frac{1}{f}\cos^2 \varphi + \frac{1}{g}\sin^2 \varphi.$$

In this paper *On solids whose surfaces can be developed onto a plane* (De solidis quarum superficiem in planum explicare licet. Petersburg, 1772) [176, vol. 28, pp. 161–186] Euler introduced the concept of a *developable surface*, that is, a surface that can be applied to a plane without folding or tearing, and proved the fundamental theorem that such a surface is a cylinder, or a cone, or a surface formed by the tangents to a space curve. His point of departure was that on such a surface an infinitesimal triangle must be congruent to the corresponding triangle in the plane to which the surface is applied. Further, he introduced on the surface curvilinear coordinates t, u equal to the rectangular coordinates of the corresponding points of the plane and, denoting the partial derivatives $\partial x/\partial t$, $\partial y/\partial t$, $\partial z/\partial t$ by l, m, n, and $\partial x/\partial u$, $\partial y/\partial u$, $\partial z/\partial u$ by λ, μ, ν, set down the developability conditions in the form $l^2 + m^2 + n^2 = 1, \lambda^2 + \mu^2 + \nu^2 = 1, l\lambda + m\mu + n\nu = 0$, that is, as the requirements of unicity and perpendicularity of the partial derivatives with respect to t and u of the radius vector of a point of the surface. An equivalent form of this condition is that the square of an arc element of a curve on a developable surface equals the sum of the squares of the differentials of the coordinates,

$$ds^2 = dt^2 + du^2,$$

that is, the arc element of a developable surface coincides with the arc element of a plane.

In a note published only in 1862 *[176, vol. 29, pp. 437–440]* Euler also established the general conditions of applicability of a surface to other surfaces. These results of Euler pertain to the so-called *intrinsic geometry of a surface*, that is, the study of properties of a surface that are unchanged when the surface is bent.

Euler's work on *geodesics* also pertains to the intrinsic geometry of surfaces. A *geodesic* is the shortest of all curves on a surface joining two of its points (in the plane the geodesics are straight lines and on a sphere they are great circles).[3]

The problem of finding geodesics on a surface was first posed in 1697 by Johann Bernoulli (1667–1748). Shortly after that, Johann Bernoulli wrote a letter to l'Hospital to the effect that he had found the general differential equation of geodesics. This result was published only in 1742 *[50, p. 364]*. Jacob Bernoulli (1654–1705) showed in 1698 that when cylindrical and conical surfaces are applied to planes, their geodesics go over into straight lines.

Euler was the first to publish the differential equation of geodesics. He did this in the paper *On the shortest curve on an arbitrary surface joining two arbitrary points* (De linea brevissima in superficie quacumque duo quaelibet puncta jungente. Petersburg, 1732) *[176, v. 25, pp. 1–12]*. Euler's equation was

$$\frac{Qddx + Pddy}{Qdx + Pdx} = \frac{dxddx + dyddy}{dt^2 + dx^2 + dy^2},$$

where t, x, y are rectangular coordinates in three-dimensional space, and the functions P and Q are determined from the differential equation of the surface $Pdx = Qdy + Rdt$. Euler returned to the problem of geodesics many times. In the second volume of his *Mechanics* (Mechanica. Petersburg) *[177, v. 2, p. 426]* he showed that, "in the absence of forces," the path of a point on a surface is a geodesic (an assumption used earlier by Bernoulli). In this connection, Euler showed that the *principal normal of a geodesic* (a straight line in the osculating plane of that curve passing through the point of contact and perpendicular to the tangent) always coincides with the normal to the surface.

---

[3] Many of Euler's papers deal with the determination of geodesics. The first was published in 1728 *[176, vol. 25, pp. 1–12]*, and the last was published posthumously *[176, vol. 25, pp. 269–279]*. Euler called geodesics *shortest curves*. The term *geodesic* first appeared in Laplace's *Celestial Mechanics* (Paris, 1799). Later it was applied to all quadrics. After the appearance of Liouville's paper (1814) it was applied to all surfaces.

We note that in the cartography paper *On the representation of the surface of a sphere in the plane* Euler compared the arc length elements of the plane and the sphere and proved the impossibility of an isometric mapping of a sphere to a plane. Also, he posed the question about the three mappings of the sphere to the plane that are of greatest importance in cartography, namely, the mapping in which the meridians and parallels are mapped onto an orthogonal system of lines, the mapping that preserves angles between curves (conformal mapping), and the mapping that preserves areas (equiareal mapping).

Independently of Euler, Monge considered developable surfaces in his *Memoir on evolutes, radii of curvature and various kinds of inflection of curves with twofold curvature* (Mémoire sur les développées, des rayons de courbure et les differents genres d'inflexion des courbes à double courbure. Paris, 1785) *[374]*. We note that Monge's book *Applications of analysis to geometry* (Applications de l'analyse à la géométrie. Paris, 1807) *[373]* contained the first systematic exposition of the theory of surfaces.

Intrinsic Geometry of Surfaces in the Work of Gauss

The general theory of the intrinsic geometry of a surface was formulated by Carl Friedrich Gauss in his *General investigations of curved surfaces* (Disquisitiones generales circa superficies curvas. Göttingen, 1828) *[196, vol. 4, pp. 217–258]*. Gauss's paper was the result of the assignment, given to him in 1820, to produce a cartographic survey of the kingdom of Hanover. This assignment made him reflect on problems of geodesy. In turn, this brought him to the theory of surfaces.

Gauss's theory unified the surface theories of Euler and Monge and the theory of quadratic forms developed by him in his *Arithmetical investigations* (Disquisitiones arithmeticae. Göttingen, 1801) *[196, vol. 1]*. In *Investigations of curved surfaces* Gauss introduced curvilinear coordinates p, q of points of the surface and defined the functions

$$a = \frac{dx}{dp}, \qquad b = \frac{dy}{dp}, \qquad c = \frac{dz}{dp}, \qquad a' = \frac{dx}{dq}, \qquad b' = \frac{dy}{dq},$$

$$c' = \frac{dz}{dq}; \qquad bc' - cb' = A, \qquad ca' - ac' = B', \qquad ab' - ba' = C;$$

$$\frac{ddx}{dp^2} = \alpha, \qquad \frac{ddx}{dpdq} = \beta, \qquad \frac{ddx}{dq^2} = \gamma, \qquad \frac{ddy}{dp^2} = \alpha',$$

$$\frac{ddy}{dpdq} = \beta', \qquad \frac{ddy}{dq^2} = \gamma', \qquad \frac{ddz}{dp^2} = \alpha'', \qquad \frac{ddz}{dpdq} = \beta'', \qquad \frac{ddz}{dq^2} = \gamma''.$$

In modern terms, the vectors $\{a, b, c\}$ and $\{a', b', c'\}$ are the partial derivatives of the radius vector of a point on the surface with respect to p and q

(Gauss denoted ordinary and partial derivatives the same way); the vector $\{A, B, C\}$ is the cross product of $\{a, b, c\}$ and $\{a', b', c'\}$ and its direction is that of the normal to the surface; and $\{\alpha, \beta, \gamma\}$, $\{\alpha', \beta', \gamma'\}$, and $\{\alpha'', \beta'', \gamma''\}$ are the second partial derivatives of the radius vector of a point on the surface with respect to p and q.

Further, Gauss introduced the quadratic forms

$$ds^2 = Edp^2 + 2Fdpdq + Gdq^2$$

and

$$Ddp^2 + D'dpdq + D''dq^2,$$

where

$$aa + bb + cc = E, \quad aa' + bb' + cc' = F, \quad a'a' + b'b' + c'c' = G,$$

$$A\alpha + B\beta + C\gamma = D, \quad A\alpha' + B\beta' + C\gamma' = D', \quad A\alpha'' + B\beta'' + C\gamma'' = D''.$$

The first of these forms, the so-called first fundamental form, gives the square of an arc length element of the surface, that is, of an element of length of an arc of a curve on the surface, and the second differs from the modern second fundamental form only by the multiplier $AA + BB + CC = EG - FF$.

Gauss introduced the all-important concept of "measure of curvature" now known as the *Gaussian curvature* of a surface. Making use of the spherical image of the surface, that is, the mapping that associates to a point P on the surface, a point P' on the unit sphere, such that the radius OP' is parallel to the normal to the surface at P (in modern terms, the end of the unit normal to the surface with beginning at a fixed point), Gauss defined the concepts of total curvature and "measure of curvature" as follows:

We shall say that a part of a curved surface bounded by a given contour has *total curvature* given by the area of the corresponding figure on the surface of the sphere. One must draw a clear distinction between the total curvature and a kind of specific curvature that we shall call the *measure of curvature*. The latter pertains to a point on the surface and is the fraction obtained by dividing the total curvature of an element of the surface adjacent to the point by the area of that element, and thus gives the ratio of corresponding infinitesimal areas on the sphere and on the curved surface. We hope that our subsequent exposition will fully explain the utility of these new concepts" *[196, vol. 4, p. 226].*

Gauss goes on to show that

the "measure of curvature" at a point of the surface is equal to a fraction whose numerator is 1 and whose denominator is the product of the two principal curvatures of the normal sections.

It is clear that the measure of curvature is positive for convex-convex or concave-concave surfaces (this is a trivial distinction) and negative for convex-concave ones. If the surface is made up of parts of both

kinds, then on their boundary the measure of curvature must vanish
[196, vol. 4, pp. 231].

Then Gauss obtains further expressions for the "measure of curvature" of
which the most important are the expression of the "measure of curvature"
k in terms of the coefficients of his two forms,

$$k = \frac{DD'' - D'D'}{EG - FF},$$

and the expression in terms of the coefficients of the first form alone and their
derivatives with respect to p and q:

$$4(EG - FF)^2 k = E\left(\frac{dE}{dq}\frac{dG}{dq} - 2\frac{dF}{dp}\frac{dG}{dq} + \left(\frac{dG}{dp}\right)^2\right) +$$

$$+ F\left(\frac{dE}{dp}\frac{dG}{dq} - \frac{dE}{dq}\frac{dG}{dp} - 2\frac{dE}{dq}\frac{dF}{dq} + \frac{dF}{dp}\frac{dF}{dq} -\right.$$

$$\left. - 2\frac{dF}{dp}\frac{dG}{dp}\right) + G\left(\frac{dE}{dp}\frac{dG}{dp} - \frac{dE}{dp}\frac{dF}{dq} + \left(\frac{dE}{dq}\right)^2\right) -$$

$$- 2(EG - FF)\left(\frac{ddE}{dq^2} - 2\frac{ddF}{dpdq} + \frac{ddG}{dp^2}\right).$$

In connection with the last formula Gauss makes the following comment:

The formula in the last section leads, of itself, to the following remarkable
  Theorem. If a curved surface is applied to any other surface then the
measure of curvature at each of its points remains unchanged.
  Also, it is clear that every finite part of a curved surface will, after
application to another surface, retain its total curvature.
  The special case to which geometers have until now limited their investigations is the case of surfaces applicable to a plane. Our theory
readily shows that the measure of curvature of such surfaces at any point
is zero. . . .
  What we have set forth in the previous section involves a special
approach to the investigation of surfaces worthy of the close attention
of geometers. Namely, when a surface is regarded not as the boundary
of a solid but as a solid with one vanishing dimension, flexible but not
stretchable, then the properties of the surface are partly dependent on
the form to which it has been reduced and in which it is being studied,
and partly independent [of it in the sense that] they remain invariant
regardless of its form under bending. The latter properties, whose investigation opens a new and fruitful area of geometry, include the measure
of curvature and total curvature as defined by us. They also include the
study of geodesic curves, as well as many other matters to be discussed
in the sequel. When this approach is adopted, then the plane, and surfaces applicable to a plane, such as cylindrical and conical surfaces,

are regarded as essentially the same, and the general method of characterizing a surface studied in this manner is based on the formula $\sqrt{Edp^2 + 2Fdpdq + Gdq^2}$ that expresses the connection of the line element of the surface with two variables p and q [196, vol. 4, pp. 237–238].

Gauss is defining the *intrinsic geometry* of a surface invariant under bending and application to another surface. Some of the properties studied in intrinsic geometry are the lengths of curves on the surface, the angles between curves, and what Gauss calls *geodesic curves*, that is, geodesics. Gauss's theorem, called by him a *remarkable theorem* (Theorema Egregium), asserts that the *measure of curvature* is a property that belongs to intrinsic geometry.

Further, Gauss considers the total curvature of a geodesic triangle (that is, a triangle whose sides are arcs of geodesics). The total curvature of such a triangle is given by $\int kd\sigma$, where $d\sigma$ is an element of the surface of the triangle. He finds that

the total curvature of a triangle is equal to the part of the spherical surface that corresponds to the triangle taken with a plus or minus sign according as the surface on which the triangle lies is concave-concave or concave-convex; as the unit of area we take a square with side one (the radius of the sphere); then the area of the sphere is 4π. It follows that the [surface of the] part of the sphere corresponding to the triangle is to the surface of the sphere as $\pm (A + B + C - \pi)$ to 4π. This theorem, that undoubtedly belongs with the most beautiful theorems of the theory of surfaces, can be stated as follows:

The excess above 180° of the sum of the angles of a triangle formed by geodesic curves on a curved concave-concave surface and the defect below 180° of the sum of the angles of a triangle formed by geodesic curves on a concave-convex surface are each measured by the area of the part of the spherical surface that corresponds to the given triangle via normal directions, provided that the whole surface (of the sphere) is taken as 720 degrees [196, vol. 4, pp. 245–246].

This theorem of Gauss implies that the Gaussian curvature at a point is equal to the limit of the ratio of the angular excess $A + B + C - \pi$ of a geodesic triangle ABC on the surface to its area as the triangle shrinks to the given point.

Minding's Theory of Surfaces of Constant Curvature

Gauss's work on the intrinsic geometry of surfaces was continued by Ferdinand Minding (1806–1885), who worked in Dorpat (now Tartu). In his *Remark on the development of curved lines on surfaces* (Bemerkung über die Abwickelung krummer Linien auf Flächen. Berlin, 1830) [366], Minding introduced the fundamental concept of *geodesic curvature* of a curve on a surface, defined as the limit of the ratio of the angle between tangent geodesics at the endpoints of an arc of the curve to the arc length as the arc shrinks to a point, and showed that this limit is invariant under bending. In the paper *How to decide whether two surfaces are mutually applicable; including remarks on surfaces of constant measure of curvature* (Wie sich entscheiden lässt ob zwei gegebene krumme Flächen auf einander abwickelbar sind, nebst Bemerkungen über die Flächen von unveränderlicher Krümmungsweise. Berlin, 1839) *[367]* Minding expressed the line element of a surface with constant "measure of curvature" as

$$ds^2 = dp^2 + \left(\frac{1}{\sqrt{k}} \sin p\sqrt{k} \right)^2 dq^2$$

and concluded that

two surfaces of equal constant measure of curvature can be applied to each other in infinitely many ways, since any two points of one can be associated to two arbitrary points of the other provided that the lengths of the shortest curves between the points of each pair are equal. This implies the following consequences:

Every surface whose measure of curvature is zero [results from] the bending of a plane—this much is well known.

Every surface whose measure of curvature (k) is constant and positive can be applied to a sphere of radius $1/\sqrt{k}$ *[367, pp. 375–376]*.

Further, Minding discovered helical surfaces and, in particular, surfaces of constant negative curvature. One such surface is given by the parametric equations

$$r = \sqrt{x^2 + y^2} = \frac{1}{\cosh \varphi}, \qquad z = \varphi - \tanh \varphi.$$

Minding comments that

This surface is the result of rotating a curve like ... *CBE* ... (Figure 102) about the zz-axis which it approaches asymptotically *[367, p. 380]*.

The curve *CBE*, which is characterized by the constancy of the length of the segment of the tangent between the point of tangency and the zz-axis, is called a *tractrix*, and the surface of revolution referred to by Minding is called a *pseudosphere*.

In *Contributions to the theory of shortest curves on curved surfaces* (Beiträge zur Theorie der kürzesten Linien auf krummen Flächen. Berlin, 1840) *[368]* Minding established the trigonometric relations in geodesic triangles on such surfaces and noted that these formulas could be obtained from the corresponding formulas of spherical geometry on a sphere of radius r by multiplying r by the complex number i.

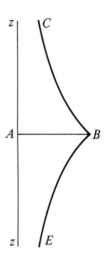

Figure 102

Interpretation of the Lobačevskian Plane on a Pseudosphere

In spite of the fact that Minding's paper appeared in the same journal (*Crelle's*) as Lobačevskiĭ's *Imaginary geometry*, neither of them noticed that the trigonometric formulas of the hyperbolic plane coincide with the trigonometric formulas of a surface of constant negative curvature.[4]

The person who *did* notice the sameness of these formulas was E. Beltrami, who considered the matter in *An attempt at an interpretation of non-Euclidean geometry [42, pp. 374–405]*. Beltrami constructed a model of the hyperbolic plane in a circle and associated to the points of the hyperbolic plane of curvature $-1/R^2$ the coordinates u, v of the corresponding points of the circle. He found that in this coordinate system the line element is given by

$$ds^2 = R^2 \frac{(a^2 - u^2)du^2 + 2uvdudv + (a^2 - v^2)dv^2}{a^2 - u^2 - v^2}.$$

Beltrami computed the total curvature of a surface with this line element and noticed that the Gaussian curvature of the hyperbolic plane is everywhere equal to the same number $-1/R^2$; that is, the hyperbolic plane can be viewed as a surface of constant negative curvature.

Thus, speaking of surfaces of constant negative curvature, Beltrami concluded that

---

[4] Lobačevskiĭ's paper appeared in vol. 17 (1837) of *Crelle's Journal* and Minding's paper in vol. 20 (1840) of that journal. B. L. Laptev *[265, p. 20]* explains that Lobačevskiĭ did not borrow the latter volume from the library of Kazan University.

the theorems of non-Euclidean planimetry apply to them. Further, most of these theorems can be thought of concretely only if we refer them to such surfaces and not to the plane *[42, p. 381]*.

That is why Beltrami suggested that all surfaces of constant negative curvature be called *pseudospherical* surfaces.

Beltrami also showed that each of the sheets of Minding's pseudosphere is isometric to the part of the hyperbolic plane enclosed between two parallel lines and a horocycle perpendicular to them. He also found parts of the hyperbolic plane isometric to other surfaces of revolution of constant negative curvature determined by Minding.

Riemannian Geometry

In the paper *On the hypotheses which lie at the foundations of geometry [122, pp. 55–71]* Riemann introduced a notion of an n-*ply extended manifold* broader than either of the similar notions introduced by Grassmann and Schläfli. After defining this manifold Riemann posed the question of the "measure-relations of which such a manifold is capable" and of the possibility "to express geometrically the calculated results." After noting that the foundations of these two parts of the question are established in Gauss's celebrated memoir *Disquisitiones generales circa superficies curvas [122, pp. 59–60]*, Riemann writes:

Position-fixing being reduced to quantity-fixings, and the position of a point in the *n*-ply extended manifold being consequently expressed by means of *n* variables $x_1, x_2, x_3, \ldots x_n$, the determination of a line comes to the giving of these quantities as functions of one variable. The problem consists then in establishing a mathematical expression for the length of a line, and to this end we must consider the quantities *x* as expressible in terms of certain units. I shall treat this problem only under certain restrictions, and I shall confine myself in the first place to lines in which the ratios of the increments *dx* of the respective variables vary continuously. We may then conceive these lines broken up into elements, within which the ratios of the quantities *dx* may be regarded as constant; and the problem is then reduced to establishing for each point a general expression for the linear element *ds* starting from that point, an expression which will thus contain the quantities *x* and the quantities *dx*. I shall suppose, secondly, that the length of the linear element, to the first order, is unaltered when all the points of this element undergo the same infinitesimal displacement, which implies at the same time that if all the quantities *dx* are increased in the same ratio, the linear element will vary also in the same ratio. On these suppositions, the linear element may be any homogeneous function of the first degree of the quantities *dx*, which is unchanged when we change the signs of all the *dx*, and in which the arbitrary constants are continuous functions of the quantities *x*. To find the simplest cases, I shall seek first an expression for manifoldness of *n* − 1 dimensions which are everywhere equidistant from the origin of the linear element; that is, I shall seek a continuous function of position whose values distinguish them from one another. In going outwords from the origin, this must either increase in all directions or decrease in all directions; I assume that it increases in all directions, and therefore has a minimum at that point. If, then, the first and second differential coefficients of this function are finite, its first differential must vanish, and the second differential cannot become negative; I assume that it is always positive. This differential expression, then, of the second order remains constant when *ds* remains constant, and increases in the duplicate ratio when the *dx*, and therefore also *ds*, increase in the same ratio; it must therefore be ds^2 multiplied by a constant, and consequently *ds* is the square root of an always positive integral homogeneous function of the second order of the quantities *dx*, in which the coefficients are continuous functions of the quantities *x*. For Space, when the position of points is expressed by rectilinear co-ordinates,

$$ds = \sqrt{\sum (dx)^2} \; [122, \, pp. \, 60–61].$$

(In Clifford's translation we changed his term n-*dimensional manifoldedness* to Riemann's *n*-ply extended manifold). By *Space* Riemann means three-dimensional Euclidean space. After considering various possible cases of the dependence of *ds* on the differentials *dx*, Riemann restricts himself to "manifolds for which the line element is given by the square root of a differential expression of the second degree," that is,

$$ds^2 = \sum_i \sum_j g_{ij} dx_i dx_j, \tag{8.2}$$

where $g_{ij} = g_{ji}$ are functions of the variables x_i.

Riemann's requirement guarantees that his *n*-dimensional space is locally Euclidean. The quadratic form (8.2) is assumed to be positive definite, that is, for all dx_i not simultaneously zero $ds^2 > 0$. Hence each pair of points with infinitesimally different coordinates x_i and $x_i + dx_i$ are a definite distance *ds* apart. By integrating this distance along different lines we determine their length. Among the different lines we can find the shortest (geodesic) lines $x_i = x_i(t)$ that are solutions of the differential equations

$$\frac{d^2 x_i}{ds^2} + \sum_j \sum_k \Gamma^i_{jk} \frac{dx_j}{ds} \frac{dx_k}{ds} = 0, \tag{8.3}$$

where the Γ^i_{jk} are point functions expressed in terms of the coefficients g_{ij} by means of the relations

$$\sum_i g_{il} \Gamma^l_{jk} = \frac{1}{2} \left(\frac{\partial g_{lk}}{\partial x_j} + \frac{\partial g_{il}}{\partial x_k} - \frac{\partial g_{jk}}{\partial x_l} \right). \tag{8.4}$$

The spaces defined in this way are called *Riemannian spaces*. Surfaces in ordinary space are particular Riemannian spaces for $n = 2$; in this case, an instance of a geometry of the kind defined by Riemann is the so-called intrinsic geometry of a surface, elaborated by Gauss in the work *General investigations of curved surfaces* alluded to by Riemann. Clearly, Euclidean space is a special case of a Riemannian space whose geodesics are Euclidean straight lines.

A more general example of a Riemannian space is a hypersurface in $(n + 1)$-dimensional Euclidean space and, in particular, a hypersphere in this space. The geodesics of a hypersphere are its great circles—sections by two-dimensional planes passing through its center. Elliptic space, obtained from a sphere by identifying antipodal points, and Lobačevskian space are also examples of Riemannian spaces.

One of the most important concepts of Riemannian geometry is the *curvature of a space*. The curvature is defined at every point in every two-dimensional direction passing through this point. To determine the curvature of a Riemannian space at a point one considers a geodesic triangle, that is, a curvilinear triangle bounded by arcs of three geodesics, one of whose vertices is at the given point and two of whose sides issuing from this vertex are tangent to the given two-dimensional direction. For this triangle ABC one computes the angular excess $A + B + C - \pi$ (as defined, the angular excess can be positive, zero, or negative), as well as the area, and forms the ratio of the angular excess over the area of the triangle. Now one lets the triangle shrink to the given vertex but insists that the sides issuing from it continue to be tangent to the given two-dimensional direction. It is obvious that in this limiting process the area of the geodesic triangle tends to zero. Since the geometry of the two-dimensional surface in which the passage to the limit is taking place is locally Euclidean and the angular excess of a Euclidean plane is zero, it follows that the angular excess of the geodesic triangle also tends to zero. But the ratio of the angular excess over the area of the triangle tends to a definite limit called the *curvature of the Riemannian space* at the given point in the given two-dimensional direction.

The curvature of n-dimensional Euclidean space, regarded as a special case of a Riemannian space, is equal to zero at all of its points; that is why a Euclidean space is called Riemannian space of *zero curvature* and Riemannian spaces different from Euclidean space are called *curved spaces*.

Riemann defined the curvature of his space in the following manner; in a neighborhood of the point at which he wanted to define curvature he introduced coordinates x_i equal to zero at the point and such that distance s of points in the neighborhood of this point was given in terms of the coordinates by the relation $s^2 = \sum_i x_i^2$. Riemann writes:

When we introduce these quantities, the square of the line-element is Σdx^2 for infinitesimal values of the x, but the term of next order in it is equal to a homogeneous function of the second order of the $\frac{1}{2}n(n - 1)$ quantities $(x_1 dx_2 - x_2 dx_1)$, $(x_1 dx_3 - x_3 dx_1)$ ... an infinitesimal, therefore, of the fourth order; so that we obtain a finite quantity on dividing this by the square of the infinitesimal triangle, whose vertices are $(0,0,0,\ldots)$, $(x_1,x_2,x_3,\ldots)$, $(dx_1,dx_2,dx_3,\ldots)$. This quantity retains the same value so long as the x and the dx are included in the same binary linear form, or so long as the two geodesics from 0 to x and from 0 to dx remain in the same surface-element; it depends therefore only on place and direction. It is obviously zero when the manifold represented is flat, i.e., when the squared line-element is reducible to $\Sigma\, dr^2$, and may therefore be regarded as the measure of the deviation of the manifold from flatness at the given point in the given surface-direction. Multiplied by $-\frac{3}{4}$ it becomes equal to the quantity which Privy Councillor Gauss has called the total curvature of a surface *[122, pp. 62–63]*.

In view of Riemann's choice of coordinates, the expansion of the square of the line element ds^2 in the neighborhood of the point he considered is given by

$$ds^2 = \sum_i dx_i^2 + \sum_i \sum_j \sum_k c_{ij,k} x_k dx_i dx_j + \sum_i \sum_j \sum_k \sum_l c_{ij,kl} x_k x_l dx_i dx_j + \ldots,$$

where $c_{ij,k}$ and $2c_{ij,kl}$ are the values of the derivatives $\dfrac{\partial g_{ij}}{\partial x_k}$ and $\dfrac{\partial^2 g_{ij}}{\partial x_k \partial x_l}$ at the point under consideration. In the first place, Riemann notes that there is no linear term in this expansion, that is, that $c_{ij,k} = 0$. The reason for this is that his coordinate lines are geodesics satisfying equation (8.3), where the coefficients Γ^i_{jk} are expressed in terms of g_{ij} by formulas (8.4). Then Riemann claims that the second-order terms in the expansion of ds^2 form a quadratic form in the $x_i dx_j - x_j dx_i$. If for the sake of uniformity we denote the infinitesimals x_i by δx_i then the latter magnitudes can be written as $\Delta x_{ij} = \delta x_i dx_j - \delta x_j dx_i$. These quantities determine a parallelogram on the vectors $\{dx_i\}$ and $\{\delta x_i\}$, so that the quadratic terms in the expansion can be written as

$$\Delta\sigma^2 = \sum_i \sum_j \sum_k \sum_l R_{ij,kl} \Delta x_{ij} \Delta x_{kl}.$$

The quantity obtained by Riemann, which he calls the fraction from the division of $\Delta\sigma^2$ by the area of the triangle on the vectors $\{dx_i\}$ and $\{\delta x_i\}$, is actually the limit of this fraction as the triangle in question shrinks to the point. Nowadays, a quantity proportional to that defined by Riemann (with a proportionality factor such that in the case of a surface it coincides with the Gaussian curvature) is called the *Riemannian curvature* of the space at a given point in a given two-dimensional direction. Riemann goes on to give our earlier geometric interpretation of the curvature he defined. He notes that on surfaces in ordinary space "the difference between the angle sum of an infinitesimal triangle and two right angles is proportional to the area of the triangle" and writes:

To give an intelligible meaning to the curvature of an n-fold manifold at a given point and in a given surface-direction through it, we must start from the fact that a geodesic proceeding from a point is entirely determined when its initial direction is given. According to this we obtain a determinate surface if we prolong all the geodesics proceeding from the given point and lying initially in the given surface-direction; this surface has at the given point a definite curvature, which is also the curvature of the n-fold continuum at the given point in the given surface-direction *[122, p. 64]*.

Riemannian geometry was applied by Riemann to the theory of differential equations in *A mathematical work containing an attempt to answer the question proposed by the most illustrious Paris Academy*: *To determine the heat state of a homogeneous solid so that a system of isothermal curves given at a certain moment in time remains a system of isothermal curves at an arbitrary moment in time and so that the temperature at a point is expressed as a function of time and two more independent variables* (Commentatio mathematica qua respondere tentatur questioni ab Illma Academia Parisensi proposita: Trouver quel doit être l'état calorifique d'un corps solide homogène pour qu'une système de courbes isothermes, à un instant donné, restent isothermes après un temps quelconque, de telle sorte que la température d'un point puisse s'exprimer en fonction du temps et de deux autres variables indépendentes), written in 1861 and published in his collected works (Leipzig, 1876) *[454, pp. 391–404]*. In this paper Riemann solved the problem of reducing the differential equation of heat conduction

$$\sum_i \frac{\partial}{\partial s_i}\left(\sum_j b_{ij}\frac{\partial u}{\partial s_j}\right) = h\frac{\partial u}{\partial t}$$

to simplest form—a problem equivalent to that of transforming the quadratic form $\sum_i \sum_j b_{ij}ds_i ds_j$ to a sum of squares. Riemann found that this can be done if and only if the expression

$$K = \frac{1}{2}\frac{\sum_i \sum_j \sum_k \sum_l (ij,kl)(ds_i\delta s_j - ds_j\delta s_i)(ds_k\delta s_l - ds_l\delta s_k)}{\left(\sum_i \sum_j b_{ij}ds_i ds_j\right)\left(\sum_i \sum_j b_{ij}\delta s_i \delta s_j\right) - \left(\sum_i \sum_j b_{ij}ds_i \delta s_j\right)^2},$$

that is invariant under a change of variables, vanishes. Of this expression, he denotes by (111), Riemann says:

The expression $\sqrt{\sum_i \sum_j b_{ij}ds_i ds_j}$ can be regarded as a line element of an n-tuply extended space that is outside the bounds of our intuition. If we lead from point $(s_1, s_2, \ldots, s_n)$ in this space all possible geodesics whose initial directions are characterized by the ratios $\alpha ds_1 + \beta\delta s_1 : \alpha ds_2 + \beta\delta s_2 : \ldots : \alpha ds_n + \beta\delta s_n$ (where α and β are arbitrary quantities) then these lines form a certain surface which we can think of as located in the ordinary space of our intuition. Then expression (111) is the curvature of this surface at the point $(s_1, s_2, \ldots, s_n)$ *[454, p. 403]*.

The quantities (ij, kl), the so-called four-index Riemann symbols, are the same quantities that we denoted earlier by $R_{ij,kl}$. We note that the quantities $R_{ij,kl}$ can be expressed in terms of the coefficients g_{ij} and their derivatives, using the quantities Γ^i_{jk} (expressed in terms of the g_{ij} and their derivatives by formulas (8.4)), by means of the formula

$$R_{ij,kl} = \left(\frac{\partial \Gamma^h_{jk}}{\partial x_i} - \frac{\partial \Gamma^h_{ik}}{\partial x_j} + \Gamma^h_{ir}\Gamma^r_{jk} - \Gamma^h_{jr}\Gamma^r_{ik} \right) g_{hl}.$$

In the paper mentioned previously Riemann also used the quantities Γ^i_{jk} but denoted them as p_{ijk}.

Riemann's investigations were continued by Elwin Bruno Christoffel (1829–1900) in the paper *On the transformation of homogeneous differential expression of second degree* (Über die Transformation der homogenen Differentialausdrücke zweiten Grades. Berlin, 1869) *[114, vol. 1, pp. 368–377]*, in which he posed the question of the conditions under which the geometry determined by a form $\sum_i \sum_j g_{ij} dx_i dx_j$ coincides with the geometry determined by a form $\sum_i \sum_j h_{ij} dy_i dy_j$. Just as Riemann's intention was to develop Gauss's theory of surfaces, so Christoffel's intention was to generalize the problem of superposition of surfaces. Christoffel's necessary condition for the coincidence of geometries turned out to be the coincidence of the differential forms $\sum_i \sum_j \sum_k \sum_l R_{ij,kl} dx_i dx_k \delta x_j \delta x_l$, computed for the two given forms. Christoffel denoted the quantities Γ^i_{jk} by $\{^{jk}_i\}$ and the quantities $\sum_l g_{il}\Gamma^l_{jk}$ by $[^{jk}_i]$. That is why these quantities are often called *Christoffel symbols of the first and second kind*, respectively.

Riemannian Spaces of Constant Curvature

After defining Riemannian spaces of variable curvature and noting that Euclidean spaces are spaces of zero curvature Riemann dwelled at length on spaces of constant nonzero curvature:

Manifolds whose curvature is constantly zero may be treated as a special case of those whose curvature is constant. The common character of these continua whose curvature is constant may be also expressed thus, that figures may be moved in them without stretching. For clearly figures could not be arbitrarily shifted and turned round in them if the curvature at each point were not the same in all directions. On the other hand, however, the measure-relations of the manifold are entirely determined by the curvature; they are therefore exactly the same in all directions at one point as at another, and consequently the same constructions can be made from it: whence it follows that in aggregates with constant curvature figures may have any arbitrary position given them. The measure-relations of these manifolds depend only on the value of the curvature, and in relation to the analytic expression it may be remarked that if this value is denoted by α, the expression for the line-element may be written

$$\frac{1}{1 + \frac{1}{4}\alpha \sum x^2} \sqrt{\sum dx^2}$$

[122, p. 65].

The latter expression generalizes the previously mentioned expression, obtained by Minding, of the line element of a surface of constant curvature in terms of its Gaussian curvature.

The simplest example of a Riemannian space of constant positive curvature is a sphere in $(n + 1)$-dimensional Euclidean space; the Riemannian curvature of a sphere of radius r, of any dimension, in all two-dimensional directions is equal to $1/r^2$, for the area of any spherical triangle is the product of r^2 and the angular excess of the triangle.

An example of an n-dimensional Riemannian space of constant negative curvature is n-dimensional Lobačevskian space, first defined by Beltrami in the paper *Fundamental theory of spaces of constant curvature* (Teoria fondamentale degli spazî di curvatura costante. Milan, 1869) *[42, pp. 406–429]* published in the same year as his previously discussed paper. This space can also be viewed as one of the sheets of a sphere of radius qi in $(n + 1)$-dimensional pseudo-Euclidean space. The Riemannian curvature of such a sphere in all two-dimensional directions is $-1/q^2$ since the area of an arbitrary spherical triangle on this sphere is the product of its angular defect by $-q^2$.

We wish to mention here a remarkable theorem of F. Schur established in the paper *On the connection between spaces of constant curvature and projective spaces* (Über dem Zusammenhang der Räume constanten Krümmungsmasses mit den projectiven Räumen. Leipzig, 1886) *[517]* which asserts that, on a Riemannian manifold, the constancy of the Riemannian curvature in all two-dimensional directions at each point implies the constancy of the Riemannian curvature at all points. Schur's proof is purely geometric and is based on a projective mapping of the bundle of linear elements at one point onto a similar bundle at another point. An analytic proof of this result was given by Luigi Bianchi (1856–1928) in the paper *On four-index symbols and on Riemannian curvature* (Sui simboli a quatro indici e sulla curvatura di Riemann. Rome, 1902) *[56]*.

Elliptic Geometry

Another form of n-dimensional Riemannian space of constant positive curvature is a sphere in $(n + 1)$-dimensional Euclidean space with identified antipodal points called n-dimensional *elliptic space* or *non-Euclidean space of Riemann*. When $(n + 1)$-dimensional Euclidean space is completed to $(n + 1)$-dimensional projective space, pairs of antipodal points of a sphere in $(n + 1)$-dimensional Euclidean space are projected from its center onto the points of its plane at infinity. Since this plane is an n-dimensional projective space, elliptic space can be viewed as a metrized projective space. The details are as follows.

An arbitrary sphere of $(n + 1)$-dimensional Euclidean space is given by an equation of the form

$$A \sum_i x_i^2 + 2 \sum_i b_i x_i + c = 0,$$

which can be written in homogeneous coordinates as

$$A \sum_i x_i^2 + 2 \sum_i b_i x_i x_0 + c(x_0)^2 = 0.$$

The latter equation shows that our sphere intersects the plane at infinity, $x_0 = 0$, in the imaginary quadric $\sum_i x_i^2 = 0$. Hence an elliptic space can be described as a projective space with a given imaginary conic. The distance ω between two points X and Y of an elliptic space of curvature $1/r^2$ is connected with the angle φ between the corresponding diameters of the sphere by the relation $\omega = \varphi r$. On the other hand, Edmond Laguerre (1834–1886) showed in the paper *On the theory of foci* (Sur la theorie des foyers. Paris, 1853) *[298, vol. 2, pp. 6–15]* that the angle φ between two straight lines in Euclidean space can be expressed in terms of the cross ratio (ij, xy) of the straight lines x, y and two *isotropic straight lines* i, j, that is, imaginary lines of zero length that lie in the plane of x and y and pass through the point of intersection of x and y, by means of the relation

$$\varphi = \frac{1}{2i} \ln(ij, xy). \tag{8.5}$$

Since the imaginary lines i, j that join the center of the sphere to the points I, J of the imaginary quadric $\sum_i x_i^2 = 0$ in the plane at infinity are, in view of the equation of the quadric, isotropic, Laguerre's formula (8.5) yields an expression for the distance ω between points X, Y of elliptic space in terms of the cross ratio (IJ, XY) of these points and the two points of intersection of the straight line XY and the imaginary quadric:

$$\omega = \frac{r}{2i} \ln(IJ, XY). \tag{8.6}$$

An elliptic metric in the projective plane was defined in 1859 by Cayley in the *Sixth memoir upon quantics [103, vol. 2, pp. 561–592]*. An elliptic metric in space, and the term *elliptic geometry*, were introduced by Klein in 1871 in the paper *On the so-called non-Euclidean geometry [282, vol. 1, pp. 254–305]*, in which he wrote that

The basis of a general projective metric in space is provided by an arbitrary *fundamental surface of the second order.* To define the distance between two points one joins them by a straight line. It intersects the fundamental surface in two new points that are in a definite cross ratio with the two given points. *The logarithm of this cross ratio multiplied by an arbitrary constant c yields what one should call the distance between the two given points [282, vol. 1, p. 300].*

Klein gave a similar definition of the angle between two planes (see p. 238). Further Klein wrote:

By a *motion* is meant the totality of linear transformations that leave the fundamental surface invariant.

By *spheres* one means quadric surfaces that meet the fundamental surface along a plane curve. The center of the sphere is the pole of the plane that contains the osculating curve. . . .

If the fundamental surface is *imaginary* then all straight lines have finite length and each pencil of planes has finite angle sum. *Elliptic* geometry comes under this case (provided that the constant c' in the definition of angles is taken as $\sqrt{-1}/2$, so that the sum of the angles in a pencil of planes is π).

We do not investigate the case when the fundamental surface is *real* and *ruled* (like a hyperboloid of one sheet) for this case is no way connected with the three geometries (elliptic, hyperbolic, parabolic) considered here.

Finally, if the fundamental surface is *real* and *nonruled* then for the interior points of the surface we obtain a metric that includes the metric of hyperbolic geometry provided that we again take the constant c' to be $\sqrt{-1}/2$.

Parabolic geometry is included as a special case of the general metric; this case arises if the fundamental surface specializes (degenerates) into an imaginary conic section. The fundamental conic section of parabolic geometry is the so-called imaginary circle at infinity *[282, vol. 1, p. 301].*

Klein's *fundamental surface* is now called by Cayley's term *the absolute.* We note that Klein's *linear transformations* are collineations, his *hyperbolic geometry* is Lobačevskian geometry, and his *parabolic geometry* is Euclidean geometry; the *imaginary circle at infinity* is the imaginary circle $\sum_i x_i^2 = 0$ in the plane at infinity of Euclidean space that plays the role of the absolute of this space. Klein did not investigate the geometry of the exterior of an oval (nonruled) quadric and the geometry of a ruled quadric. The reason for this is that, unlike elliptic geometry and Lobačevskian geometry, the geometry of these spaces is locally not Euclidean but pseudo-Euclidean.

We also note that the only difference between formula (8.6) and formula (6.27) for Lobačevskian space is that in one case we have the pure imaginary constant r/i and in the other the real constant q.

Clifford Parallels and Surfaces

The geometry of elliptic space was significantly developed by William Kingdon
Clifford (1845–1879) in his *Preliminary sketch of biquaternions* (London,
1879) *[122, pp. 181–200]*.

After defining elliptic, hyperbolic, and parabolic geometries whose absolutes
are, respectively, an imaginary quadric, a real (oval) quadric, and an imaginary
conic in a real plane, Clifford investigates *elliptic geometry*. After defining
poles and polar planes and mutually polar straight lines with respect to an
absolute and pointing out that two points that are polar conjugates with
respect to the absolute "are a *quadrant* apart," that is, are $(\pi/2)r$ apart, or, if
$r = 1$, are $\pi/2$ apart, Clifford notes that

> Through an arbitrary point can in general be drawn *one* line perpen-
> dicular to a given plane; namely, the line joining the point to the pole
> of the plane. If, however, the point *is* the pole of the plane, every line
> through it is perpendicular to the plane. Similarly, from a point not on
> the polar of a given line can be drawn one and only one perpendicular
> to the line; namely, the line through the point which meets the given line
> and its polar *[122, p. 192]*.

and further:

> *In general, two lines can be drawn so that each meets two given lines
> at right angles, and these are polars of one another.* One line may therefore
> be converted into another by rotation about two polar axes. These axes
> are determined as the lines which meet the two given lines and their
> polars. If we travel continuously along one of these lines and draw per-
> pendiculars on the other, one of these axes determines the shortest dis-
> tance between the lines, and the other the longest. If then these two are
> equal, the lines are equidistant along their whole length. Thus *there is
> a case of exception in which two lines and their polars belong to the same
> set of generators of a hyperboloid; the lines are then equidistant along their
> whole length, and meet the same two generators of one system of the
> absolute.* I shall use the word *parallel* to denote two lines so situated;
> and they shall be called *right* parallel or *left* parallel according as one
> is converted into the other by a right-handed or left-handed twist.
> Through an arbitrary point can be drawn one right parallel and one left
> parallel to a given line; the angle between them is twice the distance of
> the point from the line. There are many points of analogy between the
> *parallels* here defined and those of parabolic geometry. Thus, if a line
> meets two parallel lines, it makes equal angles with them; and a series of
> parallel lines meeting a given line constitute a ruled surface of zero
> curvature. The geometry of this surface is the same as that of a finite
> parallelogram whose opposite sides are regarded as identical *[122,
> pp. 192–193]*.

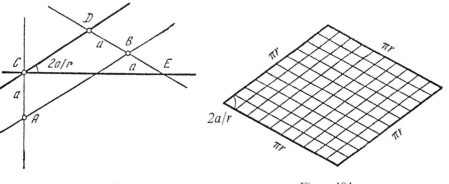

Figure 103   Figure 104

The parallels defined by Clifford are now called *paratactic straight lines*;[5] one must bear in mind that, unlike parallels in Euclidean (*parabolic*) geometry, Clifford parallels are skew straight lines. Figure 103 illustrates the construction of a right, CD, and left, CE, Clifford parallel through a point C to a straight line AB. From the point C one drops the perpendicular CA to the straight line AB and constructs its polar DBE. On the latter one lays off on either side of the point B segments BD and BE equal to the segment $CA = a$. The required parallels are the straight lines CD and CE; the angle DCE is measured in terms of the segment $DE = 2a$ and is equal to $2a/r$. The surface constructed by Clifford is now known as a *Clifford surface*. It is a ruled quadric obtained by rotating one of two paratactic straight lines about the other; its rectilinear generators in both families are paratactic to the axes; at the same time it is also a surface of revolution about the polar of the first axis. The geometry of a Clifford surface is Euclidean. Its area is finite and equal to $\pi^2 r^2 \sin 2a/r$, for it is isometric to a rhombus with sides πr and acute angle $2a/r$ (Figure 104) with sides in each pair of opposite sides glued together. To prove that the geometry of this surface is Euclidean it suffices to write down its parametric equations in homogeneous coordinates whose basis points are located on the axes of the surface,

$$x_0 = \cos\frac{a}{r}\cos u, \qquad x_1 = \cos\frac{a}{r}\sin u,$$

$$x_2 = \sin\frac{a}{r}\cos v, \qquad x_3 = \sin\frac{a}{r}\sin v,$$

and to compute the line element

$$ds^2 = \sum_i dx_i^2 = r^2\left(\cos^2\frac{a}{r}du^2 + \sin^2\frac{a}{r}dv^2\right).$$

[5] This term was introduced by E. Study in the paper *[568]*.

If in the latter we put

$$U = r\cos\frac{a}{r}\cdot u, \qquad V = r\sin\frac{a}{r}\cdot v,$$

then we have $ds^2 = dU^2 + dV^2$.

If one eliminates from the parametric equations of the Clifford surface the parameters u and v then one obtains its equation in the form

$$\sin^2\frac{a}{r}(x_0^2 + x_1^2) - \cos^2\frac{a}{r}(x_2^2 + x_3^2) = 0.$$

A Clifford surface is the simplest solution of the Clifford-Klein problem of finding spaces with a Euclidean metric that are not isometric to Euclidean space in the large.

Clifford's Idea of the Geometrization of Physics

W. K. Clifford was also interested in questions of the philosophy of space. In his philosophical work *Philosophy of pure science* (London, 1873) there are his famous words on Lobačevskian geometry!

> What Vesalius was to Galen, what Copernicus was to Ptolemy, that was Lobatchewsky to Euclid. There is, indeed, a somewhat instructive parallel between the last two cases. Copernicus and Lobatchewsky were both of Slavic origin. Each of them has brought about a revolution in scientific ideas so great that it can only be compared with that wrought by the other. And the reason of the transcendent importance of these two changes is that they are changes in the conception of the Cosmos [123, vol. 1, p. 356].

What Clifford is saying here is that Lobačevskiĭ's discovery that Euclidean geometry is not the only conceivable geometry is just as revolutionary as Vesalius' shattering of the myth of the exclusive position of man in the animal kingdom and Copernicus' discovery that the Earth is just one of the planets.

Clifford expressed his interesting thoughts on space in his posthumously published *The Common sense of the exact sciences* (London, 1885) [121]. Clifford begins the chapter on space with the words

> Geometry is a physical science [121, p. 43],

and at the end of the chapter on position he writes:

> We may conceive our space to have everywhere a nearly uniform curvature, but that slight variations of the curvature may occur from point to point, and themselves vary with the time. These variations of the curvature with the time may produce effects which we not unnaturally attribute to physical causes independent of the geometry of our space.

We might even go so far as to assign to this variation of the curvature of space "what really happens" in that phenomenon which we term the motion of matter *[121, pp. 202–203]*.

Clifford developed these ideas in greater detail in his report *On the space theory of matter* (Cambridge, 1876). First he paraphrases Riemann:

Riemann has shewn that as there are different kinds of lines and surfaces, so there are different kinds of space of three dimensions; and that we can only find out by experience to which of these kinds the space in which we live belongs. In particular, the axioms of plane geometry are true within the limits of experiment on the surface of a sheet of paper, and yet we know that the sheet is really covered with a number of small ridges and furrows, upon which (the total curvature not being zero) these axioms are not true. Similarly, he says although the axioms of solid geometry are true within the limits of experiment for finite portions of our space, yet we have no reason to conclude that they are true for very small portions; and if any help can be got thereby for the explanation of physical phenomena, we may have reason to conclude that they are not true for very small portions of space *[122, p. 21]*.

Then he states the following four principles:

(1) That small portions of space *are* in fact of a nature analogous to little hills on a surface which is on the average flat; namely, that the ordinary laws of geometry are not valid in them.
(2) That this property of being curved or distorted is continually being passed on from one portion of space to another after the manner of a wave.
(3) That this variation of the curvature of space is what really happens in that phenomenon which we call the *motion of matter*, whether ponderable or etherial.
(4) That in the physical world nothing else takes place but this variation, subject (possibly) to the law of continuity *[122, pp. 21–22]*

These principles form a program of geometrization of physics.

Riemann's Topology

In addition to the theory of Riemannian spaces Riemann also founded topology—one of the most important disciplines that significantly broadened our notions of space.

We have already pointed out that Euler interpreted Leibniz's term *geometry of position* in a topological sense, whereas Carnot and Grassmann gave it narrower interpretations. Euler's interpretation of the term was developed by the German physicist Johann Benedict Listing (1808–1862), who suggested the now generally accepted term *topology* in his *Preliminary studies on topology* (Vorstudien zur Topologie. Göttingen, 1847) *[329]*. Listing's term comes from replacing Leibniz's Latin *situs* (place) with its Greek counterpart *topos*. Listing studied

> *linear complections, that is, lines or curves or sets of such, located on a surface, such as a plane or a sphere, or arbitrarily disposed in space [329, p. 867].*

He investigated knots, chains, plaits, and other forms of mutual disposition of linear complections—later he called them *linear complexes*—with numerous examples drawn from biology and technology. Listing's work *The census of spatial complexes or a generalization of Euler's therorem on polyhedra* Der Census räumlicher Komplexe oder Verallgemeinerung des Euler'schen Satzes von den Polyedern. Göttingen, 1862) *[330]* dealt with more general *spatial complexes.*

Gauss interpreted Leibniz's ideas in the same topological sense. Gauss devoted to topological investigations, as to non-Euclidean geometry, a number of rough notes and letters (see *[196, vol. 8, pp. 407–410]* and the study of Jean Claude Pont *[435a, pp. 31–38]*). When he founded the topology of two-dimensional manifolds and laid the foundations of multidimensional manifolds Riemann also interpreted Leibniz's ideas in the same way. Riemann laid the foundations of two-dimensional topology in his *Theory of Abelian functions* (1857) *[454, pp. 82–142]*, mentioned earlier. In this work he wrote, with reference to Leibniz, that

> For the study of functions which arise as integrals of exact differentials, some theorems belonging to *analysis situs* are nearly indispensable *[454, p. 91]*.

Riemann associates to algebraic functions $f(x, y)$ in the complex variables x and y multisheeted surfaces now called *Riemann surfaces*. He divides these surfaces into

> simply connected ones, in which every closed curve bounds a region of the surface—as, for example, a disk—and multiply connected ones, for which this does not happen—as, for example, an annulus bounded by two concentric circles *[454, p. 22]*.

and points out that by means of a system of cuts it is possible to make a multiply connected surface into a simply connected one. Riemann introduces the characteristic p of a plane algebraic curve (later called by Clebsch *[119]* the genus of the curve) which he defines as half the number of cuts needed to make the corresponding multiply connected Riemann surface into a simply connected one. The number p is now called the *genus of the surface*. We note that, as S. L'Huillier showed *[323]*, for polyhedra the genus p is connected with the Euler characteristic $\chi = N_0 - N_1 + N_2$ of a polyhedron by means of the relation $\chi = 2 - 2p$. For a sphere $p = 0$, for a torus $p = 1$, and for a sphere with p handles it is equal to p. Riemann showed that for a Riemann surface the genus p, the number n of its sheets, and the number w of its branch points are connected by the relation $w - 2n = 2p - 2$.

In a survey that is a supplement to the article *On the hypotheses that lie at the foundations of geometry* Riemann adds a remark in connection with the first chapter of the article *Notion of an n-ply extended magnitude*:

> Chapter 1 is at once an introduction to investigations on *analysis situs* [454, p. 286].

Fragments dealing with *analysis situs* were published in 1876 in an edition of Riemann's collected works [454, pp. 479–482]. In these fragments Riemann generalized the topological properties of a two-dimensional surface to an n-dimensional manifold, here called an *n-stretch* (*n-Streck*). Here Riemann defines what are now called *homologous n*-stretches:

> an n-stretch A is said to be transformable into an n-stretch B if A and parts of B together form the complete boundary of an interior $(n + 1)$-stretch [454, p. 479].

Then he gives the extremely important definition:

> If in the interior of a continuously extended manifold it is possible to make every unbounded n-stretch bounding by means of m definite parts of nonbounding n-stretches then this manifold has $(m + 1)$-tuple connection of the n-th dimension.
>
> A continuously extended manifold is called *simply connected* if the connection of every dimension is simple [454, p. 479].

It is easy to verify that for a two-dimensional manifold of genus p Riemann's *connection* is $(2p + 1)$-ple. Then Riemann explains the dependence of the connection of the boundary of a manifold on the connection of the manifold itself.

These ideas of Riemann were set forth by his friend Enrico Betti in the previously mentioned paper *On spaces of an arbitrary number of dimensions* [55, vol. 2, pp. 273–290]. By "spaces" Betti meant manifolds in multidimensional Euclidean spaces. He wrote:

> If in an n-dimensional space R, bounded by one or more $(n - 1)$-dimensional spaces, every closed m-dimensional space, $m < n$, is the boundary of a part of a linear connected $(m + 1)$-dimensional space, entirely contained in R, then we shall have a connection in $m + 1$ dimensions and we shall say that R has a *simple* connection of the m-th species. If a space R has only simple connections, then we shall say that it is simply connected. If, however, one can imagine in R a number p_m of closed m-dimensional spaces which cannot form the boundary of a linear connected part of an $(m + 1)$-dimensional space, entirely contained in R, and such that every other closed m-dimensional space forms

by itself or with some or all of these spaces the boundary of a linear connected part of an $(m + 1)$-dimensional space, entirely contained in R, then we shall say that R has a connection of $(p_m + 1)$-th order of the m-th species *[55, vol. 2, p. 278]*.

These ideas were further developed by Poincaré. Already in the memoir *On curves defined by differential equations* (Sur les courbes defines par les équations differentielles. Paris, 1881–1885) *[431, vol. 1, pp. 90–161]* he made extensive use of topological properties of curves to give qualitative descriptions of solutions of differential equations. Poincaré devoted to the topology of multidimensional manifolds the large memoir *Analysis situs* (the title is taken over from Riemann) (Paris, 1895) *[431, vol. 6, pp. 193–288]* and five supplements to it (Palermo-London-Paris, 1899–1904).

Poincaré determined $(n - p)$-dimensional manifold in n-dimensional space by means of p equations and q inequalities between the n coordinates. Then he defined the boundary of a manifold, homomorphism between manifolds, and homology—the fundamental concept of combinatorial topology. Using this concept he defined the *Betti numbers* of a manifold, that is, Betti's *orders of connection* (in the twenties of this century Solomon Lefschetz (1884–1972) and James Alexander (1888–1971) proposed to call *Betti numbers* numbers b_i one less than Poincaré's *Betti numbers*; cf. *[308]*). Poincaré extended Euler's theorem to polyhedra with arbitrary Betti numbers by showing that

$$N_0 - N_1 + N_2 - \cdots + (-1)^n N_{n-1} = b_0 - b_1 + b_2 - \cdots - (-1)^n b_{n-1}.$$

The polynomial $\sum b_i t^i$ is called the *Poincaré polynomial*. In the same paper Poincaré also laid the foundation for the homotopic theory of manifolds in which a leading role is played by the noncommutative *Poincaré group* introduced by him.

Topological Spaces

At the beginning of the 20th century, in connection with the spread of the group-theoretic viewpoint, mathematicians began to study *abstract spaces* together with the study of topological invariants of manifolds that are submanifolds of Euclidean spaces or are obtained from such manifolds by identification of points (as in the case of obtaining the projective plane from a sphere) or by splitting them into points of various sheets (as in the case of constructing the Riemann surface of an algebraic function out of its domain of definition in the complex plane). One of the first works in this direction was the paper *On certain points of the functional calculus* (Sur quelques points du calcul fonctionelles. Palermo, 1906) *[184]* of the French mathematician Maurice Fréchet (1878–1973). In this paper Fréchet defines for the first time an *abstract metric space* exemplified by a space of functions or curves. To construct an abstract space Fréchet found it necessary to generalize, first of all, the theory of linear sets that has brought about so much progress in the theory of functions of a single variable,

by which theory he meant the, by that time, extensively developed theory of sets of points of the real line. Fréchet goes on to say:

If we assume this preliminary investigation of sets then there arises a difficulty. The first generalization that seems natural is that of the notion of a continuous function. But if one wishes to consider operations where the variable is an element of arbitrary nature, then we must first know what is to be meant by neighboring elements or by the limit of a sequence of elements. This seems impossible: usually one gives a special definition of limit for each category of elements under study—points, curves, and so on. I circumvented this difficulty by a method similar to that which allows one to reason in the theory of abstract groups about a composition of explicitly indeterminate form.

Now I note that almost all (but not all) classical definitions of limit can be formulated as follows: in the given category of elements one can associate to each pair of elements a number $\rho(A, B)$ whose properties are very close to the properties of distance of two points, namely that A coincides with B if $\rho(A, B) = 0$ and A tends to B if $\rho(A, B)$ tends to zero. If we accept this hypothesis, less general but nevertheless very broad, then we obtain numerous more specific results.

The approach just outlined leads to the generalizing of almost all theorems about linear sets and about continuous functions (at least those that can be formulated independently of the nature of the investigated sets) [184, pp. 1–2].

When mentioning abstract groups Fréchet emphasizes that he is applying the same methods of creating abstract mathematical concepts that have hitherto been used only in algebra, where, together with the abstract theory of groups, founded in the seventies of the 19th century, there also appeared Dedekind's abstract ring theory.

Developing his program, Fréchet defines a general metric space, which he calls *class* (V), as

a set of elements of arbitrary nature such that we know how to determine whether two given elements are identical or not and, in addition, such that to any two of them, A, B, we can associate a number $(A, B) = (B, A)$ with the following two properties:

1′ The necessary and sufficient condition that (A, B) is zero is the identity of A and B.

2′ There exists a completely determined positive function $f(\varepsilon)$ such that the inequalities $(A, B) \leq \varepsilon$ and $(B, C) \leq \varepsilon$ imply $(A, C) \leq f(\varepsilon)$ for all elements A, B, C. In other words, for (A, C) to be small it suffices that (A, B) and (B, C) are small [184, p. 18].

At present, metric spaces are defined by means of three axioms: the *identity axiom* that coincides with Frechet's axiom 1'; the *symmetry axiom* $(A, B) = (B, A)$ that Frechet regards as part of the definition of distance; and the *triangle inequality axiom* $(A, B) + (B, C) \geq (A, C)$ that replaces Fréchet's axiom 2'.

Fréchet goes on to consider sets in metric spaces, limits of sequences, and limit points of sets. All limit points of a set form its *derived set*. Sets that contain all their limit points are called *closed*, and complements of closed sets with respect to the whole space are called *open* sets. The intersection of all closed sets containing a given set M is called the *closure* $\overline{M}$ of the set M. Then, after defining the Cauchy criterion for a sequence of elements $A_1, A_2, \ldots$ of the metric space as the possibility of associating to every $\varepsilon > 0$ an integer n with the property that the inequality $(A_n, A_{n+p}) > \varepsilon$ is satisfied for every p, Frechet defines a *complete* metric space as a metric space in which every sequence satisfying the Cauchy criterion has a, necessarily unique, limit.

Fréchet restricts himself to spaces that can be regarded in at least one way as the derived sets of countable sets of their elements. He considers metric spaces of functions with different definitions of distance.

The definition of limit points and closed sets made it possible to introduce continuous mappings and homeomorphisms of the metric spaces considered by Fréchet. Fréchet's paper became the starting point for the development of a general theory of abstract spaces and for the development of the basic concepts of functional analysis.

Since distance is not needed for the study of topological properties of abstract spaces, just as it is not needed for the study of topological properties of lines and surfaces, soon after the appearance of abstract spaces Felix Hausdorff (1868–1942) in his *Foundations of set theory* (Grundzüge der Mengenlehre. Leipzig, 1914) *[217; 218]* defined an *abstract topological space*. Hausdorff defined a topological space as any collection of elements, called *points*, with a distinguished collection of subsets $\{U\}$, such that to each point x there are associated some, and, at the very least, one of the sets of the system $\{U\}$, called *neighborhoods* $U(x)$ of x; the intersection of any two neighborhoods $U_1(x)$ and $U_2(x)$ contains a neighborhood $U_3(x)$ of this point, and if y is a point of a neighborhood $U(x)$ then there exists a neighborhood $U(y)$ of y contained in $U(x)$. Prescribing neighborhoods makes it possible to define limit points of any set M as points x such that every neighborhood of x contains at least one point of M other than x. Just as in Fréchet spaces so too in a topological (Hausdorff) space one defines derived, closed, and open sets. It is easy to see that all neighborhoods are open sets.

Every Fréchet metric space is a topological space. In fact, one can take as neighborhoods the sets of points whose distances from the points of a (fixed) countable set, whose derived set is the whole space, are less than some rational numbers. It is easy to see that the number of such neighborhoods is countable, or, as one says, the space has a *countable basis*.

A further development of the Hausdorff definition is the definition of a topological space proposed by the Polish mathematician Kazimierz Kuratowski (1896–1980) in the paper *The operation $\bar{A}$ of analysis situs* (L'operation $\bar{A}$ de l'analysis situs. Warsaw, 1922) *[296]*. By *operation $\bar{A}$* Kuratowski means the transition from a set A to its closure $\bar{A}$. According to Kuratowski, a topological space is a collection of elements of arbitrary nature, called *points*, on whose subsets A there is defined an operation of *closure* satisfying the following axioms:

1' for the union $A + B$ of two sets A and B we have $\overline{A + B} = \bar{A} + \bar{B}$;
2' a set A is contained in its closure, $A \subset \bar{A}$;
3' the closure of the empty set $\varnothing$ coincides with it, $\bar{\varnothing} = \varnothing$;
4' the closure of the closure coincides with the closure, $\bar{\bar{A}} = \bar{A}$.

If for every set there is defined its closure $\bar{A}$ then the points of $\bar{A}$ not in A are called the *limit points of A*; the sets A that coincide with their closures $\bar{A}$ are called the *closed sets*, and their complements relative to the whole space are called the *open sets*.

The Soviet mathematician Pavel Sergeevič Aleksandrov [Alexandroff] (1896–1983) proposed in the paper *On the foundation of n-dimensional topology* (Zur Begründung der *n*-dimensionalen Topologie. Leipzig, 1925) *[14]* a more symmetric form of Kuratowski's definition:

1. A topological space is a set of elements of arbitrary nature, called *points*, in which certain subsets, called *open sets*, have been singled out such that:
   1' the whole space is an open set;
   2' the null set is an open set;
   3' the intersection of finitely many open sets is an open set;
   4' every union of open sets is an open set.
2. A topological space is a set of elements of arbitrary nature, called points, in which certain subsets, called *closed sets*, have been singled out such that:
   1' the whole space is a closed set;
   2' the null set is a closed set;
   3' every union of closed sets is a closed set;
   4' the intersection of finitely many closed sets is a closed set.

In the first case a closed set is defined as the complement of an open set and in the second case an open set is defined as the complement of a closed set. Neighborhoods can be defined as subsystems of open sets such that an arbitrary open set can be represented as the union of sets of this subsystem (see *[218, p. 258]*).

To different systems of closed or open subsets in the same set of points there correspond different topologies on this set. For example, our axiom systems are satisfied if

(A) the only closed (and open) sets are the whole space and the null set;
(B) the closed (and open) sets are all sets of points of the space.

In case (A) the closures of all nonempty sets, including single points, coincide with the whole space. In case (B) the closure of each set coincides with this set, and a minimal system of neighborhoods is a system in which the neighborhood of each point is that point alone. Then the space has no limit points and is called *discrete*; the space of case (A) is called *trivial*.

Hausdorff ruled out case (A) by means of the following axiom: for any two points of the space there are disjoint neighborhoods *[218, p. 260]*. Sometimes Hausdorff's axiom is replaced with a weaker one due to Frigyes Riesz, that the closures of all single points of the space are the points themselves *[456, vol. 1, pp. 155–169]*—or by an even weaker axiom introduced by the Soviet mathematician Andreĭ Nikolaevič Kolmogorov (1903–1987), that any two points of the space have different closures *[15, p. 58]*.

A very important class of topological spaces is the class of *compact spaces* such that every covering of the space by means of open sets contains a finite subcovering. The Soviet mathematicians P. S. Aleksandrov and Pavel Samuilovič Uryson [Urysohn] (1898–1924) studied such spaces in the case when the minimal cardinality of a system of neighborhoods is greater than the cardinality of a countable set and called them *bicompact [16]*. For these spaces P. S. Aleksandrov developed a homology theory analogous to the homology theory of manifolds in Euclidean spaces. By now mathematicians have found a great many topological invariants of the most varied topological spaces.

One of the most obvious topological invariants is the dimension of a topological space. Whereas it is possible to have a one-to-one correspondence between manifolds of different dimensions, a one-to-one bicontinuous correspondence can only be established between manifolds of the same dimension. If the points of a topological space have neighborhoods homeomorphic to n-dimensional Euclidean space then it is natural to say that the dimension of the space is n. In more complicated cases dimension was defined by the Dutch mathematician Luitzen Egbertus Jan Brouwer (1882–1966) in the paper *Proof of the invariance of the dimension number* (Beweis der Invarianz der Dimensionzahl. Leipzig, 1911) *[80]*. The ideas of Brouwer were developed by Urysohn in *A memoir on Cantor manifolds* (Memuar o kantorovyh mnogoobraziyah. 1928) *[599]* and by the German mathematician Karl Menger (b. 1902) in *Dimension theory* (Dimensionstheorie. Leipzig, 1932) *[362]*. They defined dimension by induction on the dimension number, beginning with the dimension -1 which they assigned to the empty set.

Urysohn and Menger also proposed another definition of dimension, equivalent to the preceding for the most important topological spaces: a nonempty space is n-dimensional if every finite covering of the space has a finite subdivision of order $\leq n$ and there exists a finite covering of the space without finite subdivisions of order $< n$. A nonempty space is infinite-dimensional if it is not n-dimensional for any nonnegative n. The dimension of the null set is again taken to be -1.

In the previously mentioned paper Brouwer proved his famous *fixed-point theorem*: every continuous mapping of an n-dimensional simplex into itself has at least one fixed point. This theorem was first proved by the Latvian mathematician Piers Bohl (1865–1921), who worked in Riga, in the paper *On the motion of a mechanical system near an equilibrium position* (Über die Bewegung eines mechanischen Systems in der Nähe einer Gleichgewichtlage. Berlin, 1904) *[64, pp. 79–125]*. The Bohl-Brouwer theorem was extended to infinite-dimensional analogues of an n-dimensional simplex by the Soviet mathematician Andrei Nikolaevič Tihonov [Tychonoff] (b. 1906) *[597]*. Bohl came to the fixed-point theorem in connection with the problem of proving the existence of a solution of a system of differential equations connected with the notion of certain mechanical systems he investigated. The Bohl-Brouwer theorem was later used to prove the existence of solutions of finite systems of ordinary differential equations, and the Tihonov theorem was used to prove the existence of solutions of infinite systems of differential equations.

The Influence of the General Theory of Relativity

We have already pointed out the importance of Einstein's discovery of the special theory of relativity for the elaboration of the concept of pseudo-Euclidean geometry. For example, the appearance of Weyl's book *Space, time, matter [626]* in which, among other matters, he set forth his axiomatization of n-dimensional Euclidean space, was connected with the general theory of relativity. But the importance of the general theory of relativity for the development of geometry is far broader and deeper. Whereas in the special theory of relativity spacetime was viewed as a pseudo-Euclidean space, in the general theory it is viewed as an analogue of a Riemannian space that stands in the same relation to such a space as pseudo-Euclidean space to Euclidean space. Such a space is now called a *pseudo-Riemannian* or *general Riemannian space*. Just as in a Riemannian space, so too at each point of this space there is given the square of a line element

$$ds^2 = \sum_i \sum_j g_{ij} dx^i dx^j,$$

where the g_{ij} are point functions and, although the quadratic form is no longer positive definite, it can be reduced at each point to the form

$$ds^2 = -(dx^0)^2 + (dx^1)^2 + (dx^2)^2 + (dx^3)^2.$$

The first sketch of the general theory of relativity was set forth by Einstein together with the German mathematician Marcel Grossmann (1878–1936) in *Outline of a generalized theory of relativity and theory of gravitation* (Entwurf einer verallgemeinerten Relativitätstheorie und Theorie der Gravitation. Leipzig, 1913), and the final theory was presented in *Foundations of the general theory of relativity* (Grundlagen der allgemeinen Relativitätstheorie. Leipzig, 1916) *[340, pp. 109–164]*.

Already in the 1913 paper Einstein and Grossmann used the *tensor calculus* created by the Italian geometer Gregorio Ricci-Curbastro (1853–1925) in *Principles of a theory of differential quadratic forms* (Principii di una theoria delle forme differenziale quadratiche. Milan, 1884) *[453, vol. 1, pp. 138–171]* and subsequently developed together with Tullio Levi-Civita (1873–1941) in *Methods of the absolute differential calculus and their applications* (Méthodes du calcul différentiel absolu et leurs applications. Leipzig, 1901) *[453, vol. 2, pp. 185–271]*. Ricci gave the name *covariant system of the first order* to the functions a_i of the coordinates x^i that change as a result of a coordinate transformation $x^{i'} = x^{i'}(x^1, \ldots, x^n)$ in accordance with the rule

$$a_{i'} = \sum_i a_i \frac{\partial x^i}{\partial x^{i'}}, \tag{8.7}$$

exemplified by the law of transformation of partial derivatives $\dfrac{\partial \varphi}{\partial x^i}$, and the name *contravariant system of the first order* to functions a^i of the coordinates x^i that transform in accordance with the opposite rule

$$a^{i'} = \sum_i a^i \frac{\partial x^{i'}}{\partial x^i}, \tag{8.8}$$

exemplified by the law of transformation of the differentials dx^i of the coordinates. Ricci considered scalars as systems of zero order.

Ricci called the functions a_{ij}, a^{ij}, and a^i_j that transform in accordance with the respective rules

$$a_{i'j'} = \sum_i \sum_j a_{ij} \frac{\partial x^i}{\partial x^{i'}} \frac{\partial x^j}{\partial x^{j'}}, \quad a^{i'j'} = \sum_i \sum_j a^{ij} \frac{\partial x^{i'}}{\partial x^i} \frac{\partial x^{j'}}{\partial x^j}, \quad a^{i'}_{j'} = \sum_i \sum_j a^i_j \frac{\partial x^{i'}}{\partial x^i} \frac{\partial x^j}{\partial x^{j'}}, \tag{8.9}$$

twice covariant, *twice contravariant*, and *mixed* systems of the second order.

Ricci defined in a similar way systems of higher order. Contravariant systems of the first order at a point can be regarded as coordinates of vectors; covariant systems of the first order at the same point can be regarded as coefficients of linear forms defined on these vectors; twice covariant systems of the second order can be regarded as coefficients of bilinear forms defined on these vectors, and mixed systems of the second order can be regarded as matrices of linear transformations defined on these vectors. Einstein and Grossmann proposed that Ricci's "systems" be called *tensors* (and thus extended to them the term *elastic tensor* and that the "orders" of the systems be called *ranks*. The Dutch geometer Jan Arnoldus Schouten (1883–1971) in his *Ricci calculus* (Der Ricci-Kalkül. Berlin, 1924) *[510]* called Ricci's systems *affinors* and thereby extended to them a term used by F. Jung for linear operators; he called "orders" of the systems *valences*, borrowing this term from chemistry.

Most mathematicians now use the term *tensor* in the sense of Einstein and *valence* in the sense of Schouten. Following Einstein's suggestion, in the tensor calculus one omits the summation sign when summing with respect to covariant and contravariant indices. We will follow this rule in the following.

The coefficients of the quadratic form that defines the metric of a Riemannian space form a twice covariant tensor g_{ij}; the quadratic form is written in tensor notation as

$$ds^2 = g_{ij}dx^i dx^j, \tag{8.10}$$

and the equation of the geodesics is written as

$$\frac{d^2 x^i}{dt^2} + \Gamma^i_{jk}\frac{dx^j}{dt}\frac{dx^k}{dt} = 0. \tag{8.11}$$

The Christoffel symbols Γ^i_{jk} do not form a tensor and transform under coordinate transformations in accordance with a more involved law. But the four-index Riemann symbols, now written as $R^{\ \ \ l}_{ij,k}$, do form a tensor of valence four called the *curvature tensor*.

Covariant differentiation as defined by Ricci associates to every tensor field in a Riemannian space—that is, to every function that defines a tensor at every point of a region in a Riemannian space—the field of a new tensor that has one additional covariant valence. The covariant derivative $\nabla_i \varphi$ of a scalar is its partial derivative $\dfrac{\partial \varphi}{\partial x^i}$ with respect to the coordinate x^i. The covariant derivatives $\nabla_j a^i$ and $\nabla_j a_i$ of vectors a^i and a_i are given by

$$\nabla_j a^i = \frac{\partial a^i}{\partial x^j} + \Gamma^i_{jk}a^k, \qquad \nabla_j a_i = \frac{\partial a_i}{\partial x^j} - \Gamma^k_{ji}a_k. \tag{8.12}$$

The covariant derivatives $\nabla_k a_{ij}$ and $\nabla_k a^i_j$ of the tensors a_{ij} and a^i_j have the form

$$\nabla_k a_{ij} = \frac{\partial a_{ij}}{\partial x^k} - \Gamma^l_{ki}a_{lj} - \Gamma^l_{kj}a_{il}, \qquad \nabla_k a^i_j = \frac{\partial a^i_j}{\partial x^k} + \Gamma^i_{kl}a^l_j - \Gamma^l_{kj}a^i_l. \tag{8.13}$$

The Geometry of the General Theory of Relativity

Einstein formulated the principle of general relativity in *The foundations of the general theory of relativity* as follows:

The general laws of nature are to be expressed by equations which hold good for all systems of co-ordinates, that is, are covariant with respect to any substitutions whatever (generally covariant) *[340, p. 117]*.

The most important element of the general relativity theory is the geometric interpretation of gravity. In this theory gravitation is linked to the curvature of space, whose definition in a pseudo-Riemannian space is the same as that in a Riemannian space; the greater the density of matter in a certain region, and thus the intensity of the gravitational field, the greater the curvature of pseudo-Riemannian spacetime. A mass point subject to gravitational forces alone moves along a geodesic which satisfies the same equations as in Riemannian space.

Light rays, too, propagate along geodesics of pseudo-Riemannian spacetime. The deflection of the trajectories of material particles and of light rays from straight lines represents the attraction of particles of matter and of light by the heavy masses that gives rise to the gravitational field. In this connection we quote from Einstein's paper *On the ether* (Über den Aether. Zürich, 1924) *[166]*:

General relativity theory sets aside one other shortcoming of classical dynamics: in the latter, inertia and weight appear as phenomena that are completely independent from one another in spite of the fact that both are conditioned by the same material constant—mass. Relativity theory surmounts this shortcoming by establishing for the dynamical behavior of an electrically neutral particle the law of geodesics in which the interaction of inertia and gravitation is already inseparable. This interaction imparts to the ether a metric that varies from point to point and properties that determine the dynamical behavior of material points. In turn, these are determined by physical factors, namely the distribution of mass or energy. Thus the ether of general relativity differs from the ether of classical mechanics in that it is no longer "absolute" but is determined in the sense of its properties, that vary in space, by the distribution of weighted matter. This definition is complete provided that the world is spatially finite and closed *[166, p. 89]*.

Here the word *ether* is used in the sense of curved space whose geometric properties are determined by matter; in other papers Einstein replaces this term by the term *space-time continuum*. Einstein's reference to the finite and closed character of the universe reflects his original view that the curvature of space implies that space is finite and closed. (Cf. the paper *[649]* of Yakov Borisovič Zel'dovič (1914–1988).)

General relativity theory has given a physical interpretation to the curvature of pseudo-Riemannian spacetime. Light rays propagate along the isotropic geodesics of this spacetime. The deviation of the trajectories of material points and light rays from geodesics represents the attraction of material points and light by the heavy masses that give rise to the gravitational field.

The primary connection between non-Euclidean geometry and general relativity is that the discovery of non-Euclidean geometry extended the range of conceivable spaces and thus prepared the ground for the discovery of spaces of variable curvature, both Riemannian and pseudo-Riemannian; the latter represents the mathematical apparatus of general relativity. However, there is also a direct link between general relativity and non-Euclidean

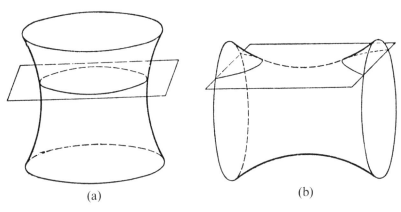

(a)   (b)

Figure 105

geometry, but what is involved is a four-dimensional rather than a three-dimensional space as in the case of special relativity. The geometry in question describes the spacetime of general relativity "in the mean," under the assumption of the uniform distribution of matter. This view was presented by Einstein already in the paper *Cosmological considerations linked to the general relativity theory* (Kosmologische Betrachtungen zur allgemeinen Relativitätstheorie. Berlin, 1917) *[340, pp. 175–189]*. Here Einstein considered a uniform distribution of matter in space alone, regarding space itself as a sphere in four-dimensional Euclidean space and spacetime as a cylinder in five-dimensional space. However, if we proceed from the assumption that we are dealing with a uniform distribution of matter throughout the spacetime continuum then spacetime is a sphere of real or imaginary radius in five-dimensional pseudo-Euclidean spaces of indices 4 and 3, respectively. If we assume, for simplicity, that the hypersurfaces t = const. are parallel hyperplanes, then, in time, the "space section" of the world decreases or increases, depending on the position of the cutting hyperplane. In the first case the curvature of the "space section" is constant and positive, whereas in the second case it is constant and negative (Figures 105a and 105b), in full agreement with astronomical observations. This confirmation shows that real spacetime, that is, a pseudo-Riemannian space of variable curvature, corresponds in the mean to the picture just outlined; the theory of an expanding universe of constant negative curvature was proposed by A. A. Friedmann in the paper *On the possibility of a world of constant negative curvature* (Über die Möglichkeit einer Welt mit konstanter negativer Krümmung des Raumes. Leipzig, 1924) *[188]*.

The view of spacetime as a four-dimensional pseudo-Riemannian space of index 3 with positive curvature was proposed by the Belgian astronomer Willem de Sitter (1872–1934) in the paper *On Einstein's theory of gravitation and its astronomical consequences* (Brussels, 1917) *[541]*. In the paper *Geometrical note on de Sitter's world* (London, 1924) *[158]* Patrick du Val showed that "de Sitter's world" is a Lobačevskian space of positive curvature, that is, a sphere of real radius in five-dimensional pseudo-Euclidean space of index 4 with identified antipodal points (see also *[129a]*).

Einstein's Philosophy of Space

The discovery of the general theory of relativity was a brilliant concretization of the connection between matter and space and time, which are an existence form of matter—an issue often addressed by the classics of Marxism.

Einstein dealt with the philosophical questions of space in many of his papers. One relevant passage is the preceding quotation from his paper *On the ether*. The following quotation is from the paper *Geometry and experience* (Geometrie und Erfahrung. Berlin, 1921) *[165]*:

Of all sciences mathematics is held in especially high esteem, for its theorems are absolutely true and indisputable while the areas of other sciences are to some extent debatable and there is always the danger of their results being overturned by new discoveries. But it is not fitting for a researcher working in some other area of science to envy mathematics, for the propositions of mathematics rest not on real objects but exclusively on objects of our imagination. In fact, it is little wonder that one attains logical agreement of deductions if one earlier reached agreement concerning the principal propositions (axioms) as well as the methods to be employed for deriving from the principal propositions other propositions. At the same time, the deep esteem for mathematics has another basis, which is that mathematics is that which gives to the exact sciences a measure of confidence which they could not otherwise attain.

In this connection there arises a riddle that has worried researchers of all times. Whence the remarkable correspondence between mathematics and real things if mathematics is just a product of human thought unrelated to any experience? Can the human mind understand the properties of real things without any experience, just by way of reflection?

To my mind, a concise answer to this question is this: to the extent to which the theorems of mathematics can be applied to reflect the real world they are not exact; they are exact to the extent to which they do not refer to reality. It seems to me that complete clarity in this matter can be achieved only by following the kind of mathematics known as axiomatics. The progress achieved through axiomatics consists in the fact that it sharply separated its formal-logical content from its objective and intuitive content. By point, straight line, and so on, in axiomatic geometry one should mean only contentless ideas. That which gives them content lies outside mathematics.

On the other hand, however, it is true that mathematics in general and geometry in particular owe to their origins the necessity of learning a little about the behavior of objects existing in reality. This is shown by the very word "geometry" which means "measurement of the earth" *[165, pp. 3–5]*.

In this paper Einstein arrives

at the viewpoint held by as original and deep a thinker as Henri Poincaré: Euclidean geometry differs from all conceivably possible axiomatic geometries by its simplicity. And since axiomatic geometry contains no statements about actual reality and can make such statements only in conjunction with physical laws, it seems possible and reasonable to adhere to Euclidean geometry regardless of the properties of reality. On the other hand, in case of an observed disagreement between theory and experience, it is easier to agree to change physical laws than axiomatic Euclidean geometry. If one forgets all about the connection between a practically rigid body and geometry, then it is not easy to refuse to admit that Euclidean geometry must be viewed as simplest.

Why did Poincaré and other researchers reject the equivalence that thrusts itself upon one between a practically rigid body of real experience and a geometric body? Simply because, when one considers them more closely, real rigid bodies in nature turn out not to be rigid at all, since their geometric behavior, that is, their possible mutual disposition, depends on temperature, external forces, and so on. Thus the initial direct connection between geometry and physical reality is destroyed and we are forced to shift to the following more general position characteristic of Poincaré's viewpoint. Geometry (G) tells us nothing about the behavior of real objects; this behavior is described only by geometry together with the totality of physical laws (Ph). In symbolic terms we can say that only the sum (G) + (Ph) lends itself to experimental verification. Thus one can choose arbitrarily (G) as well as different parts of (Ph); all of these laws represent agreements. To avoid contradictions it is necessary to choose the rest of (Ph) so that (G) and all (Ph) are experimentally verifiable *[165, pp. 7–8]*.

We see that Einstein refers here to the philosophical works of Poincaré mentioned earlier but, unlike Poincaré, who treated "objective reality" and "laws of nature" in the spirit of Mach, Einstein proceeds from "physical reality" and the possibility of "choosing arbitrarily" (G), that is, geometric axioms which, together with (Ph), that is, physical laws, must confront experience. Thus Einstein interprets Poincaré's position in materialistic terms and, like Lobačevskiĭ, believes that the question of which of the conceivable geometric systems corresponds to the geometry of the real world must be decided experimentally.

It should be pointed out that Einstein frequently refers to Mach, whom he regards as one of his predecessors. In the obituary *Ernst Mach* (Ernst Mach. Leipzig, 1916) Einstein considered Mach's critique of Newton's mechanics and wrote:

> The quoted lines show that Mach clearly understood the weak points of classical mechanics and came close to discovering the general relativity theory. And that half a century before its creation! [167].

Einstein regarded Mach's idea that the inertia of a body is explainable by its interaction with other bodies—what Einstein called *Mach's principle*—as one of the factors in the preparation of the general relativity theory. Mach's idea that "space and time cannot be completely separated from one another" [347, p. 350] is very close to the ideas of special relativity theory.

Undoubtedly, Einstein's famous *thought experiments* were influenced by Mach. At the same time Einstein did not agree with Mach's doctrine of sensations as elements of the world. He not only interpreted the terms *sensations* and *experience* in a materialist spirit but also ascribed his viewpoint to Mach when he said (apropos Mach's use of the term *sensations*) that

> frequently, as a result of insufficient familiarity with his, that is, Mach's, works, some persons tend to confuse them with the terminology of philosophical idealism and solipsism [167].

Here Einstein's opinion was based on the fact that he took Mach's materialist terminology literally and throught it inconceivable that a serious physicist could be a subjective idealist. Later Einstein changed his opinion about Mach's philosophy and during a meeting of the French philosophical society, while answering a question from Emile Meyerson, he referred to Mach as a "deplorable philosopher" (*un déplorable philosophe*) and accused him of a "myopic view of science that led him to reject the existence of atoms" [364, p. 62].

Parallel Displacement

The heightened interest in Riemannian geometry brought about by general relativity theory resulted in an extremely important discovery in this geometry, a concept known as *parallel displacement of vectors*. The discovery was made by Levi-Civita in the paper *The notion of parallelism in an arbitrary manifold and the geometric characterization of Riemannian curvature implied by it* (Nozione di parallelismo in una varieta qualunque e consequente specificazione geometrica della curvatura Riemannina. Palermo, 1917) [320].

If a vector a^i is defined at a point A of a Riemannian space then the result of its displacement to a point B of the line $x^i(t)$ is a solution of the differential equation

$$\frac{da^i}{dt} + \Gamma^i_{jk} a^j \frac{dx^k}{dt} = 0, \tag{8.14}$$

that can be written with the help of the covariant derivative as

$$\nabla_k a^i \frac{dx^k}{dt} = 0. \qquad (8.15)$$

It follows that if one defines along a coordinate line a vector field of vectors obtained by parallel displacement of a given vector then at all points of this line the covariant derivative with respect to the corresponding coordinate is equal to zero. Parallel displacement of vectors admits of a visual geometric definition. Parallel displacement of vectors from a point A of the space to an infinitely near point B is generated by mapping a neighborhood of the point A onto a neighborhood of the point B that is the result of applying first a reflection (along geodesics) in the point A in the neighborhood of A, and then a reflection in C, the midpoint of the geodesic arc from A to B, in a neighborhood of C that contains the neighborhood of A. Under this neighborhood mapping the increments of the coordinates of the point A map to the increments of the coordinates of the point B and, consequently, the differentials of the coordinates of A, that is, the vectors at A, map to the differentials of the coordinates of B, that is, the vectors at B. The name *parallel displacement* is due to the fact that the product of two reflections in two points in ordinary space is a translation of this space. Levi-Civita showed that the curvature of a Riemannian space, at a given point, in a two-dimensional direction is the limit of the ratio of the angle by which a vector is rotated as a result of parallel displacement along a closed contour passing through the given point to the area of the surface bounded by this contour when the contour shrinks to the given point (this is Levi-Civita's *geometric characterization of Riemannian curvature*).

The Problem of a Unified Field Theory

In general relativity theory the geometric properties of space are determined by the gravitational field alone, without any effect of the electromagnetic field on the geometry of space. After creating the general theory of relativity Einstein posed the problem of constructing a geometry of four-dimensional spacetime determined by the electromagnetic as well as the gravitational field. The first attempt of construction of a unified field theory was made by Hermann Weyl in his paper *Gravitation and electricity* (Gravitation und Elektrizität. Leipzig, 1918) *[340, pp. 200–216]*. Besides the metric quadratic form

$$ds^2 = \sum_{\mu, \nu} g_{\mu\nu} dx^\mu dx_\nu \qquad (2)$$

Weyl considers also the linear form

$$d\phi = \sum_\mu \phi_\mu dx_\mu \qquad (7)$$

and writes that

The internal metrical connexion of space thus depends on a linear form (7) besides the quadratic form (2)—which is determined except as to an arbitrary factor of proportionality *[340, p. 206]*.

The vector ϕ_μ is the *vector-potential* of the electromagnetic field $F_{\mu\nu}$ given by

$$F_{\mu\nu} = \partial\phi_\mu/\partial x_\nu - \partial\phi_\nu/\partial x_\mu.$$

The geometry constructed by Weyl is called the *Weyl connection*. Weyl expounded this theory also in his *Space, Time and Matter* (1918) *[626]*.

Einstein himself devoted the last 35 years of his life to the search for a unified field theory. Of other attempts at constructing a unified field theory we note, first of all, the paper of Theodor Kaluza (1885–1954) *On the unicity problem of physics* (Zum Unitätsproblem der Physik. Berlin, 1921) *[259]*. To increase the number of relevant parameters Kaluza went in the direction of increasing the number of dimensions of his space and assumed that in addition to the four dimensions of physical spacetime there is a fifth dimension without a direct physical sense. Proceeding from Kaluza's idea the mathematician Veblen and the physicist Pauli proposed in 1930–1933 a unified field theory in which five coordinates are viewed as the projective coordinates of particles in a four-dimensional space.

A detailed history of unified field theories in the first third of the 20th century is given by Vladimir Pavlovič Vizgin in the book *Unified field theories in the first third of the 20th century* (Edinye teorii polya v pervoĭ treti XX veka. Moscow, 1985) *[615]*.

We note that when Einstein posed the question of a unified field theory he assumed that the theories of interactions of the gravitational and electromagnetic fields exhaust all forms of physical interactions. The discovery, in the last few decades, of "strong" and "weak" interactions that do not coincide with these interactions deprived the problem posed by Einstein of the importance he attached to it. In his "Remarks to the Einsteinian sketch of a unified field theory" the famous German physicist Werner Heisenberg (1901–1976) wrote:

This attempt, grandiose in its conception, failed from the very beginning. At the time when Einstein concerned himself with the unified field theory new elementary particles were continuously discovered and with them their associated new fields. This meant that there existed no hard, solid, empirical basis for the implementation of the Einsteinian program and Einstein's attempts produced no conclusive results whatever. However, the failure that dogged the Einsteinian program also had deeper foundations than the uncertainty of empirical facts; these foundations have to do with the relations of Einstein's field-theoretic view of quantum theory. Einstein proceeded from the classical nonlinear field theory of matter for the metric tensor of the field that defined his geometry.

Einstein hoped that in the end it will be possible to think of atoms and elementary particles in such a theory as singular solutions of the non-linear equations of the field. In reality, however, the element of discreteness that expresses the existence of elementary particles has a far wider character. It becomes dominant as soon as we go over to the domain of atoms and elementary particles; which is why their description by means of a field theory of the classical type is out of the question. Conversely, we now know that there operate here quantum-mechanical laws whose structure was understood in the twenties. Since Einstein could not come to terms with such a structure, he did not make an attempt to approach the unified field theory using quantum laws.

Nevertheless, experiments with elementary particles carried out in connection with quantum theory contain very many arguments in favor of Einstein's program.

In the last ten years many elementary particles have been discovered and thus also many new fields. Also, it became clear that elementary particles can change into one another. If two elementary particles with very high kinetic energy collide in some way then there can arise new particles, and the laws that govern the coming into existence and the disappearance of elementary particles can, apparently, be formulated by means of relatively simple selection rules and the corresponding quantum numbers. It follows that all elementary particles "consist," so to say, of one and the same substance that can be simply called "energy" or "matter"; their structure and their ability to go over into one another should follow from a simple law for matter.

Thus, by the nature of things, a satisfactory theory of elementary particles must, at the same time, be a unified field theory of matter *[221a, pp. 120–125]*.

An attempt to create such a unified theory of elementary particles was made by Heisenberg himself.

Spaces with Affine Connection

Although the problem of a unified field theory has not yielded any substantial results for physics, it has been remarkably fruitful for geometry. Weyl's theory was further developed by Schouten, who introduced the general notion of a space with an affine connection in the paper *On different forms of connection that can be laid at the foundation of a differential geometry* (Über die verschiedenen Arten der Übertragung die einer Differentialgeometrie zugrunde gelegt werden können. Berlin, 1922) *[511]*. A detailed exposition of Schouten's theory is found in his *Ricci calculus*. A space with an affine connection is defined as a manifold of points with coordinates x^i at each of whose points there is given a point function $\Gamma^i_{jk} = \Gamma^i_{kj}$ that is not expressible in the general case in terms of the point functions g_{ij}. Because of this last condition it is not possible, in the general case, to define in a space with an affine connection the length of lines, but it *is* possible to define parallel displacement of vectors by formula (8.14). In a space with an affine connection geodesics are defined. They are solutions of the equations (8.11). For each arc of a geodesic there is defined its midpoint. In such a space one can define reflections in points along geodesics. Just as in a Riemannian space, the geometric sense of parallel displacement of vectors is that it is determined by a mapping of a neighborhood of a point of the space onto a neighborhood of an infinitely near point that is the result of two reflections in two points along geodesics. The name *affine connection* is due to the fact that the mapping of vectors at a point of space onto vectors at an infinitely near point of this space is an affine transformation of vector spaces.

It is clear that Riemannian and pseudo-Riemannian spaces are special cases of spaces with an affine connection. In such spaces the mapping of vectors at one point onto vectors at an infinitely close point under parallel displacement is an isometry of Euclidean and pseudo-Euclidean spaces, respectively.

The space introduced by Weyl in his unified field theory is also a space with an affine connection. In this space the mapping of vectors at one point onto vectors at an infinitely close point under parallel displacement is a similarity transformation.

Even more general spaces were introduced by the French mathematician Élie Cartan (1869–1951). In the papers *On manifolds with an affine connection and the generalized relativity theory* (Sur les variétés à connexion affine et la théorie de la relativité généralisée. Paris, 1923) *[96, part 3, pp. 659–746, 798–825, 921–992]*, *Spaces with a conformal connection* (Les espaces à connexion conforme. Paris, 1923) *[96, part 3, pp. 747–797]* and *On manifolds with a projective connection* (Sur les variétés à connexion projective. Paris, 1924) *[96, part 3, pp. 825–861]* Cartan introduced the notion of a *space with a homogeneous connection*. Whereas in the case of a space with an affine connection there is associated to every point a vector space that can be regarded as an affine space, and under parallel displacement of vectors the space at one point is mapped by an affine transformation onto the space at an infinitely near point, in the case of a space with a homogeneous connection there corresponds to every point of the space a homogeneous space with a definite group of transformations. Also, for every pair of infinitely near points there is specified a transformation of the spaces associated with these points that preserves their geometry. In addition to spaces with an affine connection and special cases of such spaces noted above, the most important spaces with a homogeneous connection are *spaces with a projective* and *conformal connection*, respectively. To each point of these spaces there is associated, respectively, a projective or conformal space, and for each pair of infinitely near points of a space with a connection there is specified, respectively, a projective or conformal mapping of one of the spaces associated with these points onto the other.

Differentiable Manifolds

The spaces introduced by Riemann and the groups introduced by Lie, bearing the names of their respective inventors, as well as pseudo-Riemannian spaces, Weyl spaces, and the more general spaces with an affine or other homogeneous connection were defined only locally—in the domain of a certain coordinate system. The introduction of topological manifolds made possible the study of differential-geometric spaces in the large. In every n-dimensional topological manifold one can define a homeomorphic mapping of a neighborhood U of every point onto some domain of n-dimensional Euclidean space, and the coordinate system in the Euclidean space provides coordinates for the neighborhood U. This system of coordinates in the neighborhood U is now called a *system of local coordinates* or a *local map* of the manifold, and the totality of local maps is called the *atlas* of the manifold. A manifold is called *differentiable of class v* if (1) the domains of the maps of the atlas cover the whole manifold; and (2) if U and U' are two domains of maps of the atlas with nonempty intersection and a point x in the intersection has coordinates x^i and $x^{i'}$ in the two maps then the functions $x^{i'} = f(x^1, \ldots, x^n)$ have continuous partial derivatives up to and including order v, and the jacobian $\det\left(\dfrac{\partial x^{i'}}{\partial x^j}\right)$ is different from zero throughout the intersection of the domains. The manifold is called *analytic* if the functions in (2) are analytic.

The definitions of a differentiable manifold and of an analytic manifold were formulated by Oswald Veblen (1880–1960) and by John Henry Constantine Whitehead (1904–1960) in *The foundations of differential geometry* (Cambridge, 1932) *[612]*.

With every point of a differentiable manifold there is associated a vector space with vectors $\{dx^i\}$ whose coordinates are the differentials of the coordinates x^i; these spaces are called the *tangent spaces* of the differentiable manifold. Under coordinate transformations $x^i \rightarrow x^{i'}$ the coordinates of the vectors and tensors of the tangent spaces transform according to the rules (8.7)–(8.9), and so on. If in every tangent space of a differentiable manifold there is given a metric tensor g_{ij} that determines an inner product in the Euclidean or pseudo-Euclidean space then the differentiable manifold is called, respectively, a *Riemannian* or *pseudo-Riemannian space in the large*. If in each of the tangent spaces of a differentiable manifold there is given an object Γ_{jk}^i that transforms under coordinate transformations according to the same rule as the coefficients Γ_{jk}^i of Riemannian spaces then the differentiable space is called a *space with an affine connection in the large*.

Fibrations

Spaces with connections are special cases of fiber spaces. Unlike spaces with Schouten-Cartan connections, fiber spaces are considered in the large. The initial form of the theory of fiber spaces in the large was the theory of "sphere-spaces," that is, fiber spaces whose "fibers" are spheres. This theory arose in connection with the solution of the problem of continuous mapping of an n-dimensional sphere onto an m-dimensional one for, if $n > m$, then to every point of the m-dimensional sphere there corresponds an $(n - m)$-dimensional manifold of the n-dimensional sphere, so that the latter "stratifies" or "fibrates" into these manifolds; the resulting fibres are also called *spheres*. The theory of such spaces was constructed by the American mathematician Hassler Whitney (b. 1907) in the paper *Sphere-spaces* (Boston, 1935) *[632]*. In the paper *On the theory of sphere bundles* (Boston, 1940) *[633]* Whitney calls these spaces *sphere-bundles*. The most complete account of Whitney's theory is contained in his lectures *On the topology of differentiable manifolds* (Ann Arbor, 1941) *[634]*. Together with *sphere-bundles* Whitney also considered *plane-bundles*. Whitney associated to every differentiable manifold a *tangent plane-bundle* consisting of the tangent spaces of the manifold and *tangent sphere-bundle* consisting of spheres in these tangent spaces with centers at the points of tangency, with one sphere in each tangent space. In the first case, the points of the *fibers* characterize the vectors of the differentiable manifold; in the second, they characterize the directions issuing from a point of the manifold.

Soon Whitney's *sphere-bundles* were replaced by arbitrary manifolds. Then there arose the more general concept of a *fiber bundle* or a *fiber space*. The general theory of such spaces was set forth by the American mathematician Norman Steenrod (1910–1971) in the monograph *The topology of fiber bundles* (Princeton, 1951) *[560]*, which completed the development of this theory.

The simplest fiber space is the *direct product* of two topological spaces S and T, that is, the totality of pairs (s, t) of elements s in S and t in T where the closed sets of the topological product are sets of pairs (s, t) in which both elements s and t vary over closed sets in the spaces S and T, respectively. In this case one may view the set of pairs (s, t_0) with fixed t_0 as a *fiber* and the space T of all possible elements t_0 as a basis of the fiber space. One can also view the set of pairs (s_0, t) with fixed s_0 as a *fiber* and the space S as a basis of the same fiber space. In this case the fibration is called *trivial*.

In more complicated cases, a fibre space is not simply the direct product of a *fiber* by a *basis* but also consists of homeomorphic fibers, and, if the latter are viewed as points of some new space, then this space can be regarded as a basis of the fiber space. The simplest fiber space of this kind is a *Möbius strip* obtained from the lateral surface of a right circular cylinder by identification of pairs of points that are symmetric with respect to its center, or from a rectangle (which may be thought of as the application of half of the preceding cylinder to a plane) by gluing the points of a pair of its opposite sides that are symmetric with respect to its center. (Hence the name *skew products* for general fibrations of a space used in the Russian translation of N. Steenrod's book *[560]*). One usually considers also the group of topological mappings of a fiber onto itself that preserves a certain structure of these fibers and mappings of fibers, one onto another, that preserve this structure.

The most important fiber spaces are the *tangent plane-bundles* defined by Whitney for differentiable manifolds. Now they are called simply *tangent bundles* of differentiable manifolds. The bases of these fiber spaces are the differentiable manifolds, and the fibres are their tangent spaces. To the class of fiber spaces there belong Riemannian spaces, pseudo-Riemannian spaces, and spaces with an affine connection in the large together with their tangent spaces that are, respectively, Euclidean, pseudo-Euclidean, and affine, as well as spaces with a projective, conformal, and arbitrary homogeneous connection together with the corresponding homogeneous spaces that also form *fibers*. The groups of transformations of the *fibers* of these spaces are, respectively, the groups of Euclidean and pseudo-Euclidean motions, of affine, projective, conformal, and other transformations.

Exterior Forms, Curvature and Betti Numbers

The field of a skew-symmetric covariant tensor $a_{i_1 i_2 \ldots i_k}$ in a domain of a differentiable manifold defines at every point of this domain an exterior differential form

$$\omega_k = \sum_{i_1} \cdots \sum_{i_k} a_{i_1 i_2 \ldots i_k} dx^{i_1} \wedge dx^{i_2} \wedge \cdots \wedge dx^{i_k},$$

where $\wedge$ denotes exterior multiplication. The form is invariant under even permutations of the dx^{i_α} and changes sign under odd permutations of the dx^{i_α} (in virtue of the skew symmetry of $a_{i_1 \ldots i_k}$ the ordinary differential form $\sum_{i_1} \cdots \sum_{i_k} a_{i_1 i_2 \ldots i_k} dx^{i_1} dx^{i_2} \ldots dx^{i_k}$ is equal to zero). For exterior forms we define the operations of exterior multiplication

$$\omega_{k+l} = \omega_k \wedge \omega_l$$

and exterior differentiation

$$\omega_{k+1} = \mathscr{D}\omega_k = \sum_{i_1} \cdots \sum_{i_k} da_{i_1 \ldots i_k} \wedge dx^{i_1} \wedge \cdots \wedge dx^{i_k}.$$

The differential forms in multiple integrals defining areas and volumes are exterior forms, and the Ostrogradskiĭ formula in the beginning of chapter 7 can be written as

$$\int_{\partial M} \omega = \int_M \mathscr{D}\omega,$$

where M is a bounded domain and ∂M is its boundary.

The theory of exterior forms was founded by É. Cartan in the paper *On certain differential expressions and the Pfaff problem* (Sur certaines expressions différentielles et le problème de Pfaff. Paris, 1899) *[96, part 2, pp. 303–396]*. It was extensively developed by É. Cartan (see his book *Exterior differential systems and their geometrical applications* (Les systèmes différentiels extérieurs et leurs applications géométriques. Paris, 1945) *[97]*, by the Soviet geometer Sergeĭ Pavlovič Finikov (1883–1964), and by their students.

If a moving orthonormal frame $\{e_i\}$ is associated with every point of a curve or surface in the space R_n or with every figure of a family of figures in this space, then the differentials of the radius vector x and of the vectors e_i of the frame can be written as

$$dx = e_i \omega^i, \qquad de_i = e_j \omega_i^j, \qquad (8.16)$$

where

$$\omega_j^i = -\omega_i^j, \qquad (8.17)$$

and the exterior differentials of the forms ω^i and ω_j^i satisfy the equations of structure

$$\mathscr{D}\omega^i = \sum_k \omega^k \wedge \omega_k^i, \qquad \mathscr{D}\omega_j^i = \sum_k \omega_j^k \wedge \omega_k^i. \qquad (8.18)$$

A curve, surface or family of figures is defined by a system of Pfaff equations $\theta^1 = \theta^2 = \cdots = \theta^r = 0$, where the forms θ^A are linear combinations of the forms ω^i, ω_j^i. For a curve with arc length s and tangent vector e_1 these equations are $\omega^2 = \cdots = \omega^n = 0$ and for a hypersurface with normal vector e_n the equation is $\omega^n = 0$. If for the curve we put $de_1/ds = k_1 e_2$ and choose vectors $e_3, \ldots, e_n$ so that de_i/ds is a linear combination of $e_1, \ldots, e_{i+1}$, then $\omega_i^{i+1}/ds = -\omega_{i+1}^i/ds = k_i$ are curvatures of the curve and the remaining ω_j^i are zeros. In this way we obtain the *Frenet formulas*

$$de_1/ds = k_1 e_2,$$

$$de_2/ds = -k_1 e_1 + k_2 e_3, \ldots,$$

$$de_i/ds = -k_{i-1} e_{i-1} + k_i e_{i+1}, \ldots, de_n/ds = -k_{n-1} e_{n-1}$$

discovered for R_3 in 1847 by Frederic Frenet (1816–1900) and for R_n in 1874 by C. Jordan. The curvatures k_i define the curve up to a motion of R_n.

For a hypersurface we direct the vectors $e_1, \ldots, e_{n-1}$ along curvature lines of the hypersurface and the vector e_n along its normal. Exterior differentiation of the Pfaff equation $\omega^n = 0$ gives, in view of (8.18), $\mathscr{D}\omega^n = \sum_k \omega^k \wedge \omega_k^n = 0$. Hence $\omega_\alpha^n = \sum_\beta b_{\alpha\beta}\omega^\beta$ ($\alpha, \beta = 1, 2, \ldots, n - 1$). We note that the analogues of Gauss's quadratic forms for surfaces in R_3 and for hypersurfaces in R_n are

$$dx^2 = \sum_\alpha (\omega^\alpha)^2,$$

$$e_n d^2 x = -de_n dx = -\left(-\sum_\alpha \omega_\alpha^n e_\alpha\right)\left(\sum_\beta \omega^\beta e_\beta\right) = \sum_\alpha \sum_\beta b_{\alpha\beta}\omega^\alpha \omega^\beta.$$

In our basis $b_{\alpha\beta} = b_\alpha \delta_{\alpha\beta}$, $-b_\alpha = k_\alpha$ are main curvatures of the hypersurface; they are differential invariants of the hypersurface. Exterior differentiation of other forms yields a complete system of differential invariants of the hypersurface which defines it up to a motion of the space R_n.

Formulas (8.16) and (8.18) also hold in the affine space E_n and in the pseudo-Euclidean space ${}^l R_n$. In the case of E_n the forms ω_j^i are not connected by any condition. In the case of $^l R_n$ with scalar square of a vector $\mathbf{x}^2 = \sum_i \varepsilon_i (x^i)^2$ the conditions (8.17) become

$$\varepsilon_i \omega_j^i = -\varepsilon_j \omega_i^j. \tag{8.19}$$

In the spaces P_n, S_n and $^l S_n$ the points are represented by vectors of the spaces E_{n+1}, R_{n+1} and $^l R_{n+1}$, respectively. In these spaces the role of the radius vector $\mathbf{x}$ is played by e_0. In these spaces hold the second formulas (8.16) and (8.18). In P_n the forms ω_j^i are connected by the condition $\sum_i \omega_i^i = 0$, in S_n by the condition (8.17), and in $^l R_n$ by the condition (8.19).

Analogous methods can be applied in all homogeneous spaces.

These methods can also be applied in spaces with affine connection and in Riemannian and pseudo-Riemannian spaces. In these spaces the vectors $\mathbf{e}_i$ are vectors of the tangent spaces of E_n, R_n and $^l R_n$ respectively. In these spaces hold formulas (8.16) and in Riemannian and pseudo-Riemannian spaces formulas (8.17) and (8.19), respectively. The equations of structure there have the form

$$\mathscr{D}\omega^i = \sum_k \omega^k \wedge \omega_k^i + \sum_j \sum_k S_{jk}^i \omega^j \wedge \omega^k,$$

$$\mathscr{D}\omega_j^i = \sum_k \omega_j^k \wedge \omega_k^i + \sum_k \sum_h R_{jkh}^{\cdots i} \omega^k \wedge \omega^h$$

where S_{jk}^i is the *torsion tensor* (equal to zero in spaces without torsion, in particular, in Riemannian and pseudo-Riemannian spaces), $R_{jkh}^{\cdots i}$ is the *curvature tensor* (in Riemannian spaces $R_{ijkl} = g_{hl} R_{ijk}^h$ is the Riemann tensor).

An exterior differential form ω is called *closed* if $\mathscr{D}\omega = 0$ and *exact* if $\omega = \mathscr{D}\omega'$. Every exact form is closed since $\mathscr{D}(\mathscr{D}\omega) = 0$. Exterior differentiation is similar to the taking of a boundary, except that in the latter case the dimension of a manifold decreases and in the former case the "order" of a form increases. This analogy was made use of in the paper *On analysis situs* [i.e. topology] *of n-dimensional manifolds* (Sur l'analysis situs des variétés à n-dimensions. Paris, 1931) *[452]* by the Belgian mathematician Georges de Rham (1903–1969). In this paper de Rham defined the *cohomology groups*, in a certain sense dual to Betti's *homology groups*, and applied exterior differential forms for the calculation of Betti numbers. The integral of a closed form is said to be homologous to zero, and h integrals of closed forms on a p-dimensional submanifold which are independent in the usual topological sense are said to be homologically independent if no linear combination of these integrals with constant coefficients not all zero is homologous to zero. De Rham has shown that the Betti number b_p is equal to the number of homologically independent integrals of exact differential forms of order p. This method of calculation of Betti numbers was applied by Cartan in his *Topology of compact homogeneous spaces* (La topologie des espaces homogènes clos. Moscow, 1937) *[96, part 1, pp. 1331–1338]* and by his student Charles Ehresmann in the paper *On the topology of certain homogeneous spaces* (Sur la topologie de certains espaces homogènes. Princeton, 1934) *[164]*.

The Betti numbers are connected with the Gaussian curvature of a surface: if the Gaussian integral curvature $\int K d\sigma$ is taken over a closed compact surface in R_3, then this integral is equal to $4\pi\left(1 - \dfrac{b_1}{2}\right)$, where b_1 is the first Betti number of the surface. This number is equal to double the number of handles of the surface ($b_1 = 0$ for a sphere and $= 2$ for a torus). About this problem see the book of Kentaro Yano and Salomon Bochner *Curvature and Betti numbers* (Princeton, 1953) *[645]*.

Chapter 9
Groups of Transformations

The Emergence of the Group Concept

The group concept was first defined for a certain class of concrete groups, namely *groups of substitutions*, which were studied in connection with attempts to obtain solutions in radicals of algebraic equations of degree $n \geq 5$. Permutations of roots of algebraic equations were first studied by J. L. Lagrange in his *Reflections on the solution of equations* (Réflexions sur la résolution des équations. Berlin, 1771) *[298, vol. 3, pp. 205–515]*. Lagrange noticed that if x_1, x_2, x_3 are the roots of a cubic equation, then each of the cubic radicals in the *Cardano form* can be written as $\frac{1}{3}(x_1 + \omega x_2 + \omega^2 x_3)$, where ω is a cube root of 1. Since the function $(x_1 + \omega x_2 + \omega^2 x_3)^2$ takes on two values under all possible permutations of the roots, it follows that this function is a root of a quadratic equation whose coefficients are rationally expressible in terms of the coefficients of the given equation. Lagrange also noticed that in the case of the fourth-degree equation the function $x_1 x_2 + x_3 x_4$ of the four roots of this equation takes on only three values as a result of all permutations of the roots and is therefore a root of a cubic equation whose roots are rationally expressible in terms of the coefficients of the given equation. He called this pattern

> the true principle, and, so to say, the metaphysics of the solution of an equation of third and fourth degree *[298, vol. 3, p. 357]*.

Lagrange posed the problem of the number v of values that can be taken on by a rational function V of the roots of an equation as a result of all possible permutations of the roots and showed that v is a divisor of the number $n!$ of all possible permutations of the n roots.

Permutations of the roots were also studied by Paolo Ruffini (1765–1822) in his *General theory of equations in which is demonstrated the impossibility of an algebraic solution of general equations of degree higher than the fourth* (Teoria generale delle equazioni in cui si dimostra impossibile la soluzione algebrica delle equazioni generale di grado superiore al quatro. Bologna, 1799) *[489]*. In this work Ruffini explicitly supported the point of view that it is impossible to obtain a solution in radicals of the general equation of degree $n \geq 5$. Ruffini's proof was incomplete. Essentially, it was also based on the investigation of groups of substitutions. A complete solution of the problem was obtained by Niels Henrik Abel (1802–1829) in his *Proof of the impossibility of an algebraic solution of general equations exceeding the fourth degree* (Démonstration de l'impossibilité de la résolution algébrique des équations générales qui passent quatrième degré. Berlin, 1826) *[4, vol. 1, pp. 66–94]*. The virtual identity of the titles of the works of Abel and Ruffini is an indication of their "successor" connection.

Explicit groups of substitutions were investigated by Augustin Louis Cauchy in his *Memoir on the number of values that can be taken on by a function if one permutes in all possible ways the quantities it contains* (Mémoire sur les nombre des valeures qu'une fonction peut acquérir lorsqu'on y permute de tout les manières possibles les quantités qu'elle renferme. Paris, 1815) *[100, vol. 1, pp. 64–90]*.

Cauchy called groups of substitutions *systems of conjugate substitutions* (systèmes de substitutions conjugées). In this paper Cauchy first used such now generally accepted terms as *transitive* for a system of substitutions (a system such that for any two of the permuted elements there is a permutation that permutes them), and *transposition* (a substitution that interchanges the positions of two elements and leaves each of the remaining elements fixed). Of Cauchy's many papers on the theory of groups of substitutions we mention the *Memoir on arrangements which can be formed of given letters and on permutations and substitutions by means of which one can pass from one arrangement to another* (Mémoire sur les arrangement qu l'on peut former avec les lettres données et sur les permutations et substitutions à l'aide desquelles on passe d'un arrangement à un autre) which is part of volume III of Cauchy's *Exercises in analysis and in mathematical physics* (Exercises d'analyse et de physique mathématique. Paris, 1844) *[100, vol. 13, pp. 171–282]*. One of the results proved in this paper is that a group with pq elements, p a prime, has at least one subgroup with p elements.

The term *group* first appeared in a paper by the French mathematician Evariste Galois (1811–1832), *Memoir on the conditions of solvability of equations by radicals* (Mémoire sur les conditions de resolubilité des équations par radicaux) written in 1830 and published by J. Liouville in 1846 *[194, pp. 43–71]*.

Galois introduced the term *group* only for substitutions and formulated the basic property of groups as follows:

If in such a group there are the substitutions S and T then there is the certainty of there being the substitution ST *[194, p. 47]*.

For groups of substitutions, the remaining group properties—associativity, the existence of a *neutral element*, and the existence of an *inverse* (0 and $-a$

for addition and 1 and a^{-1} for multiplication)—are automatically satisfied. Galois wrote the group operation as multiplication. He used the word *group* in a wider sense than we do. This is clear from the following phrase in Galois's letter to his friend Auguste Chevalier, written the night before his fatal duel:

> When a group G contains another group H then the group G can be decomposed into groups each of which can be obtained by applying to the substitutions of H one and the same substitution in such a way that $G = H + HS + HS' + \cdots$ [194, pp. 173–175].

This phrase shows that Galois called *groups* also what we now call right cosets of a group with respect to a subgroup. Galois defined a similar decomposition of a group into left cosets, $G = H + TH + T'H + \cdots$.

We note that since all permutations of n roots of an equation form a group and the permutations that preserve a value of a rational function V of the n roots of the equation form a subgroup of this group, Lagrange's theorem, that the number v of values that can be taken on by the function V under all possible permutations is a divisor of the number $n!$ of all possible permutations of the n roots, is a special case of the general theorem that the number of elements of a subgroup of a finite group divides the number of elements of that group. Lagrange's argument can be applied to any group G and its subgroup H. Essentially, the argument amounts to this: one considers products of elements of G not in H by all elements of H. These products form right or left cosets, depending on whether the subgroup elements are the left or right factors in the products. The theorem follows from the fact that the number of elements in each coset is the same as the number of elements in the subgroup H and each element belongs to exactly one coset. That is why the general theorem is now called *Lagrange's theorem*. The equations introduced by Abel and subsequently named for him are characterized by the fact that they admit a commutative group of permutations of their roots. That is why commutative groups are often called *Abelian groups*.

Galois Theory

The problem of solvability of equations in radicals was finally settled by Galois in his *Memoir on the conditions of solvability of equations in radicals* (mentioned previously).

In this paper Galois also introduces the notion of a *number field*. A *number field* is a set of numbers including 0 and 1 whose elements form a group under addition and whose nonzero elements form a group under multiplication. The simplest examples of fields are the fields $\mathbf{Q}$, $\mathbf{R}$, and $\mathbf{C}$ of rational, real, and complex numbers, respectively. Another example of a number field is the set of numbers of the form $a + b\sqrt{3}$, a, b rational. If the coefficients of a polynomial belong to a certain field $\mathbf{F}$ and the polynomial can be written as a product of polynomials with coefficients from that field then the polynomial is said to be reducible; otherwise it is said to be irreducible. If a number α is not in $\mathbf{F}$ then we can form the field $\mathbf{F}(\alpha)$ of numbers of the form $a + b\alpha$, where a and b are in $\mathbf{F}$. The field $\mathbf{F}(\alpha)$ is called an *extension* of the field $\mathbf{F}$ generated by α. In particular, the field of numbers of the form $a + b\sqrt{3}$ where a and b are rational is the field $\mathbf{Q}(\sqrt{3})$. If the coefficients of a polynomial of degree n belong to a field $\mathbf{F}$ then the extension $\mathbf{F}(\alpha_1, \alpha_2, \ldots, \alpha_n)$ of $\mathbf{F}$ generated by the roots of that polynomial is called its *splitting field*, for the polynomial in question can be written as a product of linear factors $(x - \alpha_1)(x - \alpha_2) \ldots (x - \alpha_n)$. In particular, the extension $\mathbf{Q}(\sqrt{3})$ of the field $\mathbf{Q}$ of rational numbers is the splitting field of the polynomial $x^2 - 3$ which is irreducible over $\mathbf{Q}$ but can be factored, $x^2 - 3 = (x - \sqrt{3})(x + \sqrt{3})$, over the field $\mathbf{Q}(\sqrt{3})$.

Galois shows that the splitting field $\mathbf{K}$ of a polynomial irreducible over $\mathbf{F}$ has the property that any polynomial irreducible over $\mathbf{F}$ with one root in $\mathbf{K}$ splits into linear factors over $\mathbf{K}$; that is, $\mathbf{K}$ contains the splitting field of that polynomial; such an extension is called a *normal extension* of the field $\mathbf{F}$. To every field $\mathbf{K}$ which is the splitting field of a polynomial that is irreducible over some field $\mathbf{F}$, Galois associates the group of bijective mappings $\alpha \to \alpha^s$ of $\mathbf{K}$ onto itself that preserve sums and products, that is, mappings such that

$$(\alpha + \beta)^s = \alpha^s + \beta^s, \qquad (\alpha\beta)^s = \alpha^s\beta^s,$$

and that fix all elements of $\mathbf{F}$; briefly, the group in question is the group of *automorphisms* of $\mathbf{K}$ that fix the elements of $\mathbf{F}$. This group is now called the *Galois group* of the field $\mathbf{K}$ over the subfield $\mathbf{F}$. If a polynomial

$$f(x) = x^n + a_1 x^{n-1} + \cdots + a_n = 0 \tag{9.1}$$

with coefficients in $\mathbf{F}$ has at least one root α in $\mathbf{K}$ then we have the equality

$$\alpha^n + a_1\alpha^{n-1} + \cdots + a_{n-1}\alpha + a_n = 0.$$

By applying to this equality the mapping $\alpha \to \alpha^s$ of the Galois group we see that

$$(\alpha^s)^n + a_1(\alpha^s)^{n-1} + \cdots + a_{n-1}\alpha^s + a_n = 0,$$

that is, α^s is also a root of the polynomial (9.1). This means that the Galois group of an irreducible polynomial (that is, the group of its splitting field) may be regarded as a group of permutations of its roots.

Galois also showed that if $\mathbf{L}$ is a field contained in the field $\mathbf{K}$ and containing the field $\mathbf{F}$ then the Galois group of $\mathbf{K}$ over $\mathbf{L}$ is a subgroup of the Galois group of $\mathbf{K}$ over $\mathbf{F}$, and that to a subgroup of the Galois group of $\mathbf{K}$ over $\mathbf{F}$ there corresponds the subfield of elements of $\mathbf{K}$ that are left fixed by the elements of that subgroup.

Galois singled out an important class of subgroups with the property that the decompositions of the group into left and right cosets with respect to the subgroup coincide. Another way of saying this is that every product gh of a group element $g \in G$ by a subgroup element $h \in H$ can be written as $h'g$ where, $h' \in H$ (in general, $h' \neq h$). The equality $gh = h'g$ can also be written as $h' = ghg^{-1}$, and one says that the subgroup is *invariant* under the transformation that consists in multiplying by an arbitrary group element g on the left and by its inverse g^{-1} on the right. In such a case, Galois called the decomposition into cosets ("into groups") a *proper decomposition*. Today we call the subgroups singled out by Galois *normal subgroups* of the group G. If N is a normal subgroup of a group G and we associate to each element g of G the coset gN consisting of all products of g by elements of N, then for two elements g_1 and g_2 from different cosets the product $(g_1n_1)(g_2n_2) = g_1(n_1g_2)n_2 = g_1g_2n_1'n_2 = g_1g_2n_3$ is in the coset that corresponds to the element g_1g_2. Therefore the cosets form a group under multiplication. This group is called the *factor group* of the group G with respect to its normal subgroup N and is denoted by G/N. If there is a mapping φ from a group A onto a group B that preserves products (that is, the element corresponding to a product is the product of the elements corresponding to its factors), then φ is called a *homomorphism* and B is said to be a *homomorphic* image of A (from the Greek words *homos*—the same—and *morphē*—form); if there is a bijective correspondence between the two groups with the above property, then we say that the groups are *isomorphic* (*isos*—equal). Thus the group G/N is a homomorphic image of the group G; the identity element of G/N is the normal subgroup N of G; that is, under the homomorphism the elements of the subgroup N of G correspond to the identity element of the group G/N. It is easy to verify that under a homomorphism between groups G and G' the set of elements of G that are mapped onto the identity of G' forms a normal subgroup of G.

Galois showed that if a field L that corresponds to a subgroup of the Galois group of K over F is a normal extension of F then the Galois groups of K and L are homomorphic, so that the subgroup of the Galois group is normal.

By means of these concepts Galois found necessary and sufficient conditions for solvability in radicals of an algebraic equation $f(x) = 0$ with coefficients in the field F. This condition is that the Galois group of the splitting field K of the polynomial $f(x)$ must have a nested sequence of subgroups

$$G, N_1, N_2, \ldots, N_j, \ldots, N_r$$

such that N_1 is normal in G (that is, is a normal subgroup of G), N_j is normal in N_{j-1}, and the factor groups $G/N_1, \ldots, N_j/N_{j+1}$ and the subgroup N_r are isomorphic to the groups of the different powers of the cycle $(a_1, a_2, \ldots, a_{n_j})$. The cycle $(a_1, a_2, \ldots, a_{n_j})$ is the permutation that sends a_k for $k \neq n_j$ into a_{k+1} and a_{n_j} into a_1. For $j \neq 0$, and $j \neq r$, n_j is the number of elements of the factor group N_j/N_{j+1}; for $j = 0$ it is the number of elements of the factor group G/N_1, and for $j = r$ it is the number of elements of the group N_r. The powers of the cycle $(a_1, a_2, \ldots, a_{n_j})$ form a group called a *cyclic group of order n_j*. (This group is isomorphic to the group of rotations of a circle through the respective angles

$0, 2\pi/n_j, 4\pi/n_j, \ldots, 2(n_j - 1)\pi/n_j$, as well as to the group of complex numbers $\omega_k = \cos 2\pi k/n_j + i \sin 2\pi k/n_j$, that is, the group of n_j-th roots of 1.) Groups with such chains of subgroups are called *solvable groups*.

Cyclic groups are the simplest solvable groups. To such groups belong the Galois groups of the splitting fields of the left sides of equations of the form $x^n - a = 0$, provided that the field $\mathbf{F}$ that contains the number a also contains a root ω_k of 1 such that k is relatively prime to n (a so-called *primitive root* of 1; ω_k has the property that every n-th root of 1 is a power of ω_k). In this case, the Galois group consists of the transformations $\alpha \to \alpha^s = \alpha \omega_k^s$.

All commutative groups are solvable. The following is a sketch of an argument that supports this assertion. Let G and G' be two commuting groups (that is, groups for which $gg' = g'g$ for all $g \in G$ and $g' \in G'$) that have no common elements other than 1, and let us define the direct product $G \times G'$ of G and G' as the group of products gg'. It can be shown that every commutative group is a direct product of cyclic groups. Since the groups G and G' are normal subgroups of the group $G \times G'$, and G and G' are, respectively, isomorphic to the factor groups $G \times G'/G'$ and $G \times G'/G$, it follows that for a commutative group that is the direct product of cyclic groups $Z_1, Z_2, \ldots, Z_r$ the role of the normal subgroups $N_1, N_2, \ldots, N_r$ is played by the nested sequence of direct products of $r - 1, r - 2, \ldots, 2$ of the groups Z_j and the last one of these groups. The Galois group of an Abelian equation is commutative.

In the case of a solvable Galois group, the solution of the corresponding equation reduces to the solution of a chain of binomial equations, and a solution of a binomial equation $x^n - a = 0$ is a radical $\sqrt[n]{a}$.

In the case of an algebraic equation of degree n the largest possible Galois group is the group of $n!$ permutations of n elements. Such a group is called the *symmetric group* of degree n and is denoted by S_n. Every group S_n has a normal subgroup consisting of the *even permutations*, that is, permutations that are products of an even number of transpositions; this subgroup is called the *alternating group* of degree n and is denoted by A_n. The factor group S_n/A_n consists of 2 elements and is a cyclic group of order 2. The group S_2 is cyclic of order 2. The group S_3 consists of 6 elements and has the normal subgroup A_3 which is a cyclic group of order 3. The group S_4 consists of 24 elements and has a normal subgroup A_4 of 12 permutations. In turn, the group A_4 has a normal subgroup B_4 whose elements are the identity permutation and the three products of cycles $(a_1 a_2)(a_3 a_4)$, $(a_1 a_3)(a_2 a_4)$, and $(a_1 a_4)(a_2 a_3)$. This group is commutative and is a direct product of two cyclic groups of order 2, and the factor group A_4/B_4 is a cyclic group of order 3. Thus in the case of equations of degree 2, 3, and 4 the maximal Galois groups are solvable, and if the Galois groups are not maximal then they are subgroups of the maximal subgroups and as such are also solvable.

The group S_n for $n \geq 5$ contains the normal subgroup A_n and the factor group S_n/A_n is cyclic of order 2. But for $n \geq 5$ the group A_n has no normal subgroups other than the trivial ones (itself and the group identity). Therefore the groups S_n are not solvable for $n \geq 5$, and so, in general, algebraic equations of degree ≥ 5 are not solvable in radicals. In particular, the equation $x^5 + x - a = 0$ is not solvable in radicals for many integer values of a, such as $a = 3, 4, 5, 7, 8, 9, 10, 11, \ldots$.

Abstract Groups

The concept of isomorphism permits us to talk about groups abstracted from various realizations as groups of substitutions, motions, and other transformations of objects of one kind or another. The concept of an abstract group was first formulated by A. Cayley in his paper *On the theory of groups, as depending on the symbolic equation $\theta^n = 1$*. London, 1854 *[103, vol. 2, pp. 123–192; vol. 4, pp. 88–91]*. In this paper was given the first axiomatic definition of a group. An important role in the propagation of the group notion among mathematicians was played by the *Treatise on substitutions and algebraic equations* (Traité sur les substitutions et des équations algébriques. Paris, 1870) *[249]* of the French mathematician Camille Jordan in which are also considered general problems of group theory.

At the end of the 19th century Heinrich Weber (1842–1913) formulated an axiomatic definition of a group in his *Algebra* (Leipzig, 1898) *[621]*, written at the same time as Peano's treatises on axiomatics and Hilbert's *Foundations of geometry*. Here a *group* is defined as a set of elements of arbitrary nature on which there is defined a group operation $a \circ b = c$. This operation is associative; that is, for any three of its elements a, b, c

$$(a \circ b) \circ c = a \circ (b \circ c),$$

there exists a neutral element e such that for every a in the group

$$e \circ a = a \circ e = a,$$

and for every element a there exists an element $\bar{a}$ such that

$$\bar{a} \circ a = a \circ \bar{a} = e.$$

One of the earliest expositions of abstract group theory is *Abstract group theory* (Abstraktnaya teoriya grupp. Kiev, 1916) by the Russian algebraist (later a famous Soviet arctic explorer) Otto Yul'evič Schmidt (1891–1956) *[536, vol. 1, pp. 17–175]*.

The Helmholtz Paper

The appearance of the paper *On the facts which lie at the foundations of geometry* (Über die Thatsachen die der Geometric zu Grunde liegen. Göttingen, 1868) *[223, pp. 618–639]* by the great German naturalist Hermann Helmholtz (1821–1894) played a vital role in the application of the group concept to geometry. The very title of this paper indicates that it was meant to be a response to Riemann's paper discussed in the previous chapter. In the fifties and sixties Helmholtz investigated the physiology of vision and hearing and, as he points out at the beginning of his paper, it was physiological findings that made him reflect on problems of space. He wrote:

> My investigations on space perceptions in the area of vision led me to investigate the question of the origin and essence of our general views on space [223, p. 618].

and continued as follows:

> As in the case of Riemann, my immediate objective was to investigate which characteristics of space belong to every manifold that depends on many variables and that changes continuously, a manifold whose differences admit of quantitative comparisons and, conversely, which of them, not conditioned by this general character, belong to space alone.
>
> I happened to have in physiological optics two examples that admitted of spatial representation of variable manifolds, namely the system of colors mentioned by Riemann and measurements of the field of vision by means of an ophthalmometer. Both manifolds represent well-known basic differences and they suggested to me a comparison [223, p. 619].

Helmholtz goes on to say that, in many things, he obtained the same results as Riemann and, although the publication of Riemann's paper took away his priority rights for a number of his results, the coincidence of these results is for him "an important guarantee of correctness" of the road he chose "in an area of problems discredited by earlier unsuccessful attempts." Helmholtz goes on to consider those of his results that do not agree with the results of Riemann. He writes:

> My investigations differ from the findings of Riemann in that I have studied more closely the effect of his restriction—a restriction that distinguishes real space from other multiply extended manifolds—on the validation of the condition that is the cornerstone of all research, namely that the square of a line element is a homogeneous quadratic function of the differentials of the coordinates. It is possible to show that if one adheres from the very beginning to the requirement of unconditionally free mobility of rigid bodies, without change of form, throughout space, then it is easy to deduce Riemann's initial hypothesis from more general assumptions.
>
> My starting point was that every initial measurement of space is based on the observation of superposition; obviously, the rectilinearity of light rays is a physical fact based on different experiences and is

meaningless for a blind person who can, nevertheless, be fully convinced of the truth of geometric propositions. As for coincidence, one cannot talk about it if rigid bodies or systems of points cannot be moved without change of form and if the coincidence of two spatial magnitudes is not a fact that exists independently of all motions. That is why I assumed from the very beginning the possibility of spatial measurement based on the verification of coincidence and set for myself the task of finding the most general analytic form of a multiply extended manifold which admits of motion of the required kind *[223, pp. 620–621]*.

Helmholtz goes on to list "the hypotheses that form the foundations of the research":

I. *A space of* n *dimensions is an* n-*ply extended manifold*, that is, every individual (Einzelne)—every point—is determined by measuring certain quantities (coordinates), n in number, that vary continuously and independently of one another. Thus every motion of a point is accompanied by continuous variation of at least one of its coordinates. If exceptions occur, where either the variation becomes discontinuous or, in spite of motion, none of the coordinates varies, then such exceptions will pertain only to certain definite locations (points, lines, a surface) determined by equations; we exclude all such locations from our investigations....

II. *One assumes the existence of mobile but immutable* (*rigid*) *bodies or systems of points*; such an assumption is necessary for equating spatial quantities by means of superposition. Since we cannot, as yet, assume any special devices for measuring spatial quantities, we can now give only the following definition of a rigid body: *The* 2n *coordinates of every pair of points of a rigid body are connected by an equation which is independent of the motion of the rigid body and is the same for all coincident pairs of points.*

Pairs of points are superposable if they coincide simultaneously or successively with the same pair of points in space.

In spite of its apparent vagueness, this definition is extremely fruitful for it implies that m points must be connected by $m(m - 1)/2$ equations, at a time when the number of unknown coordinates contained in them is mn, of which, moreover, $n(n + 1)/2$ must be available for the determination of the variable position of the immutable system. Therefore, if $m > n + 1$ then the number of equations exceeds the number of unknowns by $(\frac{1}{2})(m - n) \cdot (m - n - 1)$. It follows that the equation connecting the coordinates of any two fixed points cannot be of arbitrary form and that such equations must have special properties. This gives rise to the definite analytic problem of a more detailed determination of the form of these equations....

III. *One admits completely free mobility of rigid bodies*, that is, one assumes that each of their points can continuously shift to the position of every other, insofar as this first point is not restricted by equations between it and other points of the immutable system to which it belongs.

Thus the first point of an immutable system is absolutely mobile. If it is fixed, then the second point is subject to one equation and one of its coordinates becomes a function of the remaining $n - 1$ coordinates. If that second point is also fixed then there already are two equations for the first, and so on. Hence, all told, $n(n + 1)/2$ quantities are needed for the determination of the position of an immutable system.

From this assumption, as well as from the assumption made under II, it follows that *two immutable systems of points* A *and* B *that can be brought to superposition of corresponding points for one position of* A *can also be brought to superposition of all those points that were previously superposed for every other position of* A. In other words, the superposition of two spatial forms does not depend on their positions, so that all parts of space are mutually superposable if we ignore their boundaries, much as all parts of the same spherical surface can be superposed if we ignore their contours.

IV. Finally, we must ascribe to space one other property, analogous to the *monodromy* property of functions of a complex variable, which is that two superposable bodies remain superposable even if one of them is rotated about some axis. As for a *rotation*, it is characterized analytically by the fact that, during the period of motion, the coordinates of a certain number of points of the moving body remain fixed; as for the *inverse motion*, or *reversion*, it is characterized by the fact that the continuously varying sets of numerical values of the coordinates passed earlier are now passed in the opposite direction.

This fact can be expressed in the following manner: *if a rigid body rotates about* n − 1 *points selected so that the position of the body depends on just one independent variable, then a rotation alone returns the body to its initial position.*

We shall see that this last property of space need not necessarily obtain even if the first three conditions hold. Therefore, regardless of its complete obviousness, it must be spelled out as a special property of space.

In ordinary geometry one assumes this last property without special mention, as indicated by the fact that one regards a circle as a closed curve; one assumes postulates II and III in all propositions that involve superposition, for the existence of rigid and freely moving bodies with the abovementioned properties is the precondition of all superposability ... *[223, pp. 621–624].*

After studying various consequences of these hypotheses Helmholtz states:

Further investigation refers to the questions of the consequences that follow from the assumption, in accordance with postulate III, of superposability of finite parts of space independently of boundary and under all possible rotations. Riemann showed that just as in the case of two dimensions, when a curved surface changes into the surface of a sphere or into a surface obtained from the latter by bending without stretching, so too in the case of three and more dimensions the quantity he calls curvature remains constant. I will not here set forth the part of my investigations that is implicitly contained in Riemann's investigations.

My result is as follows.

If our assumptions I–IV hold, then the most general system of geometry would be the one obtained in accordance with the rules of our usual analytic geometry applied to a spherelike three-dimensional configuration whose equation in four-dimensional rectangular coordinates can be represented as

$$X^2 + Y^2 + Z^2 + (S + R)^2 = R^2.$$

Here X, Y, Z can become infinitely large only for $R = \infty$. The latter special case corresponds to our real geometry based on Euclid's axioms. Then X, Y, Z can take on finite values only if $S = 0$; the equation $S = 0$ is the equation of a flat configuration. Therefore, like Riemann, we must consider Euclidean space flat in comparison with spaces with a greater number of dimensions.

Finally, I note that, if we dispense with postulate IV then we obtain geometric systems very different from ours but capable of entirely consistent development. This is easiest to prove in the case of two coordinates. If the quantity θ in equation (5^b) were not equal to zero then the linear dimensions of every plane figure would increase at a constant rate upon rotation by a constant angle in one and the same direction; in this case the locus of points physically equidistant from a fixed point is a spiral.

Another easy-to-study example can be obtained if in plane analytic geometry with rectangular coordinates we suppose the y-coordinates to be imaginary. This corresponds to the assumption that h_1 and h_2 are real and $h_1 + h_2 = 0$.

In this case the locus of points equidistant from a point is an isosceles hyperbola *[223, pp. 637–638]*.

Helmholtz's final conclusions are:

My own investigations, together with those of Riemann, show that the above postulates, together with two additional stipulations, namely that

V. *Space has three dimensions*,

and that

VI. *Space can be indefinitely extended,*
provide a sufficient basis for the development of a study of space.[1] I
have already pointed out that postulates should be adopted in ordinary
geometry where they are not even mentioned; thus our postulates allow
less than what is usually assumed in geometric proofs.

Here one cannot but point out that the very possibility of our system
of spatial measurements depends, as is shown by the above exposition,
on the availability in nature of objects that are sufficiently close to our
notion of rigid bodies. The independence of superposition from the
position and direction of the superposed spatial forms and of the way
along which they are brought into coincidence is a fact that provides a
basis for the possibility of measurement of space *[223, pp. 638–639]*.

Helmholtz's hypothesis III represents the requirement of existence in space
of a group of geometric transformations that not only carry every point in
space into every other point but also every line element issuing from the first
point into every line element issuing from the second point. For Riemannian
spaces this condition is equivalent to the condition of constant curvature.
Hypothesis IV rules out pseudo-Euclidean and pseudo-Riemannian spaces.

The Norwegian mathematician Sophus Lie (1842–1899) had this to say of
the Helmholtz paper in his *Remarks on Helmholtz's paper "On the facts which
lie at the foundations of geometry"* (Bemerkungen zu v. Helmholtz' Arbeit
"Über die Thatsachen die der Geometrie zu Grunde liegen." Leipzig, 1886)
[326]:

Helmholtz's remarkable paper *On the facts which lie at the foundations
of geometry* is extremely closely linked to the new theory of groups of
transformations *[326, p. 337]*.

Lie goes on to analyze Helmholtz's hypotheses from the viewpoint of his
(Lie's) theory of continuous groups.

Groups of Geometric Transformations

We consider in greater detail continuous groups of *geometric transformations*, that is, groups of transformations of geometric spaces for whose
elements it is possible to define the notions of passing to a limit and continuity.
Felix Klein, who met C. Jordan in Paris on the eve of the Franco-Prussian
War and became familiar with his works, noticed that Euclidean motions,
non-Euclidean motions, as well as similarity, affine, projective, circular, and
conformal transformations, form groups if the product of two transformations is defined to be the result of their successive application. In 1872,
when he became a professor at the University of Erlangen, Klein gave

---

[1] They do not separate Euclidean geometry from Lobačevskian geometry (Helmholtz's remark).

a lecture entitled *Comparative overview of recent geometric investigations* (Vergleichende Betrachtungen über neuere geometrische Forschungen) known as the *Erlangen Program [282, vol. 1, pp. 460–497]*.

First Klein defines groups of transformations of *space*, that is, three-dimensional Euclidean space that can be extended to projective space:

The result of composition of an arbitrary number of transformations of space is again a transformation of space. If a given set of transformations has the property that every change resulting from the successive application of certain of the transformations that belong to the set then we call the set a *group of transformations*.

An example of a group of transformations is the set of motions (every motion is viewed as an operation on the whole space); its subgroup is, for example, the rotations about a point. The set of all collineations is a group containing the group of motions *[282, vol. 1, p. 462]*.

Klein points out that correlations ("dual transformations") do not form a group, but the totality of collineations and correlations does form a group. Then Klein considers the geometric properties of spatial figures:

By their very definition, geometric properties do not depend on the position occupied in space by the investigated figure, on its absolute magnitude or, finally, on the orientation and disposition of its parts. Therefore the properties of a spatial figure are unchanged by any motion of space, by its similarity transformations, by the process of reflection, or by any transformations that can be composed of them. We call the set of all these transformations the *principal group* of transformations of space. *Geometric properties are unaltered by the transformations of the principal group* and, conversely, one may say: *geometric properties are characterized by their unalterability under the transformations of the principal group [282, vol. 1, p. 463]*.

Thus the geometric properties of *space*, by which Klein means three-dimensional Euclidean space, are determined by "the principal group of transformations of space", that is, by its group of motions and similarities. After considering *space* Klein considers arbitrary *manifolds*.

By analogy with transformations of space we speak of transformations of a manifold; they also form *groups*. But now there is no longer a group which is distinguished from the other groups by its significance—each group is equivalent to every other group. Thus the following comprehensive problem arises as a generalization of geometry:

Given a manifold and a group of transformations on it. It is required to investigate those properties of the figures belonging to the manifold that are unchanged by the transformations of the group.

Using modern terminology, which, incidentally, is usually applied only to a certain group—the group of all linear transformations—we can put it thus:

Given a manifold and a group on it. It is required to develop the theory of invariants of this group.

This general problem includes not only ordinary geometry but also the latest geometric methods to be mentioned in the sequel and various devices for the investigation of manifolds of an arbitrary number of dimensions. Above all, one should emphasize the arbitrariness in the selection of the associated group of transformations and the resulting equality, understood in this sense, of all types of investigations that fit this general requirement *[282, vol. 1, pp. 463–464]*.

Next Klein investigates "groups of transformations of which one contains the other" in *space* as well as in various *manifolds*. In the case of *space*, such groups are the group of motions, the *principal group*, and, on the one hand, the *affine group*, the *group of collineations* and the *group of all projective transformations* (collineations and correlations), and on the other hand, *groups of inversive transformations*, that is, groups of *circular transformations* of the plane and of *conformal transformations* of space generated by inversions in circles or spheres. These transformations are also called *Möbius transformations*. After noting that the addition of correlations "implies the simultaneous introduction of points and planes as elements of space", Klein calls attention to the enlargement of

the basic group of collinear and dual transformations by the introduction into it of the corresponding *imaginary* transformations. This step calls for the preliminary extension of the class of basic elements of space by the addition of imaginary elements *[282, vol. 1, p. 467]*.

The Transfer Principles

Further Klein considers cases of isomorphic groups of transformations which enable one to interpret one geometry within another or, as Klein puts it, to *transfer* one geometry into another.

In particular, by projecting a line A to a conic A' from its point S (Figure 106) Klein notices that the group of collineations of the line is isomorphically

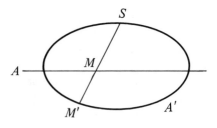

Figure 106

represented by means of the group of collineations of the plane that map the conic onto itself. In the first case, there is defined on the straight line a geometry of binary forms $\sum a_i x_i$ in two projective coordinates x_0 and x_1 of the points of the line, each of which determines a point of the line. Therefore Klein says that:

The theory of binary forms and projective geometry of systems of points on a conic are the same thing, that is, to each binary-form theorem there corresponds a theorem about such point systems and conversely [282, vol. 1, p. 469].

Further, by considering on a conic pairs of real or conjugate complex points instead of single points, and lines that determine the pairs of points on the conic instead of the pairs of points, or the poles of these lines relative to the conic, Klein arrives at the conclusion that

the theory of binary forms and the projective geometry of the plane that we study by adopting as a basis a certain conic are equivalent.

Since

projective geometry of the plane, with a conic as a basis, coincides, as a result of group equality, with projective metric geometry,

Klein also finds that

the theory of binary forms and general projective metric geometry of the plane are one and the same thing,

and adds that

instead of a conic in the plane we can take in this study a third-order curve in space and so on [282, vol. 1, p. 471].

If the conic is nondegenerate and real, then "general projective metric geometry in the plane" coincides with the projective interpretation of the Lobačevskian (hyperbolic) plane set forth by Klein in the paper *On so-called non-Euclidean geometry.* Klein points out that

the connection [established here] between plane geometry and, further, geometry of space or of manifolds of an arbitrary number of dimensions, coincides, in its essential features, with the transfer principle proposed by Hesse [282, vol. 1, pp. 471–472].

Here Klein has in mind the paper of Otto Hesse (1811–1874) *On a transfer principle* (Über ein Übertragungsprinzip. Berlin, 1866) [225]. This transfer is called the *Hesse transfer.*

After stereographically projecting an oval quadric onto a plane Klein notes that the principal group of the plane is isomorphic to the group of collineations of space that map onto themselves the quadric and the center of projection, respectively, and concludes that *plane elementary geometry and projective investigation of a surface of second order with adjunction of one of its points are the same thing [282, vol. 1, p. 470].*

Klein goes on to consider projective geometry of space, which he calls "the theory of quaternary forms." The reason for this is that planes, which by the duality principle have the same standing as points, are determined by linear forms $\sum a_i x_i$ of the four projective coordinates x_0, x_1, x_2, x_3 of points in space. After pointing out that the so-called Plücker coordinates

$$p_{ij} = x_i y_j - x_j y_i \tag{9.2}$$

of the lines passing through the points $X(x_j)$ and $Y(y_i)$, introduced by Plücker in his *New geometry of space based on considering a straight line as a spatial element [429]*, are connected by the quadratic relation

$$p_{01}p_{23} + p_{02}p_{31} + p_{03}p_{12} = 0 \tag{9.3}$$

(which is easy to obtain from the determinant whose first and third rows are the coordinates x_i and whose second and fourth rows are the coordinates y_i by expanding it by its first and second rows), and noting that as a result of collineations of space the coordinates p_{ij} also transform linearly and the relation (9.3) is preserved, Klein arrived at the conclusion that

the theory of quaternary forms coincides with the projective metric in a manifold of six homogeneous variables *[282, vol. 1, p. 472]*.

This transfer of three-dimensional projective geometry to five-dimensional hyperbolic geometry whose absolute is given by equation (9.3) is called the *Plücker transfer*. We note that equation (9.3) can be reduced to the form

$$X_0^2 + X_1^2 + X_2^2 - X_3^2 - X_4^2 - X_5^2 = 0.$$

This shows that a quadric with this equation has not only rectilinear generators but also two families of two-dimensional plane generators; here the points of the quadric represent lines in 3-space, the rectilinear generators represent *plane pencils* of lines, the plane generators of one family represent *bundles* of lines, and the plane generators of the other family represent *plane fields* of lines dual to the bundles.

Considering once more a stereographic projection of an oval quadric onto a plane Klein notes that if one adjoins to the plane a single point at infinity, and thus makes it into a conformal (inversive) plane that is in one-to-one correspondence with the quadric, then the circular transformations of the plane are represented by the collineations that map the quadric onto itself. This leads him to the conclusion that

plane inversive geometry and projective geometry on a quadric surface are identical [282, vol. 1, p. 475].

Similarly, by projecting an oval quadric in four-dimensional projective space stereographically onto a three-dimensional plane in that space Klein notes that if one adjoins to this plane a single point at infinity, and thus makes it into three-dimensional conformal (inversive) space which is in one-to-one correspondence with the quadric, then the conformal transformations of the space are represented by the collineations of the four-dimensional space that map the quadric onto itself. This makes Klein conclude that

inversive geometry in space coincides with the projective study of the manifold represented by a quadratic equation connecting five homogeneous variables [282, vol. 1, p. 475].

We note that here the conformal plane and three-dimensional conformal space are mapped onto the absolutes of three-dimensional and four-dimensional hyperbolic space, respectively, and the circles of the conformal plane and the spheres of conformal space are represented by the poles of the planes that determine their images on the respective absolutes, that is, by the points of the so-called ideal domains of three-dimensional and four-dimensional hyperbolic space. The projective coordinates of these points coincide with the *tetracyclic coordinates* of circles and *pentaspheric coordinates* of spheres introduced by Gaston Darboux (1842–1917) in the paper *On a remarkable class of algebraic curves and surfaces and on the theory of imaginaries* (Sur une classe remarquable des courbes et des surfaces algébriques et sur la théorie des imaginaries. Bordeaux, 1873) *[136]*, so that this transfer is often called the *Darboux transfer.*

Noting that the conformal plane can be viewed as the extended complex plane as well as a projective line, Klein formulates the first of the above transfers in the following alternative form:

the theory of binary forms is represented by inversive geometry in the real plane in such a way that, at the same time, the complex values of the variables are also represented [282, vol. 1, p. 476].

or, by viewing an oval quadric as a sphere, in the form:

the theory of binary forms of a complex variable is represented in projective geometry on the surface of a sphere [282, vol. 1, p. 476].

Klein extends the study of groups of transformations of *space* to *manifolds* (that is, *n*-dimensional spaces) and notes in particular that

If we consider a stereographic projection of a manifold then we obtain the well known theorem: in a multiply-extended domain (in space) there are no conformal point transformations other than the inversive transformations. On the other hand, there are arbitrarily many such transformations in the plane *[282, vol. 1, p. 479].*

Klein has in mind the relevant theorem proved by J. Liouville for three-dimensional space in an appendix to his 1850 edition of G. Monge's *Applications of analysis to geometry [373, pp. 609–616].* This theorem explains the fact that the two-dimensional analogue of conformal transformations of space is the circular transformations of the plane.

Klein considers an even larger group of transformations of spheres, namely the so-called *group of Lie's higher geometry of spheres*, introduced by Sophus Lie in his paper *On complexes, in particular line and sphere complexes, with applications to the theory of partial differential equations* (Über Komplexe, insbesondere Linien—und Kugelcomplexe mit Anwendungen auf die Theorie partieller Differentialgleichungen. Leipzig, 1872) *[325, vol. 2, part 1, pp. 1–121]*. These transformations map spheres onto spheres and preserve their tangency; here points and planes are viewed as spheres of zero and infinite radius, respectively. The subgroup of this group that maps points onto points is the group of conformal transformations that preserves angles between spheres; the subgroup of this group that maps planes onto planes is the group of *Laguerre transformations* introduced by E. Laguerre in his paper *On the geometry of direction* (Sur la géométrie de la direction. Paris, 1880) *[300, vol. 2, pp. 592–607]*. This group preserves the *tangent distance* between spheres, that is, the segment of the common tangent of two equally oriented spheres contained between the tangency points. We note that if we take the angle between two spheres as a measure of their distance then the manifold of spheres of three-dimensional Euclidean space is isometric to the ideal domain of four-dimensional hyperbolic space of curvature $+1$ (that is, a sphere of radius 1 in five-dimensional pseudo-Euclidean space having the form of a one-sheeted hyperboloid with identified antipodal points). If we take the same manifold of spheres, consider their orientations, and define the distance between two spheres to be their tangent distance, then it is isometric to four-dimensional pseudo-Euclidean space; and the same manifold of spheres, including points and planes, can be mapped in one-to-one bicontinuous manner onto a five-dimensional ruled quadric in six-dimensional projective space. Also, the groups of conformal transformations, of Laguerre transformations, and of the transformations of Lie's higher geometry of spheres are, respectively, isomorphic to the group of motions of four-dimensional hyperbolic space, the group of motions of four-dimensional pseudo-Euclidean space, and the group of motions of five-dimensional hyperbolic space whose absolute is a ruled quadric.

Finite and Discrete Groups of Transformations

Klein also studied finite groups of geometric transformations. He did this in his *Lectures on the icosahedron and the solution of equations of the fifth degree* (Vorlesungen über das Ikosaeder und die Auflösung der Gleichungen fünften Grades. Leipzig, 1884) *[281]*, where he defined the groups of symmetries of regular polyhedra, that is, the finite groups of symmetries of these polyhedra.

Since every motion of the Euclidean space R_n is a product of reflections in not more than $n + 1$ hyperplanes, every finite group of rotations is *generated*

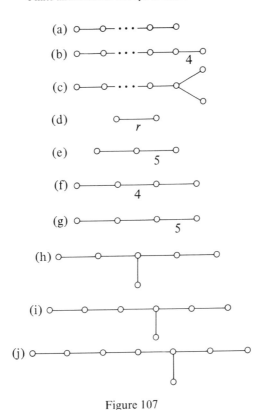

Figure 107

by reflections. A very clear classification of such groups was proposed by Harold Scott Macdonald Coxeter (b. 1907) in the paper *Discrete groups generated by reflections* (Princeton, 1934) *[128]*. In the *Coxeter diagrams* proposed in this paper each hyperplane, reflections in which generate the group, is represented by a point. These points are joined by a line without a number if the angle between the corresponding hyperplanes is equal to $\dfrac{2\pi}{3}$ and with a number k if this angle is equal to $\dfrac{2\pi}{k}$.

Figure 107 shows Coxeter diagrams for groups of symmetries of the following figures: (a) an n-dimensional simplex $\{3, 3, .., 3\}$ (for $n = 2$ a triangle $\{3\}$, for $n = 3$ a regular tetrahedron $\{3, 3\}$; (b) an n-dimensional cube and cross polyhedron $\{4, 3, \ldots, 3, 3\}$ and $\{3, 3, \ldots, 3, 4\}$ (for $n = 2$ a square $\{4\}$, for $n = 3$ a cube $\{4, 3\}$, and an octahedron $\{3, 4\}$); (c) an n-dimensional *semicube* and *semicross polyhedron* (convex polyhedra obtained from previous polyhedra by selection of one vertex on each edge and rejection of the other vertices on this edge; for $n = 3$ this polyhedron is a tetrahedron); (d) a regular polygon $\{r\}$; (e) a regular dodecahedron $\{5, 3\}$ and an icosahedron $\{3, 5\}$; (f) a regular polyhedron $\{3, 4, 3\}$; (g) regular polyhedra $\{5, 3, 3\}$ and $\{3, 3, 5\}$; (h) the polyhedron with 27 vertices and 99 faces, of which 72 are $\{3, 3, 3, 3\}$'s and 27 are $\{3, 3, 3, 4\}$'s, and the configuration of 27 straight lines on a cubic surface without double points in three-dimensional space (see *[129]*); (i) the polyhedron with 56 vertices and 702 faces of which 576 are $\{3, 3, 3, 3, 3\}$'s and 726 are $\{3, 3, 3, 3, 4\}$'s and the configuration of 28 double tangents of a plane quartic without double points; (j) the Gosset polyhedron with 240 vertices and 19440 faces of which 17280 $\{3, 3, 3, 3, 3, 3\}$'s and 2160 are $\{3, 3, 3, 3, 3, 4\}$'s; the latter figure, discovered by Thorold Gosset (see *[209]*), can also be defined with the aid of integer octaves (see *[127]*); about all these figures see also Coxeter's *Regular polytopes [130]*. Many of these groups can also be interpreted with the aid of finite geometries (see chapter 10). The orders of these groups are, respectively, equal to (a) $(n + 1)!$, (b) $2^n \cdot n!$, (c) $2^{n-1} \cdot n!$, (d) $2r$, (e) 120, (f) 1152, (g) 14,400, (h) $2^7 \cdot 3^4 \cdot 5 = 51,840$, (i) $2^{10} \cdot 3^4 \cdot 5 \cdot 7 = 2,903,040$, (j) $2^{14} \cdot 3^5 \cdot 5^2 \cdot 7 = 696,729,600$. If we restrict ourselves to symmetries preserving the orientations of simplices then the orders of the groups decrease by half.

Klein showed that the orientation-preserving group of the regular tetrahedron is isomorphic to the alternating group A_4, the groups of the cube and the octahedron are each isomorphic to the symmetric group S_4 and the groups of the icosahedron and the dodecahedron are each isomorphic to the alternating group A_5. The connection between the icosahedron and quintic equations consists in the isomorphism between its group and the group A_5 whose simplicity, as we saw, is closely related to the unsolvability in radicals of certain quintic equations.

A number of papers by Klein and Poincaré are devoted to finite groups of fractional linear transformations of the complex plane (first considered by the German mathematician Lazarus Fuchs and therefore called *Fuchsian groups*) and to functions of a complex variable whose values are preserved under the transformations of these groups—the so-called *automorphic functions*. The group of fractional linear transformations of the complex plane and its subgroup whose elements preserve a line or a circle are, respectively, isomorphic to the groups of motions of three-dimensional Lobačevskian space and the Lobačevskian plane. These isomorphisms are established by means of interpretations of the latter groups in the complex plane. They are the basis of the application of Lobačevskian geometry to the theory of automorphic functions. The title of Poincaré's paper, *The theory of Fuchsian groups*, is connected with the name of these groups. (This paper contains Poincaré's interpretation of Lobačevskian geometry.)

The discrete infinite subgroups of the group of motions of the Euclidean plane and space were studied by C. Jordan in his *Treatise on theory of substitutions* (1870) and by Leonhard Sohnke (1842–1893) in the paper *Unbounded regular point systems as foundation of a theory of crystal structures* (Die unbegränzten regelmässigen Punktsysteme als Grundlage einer Theorie der Krystallstrukturen. Leipzig, 1876). Jordan and Sohnke found all 17 plane crystallographic groups. We note that 11 of these groups were used in ornaments in the Alhambra in Grenada, and the remaining groups were used in ornaments in China and Africa.

The space crystallographic groups, of great importance in crystallography (a fact Sohnke was aware of), were determined by the Russian geologist and crystallographer Evgraf Stepanovič Fedorov (1853–1919), the German mathematician Arthur Schönflies (1853–1928) and the English mineralogist and crystallographer William Barlow (1845–1934). Fedorov, in his *Symmetry of regular systems of figures* (Symmetrii pravil'nyh system figur. Petersburg, 1890) *[181, pp. 109–255]*, Schönflies in his *Crystal systems and crystal structure* (Krystallsysteme und Krystallstruktur. Leipzig, 1891) *[508]*, and William Barlow in his paper *On geometrical properties of homogeneous rigid structures and their application to crystals* (Leipzig, 1894) *[37]* found that the number of such groups, mapping regular spatial systems of points onto themselves, is 230. Of these, 65 consist solely of motions that preserve the orientation of tetrahedra and 165 include motions that do not.

Birational Transformations

As for other geometric transformations, we mention *birational transformations*, that is, transformations of projective space such that they and their inverses are expressible by means of rational functions of projective coordinates. The simplest birational transformations are inversions in circles. Since we are dealing with the projective plane, the image of the center of a circle of inversion is not a single point at infinity of the plane but all points of the line at infinity. Another extremely simple birational transformation of the projective plane is the transformation $x_i' = 1/x_i$ studied by J. Plücker in *Analytic-geometrical studies* (Analytisch-geometrische Entwicklungen. Essen, 1828–1831). We have already mentioned inversions in an ellipse, hyperbola, and parabola in Apollonius' *Conica* and hyperbolisms in Newton's *Enumeration of curves of the third order*. A systematic study of birational transformations was undertaken by Luigi Cremona (1830–1903) in his paper *On geometric transformations of plane figures* (Sulle transformazioni geometriche delle figure plane. Bologna, 1862–1865) *[132]* and in subsequent papers, so that these transformations are often called *Cremona transformations*. Our examples of birational transformations show that these transformations do not preserve the order of algebraic curves. Cremona showed that the genus p of these curves is an invariant of birational transformations.

Inversions in circles and conics are quadratic birational transformations and so is Plücker's transformation $x_i' = 1/x_i$. We note that every quadratic birational transformation is a combination of inversions in conics, transformations of the form $x_i' = 1/x_i$ and collineations. This and many other results of the theory of birational transformations are expounded in the monograph of Hilda Hudson (1881–1965) *Cremona transformations in the plane and in space* (Cambridge, 1927) *[235]*.

Continuous Groups

The groups studied in Jordan's *Treatise on substitutions* are all finite. (We note that the groups of linear transformations investigated by Jordan that gave rise to what is now known as the *Jordan normal form of a matrix* had integer coefficients modulo an integer). But the year in which Jordan's treatise was published was also the year when the theory of continuous groups was born. Specifically, it was inspired by conversations between Jordan and Klein and Lie, who visited France on the eve of the Franco-Prussian War. All groups studied in the *Erlangen Program* are continuous groups. Lie developed the general theory of continuous groups. The main stimulus behind the creation of this theory was his desire to develop a theory of solvability in quadratures of differential equations analogous to Galois's theory of solvability in radicals of algebraic equations. Lie published his theory in *The theory of groups of transformations* (Theorie der Transformationsgruppen. Leipzig, 1886) *[325, vol. 5, pp. 9–223]* and in his *General investigations on differential equations that determine a continuous finite group* (Allgemeine Untersuchungen über Differentialgleichungen die eine continuierliche endliche Gruppe gestalten. Leipzig, 1885) *[325, vol. 6, pp. 139–223]* and gave a detailed account of it in the three-volume work *Theory of transformation groups* (Theorie der Transformationsgruppen. Leipzig, 1888–1893) *[327]*, coauthored with Friedrich Engel (1861–1941). The main object of his study were groups that Lie called *finite continuous groups* and that are now known as *finite Lie groups*.

In his previously mentioned *Remarks on Helmholtz's paper [326]* Lie replaces Helmholtz's hypotheses with simpler requirements formulated in terms of groups of geometric transformations. Lie writes:

> By rather simple means I managed to show that the equations of Euclidean as well as non-Euclidean motions of the space of three dimensions can be characterized as follows:
>
> 1′. They determine a continuous group of motions of the space of three dimensions.
> 2′. In this group there exists free mobility in the following sense: if we fix in the interior of a given domain an arbitrary point as well as an arbitrary line element passing through it then continuous motion is still possible. On the other hand, if we fix not only a point and a line element passing through it but also an element of a plane that passes through it, then all further continuous motion is impossible.
>
> I managed to show, rigorously, I think, but hardly concisely, that in spaces of more than three dimensions the totality of Euclidean and

non-Euclidean motions can be characterized in an entirely analogous manner *[326, pp. 341–342]*.

Topological Groups

In spite of the fact that, when he mentioned examples of the groups he studied, Lie often had in mind groups in the large, in his theory he studied only their neighborhoods of the identity. Lie assumed that the groups he considered were realized as groups of geometric transformations.

The theory of Lie groups in the large was created on the basis of the concept of a continuous, or topological, group—a concept that first appeared in L. E. J. Brouwer's paper *The theory of finite continuous groups independent of the axioms of Lie* (Die Theorie der endlichen continuierlichen Gruppen, unabhängig von der Axiomen von Lie. Leipzig, 1909–1910) *[81]*. The theory of general topological groups was to a large extent created by L. S. Pontryagin, who gave a full account of it in his monograph *Continuous groups* (Nepreryvnye gruppy. Moscow, 1938). An English translation appeared almost simultaneously (*Topological groups*. Princeton, 1939) *[436]*. Pontryagin defines topological groups as follows:

A set G of elements is called a *topological group* if

1) G is an abstract group,
2) G is a topological space,
3) the group operations in G are continuous in the topological space G *[436, p. 12]*.

The last condition means that if a and b are two elements of G then for every neighborhood W of ab there exist neighborhoods U and V of a and b such that the totality of products uv of the elements u and v of these neighborhoods are in W and, if a is an element of G, then for every neighborhood V of a^{-1} there exists a neighborhood U of a such that all inverses u^{-1} of the elements u of U are in V.

Brouwer indicates that he arrived at the notion of *a finite continuous group independent of the axioms of Lie* under the influence of *Hilbert's fifth problem*. In 1900, at the Second International Congress of Mathematicians in Paris, Hilbert presented a now celebrated paper entitled *Mathematical problems* (Mathematische Probleme. Leipzig, 1901) in which he formulated 23 of the (in his view) most important problems bequeathed by the 19th century to the 20th. An edition with extensive commentaries is *Die Hilbertsche Probleme* (Unter der Red. von P. S. Alexandroff. Leipzig, 1971) *[231]*. (See also *[231a]*.)

The groups considered by Brouwer are topological groups that are manifolds in Brouwer's sense; that is, their elements have neighborhoods homeomorphic to n-dimensional Euclidean space (in view of the properties of a group it suffices to require that such a neighborhood exists for the identity of the group). If x, y, and $z = xy$ are three elements in a neighborhood of the identity then, in view of the stated property, these elements have coordinates x^i, y^i, z^i, n each, and clearly, the coordinates z^i are functions of the coordinates x^i and y^i that are continuous in view of the properties of the topological group. *The axioms of Lie* imply that these functions are differentiable a certain number of times.

Hilbert's fifth problem is the question:

To what extent is Lie's concept of continuous groups of transformations useful for the solution of the posed problem (in the calculus of variations) if one does not require the functions that define the group to be differentiable *[231, p. 44]*.

The contemporary statement of Hilbert's fifth problem is formulated for groups in the large as follows: (For a suitable choice of coordinates) is every topological group which is a manifold a Lie group?

For certain classes of Lie groups this question was answered in the affirmative by John von Neumann in his paper *Introduction of analytic parameters in topological groups* (Die Einführung analytischer Parameter in topologischen Gruppen. Leipzig, 1933) *[385, vol. 2, pp. 366–386]* and by Pontryagin in the paper *The theory of topological commutative groups*. Baltimore, 1934) *[437]*. The solutions of these problems are given in detail in Pontryagin's *Topological groups [436, pp. 153–171, 212–216]*. Later the French mathematician Claude Chevalley (b. 1909) answered Hilbert's fifth problem affirmatively for solvable Lie groups in the paper *Two theorems on solvable topological groups* (Ann Arbor, 1941) *[111]*. In the paper *Solvable topological groups* (Topologičeskie razrešimye gruppy. Moscow, 1946) *[351]* the Soviet algebraist Anatoliĭ Ivanovič Mal'cev (1909–1967) showed that solvable topological groups belonging to a wider class are Lie groups. Finally, Hilbert's fifth problem was solved by the American mathematicians Andrew Gleason (b. 1921) in the paper *Groups without small subgroups*. (Baltimore, 1952) *[206]*, and Dean Montgomery and Leo Zippin in the paper *Small subgroups in finite dimensional groups* (Baltimore, 1952) *[375]*. This proof is set forth in the book by Montgomery and Zippin entitled *Topological transformation groups* (New York, 1955) *[376]*. Modified versions of that proof were presented by the Japanese mathematician H. Yamabe in the paper *On the conjecture of Iwasawa and Gleason* (Baltimore, 1953) *[641]* and by the Soviet mathematician Viktor Mihaĭlovič Gluškov (1923–1982) in the paper *The structure of locally bicompact groups and Hilbert's fifth problem* (Stroenie lokal'no bikompaktnyh grupp i pyataya problema Gil'berta. Moscow, 1957) *[207]*. In the paper *[642]* Yamabe has given a generalization of Gleason's theorem.

Lie Groups in the Large

After the affirmative answer to Hilbert's fifth problem, Lie groups in the large can be defined as topological groups that are manifolds. We note that in Lie groups it is always possible to choose coordinates in a neighborhood of the identity in such a way that the coordinates z^i of the product $z = xy$ can be expressed in terms of the coordinates x^i and y^i of the factors by means of functions that are not only differentiable an arbitrary number of times but also analytic. Therefore, to every real group G there corresponds a complex group CG called a *complex extension* of the group G for which the functions that express the dependence of the coordinates z^i of the product $z = xy$ on the coordinates x^i and y^i of the factors are the same analytic functions as in the group G. Also, to one and the same complex form there corresponds a certain number of different real functions.

If two Lie groups have the same neighborhoods of the identity then the two groups are said to be *locally isomorphic*. Therefore we can say that Lie considered Lie groups up to a local isomorphism.

Lie Algebras

The analyticity of the functions that define the group operation implies that in order to characterize Lie groups up to local isomorphism it suffices to consider the group elements in an infinitely small neighborhood of the identity. Lie called such elements *infinitesimal transformations*. Giving an infinitesimal transformation is equivalent to giving the derivatives

$$X^i = \left(\frac{dx^i}{dt}\right)_0$$

of the functions $x^i = x^i(t)$ that define in the group a curve that passes at $t = 0$ through the group identity. The numbers X^i can be viewed as the coordinates of a vector X tangent to the curve at the group identity. These vectors form a linear space in which the group operations induce a multiplication $Z = [XY]$ that is neither commutative nor associative but satisfies the identities $[XY] = -[YX]$ and $[[XY]Z] + [[YZ]X] + [[ZX]Y] = 0$. The latter is called *the Jacobi identity*, for it was C. G. J. Jacobi who first proved it for *Poisson brackets*. A linear space with an operation $Z = [XY]$ satisfying the above identities is now called a *Lie algebra*. An example of a Lie algebra is the set of vectors in three-dimensional space with the usual cross product operation. Instead of vectors, Lie considered the differential operators

$\sum_i X^i \frac{\partial}{\partial x^i}$ and called Lie algebras *infinitesimal groups*. Lie showed that every Lie group has a Lie algebra and is determined by it (of course, to within a local isomorphism). (We note that the Lie algebra of vectors in three-dimensional space determines its group of rotations.) Therefore the study of Lie groups is often reduced to the study of the corresponding Lie algebras.

If the Lie group is a group of matrices U and we consider the *one-parameter subgroups* $U(t)$ of this group with *canonical parameter* t: $U(t_1 + t_2) = U(t_1)U(t_2)$, then the Lie algebra of this group consists of the matrices $A = \frac{dU}{dt}\Big|_{t=0}$. The Lie algebra of the group SL_n of *unimodular matrices*

(matrices with determinant 1) consists of *traceless matrices*: if we consider the columns $\mathbf{u}_1, \mathbf{u}_2, \ldots, \mathbf{u}_n$ of the matrix U as vectors, then the determinant $\det U$ can be viewed as a skew product $[\mathbf{u}_1, \mathbf{u}_2, \ldots, \mathbf{u}_n]$, and the differentiation of this product for $t = 0$ when $U(0) = I = (\delta_{ij})$ gives for $A = (a_{ii}) = \dfrac{dU}{dt}\bigg|_{t=0}$ the value $\text{Tr}A = \sum a_{ii} = 0$. For the group O_n of *orthogonal matrices* we have the equality $U^T U = I$ (T denotes transposition). Differentiation of this equality gives $A^T + A = 0$, i.e., the Lie algebra of the group O_n consists of *skew-symmetric matrices*. Similarly, for the group ${}^l O_n$ of *pseudoorthogonal matrices* of linear transformations preserving the quadratic form $\sum_i \varepsilon_i x_i^2$ of index l, where $\varepsilon_i = -1$ if $i \leq l$ and $\varepsilon_i = 1$ if $i > l$, we have the equality $U^T E_l U = E_l$, where $E_l = (\varepsilon_i \delta_{ij})$, and its Lie algebra consists of matrices A with

$$A^T E_l + E_l A = 0. \tag{9.4}$$

Again, for the group Sy_{2n} of *symplectic matrices* of linear transformations preserving a skew bilinear form $\sum (x^i y^{n+i} - x^{n+i} y^i)$ $(i = 1, 2, \ldots, n)$ we have the equality $U^T JE = J$, where J is the block matrix

$$J = \begin{array}{|c|c|} \hline 0 & I \\ \hline -I & 0 \\ \hline \end{array},$$

and its Lie algebra consists of matrices A for which

$$A^T J + JA = 0. \tag{9.5}$$

If we replace the multiplication AB in the algebras $\mathbf{R}_n$, $\mathbf{C}_n$ and $\mathbf{H}_n$ of real, complex, and quaternion matrices by *commutation* $[AB] = AB - BA$, then we obtain the Lie algebras $\mathbf{R}_n^-$, $\mathbf{C}_n^-$ and $\mathbf{H}_n^-$. If we designate the Lie algebras of traceless matrices by $P\mathbf{R}_n^-$ and $P\mathbf{C}_n^-$, the Lie algebras of skew-symmetric matrices by $S\mathbf{R}_n^-$ and $S\mathbf{C}_n^-$ and the Lie algebras of hermitian skew-symmetric matrices by $\overline{S}\mathbf{C}_n^-$ and ${}^l\overline{S}\mathbf{H}_n^-$, then the Lie algebras satisfying the conditions (9.4) and (9.5) are, respectively, ${}^l S\mathbf{R}_n^-$, ${}^l S\mathbf{C}_n^-$ and $Sp\mathbf{R}_n^-$, $Sp\mathbf{C}_n^-$. If we replace A^T with $\bar{A}^T$ in (9.4) and (9.5), then we obtain the Lie algebras ${}^l\overline{S}\mathbf{C}_n^-$, ${}^l\overline{S}\mathbf{H}_n^-$ and $\overline{Sp}\mathbf{H}_n^-$.

Solvable and Semisimple Lie Groups

The most important class of Lie groups—the solvable ones—was defined by Lie by analogy with Galois' solvable finite groups. Just as Galois created the group concept to explain when an algebraic equation is solvable in radicals and gave as the required criterion the solvability of the group of permutations of the roots of the equation, so too Lie posed the question of when the integral of a differential equation is expressible by means of quadratures. This is reflected in the name of the paper, mentioned earlier, in which Lie built the foundations of the theory of Lie groups—*General investigations on differential equations that determine a continuous finite group* (Lie called Lie groups finite continuous groups). Like a finite solvable group, a solvable Lie group is defined as a group G that contains a nested sequence of normal subgroups $H_1 \supset H_2 \supset \ldots \supset H_r \supset e$ such that the factor groups G/H_1, H_1/H_2, ..., H_{r-1}/H_r as well as H_r are commutative. We note that with every Lie group there is associated the so-called commutator subgroup—a Lie group whose Lie algebra is the subalgebra of the Lie algebra of the given group consisting of the commutators $[XY]$ of that algebra. It is clear that the subalgebra of a Lie algebra corresponding to the commutator subgroup is an ideal of the Lie algebra, so that the commutator subgroup of a Lie algebra is its normal subgroup. From the definition of the commutator subgroup it follows that the factor group of a group by its commutator subgroup is commutative. If the Lie group is itself commutative then its commutator subgroup consists of the identity element of the group; if the group is solvable then its subgroup H_1 is its commutator subgroup and each subgroup H_i is the commutator subgroup of H_{i-1}. Lie proved that if a linear partial differential equation $\sum_i A_i \partial f/\partial x_i = 0$ in n variables $x_1, x_2, \ldots, x_n$ admits an $(n-1)$-dimensional solvable group whose infinitesimal transformations, together with the infinitesimal transformation determined in the domain of the variables x_i by the differential operator $\sum_i A_i \partial/\partial x_i$, form an independent system, then the differential equation can be integrated by quadratures; that is why solvable Lie groups were originally called integrable groups.

A Lie group without solvable normal subgroups is called *semisimple*. This notion was introduced by É. Cartan in his dissertation *On the structure of finite continuous groups* (Sur la structure des groupes finis et continus. Paris, 1894) *[96, part 1, vol. 1, pp. 137–287]*, in which he gave criteria for the solvability and semisimplicity of groups (the term *semisimple* (halbeinfach) was introduced by the German mathematician Wilhelm Killing (1847–1923) in the paper *The structure of continuous finite groups of transformations* (Zusammensetzung der stetigen endlichen Transformationsgruppen. Leipzig, 1888–1890) *[278a]*). Knowledge of the structure of solvable and semisimple groups enables one to establish the structure of an arbitrary Lie group. This is so because of Levi's theorem, established by Elia Levi (1883–1917) in a paper of the same title as Cartan's dissertation (Sulla struttura dei gruppi finiti e continui. Turin, 1905) *[319]*, to the effect that if a Lie group has a normal subgroup with semisimple factor group then it also has a subgroup isomorphic to that factor group. The term *semisimple Lie groups* is due to their connection with *simple Lie groups*, that is, Lie groups without normal subgroups of smaller dimension. Simple Lie groups were introduced by Lie himself in his *General investigations on differential equations that determine a finite continuous group* (1885). In his dissertation, Cartan showed that every semisimple group is isomorphic, or locally isomorphic, to the direct product of a certain number of simple Lie groups, that is, to a group consisting of the products of the elements of a number of noncommutative simple groups such that the elements from different groups commute; each of these simple groups is a normal subgroup of the direct product (this property was named semisimplicity by Killing).

The Classification of Complex Simple Lie Groups

In the abovementioned paper Lie showed that there are four infinite series of simple complex groups—the group CSL_{n+1} of complex *unimodular matrices*; the groups CO_{2n+1} and CO_{2n} of complex *orthogonal matrices* of odd and even order; and the group CSy_{2n} of complex *symplectic matrices*. The latter group was originally called the *complex-group*, and later, to avoid confusion with the term *complex group*, the Latin word *complexus* (complicated) was replaced by its Greek equivalent *symplektikos*. (The term *symplectic group* was first used by H. Weyl in the book *[628]*.)

Cartan called simple Lie groups, locally isomorphic to the groups CSL_{n+1}, CO_{2n+1}, CSy_{2n} and CO_{2n}, complex simple groups of the four *large classes* A_n, B_n, C_n and D_n, respectively; the subscript n is the *rank* of the group and is equal to the dimension of the set of elements of the group that commute with a regular element—an element of common state. This set is a commutative subgroup called the *Cartan subgroup* and the corresponding subalgebra of the Lie algebra of the group is the *Cartan subalgebra*.

For example, for the groups SL_{n+1} and CSL_{n+1} the regular elements are the matrices with distinct eigenvalues. If such a matrix has the form $U = (u_i \delta_{ij})$, then the condition $\det U = 1$ can be written as $\Pi_i u_i = 1$. The Cartan subgroup of these elements consists of matrices of the same form and the Cartan subalgebra of these elements consists of the matrices of the form $H = (h_i \delta_{ij})$ for which $\sum_i h_i = 0$. If we denote the matrix with 1 at the intersection of the *i*-th row and *j*-th column and with 0 in all remaining places by E_{ij}, then the matrices of the Cartan subalgebra can be written in the form $H = \sum_i h_i E_{ii}$.

In the previously mentioned paper *The structure of continuous finite groups of transformations [278a]*, Wilhelm Killing showed that in addition to these complex simple groups there are also certain *exceptional* simple groups.

In his dissertation Cartan showed that all simple complex Lie groups belong either to the four *large classes* or to the five *exceptional classes* with respective dimensions 14, 52, 78, 133 and 248 and called the latter classes G_2, F_4, E_6, E_7, E_8 respectively.

In order to classify simple Lie groups Cartan considered a map $A \to [AH] = AH - HA$ of the Lie algebra of a simple group of matrices onto itself. This map is a linear transformation. For $A = E_{ij}$ this map takes the form $E_{ij} \to E_{ij}(\sum_k h_k E_{kk}) - (\sum_k h_k E_{kk})E_{ij} = (h_j - h_i)E_{ij}$. This shows that the matrices E_{ij} $(i \neq j)$ play the role of eigenvectors of this transformation and that the corresponding eigenvalues are the linear forms $h_j - h_i$ defined on the Cartan subalgebra. These forms are called *roots* and can be considered as inner products of elements H, R of this subalgebra, viewed as vectors. The vectors R are called *vector roots*. For the groups SL_{n+1} and CSL_{n+1} these vectors are $E_i - E_j$ where E_i stands for E_{ii}. Analogous vector roots are defined for all the remaining simple Lie groups.

Since among all real simple Lie groups with common complex form there is a single compact one, the classification of complex simple Lie groups is equivalent to the classification of compact real simple Lie groups.

The classification of real compact simple Lie groups was greatly simplified by B. L. van der Waerden in *The classification of simple Lie groups* (Die Klassifikation der einfachen Lieschen Gruppen. Berlin, 1933) *[610]* and by Eugene Dynkin (b. 1925; worked in Moscow and in Ithaca) in *The structure of simple Lie groups* (Struktura prostyh grupp Li. Moscow, 1946) *[159]*. Both made use of the study of vector roots of compact simple Lie groups by Hermann Weyl in his paper *The theory of representations of continuous semisimple groups by linear transformations* (Theorie der Darstellung kontinuierlicher halbeinfacher Gruppen durch lineare Transformationen. Berlin, 1925) *[625, pp. 262–366]*.

Cartan noted that all vector roots are linear combinations with integer coefficients of several *fundamental roots* equal in number to the rank of the group. Weyl showed that the vector roots form regular figures, and van der Waerden noted that these figures have the following property: if **a** and **b** are two vectors of such a figure, then the numbers $2\mathbf{ab}/\mathbf{a}^2$ are integers. Therefore the angles between these vectors can only be $\dfrac{\pi}{6}, \dfrac{\pi}{4}, \dfrac{\pi}{3}, \dfrac{\pi}{2}, \dfrac{2\pi}{3}, \dfrac{3\pi}{4}, \dfrac{5\pi}{6}, \pi$. The lengths of vectors forming angles $\dfrac{\pi}{3}$ and $\dfrac{2\pi}{3}$ are equal, the lengths of vectors forming angles $\dfrac{\pi}{4}$ and $\dfrac{3\pi}{4}$ are in the ratio $\sqrt{2}:1$, and the lengths of vectors forming angles $\dfrac{\pi}{6}$ and $\dfrac{5\pi}{6}$ are in the ratio $\sqrt{3}:1$. For rank 2 these figures have the forms shown in Figure 108.

The systems of vector roots for simple Lie groups are:

A_n: $\mathbf{E}_j - \mathbf{E}_k$, $\quad j \neq k$, $\quad j, k = 0, 1, \ldots, n$.

B_n: $\pm \mathbf{E}_j$, $\quad j = 1, 2, \ldots, n$; $\quad \pm \mathbf{E}_j \pm \mathbf{E}_k$, $\quad j, k = 1, 2, \ldots, n$.

C_n: $\pm 2\mathbf{E}_j$, $\quad j = 1, 2, \ldots, n$; $\quad \pm \mathbf{E}_j \pm \mathbf{E}_k$, $\quad j, k = 1, 2, \ldots, n$.

D_n: $\pm \mathbf{E}_j \pm \mathbf{E}_k$, $\quad j < k$, $\quad j, k = 1, 2, \ldots, n$.

G_2: $\mathbf{E}_i - \mathbf{E}_j$, $\pm \left(\sum_i \mathbf{E}_i - 3\mathbf{E}_j \right)$ $\quad i, j = 0, 1, 2$.

F_4: $\pm \mathbf{E}_i$, $\pm \mathbf{E}_i \pm \mathbf{E}_j$, $\dfrac{1}{2}(\pm \mathbf{E}_1 \pm \mathbf{E}_2 \pm \mathbf{E}_3 \pm \mathbf{E}_4)$, $\quad i, j = 1, 2, 3, 4$.

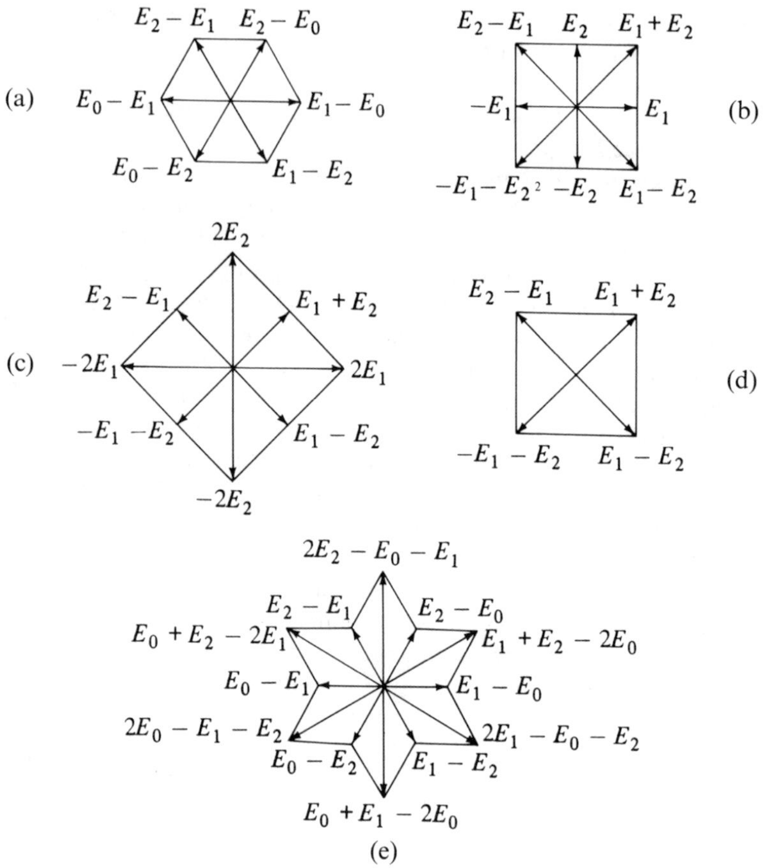

(a)

$E_2 - E_1 \quad E_2 - E_0$

$E_0 - E_1 \qquad\qquad E_1 - E_0$

$E_0 - E_2 \qquad E_1 - E_2$

(b)

$E_2 - E_1 \quad E_2 \quad E_1 + E_2$

$-E_1 \qquad\qquad E_1$

$-E_1 - E_2{}^2 \quad -E_2 \quad E_1 - E_2$

(c)

$2E_2$

$E_2 - E_1 \qquad E_1 + E_2$

$-2E_1 \qquad\qquad 2E_1$

$-E_1 - E_2 \qquad E_1 - E_2$

$-2E_2$

(d)

$E_2 - E_1 \quad E_1 + E_2$

$-E_1 - E_2 \quad E_1 - E_2$

(e)

$2E_2 - E_0 - E_1$

$E_2 - E_1 \qquad E_2 - E_0$

$E_0 + E_2 - 2E_1 \qquad\qquad E_1 + E_2 - 2E_0$

$E_0 - E_1 \qquad\qquad E_1 - E_0$

$2E_0 - E_1 - E_2 \qquad\qquad 2E_1 - E_0 - E_2$

$E_0 - E_2 \qquad E_1 - E_2$

$E_0 + E_1 - 2E_0$

Figure 108

$$E_6: \mathbf{E}_i - \mathbf{E}_j, \; \pm\sqrt{2}\mathbf{E}_0, \frac{1}{2}\left(\frac{\sqrt{2}}{2}\mathbf{E}_0 + \frac{1}{2}\sum_i \mathbf{E}_i - \mathbf{E}_j - \mathbf{E}_k - \mathbf{E}_l\right),$$

$i, j, k, l = 1, 2, 3, 4, 5, 6.$

$$E_7: \mathbf{E}_i - \mathbf{E}_j, \; \pm\left(\frac{1}{2}\sum_i \mathbf{E}_i - \mathbf{E}_h - \mathbf{E}_j - \mathbf{E}_k - \mathbf{E}_l\right),$$

$h, i, j, k, l = 0, 1, 2, 3, 4, 5, 6. 7.$

$$E_8: \pm\mathbf{E}_i - \mathbf{E}_j, \; \pm\sum_i \mathbf{E}_i, \; \pm\left(\frac{1}{2}\mathbf{E}_i - \mathbf{E}_j - \mathbf{E}_k\right), \; \pm\left(\frac{1}{2}\mathbf{E}_i - \mathbf{E}_h - \mathbf{E}_j - \mathbf{E}_k - \mathbf{E}_l\right),$$

$h, i, j, k, l = 1, 2, 3, 4, 5, 6, 7, 8.$

Figures 108a, 108b, 108c, and 108d represent the systems of vector roots of the groups A_2, B_2, C_2, D_2, and G_2, respectively. Dynkin suggested that one should take as fundamental roots the *simple positive roots*: if vector roots are written in the form $\sum_i a_i \mathbf{E}_i$, where the $\mathbf{E}_i$ are a basis of a Cartan subalgebra, then call a root positive if the first $a_i \neq 0$ is positive: if $\mathbf{a} - \mathbf{b}$ is positive then the root $\mathbf{a} = \sum_i a_i \mathbf{E}_i$ is longer than $\mathbf{b} = \sum_i b_i \mathbf{E}_i$.

The systems of simple positive roots for simple Lie groups are:

A_n: $\mathbf{E}_0 - \mathbf{E}_1, \mathbf{E}_1 - \mathbf{E}_2, \ldots, \mathbf{E}_{n-1} - \mathbf{E}_n$.

B_n: $\mathbf{E}_1 - \mathbf{E}_2, \mathbf{E}_2 - \mathbf{E}_3, \ldots, \mathbf{E}_{n-1} - \mathbf{E}_n, \mathbf{E}_n$.

C_n: $\mathbf{E}_1 - \mathbf{E}_2, \mathbf{E}_2 - \mathbf{E}_3, \ldots, \mathbf{E}_{n-1} - \mathbf{E}_n, 2\mathbf{E}_n$.

D_n: $\mathbf{E}_1 - \mathbf{E}_2, \mathbf{E}_2 - \mathbf{E}_3, \ldots, \mathbf{E}_{n-1} - \mathbf{E}_n, \mathbf{E}_{n-1} + \mathbf{E}_n$.

G_2: $\mathbf{E}_2 - \mathbf{E}_3, \mathbf{E}_1 + \mathbf{E}_3 - 2\mathbf{E}_2$.

F_4: $\mathbf{E}_2 - \mathbf{E}_3, \mathbf{E}_3 - \mathbf{E}_4, \mathbf{E}_4, \dfrac{1}{2}(\mathbf{E}_1 - \mathbf{E}_2 - \mathbf{E}_3 - \mathbf{E}_4)$.

E_6: $\mathbf{E}_1 - \mathbf{E}_2, \mathbf{E}_2 - \mathbf{E}_3, \mathbf{E}_3 - \mathbf{E}_4, \mathbf{E}_4 - \mathbf{E}_5, \mathbf{E}_5 - \mathbf{E}_6,$

$\dfrac{\sqrt{2}}{2}\mathbf{E}_0 - \dfrac{1}{2}(\mathbf{E}_1 + \mathbf{E}_2 + \mathbf{E}_3 - \mathbf{E}_4 - \mathbf{E}_5 - \mathbf{E}_6)$.

E_7: $\mathbf{E}_1 - \mathbf{E}_2, \mathbf{E}_2 - \mathbf{E}_3, \mathbf{E}_3 - \mathbf{E}_4, \mathbf{E}_4 - \mathbf{E}_5, \mathbf{E}_5 - \mathbf{E}_6, \mathbf{E}_6 - \mathbf{E}_7,$

$\dfrac{1}{2}(\mathbf{E}_0 - \mathbf{E}_1 - \mathbf{E}_2 - \mathbf{E}_3 - \mathbf{E}_4 + \mathbf{E}_5 + \mathbf{E}_6 + \mathbf{E}_7)$.

E_8: $\mathbf{E}_2 - \mathbf{E}_3, \mathbf{E}_3 - \mathbf{E}_4, \mathbf{E}_4 - \mathbf{E}_5, \mathbf{E}_5 - \mathbf{E}_6, \mathbf{E}_6 - \mathbf{E}_7, \mathbf{E}_7 - \mathbf{E}_8,$

$\dfrac{1}{2}(\mathbf{E}_1 - \mathbf{E}_2 - \mathbf{E}_3 - \mathbf{E}_4 - \mathbf{E}_5 - \mathbf{E}_6 - \mathbf{E}_7 + \mathbf{E}_8), \mathbf{E}_7 + \mathbf{E}_8$.

Dynkin also suggested that these systems of simple roots be represented by means of certain diagrams now named *Dynkin diagrams*. In these diagrams, similar to Coxeter diagrams for groups generated by reflections, each simple positive root is represented by a point. These points are joined by a single line if the angle between corresponding roots is equal to $\dfrac{2\pi}{3}$, by a double line if it is $\dfrac{3\pi}{4}$, and by a triple line if it is $\dfrac{5\pi}{6}$. If two points, corresponding to two roots are joined by a single, double, or triple line and the first root is longer that the second, there is an arrow from the first point to the second. The Dynkin diagrams for simple Lie groups are shown in Figure 109.

The similarity between the Dynkin diagrams for the groups A_n, B_n, C_n, D_n, G_2, F_4, E_6, E_7, E_8, and the Coxeter diagrams (a), (b), (c), (d) for $r = 6$, (f), (h), (i), (j) is explained by the close connection between the groups of symmetries of systems of vector roots and the groups generated by reflections; we note that the vector roots $\pm \mathbf{E}_i, \frac{1}{2}(\pm \mathbf{E}_1 \pm \mathbf{E}_2 \pm \mathbf{E}_3 \pm \mathbf{E}_4)$ of the group F_4 are radius vectors of vertices of a regular polyhedron $\{3, 4, 3\}$ in R_4.

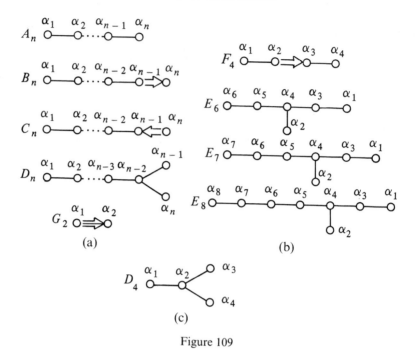

(a)   (b)

(c)

Figure 109

The Classification of Real Simple Lie Groups

In the paper *Real simple finite continuous groups* (Les groupes réels simples finis et continus. Paris, 1914) *[96, part 1, vol. 1, pp. 399–530]* Cartan determined all real simple Lie groups. The number of such groups is considerably larger than that of the complex groups, for a number of nonisomorphic real groups can have the same complex form. The real groups in the class A_n are the group SL_{n+1} of unimodular real matrices, the groups CSU_{n+1} and $C^l SU_{n+1}$ of unimodular complex unitary and pseudounitary matrices (that preserve Hermitian positive-definite and indefinite forms, respectively) and the group $HSL_{(n+1)/2}$ of unimodular quaternion matrices. The real groups in the class B_n are the groups O_{2n+1} and $^l O_{2n+1}$ of real orthogonal and pseudoorthogonal matrices (that preserve positive definite and indefinite quadratic forms, respectively). The real groups in the class C_n are the group Sy_{2n} of real sympletic matrices and the groups HU_n and $H^l U_n$ of quaternion unitary and pseudounitary matrices. The real groups in the class D_n are the groups O_{2n} and $^l O_{2n}$ of real orthogonal and pseudoorthogonal matrices and the group HSy_n of quaternion symplectic matrices. The letter l in the symbols for groups of pseudoorthogonal and pseudounitary matrices denotes the number of minus signs in the canonical expressions $\sum_i \pm (x^i)^2$ and $\sum_i \pm \bar{x}^i x^i$ of the corresponding quadratic and Hermitian forms. Later, the classification of noncompact real simple Lie groups was simplified by Cartan himself in

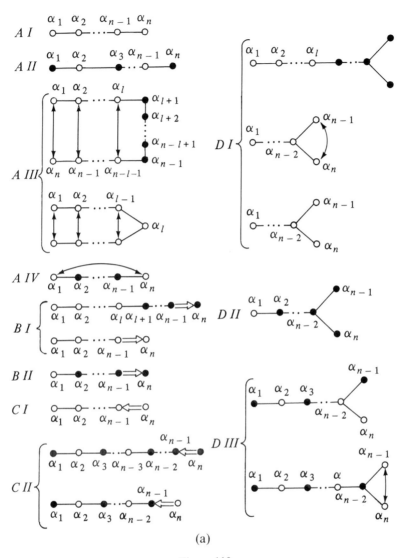

(a)

Figure 110

the paper *Compact and noncompact simple groups and Riemannian geometry* (Groupes simples clos et ouverts et géométrie riemannienne. Paris, 1929) *[96, part 1, pp. 1011–1043]*, where this classification was connected with the theory of symmetric spaces.

Real simple Lie groups are defined by their *Satake diagrams*. As complex simple Lie groups, these groups have vector roots. Some of these roots are real and some are imaginary. All vector roots of compact simple Lie groups are imaginary. The real Lie groups all of whose vector roots are real are called

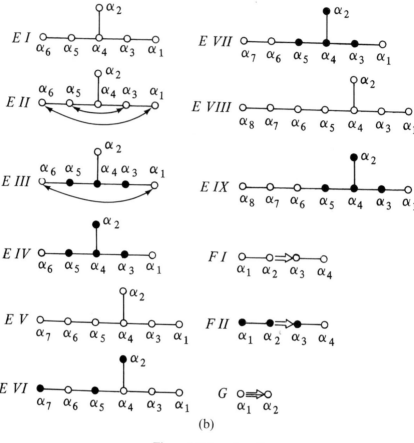

(b)

Figure 110 (*continued*)

split simple groups (or *anticompact simple groups*). The remaining real simple Lie groups have both real and imaginary roots.

Satake diagrams, introduced by I. Satake in the paper *On representations and compactifications of symmetric Riemannian spaces*. Princeton, 1960 *[503]*, have the same form as Dynkin diagrams, but white points in these diagrams correspond to real roots, and black points to imaginary roots, and pairs of white points are joined by a double arrow to pairs of imaginary conjugate roots. Satake diagrams of all noncompact real simple Lie groups are represented in Figure 110. The designations *A I, A II, ...* of noncompact simple Lie groups were introduced by É. Cartan.

In the Table I below, the first line gives the Cartan designations of the noncompact simple Lie groups in Figure 110 and analogous designations *A O, B O, C O, D O* for compact simple Lie groups in the same classes; on the second line gives the corresponding Lie algebras, and the third line gives the spaces whose fundamental groups are these groups:

Table I

A O	A I	A II	A III	A IV	B O	B I	B II
$P\overline{S}C^-_{n+1}$	PR^-_{n+1}	$PH^-_{(n+1)/2}$	$P^l\overline{S}C^-_{n+1}$	$P^1\overline{S}C^-_{n+1}$	SR^-_{2n+1}	$^lSR^-_{2n+1}$	$^1SR^-_{2n+1}$
$C\overline{S}_n$	P_n	$HP_{(n-1)/2}$	$C^l\overline{S}_n$	$C^1\overline{S}_n$	S_{2n}	$^lS_{2n}$	$^1S_{2n}$

C O	C I	C II	D O	D I	D II	D III
$\overline{S}H^-_n$	$Sp R^-_{2n}$	$^l\overline{S}H^-_n$	SR^-_{2n}	$^lSR^-_{2n}$	$^1SR^-_{2n}$	$Sp H^-_n$
$H\overline{S}_{n-1}$	Sp_{2n-1}	$H^l\overline{S}_{n-1}$	S_{2n-1}	$^lS_{2n-1}$	$^1S_{2n-1}$	HSp_{n-1}

Geometric Interpretation of Simple Lie Groups

Since the simple groups O_{n+1} are isomorphic to groups of rotations of $(n + 1)$-dimensional Euclidean spaces, they are locally isomorphic to the groups of motions of n-dimensional *elliptic spaces* S_n; since the simple groups $^lO_{n+1}$ are isomorphic to groups of rotations of $(n + 1)$-dimensional pseudo-Euclidean spaces of index $n - l + 1$, they are locally isomorphic to groups of motions of n-dimensional *hyperbolic spaces* ^lS_n of index l (for $n = 1$ these are the *Lobačevskian spaces* ^1S_n). Similarly, the simple groups SL_{n+1} and Sy_{2n} are locally isomorphic to groups of collineations of n-dimensional *projective spaces* P_n and to groups of symplectic transformations of $(2n - 1)$-dimensional *symplectic spaces* Sp_{2n-1}, that is, projective spaces P_{2n-1} in which there is given an invariant complex of $(n - 1)$-dimensional *isotropic planes* (for $n = 2$ an invariant linear complex of isotropic *lines*). In view of the *Darboux transfer*, the groups of conformal transformations of n-dimensional *conformal spaces* C_n are locally isomorphic to the groups $^1O_{n+2}$. Thus the fundamental groups of projective, elliptic, symplectic, and conformal spaces are simple Lie groups.[2]

The space whose fundamental group is the compact group CSU_{n+1} in the class A_n was first considered at the beginning of the 20th century by the Italian geometer Guido Ghirin Fubini (1879–1943) in the paper *On definite metrics of a Hermitian form* (Sulle metriche definite da una forma Hermitiana. Venice, 1903) *[191]* and by the German geometer Eduard Study (1862–1930) in the paper *Shortest paths in the complex domain* (Kürzeste Wege im komplexen Gebiete. Leipzig, 1905) *[566]*. These papers dealt with complex projective space with a real metric defined by the relation

$$\cos^2\frac{\omega}{r} = \frac{\overline{\mathbf{x}}\mathbf{y}\cdot\overline{\mathbf{y}}\mathbf{x}}{\overline{\mathbf{x}}\mathbf{x}\cdot\overline{\mathbf{y}}\mathbf{y}}. \qquad (9.6)$$

---

[2] A detailed account of the geometry of these spaces is given in chapters IX and XI of the author's book *[464]* and in chapters II–IV of his books *[465; 466]*.

The right side of this formula is the cross ratio of the points $X(\mathbf{x})$ and $Y(\mathbf{y})$ and their polar hyperplanes relative to the imaginary Hermitian quadric $\bar{\mathbf{x}}\mathbf{x} = 0$. In the same paper Study considered the space with the imaginary Hermitian quadric replaced by a real Hermitian quadric of index l. At present, n-dimensional spaces of this type defined by an imaginary Hermitian quadric and a real Hermitian quadric of index l are called, respectively, *complex Hermitian elliptic and hyperbolic spaces* $\mathbf{C}\bar{S}_n$ and $\mathbf{C}^l\bar{S}_n$. For $n = 3$ these spaces are described in *Geometry of the complex domain* (Oxford, 1924) *[125]* by Julian Lowell Coolidge (1873–1958) and in *Lectures on complex projective geometry* (Leçons sur la géométrie projective complexe. Paris, 1931) *[94]* by É. Cartan. In the case of Hermitian quadrics of arbitrary index l the groups of motions are locally isomorphic to the noncompact groups $\mathbf{C}^l SU_{n+1}$ of the same class. On the basis of a geometric interpretation of the groups $\mathbf{H}U_n$ given by C. Chevalley in *The theory of Lie groups* (Princeton, 1946) *[112]*, a similar interpretation of the compact and noncompact groups $\mathbf{H}U_{n+1}$ and $\mathbf{H}^l U_{n+1}$ of the class C_{n+1} as quaternion Hermitian elliptic and hyperbolic spaces $\mathbf{H}\bar{S}_n$ and $\mathbf{H}^l\bar{S}_n$ was given in the forties of this century. One of the earliest definitions was given by the author in an appendix to his book of Russian translations of É. Cartan's papers, entitled *Geometry of Lie groups and symmetric spaces* (Geometriya grupp Li i simmetričeskie prostranstva, Moscow, 1949 *[268, pp. 331–368]*.[3]

The group $\mathbf{H}SL_{n+1}$ admits an interpretation, analogous to the interpretation of the group SL_{n+1}, as the group of collineations of a *quaternion projective space* and the group $\mathbf{H}Sy_{n+1}$ as the group of symplectic transformations of a *quaternion Hermitian symplectic space*. The group $\mathbf{H}P_n$ was considered for $n = 3$ by É. Cartan in *[94]*. The group $\mathbf{H}\bar{S}p_n$ was considered by Lyudmila Viktorovna Rumyanceva in *Quaternion symplectic geometry* (Kvaternionnaya simplektičeskaya geometriya. Moscow, 1963) *[490]*.

Isomorphisms of Simple Lie Groups and Transfers

The following isomorphisms hold for a simple Lie groups of low dimensions: $A_1 = B_1 = C_1$, $B_2 = C_2$, $A_3 = D_3$ and $D_2 = B_1 \otimes B_1$. These isomorphisms determine geometric interpretations of the corresponding spaces in one another. On the isomorphism $A_1 = B_1$ are based the isometricity of a complex Hermitian elliptic line $\mathbf{C}\bar{S}_1$ and a two-dimensional real sphere covering twofold the elliptic plane S_2, the Poincaré interpretation of the Lobačevskian plane S_2 in the complex plane with the metric of a complex Hermitian hyperbolic line $\mathbf{C}^1\bar{S}_1$, and the *Hesse transfer* of the projective line P_1 in the plane $^1S_2$.

On the isomorphism $B_2 = C_2$ are based the analogous isometricity of

---

[3] See also chapter V of the author's book *[465]*.

a quaternion Hermitian elliptic line $H\bar{S}_1$ and a four-dimensional real sphere covering twofold the elliptic space S_4, the Poincaré interpretation of the Lobačevskian space $^1S_4$, and the interpretation of the manifold of lines of the symplectic space Sp_3 in the hyperbolic space $^2S_4$.

On the isomorphism $A_3 = D_3$ are based a number of interpretations of which we mention the *Plücker transfer*, in which the manifold of lines in the projective space P_3 is represented by the absolute of the hyperbolic space $^3S_5$, the interpretation (by E. Study in the paper *Companion to the theory of linear transformations of a complex variable* (Ein Seitenstück zur Theorie der linearen transformationen einer komplexen Veränderlichen. Berlin, 1923–1924) *[567]*) of the quaternion projective line HP_1 by the absolute of the Lobačevskian space $^1S_5$, the interperation of the manifold of lines of $C\bar{S}_3$ in the analogous manifold of S_5, the analogous interpetation of the line manifolds of $C^2\bar{S}_3$ and $^2S_5$, and the interpretation of the manifold of lines of $C^1\bar{S}_3$ in the quaternion symplectic plane $H\bar{S}p_2$.

On the isomorphism $D_2 = B_1 \otimes B_1$ are based the interpretation (by G. G. Fubini in the paper *Clifford parallelism in elliptic spaces* (Il rapallelismo di Glifford negli spazî ellittici. Pisa, 1900) *[192]*) of the manifold of lines of the elliptic space S_3 in the manifold of pairs of points of two spheres covering twofold the plane S_2, the *Kotel'nikov-Study transfers* (by A. P. Kotel'nikov in his *Projective theory of vectors* (Proektivnaya teoriya vektorov, Kazan, 1899) *[289]* and by E. Study in his *Geometry of Dynames* (Geometrie der Dynamen. Leipzig, 1903) *[565]*) of the manifolds of lines of S_3 and $^1S_3$ to the spheres covering the planes $^1CS_2$ and C^1S_2 over the respective algebras ^1C of split complex numbers $a + be$, $e^2 = 1$ (a, b real) and the field C of complex numbers, the interpretation (by Klein in his *Lectures on non-Euclidean geometry* (Vorlesungen über nicht-Euklidische Geometrie. Berlin, 1928) *[284]*) of the complex projective line CP_1 by the absolute of $^1S_3$, an analogous interpretation of the absolute of $^2S_3$ in the split complex projective line $^1CP_1$, and the interpretation of the quaternion symplectic line HSp_1 in the manifold of pairs of points of the planes S_2 and $^1S_2$ or by the "product" of these planes.

We shall discuss the interpretations by spaces over algebras in the next chapter.[4]

Symmetric Spaces

There is a close connection between Lie groups and symmetric spaces. *Symmetric Riemannian spaces* were introduced by the Soviet geometer Petr Alekseevič Širokov (1895–1944) in the paper *Constant fields of vectors and second-order tensors in Riemannian spaces* (Postoyannye polya vektorov i

---

[4] For details of these interpretations see the author's books *[464–466]* and the paper *[490]*.

tenzorov vtorogo poryadka v rimanovyh prostranstvah. Kazan, 1925) *[533, pp. 256–280]* and by É. Cartan in the paper *On a certain remarkable class of Riemannian spaces* (Sur une classe remarquable d'espaces de Riemann. Paris, 1926) *[96, part. 1, pp. 587–659]*; see also the books *[222]* of Sigurdur Helgason and *[337]* of Ottmar Loos. These spaces can be defined as Riemannian spaces in which reflections in points along geodesics preserve the metric of the space. A necessary and sufficient condition for this is the vanishing of the covariant derivative of the curvature tensor ($\nabla_l R_{ij,k}^{\;\;h} = 0$, spaces satisfying this condition were studied by P. A. Širokov). The simplest examples of symmetric spaces are Euclidean and non-Euclidean spaces whose curvature in all two-dimensional directions is constant (and equal to zero in the case of Euclidean spaces).

In *The geometry of groups of transformations* (La géométrie des groupes de transformations. Paris, 1927) *[96, part. 1, pp. 673–791]*. É. Cartan introduced the notion of *symmetric spaces with an affine connection* in which reflections in points along geodesics preserve the affine connection (that is, map geodesics to geodesics with preservation of the affine parameter). As in the case of Riemannian spaces, a necessary and sufficient condition for a space with an affine connection to be symmetric is that $\nabla_l R_{ij,k}^{\;\;h} = 0$.

In the same work Cartan also showed that every Lie group is a symmetric space with an affine connection if one takes as geodesics of the Lie group its one-parameter subgroups and their cosets, and if one takes as an affine parameter on them the canonical parameter of the one-parameter subgroups (the parameter t such that $g(t_1)g(t_2) = g(t_1 + t_2)$) and the corresponding parameters of the cosets. Reflections in points of a symmetric space generate a subgroup of the group of isometries of the space that is a Lie group. If we associate to every point of a symmetric space the reflection σ in that point then, as Cartan showed, the products $\sigma\sigma_0$, where σ_0 is a reflection in a fixed point, form a totally geodesic surface in that Lie group with an affine connection defined by Cartan.

Symmetric spaces admit geometric interpretations as manifolds of *symmetry figures* in spaces with the same fundamental groups, in particular, in the cases when these groups are simple Lie groups in the projective, elliptic, hyperbolic, symplectic, and conformal spaces defined above.

The classification of symmetric spaces with simple groups of motions is equivalent to the classification of the involutory automorphisms of non-compact simple Lie groups carried out by Marcel Berger in the papers *Classification of irreducible symmetric homogeneous spaces* (Classification des espaces homogènes symétriques irréductible. Paris, 1955) *[45]* and *Structure and classification of symmetric spaces with semisimple isometry group* (Structure et classification des espaces symétriques à groupe d'isométrie semisimple. Paris, 1955) *[46]* (see also *[47]*), and by Anatoliĭ Semenovič Fedenko (b. 1929) in the paper *Symmetric spaces with simple noncompact fundamental groups* (Simmetričeskie prostranstva s prostymi nekompaktnymi fundamental'nymi gruppami. Moscow, 1956) *[179]* (see also *[180]*).

The first column in Table II below gives the class designation (*A, B, C* or

Table II

					$^1C\bar{S}_n = P_n$
A	$C\bar{S}_n = (\mathbf{H} \otimes \mathbf{C})\tilde{\bar{S}}_{(n-1)/2}$	$'x^i = \bar{x}^i$	normal n-dimensional chain	$A\,I$	
		$'x^{2i} = \bar{x}^{2i+1},$ $'x^{2i+1} = \bar{x}^{2i}$	paratactic line congruence	$A\,II$	$(\mathbf{H} \otimes {}^1\mathbf{C})\tilde{\bar{S}}_{(n-1)/2} = \mathbf{H}P_{(n-1)/2}$
		$'x^a = x^a,$ $'x^u = -x^u$	$(l-1)$-dimensional plane	$A\,III$	$\mathbf{C}^l\bar{S}_n$
		$'x^0 = x^0,$ $'x^i = -x^i$	point	$A\,IV$	$\mathbf{C}^1\bar{S}_n$
B	S_{2n}	$'x^a = x^a,$ $'x^u = -x^u$	$(l-1)$-dimensional plane	$B\,I$	$^lS_{2n}$
		$'x^0 = x^0,$ $'x^i = -x^i$	point	$B\,II$	$^1S_{2n}$
C	$\mathbf{H}S_{n-1}$	$'x^i = ix^{i^{-1}}$	normal complex n-dimensional chain	$C\,I$	$^1\mathbf{H}\bar{S}_{n-1} = Sp_{2n-1}$
		$'x^a = x^a,$ $'x^u = -x^u$	$(l-1)$-dimensional plane	$C\,II$	$\mathbf{H}^l\bar{S}_{n-1}$
D	$S_{2n-1} = {}^1\mathbf{H}Sp_{n-1}$	$'x^a = x^a,$ $'x^u = -x^u$	$(l-1)$-dimensional plane	$D\,I$	$^lS_{2n-1}$
		$'x^0 = x^0,$ $'x^i = -x^i$	point	$D\,II$	$^1S_{2n-1}$
		$'x^{2i} = x^{2i+1},$ $'x^{2i+1} = -x^{2i}$	paratactic line congruence	$D\,III$	$\mathbf{H}Sp_{n-1}$

D). The second gives the space whose fundamental group is the compact simple Lie group of this class. The third gives the involutory motion of the group which generates its involutory automorphism. The fourth gives the corresponding symmetry of the space. The fifth gives Cartan's designation of the corresponding noncompact group and symmetric space. The sixth column gives the space whose fundamental group is the corresponding group in column five.

The calculation of the Poincaré polynomial $\sum b_i t^i$ for symmetric spaces is easier than for general homogeneous spaces. These polynomials and the Betti numbers b_i for compact simple Lie groups were calculated by L. S. Pontryagin in the paper *On Betti numbers of compact Lie groups* (Moscow, 1935) *[439]* (see also *[113]*: these polynomials have the form

$$f(t) = \prod_i (t^{2a_i-1} - 1), \tag{9.7}$$

where for the group A_l, $a_i = 2, 3, \ldots, l$; for B_l and C_n, $a_i = 2, 4, \ldots, 2l$; for D_n, $a_i = 2, 4, \ldots, 2l - 2, l$; for G_2, $a_i = 2, 6$; for F_4, $a_i = 2, 6, 8, 12$; for E_6, $a_i = 2, 5, 6, 8, 9, 12$; for E_7, $a_i = 2, 6, 8, 10, 12, 14, 18$; and for E_8, $a_i = 2, 8, 12, 14, 18, 20, 24, 30$.

Reductive and Parabolic Spaces

Reductive spaces are a generalization of symmetric spaces. They were introduced by the author of the well-known monograph *Riemannian geometry and tensor analysis* (Rimanova geometriya i tenzornyĭ analiz. Moscow, 1953) *[447]* Petr Konstantinovič Raševskiĭ (1907–1983) in the paper *Symmetric spaces with an affine connection with torsion* (Simmetričeskie prostranstva affinnoĭ svyaznosti s kručeniem. Moscow, 1950) *[449]*, and by Kakumi Nomizu in the paper *Invariant affine connections on homogeneous spaces* (Baltimore, 1954) *[392]*; the term *reductive* is due to Nomizu. Whereas symmetric spaces are spaces with an affine connection without torsion and a covariantly constant curvature tensor, in reductive spaces both the curvature and torsion tensors are covariantly constant. Reductive spaces also admit of interesting geometric interpretations.

The term *parabolic space* which in 19th century was equivalent to the term *Euclidean space* (intermediate between elliptic and hyperbolic spaces) has now received a new meaning: *parabolic spaces* are homogeneous spaces whose fundamental groups are simple Lie groups and whose stabilizer subgroups are parabolic subgroups of these groups, i.e., subgroups containing the maximal solvable subgroup of this group. These spaces were first considered by Charles Ehresmann (1905–1979) in his paper *On the topology of certain homogeneous spaces* (Sur la topologie des certains spaces homogènes. Princeton, 1934) *[164]*. Ehresmann found topological invariants of these spaces. Izrail Moiseevič Gel'fand (b. 1913) and Mark Aronovič Naĭmark

(1909–1980) used these spaces in their paper *Unitary representations of the classical groups [199]*. Jacques Tits considered these spaces in the paper *On certain classes of homogeneous spaces* (Sur certaines classes d'espaces homogènes. Brussels, 1955) *[579]*, introduced these spaces and named them R-spaces (*R*-espaces). J. Tits studied the cases of these spaces with maximal nonsemisimple stabilizer subgroups in the papers *Exceptional Lie groups and their geometric interpretation* (Les groupes de Lie exceptionnels et leur interprétation géométrique. Brussels, 1955) *[580]*, and *On the geometry of R-spaces* (Sur la géometrié des *R*-espaces. Paris, 1957) *[581]*.

Each parabolic subgroup of a simple Lie group is determined by one or several simple roots of this group. The maximal solvable subgroup is determined by all these roots. The parabolic subgroups determined by one simple root are maximal nonsemisimple subgroups of this group. Tits named the figures of the spaces with simple fundamental group with maximal nonsemisimple stabilizer subgroup *fundamental elements* of these spaces. In his paper *Simplicity and semisimplicity figures* (Obrazy prostoty i poluprostoty. Moscow, 1963) *[470]* the author called these figures *simplicity figures* and other elements of *R*-spaces *semisimplicity figures*. J. A. Wolf calls parabolic spaces *flag manifolds*. All maximal nonsemisimple subgroups of complex simple Lie groups were found in the thesis of Vladimir Vladimirovič Morozov (1910–1975) *On nonsemisimple maximal subgroups of the simple groups* (O nepoluprostyh maksimal'nyh podgruppah prostyh grupp. Kazan, 1943) *[378]* (see also the paper *[263]* by I. L. Kantor).

We have seen previously that in the Lie algebra $\mathbf{g}$ of a simple Lie group G there exists a basis consisting of the basis of the Cartan subalgebra of this algebra and of eigenvectors of the linear transformation $A \to [AH]$, where H is an element of the Cartan subalgebra. Therefore, for each simple root there is a decomposition of the Lie algebra into the direct sum of linear subspaces

$$\mathbf{g} = \mathbf{g}_{-\lambda} \oplus \mathbf{g}_{-\lambda+1} \oplus \cdots \oplus \mathbf{g}_{-1} \oplus \mathbf{g}_0 \oplus \mathbf{g}_1 \oplus \cdots \oplus \mathbf{g}_{\lambda-1} \oplus \mathbf{g}_\lambda, \quad (9.8)$$

where each $\mathbf{g}_\alpha$ for $\alpha \neq 0$ is a linear combination of eigenvectors such that the corresponding vector roots have the coefficient α at a selected simple root and $\mathbf{g}_0$ is the direct sum of analogous linear combinations and the Cartan subalgebra. For example, the eigenvectors in the Lie algebra of the group SL_{n+1} are the matrices E_{ij} and the corresponding vector roots are $\mathbf{E}_i - \mathbf{E}_j$. The simple roots of this group being those of A_n, they are, as was shown above, the vectors $\alpha_1 = \mathbf{E}_0 - \mathbf{E}_1$, $\alpha_2 = \mathbf{E}_1 - \mathbf{E}_2$, ..., $\alpha_n = \mathbf{E}_{n-1} - \mathbf{E}_n$. Therefore, if we put in place of the element a_{ij} in the matrix of this algebra the linear combination of the roots α_i corresponding to the eigenvectors E_{ij}, we obtain for $n = 3$ the matrix

$$\begin{pmatrix} 0 & \alpha_1 & \alpha_1 + \alpha_2 & \alpha_1 + \alpha_2 + \alpha_3 \\ -\alpha_1 & 0 & \alpha_2 & \alpha_2 + \alpha_3 \\ -\alpha_1 - \alpha_2 & -\alpha_2 & 0 & \alpha_3 \\ -\alpha_1 - \alpha_2 - \alpha_3 & -\alpha_2 - \alpha_3 & -\alpha_3 & 0 \end{pmatrix}.$$

In this case in the formula (9.8) $\lambda = 1$, the decomposition has the form $\mathbf{g} = \mathbf{g}_{-1} \oplus \mathbf{g}_0 \oplus \mathbf{g}_1$, and, with $\times$ a nonzero element, we have for the root α_1:

$$\mathbf{g}_{-1} = \begin{pmatrix} 0 & 0 & 0 & 0 \\ \times & 0 & 0 & 0 \\ \times & 0 & 0 & 0 \\ \times & 0 & 0 & 0 \end{pmatrix}, \quad \mathbf{g}_0 = \begin{pmatrix} \times & 0 & 0 & 0 \\ 0 & \times & \times & \times \\ 0 & \times & \times & \times \\ 0 & \times & \times & \times \end{pmatrix}, \quad \mathbf{g}_1 = \begin{pmatrix} 0 & \times & \times & \times \\ 0 & 0 & 0 & 0 \\ 0 & 0 & 0 & 0 \\ 0 & 0 & 0 & 0 \end{pmatrix};$$

for the root α_2:

$$\mathbf{g}_{+1} = \begin{pmatrix} 0 & 0 & 0 & 0 \\ 0 & 0 & 0 & 0 \\ \times & \times & 0 & 0 \\ \times & \times & 0 & 0 \end{pmatrix}, \quad \mathbf{g}_0 = \begin{pmatrix} \times & \times & 0 & 0 \\ \times & \times & 0 & 0 \\ 0 & 0 & \times & \times \\ 0 & 0 & \times & \times \end{pmatrix}, \quad \mathbf{g}_1 = \begin{pmatrix} 0 & 0 & \times & \times \\ 0 & 0 & \times & \times \\ 0 & 0 & 0 & 0 \\ 0 & 0 & 0 & 0 \end{pmatrix};$$

for the root α_3:

$$\mathbf{g}_{-1} = \begin{pmatrix} 0 & 0 & 0 & 0 \\ 0 & 0 & 0 & 0 \\ 0 & 0 & 0 & 0 \\ \times & \times & \times & 0 \end{pmatrix}, \quad \mathbf{g}_0 = \begin{pmatrix} \times & \times & \times & 0 \\ \times & \times & \times & 0 \\ \times & \times & \times & 0 \\ 0 & 0 & 0 & \times \end{pmatrix}, \quad \mathbf{g}_1 = \begin{pmatrix} 0 & 0 & 0 & \times \\ 0 & 0 & 0 & \times \\ 0 & 0 & 0 & \times \\ 0 & 0 & 0 & 0 \end{pmatrix}.$$

The stabilizer subgroup is determined by the subalgebra

$$\mathbf{f} = \mathbf{g}_0 \oplus \mathbf{g}_1 \oplus \cdots \oplus \mathbf{g}_{\lambda-1} \oplus \mathbf{g}_\lambda \tag{9.9}$$

and the supplementary direct sum

$$\mathbf{l} = \mathbf{g}_{-\lambda} \oplus \mathbf{g}_{-\lambda+1} \oplus \cdots \oplus \mathbf{g}_{-1} \tag{9.10}$$

can be considered the tangent space of the parabolic space. Therefore for the determination of the dimension of the parabolic space it is sufficient to find the dimension of the direct sum (9.8).

The decomposition (9.8) is also defined for an arbitrary set of simple roots. In that case, $\mathbf{g}_\alpha$ for $\alpha \neq 0$ is a linear combination of eigenvectors such that the corresponding vector roots have coefficients at selected simple roots whose sum is α, and $\mathbf{g}_0$ is defined as in the case of a single root. In our example, for two selected roots $\lambda = 2$ and for the roots α_1, α_2 we have

$$\mathbf{g}_{-2} = \begin{pmatrix} 0 & 0 & 0 & 0 \\ 0 & 0 & 0 & 0 \\ \times & 0 & 0 & 0 \\ \times & 0 & 0 & 0 \end{pmatrix}, \quad \mathbf{g}_{-1} = \begin{pmatrix} 0 & 0 & 0 & 0 \\ \times & 0 & 0 & 0 \\ 0 & \times & 0 & 0 \\ 0 & \times & 0 & 0 \end{pmatrix}, \quad \mathbf{g}_0 = \begin{pmatrix} \times & 0 & 0 & 0 \\ 0 & \times & 0 & 0 \\ 0 & 0 & \times & \times \\ 0 & 0 & \times & \times \end{pmatrix},$$

$$\mathbf{g}_1 = \begin{pmatrix} 0 & \times & 0 & 0 \\ 0 & 0 & \times & \times \\ 0 & 0 & 0 & 0 \\ 0 & 0 & 0 & 0 \end{pmatrix}, \quad \mathbf{g}_2 = \begin{pmatrix} 0 & 0 & \times & \times \\ 0 & 0 & 0 & 0 \\ 0 & 0 & 0 & 0 \\ 0 & 0 & 0 & 0 \end{pmatrix},$$

and for all three roots $\lambda = 3$ and

$$\mathbf{g}_{-3} = \begin{pmatrix} 0 & 0 & 0 & 0 \\ 0 & 0 & 0 & 0 \\ 0 & 0 & 0 & 0 \\ \times & 0 & 0 & 0 \end{pmatrix}, \quad \mathbf{g}_{-2} = \begin{pmatrix} 0 & 0 & 0 & 0 \\ 0 & 0 & 0 & 0 \\ \times & 0 & 0 & 0 \\ 0 & \times & 0 & 0 \end{pmatrix}, \quad \mathbf{g}_{-1} = \begin{pmatrix} 0 & 0 & 0 & 0 \\ \times & 0 & 0 & 0 \\ 0 & \times & 0 & 0 \\ 0 & 0 & \times & 0 \end{pmatrix},$$

$$\mathbf{g}_0 = \begin{pmatrix} \times & 0 & 0 & 0 \\ 0 & \times & 0 & 0 \\ 0 & 0 & \times & 0 \\ 0 & 0 & 0 & \times \end{pmatrix}, \quad \mathbf{g}_1 = \begin{pmatrix} 0 & \times & 0 & 0 \\ 0 & 0 & \times & 0 \\ 0 & 0 & 0 & \times \\ 0 & 0 & 0 & 0 \end{pmatrix},$$

$$\mathbf{g}_2 = \begin{pmatrix} 0 & 0 & \times & 0 \\ 0 & 0 & 0 & \times \\ 0 & 0 & 0 & 0 \\ 0 & 0 & 0 & 0 \end{pmatrix}, \quad \mathbf{g}_3 = \begin{pmatrix} 0 & 0 & 0 & \times \\ 0 & 0 & 0 & 0 \\ 0 & 0 & 0 & 0 \\ 0 & 0 & 0 & 0 \end{pmatrix}.$$

Let us call the figure corresponding to the simple root α_i the α_i-*figure* and the figure corresponding to the set α_{i_1}, α_{i_2}, ..., α_{i_k} of simple roots the $(\alpha_{i_1}, \alpha_{i_2}, \ldots, \alpha_{i_k})$-figure. If we consider the matrices of the group SL_{n+1} as matrices of collineations of the projective space P_3 and determine the stabilizer groups corresponding to the subalgebras $\mathbf{g}_0 \oplus \mathbf{g}_1$ for each root α_i, $\mathbf{g}_0 \oplus \mathbf{g}_1 \oplus \mathbf{g}_2$ for each pair $(\alpha_{i_1}, \alpha_{i_2})$ and $\mathbf{g}_0 \oplus \mathbf{g}_1 \oplus \mathbf{g}_2 \oplus \mathbf{g}_3$ for the triple $(\alpha_1, \alpha_2, \alpha_3)$, then we find that in P_3 the α_1-figure is a *point*, the α_2-figure is a *line*, the α_3-figure is a *plane*, the (α_1, α_2)-figure is a flag consisting of a point and a line through this point, the (α_1, α_3)-figure is a flag consisting of a point and a plane through this point, the (α_2, α_3)-figure is a flag consisting of a line and a plane through this line, and the $(\alpha_1, \alpha_2, \alpha_3)$-figure is a flag consisting of a point, a line through it, and a plane through this line.

Similarly, we find that the α_i-figures of the spaces P_n, $^nS_{2n-1}$ and $^nS_{2n}$, Sp_{2n-1} whose fundamental groups are split real simple Lie groups are as shown in Table III. The first column of this table gives class, the second the space, the third the simple roots, and the fourth the α_i-figures. The $(\alpha_{i_1}, \alpha_{i_2}, \ldots, \alpha_{i_k})$-figures of these spaces are flags consisting of the corresponding $\alpha_{i_1} -, \ldots, \alpha_{i_k}$-figures. This fact explains the name *flag manifold* for parabolic spaces.

An α_i-figure and an α_j-figure are called *incident* if they are contained in an (α_i, α_j)-figure. Note that the $(n-2)$-dimensional planes of the absolute of $^nS_{2n-1}$ are (α_{n-1}, α_n)-figures, for they are intersections of two $(n-1)$-dimensional planes of different families on the absolute.

The Dynkin diagrams of complex and compact real groups, shown in the Figure 109a and the Satake diagrams of split real groups that coincide with them can be considered as diagrams of α_i-figures of the spaces with these fundamental groups. The bilateral symmetry of the diagram for A_n corresponds to the *principle of duality* of the space P_n and the bilateral

Table III

A	P_n	α_1 α_2	Point Line
		α_i	$(i-1)$-dimensional plane
		α_n	Hyperplane
B	$^nS_{2n}$	α_1 α_2	Point of the absolute Line of the absolute
		α_i	$(i-1)$-dimensional plane on the absolute
		α_n	$(n-1)$-dimensional plane on the absolute
C	Sp_{2n+1}	α_1 α_2	Point Isotropic line
		α_i	Isotropic $(i-1)$-dimensional plane
		α_n	Isotropic $(n-1)$-dimensional plane
D	$^nS_{2n-1}$	α_1 α_2	Point of the absolute Line on the absolute
		α_i	$(i-1)$-dimensional plane on the absolute
		α_{n-1} α_n	$(n-1)$-dimensional plane of first family on the absolute $(n-1)$-dimensional plane of second family on the absolute

symmetry of the roots α_{n-1} and α_n for D_n corresponds to the equality of two families of $(n-1)$-dimensional planes of the absolute of the space $^nS_{2n-1}$. All α_i-figures, and therefore all $(\alpha_{i_1}, \alpha_{i_2}, \ldots, \alpha_{i_n})$-figures of the spaces with split real fundamental groups, are real, all figures of the spaces with compact real fundamental groups are imaginary; such are points, lines and planes on the absolutes of the elliptic spaces S_{2n} and S_{2n-1}.

The Satake diagrams of noncompact and nonsplit real groups shown in Figure 110a can also be considered as diagrams of α_i-figures of the corresponding spaces, but the α_i-figures represented by white points are real, the α_i-figures represented by black points are imaginary, and the α_i-figures represented by white points joined by double arrows are conjugate imaginary (if an α_i-figure and an α_j-figure are such imaginary figures, then the (α_i, α_j)-figure is real). For example, the α_i-figures of the spaces $^lS_{2n}$ and $^lS_{2n-1}$ (points, lines and i-dimensional planes on the absolute) are real for $i \le l$ and imaginary for $i > l$, and the α_{n-1} and α_n-figures of $^{n-1}S_{2n-1}$ are conjugate imaginary and determine real $(n-2)$-dimensional planes on the absolute (in

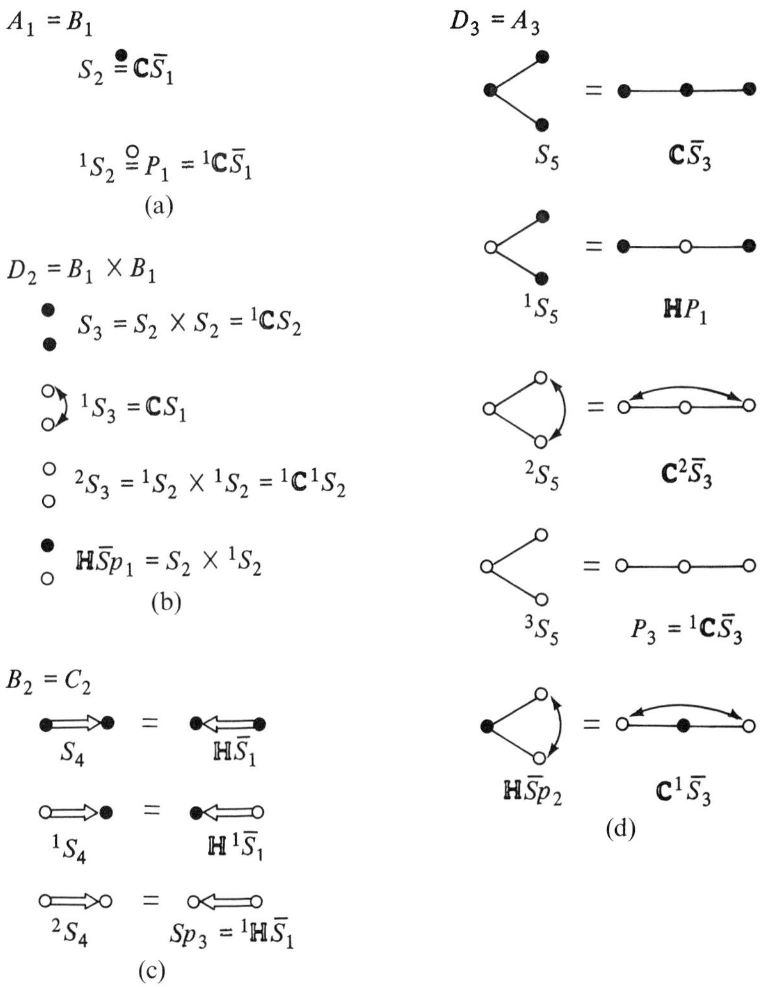

$A_1 = B_1$

$S_2 \overset{\bullet}{=} \mathbf{C}\bar{S}_1$

$^1S_2 \overset{\circ}{=} P_1 = {}^1\mathbf{C}\bar{S}_1$

(a)

$D_2 = B_1 \times B_1$

$S_3 = S_2 \times S_2 = {}^1\mathbf{C}S_2$

$^1S_3 = \mathbf{C}S_1$

$^2S_3 = {}^1S_2 \times {}^1S_2 = {}^1\mathbf{C}{}^1S_2$

$\mathbf{H}\bar{S}p_1 = S_2 \times {}^1S_2$

(b)

$B_2 = C_2$

$\underset{S_4}{\bullet\!\!\Rightarrow\!\!\bullet} \quad = \quad \underset{\mathbf{H}\bar{S}_1}{\bullet\!\!\Leftarrow\!\!\bullet}$

$\underset{{}^1S_4}{\circ\!\!\Rightarrow\!\!\bullet} \quad = \quad \underset{\mathbf{H}\,{}^1\bar{S}_1}{\bullet\!\!\Leftarrow\!\!\circ}$

$\underset{{}^2S_4}{\circ\!\!\Rightarrow\!\!\circ} \quad = \quad \underset{Sp_3 = {}^1\mathbf{H}\bar{S}_1}{\circ\!\!\Leftarrow\!\!\circ}$

(c)

$D_3 = A_3$

$S_5 \qquad \mathbf{C}\bar{S}_3$

$^1S_5 \qquad \mathbf{H}P_1$

$^2S_5 \qquad \mathbf{C}^2\bar{S}_3$

$^3S_5 \qquad P_3 = {}^1\mathbf{C}\bar{S}_3$

$\mathbf{H}\bar{S}p_2 \qquad \mathbf{C}^1\bar{S}_3$

(d)

Figure 111

particular, for $n - 2$ the α_1 and α_2-figures of $^1S_3$ are conjugate imaginary lines intersecting in real points).

The Dynkin and Satake diagrams of isomorphic groups are similar. Such diagrams are shown in Figures 111 a, b, c and d.

Quasisimple and k-Quasisimple Lie Groups

The group of motions of the Euclidean space R_n is not simple for it contains a normal subgroup, namely the group of translations. However, Euclidean space can be obtained from elliptic space S_n and from Lobačevskian space ^1S_n by passage to the limit, and its group of motions can be obtained from the groups of motions of these spaces in the same way. Similarly, the group of motions of the pseudo-Euclidean space ^lR_n can be obtained from the groups of motions of the hyperbolic spaces ^lS_n and $^{l+1}S_n$ by passage to the limit.

F. Klein, in his paper *On the geometric foundations of the Lorentz group* (Über die geometrische Grundlagen der Lorentzgruppe, Leipzig, 1910) *[282, vol. 1, pp. 533–552]* devoted to the study of the pseudo-Euclidean space $^1R_4$ used in the special theory of relativity, put forward the idea of arbitrary *projective metrics*. He noted that the metrics of the elliptic plane S_2, the Lobačevskian plane $^1S_2$, the Euclidean plane R_2 and the pseudo-Euclidean plane $^1R_2$ can be considered as geometries of the projective plane P_2 with absolutes (1) $u_0^2 + u_1^2 + u_2^2 = 0$ (imaginary conic), (2) $u_0^2 + u_1^2 - u_2^2 = 0$ (real conic), (3) $u_0^2 + u_1^2 = 0$ (imaginary pair of points), and (4) $u_0^2 - u_1^2 = 0$ (real pair of points). In adding to these 4 cases the 5th case (5) $u_0^2 = 0$ (double point), Klein wrote:

We obtain five kinds (*and only five kinds*) *of metric geometries in the plane from which only one, corresponding to the imaginary pair of points, is known to us from the example of the elementary metric.* We call the totality of theories occurring in this way the *general theory of projective metrics [282, vol. 1, p. 540]*.

This general theory of projective metrics was constructed by Duncan Maclaren Young Sommerville (1879–1934), author of the well-known bibliography of non-Euclidean geometry *[549]* and of an introduction to *n*-dimensional geometry *[550]*, in the paper *Classification of geometries with projective metrics* (Edinburgh, 1910–1911) *[551]*.

First Sommerville considers three-dimensional geometries. We quote:

Let us now investigate the different systems of geometry. We have three constants to fix, and any of them may be infinite, real or imaginary, hence there are 27 possible systems. These depend upon the form of the absolute and the conditions laid down with regard to the actual* and ideal elements. We shall make the following assumptions:

1. *An actual geometric form contains actual elements.*

2. *The distance between two actual elements of an actual geometric form is real.*

Having fixed upon one plane α as an actual plane, a line a in α as an actual line, and a point A in a as an actual point, all points at a real finite distance from A are actual points.** A line is actual if it makes a real angle with a, or if it makes a real angle with an actual line; and similarly for planes. The actual

---

*The term "actual" here is opposed to "ideal," and is preferred to real, which is opposed to "imaginary." (Sommerville's note.)

**It may happen that the harmonic conjugate A' of A is at a real finite distance from A, but points of AA' in the vicinity of A are ideal. In this case A' is ideal. (Sommerville's note.)

points are separated from the ideal points by the absolute. The actual elements of an actual sheaf (*e.g.* a sheaf of lines passing through an actual point and lying in an actual plane) are separated from the ideal elements by the two absolute elements of the sheaf.

The values of the constants are as follows:—

K, *k* or κ is infinite if the absolute degenerates to two coincident planes, lines or points.

K is real or imaginary, according as actual lines do or do not cut the absolute.

κ is real or imaginary, according as actual lines do or do not project the absolute.

k is real or imaginary according as actual points in actual planes do or do not project the section of the absolute.

When K, *k*, κ is infinite, real, imaginary, the measure of distance, plane angle, dihedral angle is parabolic, hyperbolic, elliptic. In ordinary geometry, in hyperbolic geometry, and in elliptic geometry the measure of angles, plane and dihedral, is elliptic; *k* and κ are both imaginary, while K is infinite, real, or imaginary.

The forms of the absolute and the various geometries are discussed as follows:—

A. *Absolute a proper quadric.*
  I. *Imaginary.*
     K, *k*, κ all imaginary. Distances and angles are always real and periodic. (ELLIPTIC GEOMETRY.)
  II. Real and not ruled.
     The absolute divides space into an actual and an ideal region of points, lines, and planes, and possesses an interior and an exterior. A line projects the quadric if it does not cut it.
     1. Actual points within. Actual lines and planes cut the quadric. K real, *k* and κ imaginary. (HYPERBOLIC GEOMETRY.)
     2. Actual points outside. Actual lines and planes cut the quadric. K and *k* real, κ imaginary.
     3. Actual points outside. Actual planes cut the quadric, but actual lines do not. K imaginary, *k* and κ real.
     4. Actual points outside. Actual lines and planes do not cut the quadric. K and *k* imaginary, κ real.
  III. *Ruled.*
     There is no point from which real tangent lines and planes may not be drawn to the quadric, and every plane cuts the quadric. A line projects the quadric if it cuts it.
     1. Actual lines cut the quadric.
        Take any such line and draw an arbitrary plane through it, cutting the quadric in a conic S. Let this plane be actual. Then there are two cases.

(a) Points within S are actual.
   K real, k imaginary, κ real.
(b) Points outside S are actual.
   K, k, κ real.
2. Actual lines do not cut the quadric.
   K imaginary, k real, κ imaginary.

In the case of a ruled quadric there are two systems of lines which do not cut the quadric, and these are separated by the quadric. If, therefore, we fix upon one line as actual, all lines of the other system are ideal since they contain no actual points. The absolute divides the points of space into two sets, and it is arbitrary which set we agree to take as actual.

B. *Absolute a simply degenerate quadric.*
  I. *A cone, two coincident points.* $\kappa = \infty$.
   1. *Imaginary cone.*
      K and k imaginary.
   2. *Real cone.*
      (a) Actual points within. K real, k imaginary.
      (b) Actual points outside. Actual lines cut the cone. K, k real.
      (c) Actual points outside. Actual lines do not cut the cone. K imaginary, k real.
  II. *Two coincident planes, proper conic.* $K = \infty$.
   1. *Imaginary conic.*
      k, κ imaginary. (PARABOLIC GEOMETRY).
   2. *Real conic.*
      (a) Actual lines pass within the conic. k real, κ imaginary.
      (b) Actual lines and planes pass outside the conic. k imaginary, κ real.
      (c) Actual lines pass outside, actual planes cut the conic. k, κ real.
  III. *Two planes, two coincident lines, two points.* $k = \infty$.
   1. *Imaginary planes, imaginary points.*
      K, κ imaginary.
   2. *Imaginary planes, real points.*
      K imaginary, κ real.
   3. *Real planes, imaginary points.*
      K real, κ imaginary.
   4. *Real planes, real points.*
      K, κ real.
C. *Absolute a doubly degenerate quadric.*
  I. *Two coincident planes, two coincident lines, two points.* K, $k = \infty$.
   1. *Imaginary points.* κ imaginary.
   2. *Real points.* κ real.
  II. *Two coincident planes, two lines, two coincident points.* K, $\kappa = \infty$.
   1. *Imaginary lines.* κ imaginary.
   2. *Real lines.* κ real.

III. *Two planes, two coincident lines, two coincident points.* k, $\kappa = \infty$.
  1. *Imaginary planes.*   K imaginary.
  2. *Real planes.*   K real.
D. *Absolute a triply degenerate quadric.*
Two coincident planes, two coincident lines, two coincident points. K, k,
$\kappa = \infty$. *[551, pp. 28–31].*

Sommerville considers geometries with different kinds of points, lines and planes as different geometries. His geometry *A I* is the geometry of the *elliptic space* S_3. His geometries *A II* are geometries of the *Lobačevskian space* $^1S_3$: (1) with points inside the absolute, hyperbolic lines and planes, (2) with points outside the absolute and hyperbolic lines and planes, (3) with the same points, elliptic lines, and hyperbolic planes, (4) with the same points and elliptic lines and planes. Sommerville's geometries *A III* are geometries of the *hyperbolic space* $^2S_3$; the absolute divides this space into two equal parts and he distinguishes between two cases: (1) geometries with hyperbolic lines and (2) geometries with elliptic lines. Sommerville's geometries *B I* are geometries of (1) the *co-Euclidean space* R_3^* dual to the Euclidean space R_3, (2) the *copseudo-Euclidean space* $^1R_3^*$ dual to the pseudo-Euclidean space $^1R_3$: (a) with points inside the cone and hyperbolic lines, (b) with points outside the cone and hyperbolic lines, (c) with points outside the cone and elliptic lines. His geometries *B III* are geometries of (1) the *Euclidean space* R_3, (2) the *pseudo-Euclidean space* $^1R_3$: (a) with timelike lines, (b) with spacelike lines and pseudo-Euclidean planes, (c) with spacelike lines and Euclidean planes. His geometries *B III* are geometries (1) of the *quasielliptic space* S_3^1 with absolute consisting of two imaginary planes and two imaginary points on their intersection line, (2) of the *quasihyperbolic space* $^{01}S_3^1$ with imaginary planes and real points of the absolute, (3) of the *quasihyperbolic space* $^{10}S_3^1$ dual to previous one, (4) of the *quasihyperbolic space* $^{11}S_3^1$ with real planes and points of the absolute. His geometries *D I* are geometries (1) of the *Galilean space* Γ_3 and (2) of the *pseudo-Galilean space* $^1\Gamma_3$; *D II* are the geometries (1) of the *isotropic space* I_3 and (2) of the *pseudoisotropic space* $^1S_3$; *D III* are geometries (1) of the *co-Galilean space* Γ_3^* dual to Γ_3 and (2) of the *copseudo-Galilean space* $^1\Gamma_3^*$ dual to $^1\Gamma_3$. Sommerville's geometry *D* is the geometry of the *flag space* F_3.

The quasielliptic space S_3^1 was first investigated by Wilhelm Blaschke (1885–1962) in the paper *Euclidean kinematics and non-Euclidean geometry* (Euklidische Kinematik und nichteuklidische Geometrie. Berlin, 1911) *[61]* as the space of the group of motions of the Euclidean plane R_2: if the product AB^{-1} of the motion A and the motion inverse to the motion B is a rotation with angle φ then the distance $d_1(A, B)$ is φ. If this product is a translation with distance d, then the distance $d_1 = 0$ but there is also the distance $d_2(A, B) = d$. Blaschke introduced the term *quasielliptic space*. The Galilean space Γ_4—the four-dimensional analogue of the space Γ_3—is the spacetime geometry of classical mechanics of Galileo and Newton (hence its name). In connection with this interpretation, this space was investigated by Ludwik

Silberstein in his *Projective geometry of Galilean space-time* (London, 1925) *[538]* and by Aleksandr Petrovič Kotel'nikov (1865–1944) in the paper *The principle of relativity and the Lobačevskian geometry* (Princip otnositel'nosti i geometriya Lobačevskogo. Kazan, 1927) *[290]*. Another form of two-quasielliptic space is the *isotropic* space I_3, studied by Karl Strubecker in *Differential geometry of isotropic space* (Differentialgeometrie des isotropen Raumes. Vienna, 1941) *[564]*. The flag space F_3 was first studied by Ivan Vasil'evič Parnasskiĭ (b. 1923) in the paper *Axiomatic construction of three-dimensional parabolic geometry* (Aksiomatičeskoe postroenie trehmernoĭ paraboličeskoĭ geometrii. Moscow, 1956) *[405]* and by G. W. M. Kallenberg in the paper *Differential geometry of a particular group of projective transformations* (Amsterdam, 1957) *[258]*. In the same paper Sommerville extends his investigations to n-dimensional spaces. We quote:

Here there are n constants, $k_0, k_1, \ldots, k_{n-1}$, and therefore 3^n geometries. The absolute takes the following forms:

A_0. *A proper hyperquadric of n dimensions* ("*n-quadric*").
  I. *Imaginary.*
  II. *Real and not ruled.*
  III. *Ruled.*
A_1. *Simply degenerate.*
  (*r*)*A hypercone of species r of n dimensions* ("*(n, r)-cone*"). This is formed by joining the points of a proper $(n - r)$-quadric to the points of an $(r - 1)$-flat (the *axis*), and in the axis is taken a proper $(r - 1)$-quadric. An (n, n)-cone consists of two coincident $(n - 1)$-flats with an $(n - 1)$-quadric; an $(n, n - 1)$-cone consists of two $(n - 1)$-flats with an $(n - 2)$-quadric; and an $(n, 0)$-cone is a proper n-quadric.
A_l. *l-ply degenerate.*
  $(r_1, r_2, \ldots, r_l)$. Two coincident $(r_1 - 1)$-, $(r_2 - 1)$-, $\ldots$, $(r_l - 1)$-flats $(r_1 < r_2 < \cdots < r_l)$. An (n, r_l)-cone with an (r_l, r_{l-1})-cone in its axis, and an (r_{l-1}, r_{l-2})-cone in the axis of the second hypercone, and so on, and finally an $(r_1 - 1)$-quadric in the axis of the last hypercone.

The number of geometries with non-degenerate absolute is 2^n. With an l-ply degenerate absolute with the symbol $(r_1, r_2, \ldots, r_l)$ there are 2^{n-l} geometries; and there are $_nC_l$ different l-ply degenerate absolutes *[551, p. 36]*.

Sommerville's spaces $A_0 I$ are the elliptic spaces S_n, his spaces $A_0 II$ are the Lobačevskian spaces ^1S_n, and his spaces $A_0 III$ are the hyperbolic spaces ^1S_n for $n > 1$. His spaces $A_1(r)$ are the quasielliptic spaces S_n^m ($m = n - r$) with absolute consisting of an imaginary degenerate quadric which is an imaginary cone with real $(r - 1)$-dimensional vertex (called *axis* by Sommerville) and an imaginary nondegenerate quadric on this vertex, and the quasihyperbolic spaces $^{k_1 k_2}S_n^m$ with analogous absolute with real cone and quadric or with real cone or quadric. His spaces $A_l(r_1, r_2, \ldots, r_l)$ are l-quasielliptic spaces $S_n^{m_1 m_2 \cdots m_l}$ or l-quasihyperbolic spaces $^{k_1 k_2 \cdots k_{l+1}} S_n^{m_1 m_2 \cdots m_l}$ ($m_i = n - r_i$) with absolutes consisting of a degenerate quadric with $(r_1 - 1)$-dimensional vertex and on this vertex a second degenerate quadric with $(r_2 - 1)$-dimensional vertex, and so on, and of a nondegenerate quadric on the $(r_l - 1)$-dimensional vertex. The *Euclidean* and *pseudo-Euclidean* spaces R_n and $^l R_n$ are, respectively, the spaces S_n^0 and $^{0l}S_n^0$, the dual spaces R_n^* and $^l R_n^*$ are the spaces S_n^{n-1} and $^{l0}S_n^{n-1}$, respectively, the *Galilean* and *pseudo-Galilean* spaces Γ_n and $^l\Gamma_n$ are, respectively, the spaces S_n^{01} and $^{00l}S_n^{01}$, the *isotropic* and *pseudoisotropic* spaces I_n and $^l I_n$ are, respectively, the spaces $S_0^{0,n-1}$ and $^{00l}S_n^{0,n-1}$, and the *flag space* F_n is the space $S_0^{012\cdots n-1}$.

The groups of motions of the Euclidean and pseudo-Euclidean spaces R_n and $^l R_n$ and of all quasielliptic and quasihyperbolic spaces S_n^m and $^{k_0 k_1} S_n^m$ are instances of *quasisimple Lie groups* obtained from simple groups by passage to the limit. This transition, by passage to the limit, from simple Lie groups to quasisimple Lie groups is an example of the process of *contraction of Lie groups*, introduced by the American physicist Eugene Wigner (b. 1902) and his student Erdal Inönü in the paper *On the contraction of groups and their representations* (Washington, 1953) *[245]*. Similar groups were considered on a number of occasions by the Soviet mathematician Izrail Moiseevič Gelfand (b. 1913) and his students, in particular in the papers of Feliks Aleksandrovič Berezin (1931–1980) and I. M. Gel'fand *Some remarks on the theory of spherical functions on symmetric riemannian manifolds* (Neskol'ko zamečaniĭ k teorii sferičeskih funkeiĭ na simmetričeskih rimanovyh mnogoobrariyah. Moscow, 1956) *[44]* and *Laplace operators on semisimple Lie groups and on some symmetric spaces* (Operatory Laplasa na poluprostyh gruppah Li i nekotoryh simmetričeskih prostranstvah. Moscow, 1957) *[43]*. A complete classification of all quasisimple Lie groups was given by the author and by Lyudmila Mihaĭlovna Karpova (b. 1934) in the paper *Flag groups and contraction of Lie groups* (Flagovye gruppy i sžatie grupp Li. Moscow, 1966) *[479]*. In the same paper the authors introduced the notion of *k-quasisimple Lie groups* obtained from simple Lie groups by passage to the limit, in the same way, k times, instances of which are the groups of motions of the k-quasielliptic and k-quasihyperbolic spaces $S_n^{m_1 m_2 \cdots m_{k-1}}$ and $^{l_0 l_1 \cdots l_k} S_n^{m_1 m_2 \cdots m_{k-1}}$.

In the Soviet Union, the initiator of the study of k-quasielliptic and k-quasihyperbolic spaces was Isaak Moiseevič Yaglom (b. 1921),[5] author of the well-known books *[636]* and *[639]*. By now, this theory has been substantially developed.[6]

Quasielliptic and quasihyperbolic spaces S_n^m and $^{l_0 l_1} S_n^m$ are special cases of *quasi-Riemannian* and *quasipseudo-Riemannian spaces*, and k-quasielliptic and k-quasihyperbolic spaces $S_n^{m_0 m_1 \cdots m_{k-1}}$ and $^{l_0 l_1 \cdots l_k} S_n^{m_0 m_1 \cdots m_{k-1}}$ are special

---

[5] See I. M. Yaglom's book *[637]* and other publications by him and his students.
[6] See the papers of Iraida Ivanovna Železina (b. 1931) *[648]*, Tamara Grigor'evna Čahlenkova (b.1932) *[104]*, Evgeniya Ustinovna Yasinskaya (b. 1929) *[646]*, and also the survey paper of Yaglom, the author, and Yasinskaya *[640]* and chapter V of the author's book *[466]*.

cases of k-*quasi-Riemannian* and k-*quasipseudo-Riemannian spaces*, which are spaces with a homogeneous connection in whose tangent spaces there are defined corresponding projective metrics. Such spaces were first considered by Nil Aleksandrovič Glagolev (1886–1945) in the paper *Riemannian manifolds of projective structure* (Rimanovy mnogoobraziya proektivnoy struktury. Moscow, 1925) *[205]*, by Enea Bortolotti (1896–1942) in the paper *On specialized quadratic forms* (Sulle forme differenziale spezializzate. Rome, 1930) *[74]*, by Grigore Constantin Moisil (1906–1973) in the paper *On geodesics of singular Riemannian surfaces* (Sur les géodésiques des espaces de Riemann singuliers. Bucharest, 1940) *[370]*, and by Aleksandr Petrovič Norden (b. 1904), author of the well-known monograph *Spaces with an affine connection* (Prostranstva affinnoĭ svyaznosti. Moscow-Leningrad, 1950) *[393]*, in the papers *On the interpretation of spaces with a degenerate metric* (Ob istolkovaniĭ prostranstva s vyroždayušeĭsya metrikoĭ. Moscow, 1945) *[394]* and *Generalized geometry of a two-dimensional line space* (Obobščennaya geometriya dvumernogo lineĭčatogo prostranstva. Moscow, 1946) *[395]*. In connection with physical problems, similar spaces were investigated by P. Defrise in the paper *Geometric analysis of the kinematics of continuous media* (Analyse géométrique de la cinematique des milieux continus. Brussels, 1953) *[139]*, Czesław Jankiewicz (Yankevič) in the paper *On degenerate Riemannian geometries* (O vyroždayuščihsya rimanovyh geometriyah. Warsaw, 1954) *[644]*, and R. A. Toupin in the paper *World invariant kinematics* (New York, 1958) *[589]*. The general case of a quasi-Riemannian and quasipseudo-Riemannian space was defined by I. V. Parnasskiĭ in the paper *On degenerate Riemannian geometries* (O vyroždayuščihsya rimanovyh geometriyah. Kuĭbyšev, 1962) *[406]*. Revolt Ivanovič Pimenov (b. 1931) considered the physical applications of this theory in the paper *Application of semi-Riemannian geometry to unified field theory* (Primenenie polurimanovoĭ geometri k edinoĭ teorii polya. Moscow, 1964) *[421]*; we also note his paper *On the definition of semi-Riemannian spaces* (K opredeleniyu polurimanovyh prostranstv. Leningrad, 1965) *[422]*. In the paper *Symmetric semi-Riemannian spaces* (Simmetričeskie polurimanovy prostranstva. Kazan, 1964) *[478]*,[7] L. M. Karpova investigated symmetric quasi-Riemannian and k-quasi-Riemannian spaces and, in particular, has defined invariant quasi-Riemannian and k-quasi-Riemannian metrics in quasisimple and k-quasisimple groups obtained from simple groups by passage to the limit. A special case of this metric is the metric of quasielliptic space in the group of motions of the Euclidean plane introduced by W. Blaschke in the paper mentioned previously.

While quasisimple groups, obtained by passing to the limit from the simple groups O_n and ${}^l O_n$, are interpreted as groups of motions of real quasielliptic and quasihyperbolic spaces, quasisimple groups, obtained by passage to the limit from the simple groups CSU_n and $C^l SU_n$, are interpreted as groups of motions of complex Hermitian quasielliptic and quasihyperbolic spaces, of

---

[7] See also chapter VI of the author's book *[466]*.

which complex Hermitian Euclidean and pseudo-Euclidean spaces are special cases, and quasisimple groups, obtained by passage to the limit from the simple groups HU_n and H^lU_n are interpreted as analogous quaternion spaces. Complex Hermitian Euclidean space $C\bar{R}_n$ (*unitary space*) was introduced, together with its infinite dimensional analogue, by John von Neumann in the paper *Mathematical foundations of quantum mechanics [386, pp. 35–42]*; this space is often used in linear algebra (see, for example, *[198, 350]*). *Complex* and *quaternion Hermitian quasielliptic* spaces $C\bar{S}_n^m$ and $H\bar{S}_n^m$ and analogous k-*elliptic spaces* $C\bar{S}_n^{m_0 m_1 \cdots m_{k-1}}$ and $H\bar{S}_n^{m_0 m_1 \cdots m_{k-1}}$ were first studied by Tamara Mihaĭlovna Klimanova (b. 1937) in the paper *Unitary semielliptic spaces* (Unitarnye poluelliptičeskie prostranstva. Baku, 1963) *[285]*. In the paper *The algebra and group deformations* $I^m[SO(n) \otimes SO(m) \Rightarrow SO(n, m)$, $I^m[U(n) \otimes U(m)] = U(n, m)$ and $I^m[Sp(n) \otimes Sp(m)] \Rightarrow Sp(n, m)$ *for* $1 < m < n - 1$, 1974) *[635]*, K. B. Wolf and C. B. Boyer, independently of Soviet authors, also arrived at the notions of real, complex and quaternion matrices representing groups of motions of real, complex and quaternion quasielliptic spaces, respectively. Wolf and Boyer denote the groups O_n, CU_n and HU_n by $O(n)$, $U(n)$, and $Sp(n)$, respectively, the groups $^mO_{n+m}$, C^mU_{n+m} and H^mU_{n+m} by $O(n, m)$, $U(n, m)$ and $Sp(n, m)$, respectively, and use the symbol I^m (from *inhomogeneous*) for groups of matrices of motions of the corresponding quasielliptic spaces. In their paper, Wolf and Boyer consider the transition from these groups to the groups of motions of the corresponding hyperbolic spaces. Other classes of quasisimple and k-quasisimple groups and their geometric interpretations have been considered by Lyudmila Pavlovna Andreeva (b. 1937) and L. V. Šestyreva (= Rumyanceva) in *Limit symplectic spaces* (Predel'nye simplektičeskie prostranstva. Kolomna, 1964) *[22]* and by

Table IV

A				B	
$C\bar{S}_n = (\mathbf{H} \otimes \mathbf{C})\tilde{\bar{S}}_{(n-1)/2}$				S_{2n}	
A I	*A II*	*A III*	*A IV*	*B I*	*B II*
P_n	$HP_{(n-1)/2}$	$C^l\bar{S}_n$	$C^1\bar{S}_n$	$^lS_{2n}$	$^1S_{2n}$
$^0C\bar{S}_n$	$(\mathbf{H} \otimes {}^0\mathbf{C})\tilde{\bar{S}}_{(n-1)/2}$	$C\bar{S}_n^{l-1}$	$C\bar{R}_n$	S_{2n}^{l-1}	R_{2n}

C		D		
$H\bar{S}_{n-1}$		$S_{2n-1} = {}^1HS\bar{p}_{n-1}$		
C I	*C II*	*D I*	*D II*	*D III*
Sp_{2n-1}	$H^l\bar{S}_{n-1}$	$H^1\bar{S}_{n-1}$	$^lS_{2n-1}$	$^1S_{2n-1}$	$HS\bar{p}_{n-1}$
$^0H\bar{S}_{n-1}$	$H\bar{S}_{n-1}^{l-1}$	$H\bar{R}_{n-1}$	S_{2n-1}^{l-1}	R_{2n-1}	$^0HS\bar{p}_{n-1}$

Irina Nikolaevna Semenova (b. 1928) in *Limit projective spaces* (Predel'nye proektivnye prostranstva. Kolomna, 1964) *[521]*. *Dual hermitian non-Euclidean spaces* (Dual'nye érmitovy neevklidovy prostranstva. Moscow, 1969) *[354]* have been considered by Larisa Mihailovna Markina (b. 1940), *Semiantiquaternion spaces* (Poluantikvaternionnye prostranstva. Moscow, 1969) *[389]* have been considered by Lyudmila Sergeevna Nikitina (b. 1945).

The first four lines of Table IV coincide with the first, second, fifth and sixth columns of Table II. The fifth line of this table shows the spaces whose fundamental groups are quasisimple groups intermediate between the corresponding compact and noncompact groups.

Symmetry, Duality and Stability

The application of the theory of continuous groups to the theory of differential equations initiated by Sophus Lie is today an important branch of the theory of differential equations having varied mechanical and physical applications. At the end of the 19th century and in the beginning of the 20th century Émile Picard (1856–1941), one of the teachers of É. Cartan, worked in this area. The present state of the studies in this area is described in the book of Lev Vasil'evic Ovsyannikov (b. 1919) *Group analysis of differential equations* (Gruppovoĭ analiz differencial'nyh uravneniĭ. Moscow, 1978) *[402]*. Studies in *soliton theory* (see the book *Solitons* (*[76]*) by R. K. Boullough, P. J. Caudry, S. P. Novikov, and others; see also *[648]*) show that the integrability of differential equations defining mechanical and physical systems is closely connected with the inner symmetry of the system and with its stability. The term *soliton* originally denoted a stable solitary wave in hydrodynamics, but later this term came to denote stable dynamic structures in different domains of physics defined by similar differential equations. Soliton structures in plasma are especially important.

The connection between symmetry and stability was known in antiquity. In his book *Symmetry [629]* Hermann Weyl compares Babylonian, Greek and medieval images of gods and prayers and notes that all these images have bilateral symmetry and that divinity in antiquity and in the Middle Ages was a synonym of stability. A deeper bilateral symmetry, which may be called *duality*, is found in mechanics: it is well known that the kinetic energy of a conservative mechanical system characterized by generalized coordinates q_i and generalized impulses p_i has the form $T = \sum_i \sum_j a_{ij} p_i p_j$ where the quadratic form T is positive definite, and that the potential energy of such a system has the analogous form $U = \sum_i \sum_j c_{ij} q_i q_j$ only in the case of stability of the system (see P. L. Dirichlet's *On the stability of equilibrium* (Über die Stabilität des Gleichgewichts) *[154, vol. 2, pp. 3–8]*. In this case the canonical Hamilton equations of the system $\dfrac{dq_i}{dt} = \dfrac{\partial H}{\partial p_i}, \dfrac{dp_i}{dt} = -\dfrac{dH}{\partial q_i}$, where $H = T + U$ is

Hamilton's function, have the form $\dfrac{dq_i}{dt} = 2\sum_i a_{ij}p_j$, $\dfrac{dp_i}{dt} = -2\sum_i c_{ij}q_j$ and the motion of the system has an oscillatory character. The coefficients a_{ij} define the inertia of the system, and the c_{ij} its elasticity. The kinetic and potential energies T and U of a stable electric system have similar form. For such a system the q_i are quantities of electricity, the p_i are electric tensions in the elements of the system, the a_{ij} determine the impedance, and the c_{ij} the electric capacity of the system. The physical meaning of the stability of such systems consists in the accumulation of the energy in its potential form, the setting of the system in motion, and the return of the energy to potential form. This oscillatory motion has stable character. In the simplest cases this motion is a harmonic oscillation—mechanical or electric. An analogous instance is the inner motion of hydrodynamic or electromagnetic *solitons* and of stable phenomena on different levels of development of matter: the *atom* as the junction of the electric field of electrons and the positive charges of the nucleus with the inert mass of the nucleus, the *living cell* as the junction of the nucleus capable of accumulating energy and information with inert protoplasm, etc. To this harmony undoubtedly relate the previously mentioned words of Poincaré "it is this harmony then which is the sole objective reality" *[433, p. 209]*, but here the words *objective reality* must be understood literally and not in the conventionalist sense of Poincaré.

We note that certain soliton resolutions of differential equations are connected with geometric problems. For example, the *Sine–Gordon equation* $z_{xy} = \sin z$ is connected with the problem of determination of surfaces of constant negative curvature in the Euclidean space R_3, the *hyperbolic Sine–Gordon equation* $z_{xy} = \sinh z$ with the analogous problem in the pseudo-Euclidean space $^1R_3$, and the *Klein–Gordon equation* $z_{xy} = mz$ with the analogous problem in the Galilean space Γ_3. Many physical problems connected with non-Euclidean geometry are described in the book of Anatoliĭ Kuz'mič Lapkovskiĭ *Relativistic kinematics, non-Euclidean spaces and the exponential mapping* (Relativistkaya kinematika, neevklidovy prostranstva i eksponencyal'noe otobraženie. Minsk, 1985) *[302]*.

Chapter 10
Application of Algebras

Attempts to Extend Complex Numbers to Space

The geometric interpretation of complex numbers as points of the plane appeared for the first time in the 18th century. After that there arose the natural idea of generalizing complex numbers in such a way that they could be interpreted as points of three-dimensional space. One of the earliest attempts of this kind was due to Caspar Wessel. It appeared in his previously mentioned *Attempt to represent direction [624]*. Having thought of the operation of multiplication of complex numbers in geometric terms, Wessel associated to a point in space with rectangular coordinates x, y, z the expression $x + y\varepsilon + z\eta$, where ε and η are two different imaginary units, and interpreted by means of these numbers rotations about the Oy- and Oz-axes. Wessel used his "algebra" to solve problems involving spherical polygons.

Further attempts to construct a three-dimensional analog of the complex numbers were made by English algebraists. In their case, the problem arose in connection with the publication of *The theory of conjugate functions or algebraical couples* (Dublin, 1835) *[214, vol. 3, pp. 1–96]* by the Irish mathematician and mechanist William Rowan Hamilton (1805–1865). This paper contained a rigorous justification of the complex numbers based on their representation as pairs of real numbers, equivalent to the view of them as vectors in the plane. Between 1837 and 1838 Hamilton tried to construct an analogous theory for triples of real numbers *[214, vol. 3, pp. 106–110]*, but all systems of numbers of this kind constructed by him contained divisors of zero, that is, pairs of numbers α, β such that

$$\alpha \neq 0, \quad \beta \neq 0, \quad \alpha\beta = 0. \tag{10.1}$$

In the fourth part of his treatise *On the foundation of algebra*, entitled *On triple algebra* (Cambridge, 1847), *[377]*, Augustus de Morgan (1806–1871) considered numbers of the form $a\xi + b\eta + c\zeta$. De Morgan investigated dif-

ferent algebras of this type, including the algebra with the following multiplication table of the basis elements:

	ξ	η	ζ
ξ	ξ	η	ζ
η	η	$-\zeta$	ξ
ζ	ζ	ξ	$-\eta$

In this algebra the basis element ξ plays the role of 1 and the elements η and
ζ are connected by the relations $\eta^3 = \zeta^3 = -\xi$. If we replace ξ, $-\eta$, and $-\zeta$
with, respectively, 1, e, and e^2, then we can write the elements of this algebra
as $a + be + ce^2$ with $e^3 = 1$, which is how this algebra was introduced by
Charles Graves (1810–1860) in the paper *On algebraic triplets* (Dublin,
1847) *[212]*. *Triplets* is what Graves called the elements of his algebra. Graves
showed that with every triplet there are associated three moduli

$$|\alpha| = \sqrt[3]{a^3 + b^3 + c^3 - 3abc}, \qquad |\alpha|_S = \sqrt{a^2 + b^2 + c^2 - ab - bc - ac},$$
$$|\alpha|_A = |a + b + c|, \tag{10.2}$$

such that for each of three moduli the modulus of a product is equal to
the product of the moduli of the factors, and such that $|\alpha|^3 = |\alpha|_S^2 \cdot |\alpha|_A$.
Graves showed that each triplet has an exponential representation which
is the product of the first modulus by $\exp(\varphi e + \psi e^2)$ (we recall that $\exp \alpha =$
$1 + \alpha + \alpha^2/2! + \alpha^3/3! + \cdots$), where the "amplitudes" φ and ψ of a product
of two triplets are the sums of the corresponding amplitudes of the factors.
Graves' triplets include zero divisors characterized by the vanishing of the
second or third (and therefore also the first) modulus. The equality (10.1)
holds if the second (third) modulus of α and the third (second) modulus of β
vanish.

Graves identified a triplet $x + ye + ze^2$ with a point in space whose rectangular coordinates are x, y, and z and proposed the following interpretation
of the product of two triplets. He drew a sphere with center at the origin and
denoted by l, m, n its points of intersection with the respective positive
coordinate axes (Figure 112). Next he drew the circle determined by the points
l, m, n, the *symmetric axis OA* through the center of that circle and the
symmetric plane S passing through O and perpendicular to the axis OA. Then
he considered the projections of triplets, which he here called *lines*, on the axis
OA and on the plane S and noted that

the projections of the lines of a product, of the lines of the factors, and
of the unit line on the symmetric axis form a proportion in the sense of
Euclid,

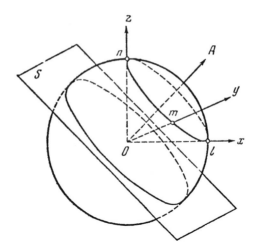

Figure 112

that is, a proportion of real numbers,

and the projections of the same lines on the symmetric plane form a proportion in which one considers lengths as well as directions of these proportions,

that is, a proportion of complex numbers *[212, p. 74]*. Graves was saying that every triplet can be represented as a sum of a real number—its projection on the axis *OA*—and a complex number—its projection on the plane *S*. Also, multiplication of triplets reduces to the multiplication of these real and complex numbers. It is not difficult to check that, similarly, addition of triplets reduces to the addition of these numbers. Using the language of modern algebra we can say that this representation shows that the algebra of triplets is the direct sum of the field of real numbers and the field of complex numbers. It is not difficult to check that the triplets e_A and e_S, the "unit lines" of the *OA* axis and the plane *S*, are given in terms of the basis elements of the algebra by the formulas

$$e_A = \frac{1 + e + e^2}{3}, \qquad e_S = \frac{1 + \omega e + \omega^2 e^2}{3},$$

and that every triplet is a linear combination with real coefficients of the triplets e_A, e_S, and its conjugate complex $\bar{e}_S = \frac{1}{3}(1 + \omega^2 e + \omega e^2)$, or, equivalently, of the triplet e_A with a real coefficient and the triplet e_S with a complex coefficient (see *[30]*). Also, it is easy to check that the coefficients in the latter representation of a triplet $a + be + ce^2$ are, respectively, the third and second moduli in (10.2). If we denote the fields of real and complex numbers by **R** and **C**, then we can write the algebra **T** of triplets as the direct sum **R** ⊕ **C** of these two fields.

Quaternions

Upon discovering that all *triple algebras* he investigated had divisors of zero, Hamilton decided to look for algebras without divisors of zero among *quadruple algebras*. He found a quadruple algebra without divisors of zero that had all the properties of real and complex numbers other than commutativity of multiplication. Hamilton called this generalization of the complex numbers *quaternions* (from the Latin *quaternus*—quadruple). He first presented his theory in the paper *On quaternions, or On a new system of imaginaries in algebra* (Dublin, 1844–1850) *[214, vol. 3]* and then in his *Lectures on quaternions* (Dublin, 1853) *[215]* (see also *[216]*). Hamilton writes quaternions as sums of the form $a + bi + cj + dk$ that are added and multiplied like polynomials subject to the conditions $i^2 = j^2 = -1$, $ij = -ji = k$. It is easy to check that these imply the relations $k^2 = -1$, $jk = -kj = i$, $ki = -ik = j$. The quaternions form a commutative group under addition, and the nonzero quaternions form a noncommutative group under multiplication. Also, multiplication is distributive over addition. Thus the quaternions form a *skew field*. This skew field is denoted by **H** (in honor of Hamilton).

If we associate with each quaternion $\alpha = a + bi + cj + dk$ the conjugate quaternion $\bar{\alpha} = a - bi - cj - dk$, then it is easy to check that

$$\overline{\alpha\beta} = \bar{\beta}\bar{\alpha} \tag{10.3}$$

The product $\alpha\bar{\alpha}$ is the nonnegative real number $a^2 + b^2 + c^2 + d^2$. Formula (10.3) implies that

$$|\alpha|^2 |\beta|^2 = |\alpha\beta|^2. \tag{10.4}$$

This implies that, given two quadruples of numbers, we can use the multiplication rule for quaternions to obtain a third quadruple such that the sum of the squares of its entries is the product of the sums of the squares of the entries comprising the given quadruples.

In 1748, in a letter to Goldbach, Euler mentioned his discovery of a law, very similar to that given previously, for transition from two quadruples to a third. Specifically, Euler's law corresponds to the transition from quaternions α, β to the quaternion $\bar{\alpha}\beta$. Euler's result appeared in the paper *Proof of Fermat's theorem on the representation of all numbers, integers as well as fractions, as sum of at most four squares* (Demonstratio theorematis Fermatiani omne numerum sive integrum sive fractum esse summam quattuor pauciorumque quadratorum. Petersburg, 1760) *[176, vol. 2, pp. 338–372]*. Hamilton called expressions of the forms $xi + yi + zk$ *vectors* (from the Latin *vector*—carrier) and viewed quaternions as sums of real numbers (*scalars*, from *scala*—ladder) and vectors. In his *Lectures on quaternions* there are defined all operations of vector algebra: the sum of two vectors $\alpha = xi + yj + zk$ and $\beta = x'i + y'j + z'k$ yields a new vector $\alpha + \beta$, and the product yields a general quaternion $\alpha\beta$ whose scalar part $S\alpha\beta$ Hamilton called the *scalar product* of the vectors α, β and whose vector part $V\alpha\beta$ he called their *vector product* (Hamilton's scalar product differs from our scalar, or inner, product in sign, but his vector product coincides with our vector, or cross, product). Hamilton also considered a quaternion of the form $\beta\alpha^{-1}$, which he called a "quotient of division of two vectors"; it is easy to see that

$$\beta\alpha^{-1} = \frac{|\beta|}{|\alpha|}(\cos\varphi + \varepsilon\sin\varphi), \tag{10.5}$$

where φ is the angle between the vectors α and β and ε is a unit vector perpendicular to α and β.

We note that just as the modulus $|\beta - \alpha|$ of the difference $\beta - \alpha$ of two numbers can be used to introduce the Euclidean metric of R_2 in the field $\mathbf{C}$ (the complex plane), so too can it be used to introduce the Euclidean metric of R_4 in the skew field $\mathbf{H}$. Also, an arbitrary rotation of that space is given by

$$\xi' = \alpha\xi\beta \quad \text{or} \quad \xi' = \alpha\bar{\xi}\beta, \qquad |\alpha| = |\beta| = 1,$$

and an arbitrary rotation of that space that preserves the real axis, that is, an arbitrary rotation of the Euclidean space R_3 is given by

$$\xi' = \alpha^{-1}\xi\alpha. \tag{10.6}$$

Quaternions can also be represented as pairs (α, β) of complex numbers $\alpha = a + bi$, $\beta = c + di$ that are multiplied according to the rule

$$(\alpha, \beta)(\gamma, \delta) = (\alpha\gamma - \beta\bar{\delta}, \alpha\delta + \beta\gamma). \tag{10.7}$$

Such pairs of complex numbers, as well as their application to motions of space equivalent to the application of the relation (10.6), were considered by Gauss in the posthumously published note *Mutations of space* (Mutationen des Raumes) *[196, vol. 8, pp. 357–362]*.

Cayley Numbers or Octaves

Soon after the appearance of quaternions, Arthur Cayley discovered their generalization, the so-called *Cayley numbers* or *octaves*, introduced in the paper *On Jacobi's elliptic functions and on quaternions* (London, 1845) *[103, vol. 1, p. 127]*. Octaves are defined as expressions of the form $a + bi + cj + dk + xl + yp + zq + tr$ that are added and multiplied like polynomials subject to the conditions $i^2 = j^2 = l^2 = -1$, $ij = -ji = k$, $il = -li = p$, $lj = -jl = q$, $kl = -lk = r$. It is easy to see that these imply the relations $k^2 = p^2 = q^2 = r^2 = -1$, $iq = -qi = r$, $jp = -pj = r$, $kp = -pk = q$, as well as the relations obtained from them by cyclic permutation of triples of elements. The reason for the name *octave* (from the Latin *octo*—eight) is that an octave is given by eight real numbers.

The name *octave* first turned up in Hamilton's note on the papers of J. T.

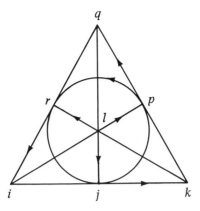

Figure 113

Graves (1848) (see *[63]*) in which it was pointed out that John Thomas Graves (1806–1870), brother of Charles Graves, had discovered these numbers as early as 1843, which is why octaves are also called *Graves-Cayley numbers*. It is easy to verify that, except for associativity of multiplication, addition and multiplication of octaves have the same properties as addition and multiplication of quaternions. Octaves form a commutative group under addition, and the nonzero octaves form a *loop* under multiplication. The multiplication of octaves has the property of *alternativity*; that is, each two octaves generate a skew field. Therefore octaves form an alternative skew field. This skew field is denoted by **O** (sometimes also by **Ca** or **Cay**, in honor of Cayley).

It is convenient to represent multiplication of the units of the algebra **O** graphically using the diagram suggested by the Dutch mathematician Hans Freudenthal (b. 1905) (Figure 113). The product of two units, represented on the oriented sides and medians of a triangle and on its inscribed circle, is equal to the third unit with a plus or minus sign according as the transition from the first to the second element agrees or disagrees with the orientation of the path determined by the two units.

With every octave α there is associated the conjugate octave $\bar{\alpha}$ obtained from α by changing the signs of b, c, d, x, y, z, t. It is easy to see that the rule (10.3) holds. The product $\alpha\bar{\alpha}$ is equal to the real number $a^2 + b^2 + \cdots + t^2$. Here (10.3) also implies (10.4); that is, given two octaves, the law of multiplication of octaves enables us to find a third octave such that the sum of the squares of its entries is equal to the product of the sums of the squares of the entries of the given octaves. That is why octaves, like quaternions, are used in the theory of numbers. Independently of Cayley, Francesco Brioschi (1824–1897), in the paper *On an analogy between a class of determinants of even order and binary determinants* (Sur l'analogie entre une classe de d'ordre pair et les déterminants binaires. Berlin, 1856) *[79, vol. 5, pp. 511–520]*, devoted to problems of number theory, discovered a law of transition from two eight-tuples to a third equivalent to the law of multiplication of octaves. That is why the formulas that express the coordinates of the octave $\alpha\beta$ in terms of the coordinates of the octaves α and β are sometimes called *Brioschi formulas*.

Just as in the space of complex numbers and in the space of quaternions, so too in the space of octaves one can introduce a Euclidean metric by defining the distance between two octaves α and β as the modulus $|\beta - \alpha|$ of their difference.

Octaves can be represented as pairs of quaternions (α, β), $\alpha = a + bi + cj + dk$, $\beta = x + yi + zj + tk$, multiplied in accordance with the rule

$$(\alpha, \beta)(\gamma, \delta) = (\alpha\gamma - \bar{\delta}\beta, \delta\alpha + \beta\bar{\gamma}), \tag{10.8}$$

which generalizes formula (10.7).

Matrices

In *A memoir on the theory of matrices* (London, 1858) *[103, vol. 1, pp. 475–496]*, Arthur Cayley introduced an algebra of square tables of numbers that he called *matrices* (from the Latin *matrix*—list, register). A table of n^2 numbers is called a *matrix of order* n. The sum of matrices $A = (a_{ij})$ and $B = (b_{ij})$ is the matrix $A + B = (a_{ij} + b_{ij})$ and their product is the matrix $C = (c_{ij})$ with

$$c_{ij} = \sum_k a_{ik}b_{kj}. \tag{10.9}$$

If we associate to the matrix $A = (a_{ij})$ the linear transformation

$$x'_i = \sum_j a_{ij}x_j, \tag{10.10}$$

then the product of the matrices A and B corresponds to the linear transformation that is the result of successive application of the linear transformations whose matrics are B and A. The former linear transformation is called the product of the latter linear transformations.

If $I = (\delta_{ij})$, where $\delta_{ii} = 1$ and $\delta_{ij} = 0$ for $i \neq j$, and if A is any matrix (of the same order as I) then $IA = AI = I$. Hence I is called a *unit matrix*. The inverse A^{-1} of a matrix A is a matrix such that $A^{-1}A = I$. The elements of the matrix A^{-1} are the solutions b_{ij} of the system of equations (10.9), where $c_{ij} = \delta_{ij}$.

If we denote by $|A|$ the determinant of a matrix A then we have

$$|AB| = |A||B| \tag{10.11}$$

A special case of matrices of orders three and four was considered by Euler in the paper *An algebraic problem worthy of mention because of its utterly unique consequences* (Problema algebraicum ob affectiones prorsus singulares memorabile. Petersburg, 1771) *[176, vol. 6, pp. 287–315]*. Euler called matrices *squares* and regarded them as generalizations of the *magic squares* that were very popular in the Middle Ages and during the Renaissance. Euler considered matrices that correspond to transformations of rectangular coordinates now known as *orthogonal* matrices. Let $A = (a_{ij})$ and define A^T, the transpose of A, to be the matrix $A^T = (a_{ji})$. Then an orthogonal matrix A has the property $A^T A = I$ or, equivalently, the property $A^T = A^{-1}$. Euler determined such matrices by the requirement $A^T A = I$, spelled out for all elements of the matrix $A^T A$. We denote the algebra of real matrices by $\mathbf{R}_n$. The analogous algebra $\mathbf{C}_n$ of complex matrices was also considered by Cayley.

In the same *Memoir on the theory of matrices* Cayley showed that if one associates to a quaternion $\alpha = a + bi + cj + dk$ the complex matrix

$$\begin{pmatrix} A & B \\ -\bar{B} & \bar{A} \end{pmatrix} = \begin{pmatrix} a + id & b + ic \\ -b + ic & a - id \end{pmatrix}, \tag{10.12}$$

then to the sum and product of quaternions there correspond the sum and product of the corresponding matrices, and the number $\alpha\bar{\alpha} = |\alpha|^2$ is equal to the determinant of the corresponding matrix.

Note that if one writes an arbitrary real matrix of order two in the form

$$\begin{pmatrix} A & B \\ C & D \end{pmatrix} = \begin{pmatrix} a + d & b + c \\ -b + c & a - d \end{pmatrix}, \tag{10.13}$$

then we can also write it as $\alpha = a \cdot 1 + bi + ce + df$, where 1 stands for the unit matrix $I = \begin{pmatrix} 1 & 0 \\ 0 & 1 \end{pmatrix}$, i for the matrix $\begin{pmatrix} 0 & 1 \\ -1 & 0 \end{pmatrix}$ whose square is $-I$, e for the matrix $\begin{pmatrix} 0 & 1 \\ 1 & 0 \end{pmatrix}$, and f for the matrix $\begin{pmatrix} 1 & 0 \\ 0 & -0 \end{pmatrix}$. Expressions of this kind resemble quaternions and are called *split quaternions*. The split quaternions i, e, f satisfy the relations $i^2 = -1$, $e^2 = 1$, $ie = -ei = f$, which imply the relations $f^2 = 1$, $ef = -fe = -i$, $fi = -if = e$. Split quaternions are also called *antiquaternions* and *pseudoquaternions*. If we associate to each split quaternion α the conjugate split quaternion $\bar{\alpha} = a - bi - ce - df$, then condition (10.3) holds and the product $\alpha\bar{\alpha}$ is equal to the real number $|\alpha|^2 = a^2 + b^2 - c^2 - d^2$, which is equal to the determinant of the matrix in (10.13); here (10.11) is equivalent to (10.4). We denote the algebra of split quaternions by $^1\mathbf{H}$. The correspondence (10.13) shows that the algebra $^1\mathbf{H}$ is isomorphic to the algebra $\mathbf{R}_2$.

Matrices with zero determinants have the property that for such a matrix one can always find another matrix with zero determinant such that the product of the two is the zero matrix. In the case of matrices of order two, if $\begin{pmatrix} a & b \\ c & d \end{pmatrix}$ has determinant zero then so do the matrices $\begin{pmatrix} d & -b \\ -c & a \end{pmatrix}$ and $\begin{pmatrix} a & b \\ c & d \end{pmatrix}$. $\begin{pmatrix} d & -b \\ -c & a \end{pmatrix} = (ad - bc)I$. In other words, matrices with zero determinant are divisors of zero.

The vector form of the transformation (10.10) is

$$\mathbf{x}' = A\mathbf{x}. \tag{10.14}$$

The symbol A is called a *linear operator*. This term, which became popular in the twenties and thirties of this century, was first introduced by the English physicist Oliver Heaviside (1850–1925) in his *Electromagnetic theory* (London, 1893) *[220]*, which is largely devoted to the exposition of the calculus of vectors.

With every linear transformation A there are associated its so-called eigenvectors, that is, vectors that the linear transformation multiplies by real numbers:

$$A\mathbf{x} = \lambda\mathbf{x}, \tag{10.15}$$

or

$$(A - \lambda I)\mathbf{x} = 0. \tag{10.16}$$

The symbol I in (10.16) stands for the identity operator $I\mathbf{x} = \mathbf{x}$. The factors λ are called eigenvalues of A. The vector equality (10.16) can be written in terms of the coordinates of $\mathbf{x}$ as

$$(a_{11} - \lambda)x_1 + a_{12}x_2 + \cdots + a_{1n}x_n = 0,$$
$$a_{21}x_1 + (a_{22} - \lambda)x_2 + \cdots + a_{2n}x_n = 0,$$
$$\cdots\cdots\cdots\cdots\cdots\cdots\cdots\cdots\cdots\cdots\cdots\cdots\cdots \tag{10.17}$$
$$a_{n1}x_1 + a_{n2}x_2 + \cdots + (a_{nn} - \lambda)x_n = 0.$$

The system of equations (10.17) has nontrivial solutions if and only if the determinant of the coefficient matrix is zero, that is, if and only if

$$\begin{vmatrix} a_{11} - \lambda & a_{12} & \cdots & a_{1n} \\ a_{21} & a_{22} - \lambda & \cdots & a_{2n} \\ \vdots & \vdots & \ddots & \vdots \\ a_{n1} & a_{n2} & & a_{nn} - \lambda \end{vmatrix} = 0. \tag{10.18}$$

The solutions of equation (10.18) are the eigenvalues of the matrix (a_{ij}). If one of them is put in place of λ in (10.17), then the entries $x_1, x_2, \ldots, x_n$ of a solution of the system of equations (10.17) are the coordinates of an eigenvector that "belongs" to the eigenvalue in question. Eigenvalues of linear transformations appeared for the first time in the papers of Lagrange on systems of linear differential equations with constant coefficients and in the papers of the French mathematician and mechanist Pierre Simon Laplace (1749–1827) on the theory of small oscillations and "secular inequalities" of the motions of planets; hence the same *secular equation* that is sometimes used in connection with the equation (10.18). Eigenvalues of linear transformations of three-dimensional space were used implicitly by Euler in the *Introduction to infinitesimal analysis* (1748) for the determination of the principal axes of a quadric *[176, vol. 9, p. 385]* and in the *Theory of motion of solid or rigid bodies* (Theoria motus corporum solidorum seu rigidorum. Greifswald, 1765) *[177, vol. 3, pp. 215–243]* for the determination of the principal axes of inertia of a rigid body. The directions of the principal axes of a quadric are those of the eigenvectors of the linear transformation that maps a vector l into a vector perpendicular to the plane that contains the locus of the midpoints of the chords of the surface parallel to l, and the directions of the principal axes of inertia of a rigid body are those of the eigenvectors of the linear transformation that maps a vector whose direction is that of the instantaneous axis of rotation of the body into a vector that gives the direction of the corresponding kinetic moment of the body. Augustin Cauchy computed the eigenvalues of linear transformations of n-dimensional space in the paper with a traditional title *On the equation which enables one to determine the secular inequalities of the motions of planets* (Sur l'equation à l'aide de laquelle on détermine des inégalités séculaires de mouvements des planètes. Paris, 1826) *[100, vol. 9, pp. 174–195]*.

If we go from a basis $e_1, e_2, \ldots, e_n$ to a basis $e'_1, e'_2, \ldots, e'_n$ then the linear transformation (10.14) is expressed by means of a different matrix A' said to be similar to the matrix A. If all eigenvalues $\lambda_1, \lambda_2, \ldots, \lambda_n$ are real and distinct then there exists a basis of eigenvectors. Relative to such a basis the matrix of the linear transformation is the diagonal matrix

$$A = \begin{pmatrix} \lambda_1 & & & 0 \\ & \lambda_2 & & \\ & & \ddots & \\ 0 & & & \lambda_n \end{pmatrix}. \tag{10.19}$$

It is clear that matrices that are reducible to the same diagonal form are similar. Euler implicitly reduced to diagonal form matrices made up of the coefficients of the equation of a quadric and of the moments of inertia $\int x^2 \, dM$, $\int y^2 \, dM, \int z^2 \, dM, \int xy \, dM, \int xz \, dM, \int yz \, dM$, where dM is an element of the mass of the body and one integrates over the volume of the body. In the cases dealt with by Euler the eigenvalues are the coefficients of the squares of the coordinates in the equation of the surface when the coordinate axes coincide with the principal axes and the principal moments of inertia. General reduction of matrices to diagonal form was carried out by Cayley in the *Memoir on the theory of matrices* (1858).

The problem of deciding whether or not two matrices, possibly with repeated eigenvalues, are similar was solved by Karl Weierstrass in the paper *On the theory of bilinear and quadratic forms* (Zur Theorie der bilinearen und quadratischen Formen. Berlin, 1868) *[622, vol. 2, pp. 19–44]*. In this paper Weierstrass introduced the notion of so-called *elementary divisors* of the determinant of the matrix $A - \lambda I$, which is a polynomial in λ. If all elementary divisors are linear functions of λ then all eigenvalues of the matrix A are different. If there are elementary divisors of degree $l > 1$ then there may be vectors such that

$$(A - \lambda I)\mathbf{x} \neq 0, \qquad (A - \lambda I)^2\mathbf{x} \neq 0, \ldots, (A - \lambda I)^{l-1}\mathbf{x} \neq 0, \quad (10.20)$$

but

$$(A - \lambda I)^l\mathbf{x} = 0. \tag{10.21}$$

Weierstrass showed that two matrices are similar if and only if they have the same elementary divisors. Two similar matrices with complex entries can be reduced to the same *canonical form* in which the elements on the principal diagonal are the eigenvalues of the matrix, the elements on the nearest diagonal below it are 0 or 1, and the remaining elements are all zero:

$$(10.22)$$

In his *Treatise on substitutions* (1870) *[249]*, C. Jordan first found the canonical form (10.22) for linear transformations with elements in a finite field. This is why the canonical form (10.22) is called the *Jordan normal form* and the submatrices on the principal diagonal are called *Jordan blocks*. To every Jordan block of order *l* there corresponds an *l*-dimensional invariant subspace of the linear space, that is, a subspace mapped into itself by the linear transformation, and in that subspace there is a nested sequence of invariant subspaces whose respective dimensions are $1, 2, \ldots, l - 1$; the one-dimensional invariant subspace is determined by an eigenvector that belongs to the eigenvalue on the principal diagonal of the Jordan matrix. The Jordan normal forms of complex matrices determine the Jordan normal forms of real matrices. The classification of all linear transformations with different Jordan normal forms was carried out by the Italian mathematician Corrado Segre (1863–1924) in the paper *On the theory and classification of homographies on a linear space of an arbitrary number of dimensions* (Sulla teoria e sulle classificazioni delle omografie in uno spazio lineare ad un numero qualunque di dimensioni. Rome, 1884) *[520]*.

The term *homography* (homographie, omografia), introduced by Michel Chasles (1793–1880) *[107, p. 67]* for denoting collineations, and used in this sense by Segre in his paper, was also used by Cesare Burali-Forti (1861–1931) in, for example, his *Foundations of the differential geometry of surfaces couched in terms of a general vector method* (Fondamenti per la geometria differenziale di una superficie col metodo vettoriale generale. Palermo, 1912) *[83]*, and by other Italian mathematicians, as a term for a linear operator. Another term for a linear operator is *affinor*, introduced by the German geometer F. Jung in the paper *Formation of derivatives in the spatial magnitude field* (Ableitungsbildung im räumlichen Grössenfelde. Berlin, 1908) *[253]*. It is derived from *affine transformation* and justified by the fact that affine transformations are expressed in terms of linear operators by means of the formula

$$\mathbf{x}' = A\mathbf{x} + \mathbf{b} \qquad (10.23)$$

A more frequently used term for a linear operator is *tensor*, from the Latin *tensio*—tension. The reason for this term is that one of the earliest examples of a linear operator was the so-called *elastic tensor* that characterizes the state of tension of an elastic body. If one singles out in a body that is in a state of tension an element of volume containing a certain point then to every plane passing through this point there corresponds the force that must be applied to the section of the volume element by this plane in order that the remaining part of the element of volume of the body remains in equilibrium. The ratio $\mathbf{p}$ of this force to the area of the section is called the *tension* that acts in this section. Tension is a linear vector function of the unit normal vector $\mathbf{n}$ of the plane of the section,

$$\mathbf{p} = T\mathbf{n}.$$

In elasticity theory, the diagonal elements of a matrix of the operator with respect to rectangular coordinates x, y, z are denoted by σ_x, σ_y, σ_z and are called *normal tensions*, and the remaining elements are denoted by τ_{xy}, τ_{xz}, τ_{yz} and are called *tangent tensions*.

We note that the American physicist Josiah Willard Gibbs (1839–1903), who in his *Elements of vector analysis* (New Haven, 1881–1884) *[202]* combined the vector calculi of Hamilton and Grassmann and gave the calculus of vectors its modern form, not only introduced the inner and cross products of vectors α and β, denoting them by $\alpha \cdot \beta$ and $\alpha \times \beta$, but also introduced a product that he denoted by $\alpha\beta$ and called their *dyadic product* or *dyad*. The dyad $\alpha\beta$ is an operator that maps a vector ξ on the vector $(\beta \cdot \xi)\alpha$. Gibbs showed that every linear vector function on three-dimensional space is the sum of three dyads.

The involvement of the physicists Gibbs and Heaviside in the elaboration of the calculus of vectors and of the theory of linear operators explains the exceptional suitability of the calculus of vectors, and especially the theory of vector fields, for physics. Vector analysis, even in its early Hamilton form, was profitably applied to electromagnetic field theory by the English physicist James Clerk Maxwell (1831–1879) in his famous work *A Treatise on Electricity and Magnetism* (Cambridge, 1873) *[359]*, in which he anticipated the existence of electromagnetic waves subsequently confirmed by the invention of wireless communication.

Grassmann and Clifford Numbers

H. Grassmann, who introduced the notion of an n-dimensional space of vectors in his *The science of linear extension* (1844) *[211]*, defined on it skew-symmetric *outer products*

$$[x_1 x_2] = -[x_2 x_1], [x_1 x_2 x_3] = -[x_2 x_1 x_3] = \cdots = -[x_3 x_2 x_1],$$

which vanish for linearly dependent vectors. For linearly independent vectors $x_1, \ldots, x_m, x_{m+1}, \ldots, x_p$ he also defined the product

$$[x_1 \ldots x_m][x_{m+1} \ldots x_p] = [x_1 \ldots x_m x_{m+1} \ldots x_p];$$

when the vectors $x_1, \ldots, x_p$ are linearly dependent the value of this product is to be zero. In this way Grassmann defined an algebra of expressions of the form

$$\alpha = a + \sum_i a_i e_i + \sum_i \sum_j a_{ij}[e_i e_j] + \cdots$$
$$\cdots + \sum_{i_1} \sum_{i_2} \cdots \sum_{i_r} a_{i_1 i_2 \ldots i_r}[e_{i_1} e_{i_2} \ldots e_{i_r}] + \cdots + a_{12\ldots n}[e_1 e_2 \ldots e_n],$$
$$(10.24)$$

which he called *extensive magnitudes* and which are now called *Grassmann numbers of order* n. These numbers form an algebra which we denoted by G_n. Since in the expression (10.24) there is one coordinate of each of the types a and $a_{12\ldots n}$, n coordinates of each of the types a_i and $a_{12\ldots i-1, i+1\ldots n}$ and $\binom{n}{r}$ coordinates of each of the types $a_{i_1 i_2 \ldots i_r}$ and $a_{1\ldots i_1-1, i_1+1\ldots i_r-1, i_r+1\ldots n}$, it follows that there are, in all,

$$1 + n + \binom{n}{2} + \cdots + \binom{n}{r} + \cdots + \binom{n}{2} + n + 1 = (1 + 1)^n = 2^n$$

coordinates of a Grassmann number. It is easy to see that the multiplication of Grassmann numbers is associative.

In the paper *Applications of Grassmann's extensive algebra* (Baltimore, 1878) *[122, pp. 266–276]*, W. K. Clifford introduced the following modified version of Grassmann's algebra. Like Grassmann, Clifford considers linear combinations of n vectors $e_1, e_2, \ldots, e_n$ and of products $e_{i_1} e_{i_2} \ldots e_{i_r}$, which, for distinct factors $e_{i_1}, e_{i_2}, \ldots, e_{i_r}$, are defined like Grassmann's outer products

$[\mathbf{e}_{i_1}\mathbf{e}_{i_2}\dots\mathbf{e}_{i_r}]$ but are denoted by $\mathbf{e}_{i_1 i_2\dots i_r}$. However, if the latter products contain repeated factors then they are not put equal to zero but are computed using the rule $\mathbf{e}_i^2 = -1$ (for example, $\mathbf{e}_1\mathbf{e}_2\mathbf{e}_1 = -\mathbf{e}_1^2\mathbf{e}_2 = \mathbf{e}_2$). In this way Clifford defined an algebra of expressions

$$\alpha = a + \sum_i a_i\mathbf{e}_i + \sum_i\sum_j a_{ij}\mathbf{e}_{ij} + \cdots$$

$$\cdots + \sum_{i_1}\sum_{i_2}\cdots\sum_{i_r} a_{i_1 i_2\dots i_r}\mathbf{e}_{i_1 i_2\dots i_r} + \cdots + a_{12\dots n}\mathbf{e}_{12\dots n}, \qquad (10.25)$$

now known as *Clifford numbers of order* n. These numbers form an algebra which we denoted by $\mathbf{K}_n$. The number of coordinates of a Clifford number is also 2^n. It is easy to see that the only basis elements that commute with all Clifford numbers are 1 and the element $\mathbf{e}_{12\dots n}$ for odd n. The algebra $\mathbf{K}_n$ of Clifford numbers coincides for $n = 1$ with the field $\mathbf{C}$ of complex numbers and for $n = 2$ with the skew field $\mathbf{H}$ of quaternions; for $n > 2$ the Clifford numbers are generalizations of the quaternions. It is easy to verify that multiplication of Clifford numbers is associative.

In the paper *A preliminary sketch of biquaternions* (1873) *[122, pp. 181–200]*, Clifford introduced two modifications of the complex numbers now known, respectively, as *split complex numbers* and *dual numbers*. Split complex numbers are of the form $a + be$ with $e^2 = 1$, and dual numbers are of the form $a + b\varepsilon$ with $\varepsilon^2 = 0$. In the same paper Clifford suggested that along with *hyperbolic biquaternions*, that is, quaternions with the usual complex coordinates (introduced by Hamilton and called by him just *biquaterions*), one should study *elliptic biquaternions* whose coordinates are split complex numbers, and *parabolic biquaternions* whose coordinates are dual numbers; this explains the title of Clifford's paper. Split complex and dual numbers form algebras which we denote by $^1\mathbf{C}$ and $^\circ\mathbf{C}$, respecitvely. Unlike complex numbers, split complex and dual numbers have divisors of zero. In the case of dual numbers the zero divisors are of the form $a\varepsilon$ and in the case of split complex numbers they are of the form $a(1 \pm e)$ (it is easy to see that $\left(\dfrac{1 \pm e}{2}\right)^2 = \dfrac{1 \pm e}{2}$ and that $(1 + e)(1 - e) = 0$). Split complex numbers are also called *paracomplex* and *double numbers*.

Since in the case of the algebra $\mathbf{K}_3$ the element $\mathbf{e}_{123}$ commutes with all elements of the form $a + a_1\mathbf{e}_1 + a_1\mathbf{e}_2 + a_{12}\mathbf{e}_{12}$, which may be regarded as quaternions, and since $\mathbf{e}_{123}^2 = 1$, it follows that the Clifford numbers of order 3 coincide with the elliptic biquaternions. The Clifford numbers of order $n \geq 3$ contain divisors of zero: such are, for example, the numbers $a(1 \pm \mathbf{e}_{123})$. Thus Clifford numbers are generalizations of the quaternions in a direction other than that of the octaves: the octaves are nonassociative under multiplication but have no divisors of zero whereas the Clifford numbers of order $n \geq 3$ are associative under multiplication but have divisors of zero.

In the paper *Complex numbers* (Komplexe. Zahlen Leipzig, 1902) *[96, part 2, pp. 107–246]* E. Study proposed the generalization of Clifford numbers in which the condition $\mathbf{e}_i^2 = -1$ is replaced with $\mathbf{e}_a^2 = 1 (a = 1,\dots,l)$, $\mathbf{e}_u^2 =$

$-1(u = l + 1, \ldots, n)$ *[96, part 2, p. 242]*. These numbers form an algebra which we denote by $^l\mathbf{K}_n$. The algebra $^l\mathbf{K}_n$ of Clifford-Study numbers coincides with the algebra $^1\mathbf{C}$ of split complex numbers for $n = l = 1$, with the algebra $^\prime\mathbf{H}$ of split quaternions for $n = 2, l = 1$ and for $n = l = 2$, and with the hyperbolic biquaternions for $n = 3, l = 1$. The algebra $\mathbf{G}_n$ of Grassmann coincides for $n = 1$ with the algebra $^\circ\mathbf{C}$ of dual numbers.

That Grassmann, Clifford and Clifford-Study numbers were introduced by geometers is due to the fact that, from the beginning, these numbers were linked to geometric problems: Grassmann numbers to the determination of volumes of multidimensional parallelepipeds and to Cartan's method of exterior forms, split complex numbers, dual numbers, biquaternions, Clifford and Clifford-Study numbers to the geometry of Euclidean and non-Euclidean spaces.

In 1886, soon after the appearance of the Clifford numbers, there appeared the paper *Investigations of sums of squares* (Untersuchungen über die Summen von Quadraten. Bonn, 1886) *[328]* by the German mathematician Rudolf Lipschitz (1832–1903) in which he showed that, quite generally, there is a close connection between the Clifford numbers and the study of groups of rotations of multidimensional spaces. Specifically, if in the algebra $\mathbf{K}_n$ one defines the *conjugate* $\bar{\alpha}$ of an element $\alpha = \sum_r a_{i_1 \ldots i_r} \mathbf{e}_{i_1 \ldots i_r}$ to be the element $\bar{\alpha} = \sum_r (-1)^r a_{i_1 \ldots i_r} \mathbf{e}_{i_r i_{r-1} \ldots i_1} = \sum_r (-1)^{r(r+1)/2} a_{i_1 \ldots i_r} \mathbf{e}_{i_1 \ldots i_r}$, then, as is easily shown, relation (10.3) holds and the coefficient of 1 in $\alpha\bar{\alpha}$ is $\sum_r a_{i_1 \ldots i_r}^2$. If we denote this coefficient by $|\alpha|^2$ then relation (10.4) also holds. Thus it is natural to call $|\alpha|$ the *modulus* of α. Just as in the field $\mathbf{C}$ of complex numbers and in the skew field $\mathbf{H}$ of quaternions so too in the algebra $\mathbf{K}_n$ of Clifford numbers one can introduce the metric of the Euclidean space R_{2^n} by defining the distance between two elements α and β to be the modulus of their difference. Lipschitz also notes that the algebra $\mathbf{K}_n$ of Clifford numbers of order n can be represented by those Clifford numbers of the algebra $\mathbf{K}_{n+1}$ that are linear combinations of basis elements with *even* indices. Then the transformation (10.6), where ξ and ξ' are Clifford numbers of $\mathbf{K}_{n+1}$ (for they are linear combinations of the elements $\mathbf{e}_i(i = 0, 1, \ldots, n)$) and α is a Clifford number of the same algebra representing a Clifford number of the algebra $\mathbf{K}_n$ such that $\alpha^{-1}\xi\alpha$ is again a linear combination of the elements $\mathbf{e}_i$, is a rotation of the Euclidean space R_{n+1}. (In the case of quaternions, the transformation (10.6) can also be represented in this manner.) The elements α satisfying the equality (10.6) are subject to the conditions

$$\sum_k a_{i_1 i_2 \ldots i_{2k}}^2 = 1$$

and

$$\left.\begin{array}{l} aa_{i_1 i_2 i_3 i_4} = 3!! \, a_{[i_1 i_2} a_{i_3 i_4]} \\ aa_{i_1 i_2 i_3 \ldots i_6} = 5!! \, a_{[i_1 i_2} a_{i_3 \ldots i_6]}, \\ \cdots\cdots\cdots\cdots\cdots\cdots\cdots\cdots\cdots\cdots \\ aa_{i_1 i_2 i_3 \ldots i_{2k}} = (2k-1)!! \, a_{[i_1 i_2} a_{i_3 \ldots i_{2k}]}, \end{array}\right\}$$

where $(2k - 1)!!$ is the product of all odd numbers from 1 to $2k - 1$, and [ ] is the alternation symbol that denotes the algebraic sum of the expressions within the brackets obtained by all possible permutations of the indices where an expression is to be preceded by a plus sign if the permutation is even and by a minus sign if it is odd. These conditions enable us to express all coordinates of an element α in terms of its coordinates a_{ij}. Formula (10.6) shows that to every rotation of Euclidean space R_n there correspond two Clifford numbers of the algebra $\mathbf{K}_{n-1}$, namely α and $-\alpha$. The transformations (10.6) of this form define the so-called *spinor representations* of the rotation groups of the Euclidean space R_{n+1} and the groups of motions of the elliptic space S_n. By means of the Clifford-Study numbers of the algebra $^l\mathbf{K}_n$ one also defines the spinor representations of the groups of rotations of the pseudo-Euclidean spaces $^lR_{n+1}$ and of the groups of motions of the hyperbolic spaces ^lS_n.

The Kotel'nikov-Study Transfer

After defining in *Preliminary sketch of biquaternions* elliptic and parabolic biquaternions as well as elliptic and parabolic bivectors, that is, vectors for which the scalars are elements of the algebras of split complex and dual numbers, respectively, Clifford connected them with the twist motions of elliptic and Euclidean space. These ideas of Clifford provided the foundation for the *twist calculus* of the Russian mathematician and mechanist A. P. Kotel'nikov and E. Study, expounded in Kotel'nikov's master's dissertation *Twist calculus and some of its applications to geometry and mechanics* (Vintovoe sčislenie i nekotorye ego priloženiya k geometrii i mehanike. Kazan, 1895) *[288]* and in his (previously mentioned) doctoral dissertation *Projective theory of vectors* (Proéktivnaya teoriya vektorov) *[289]* and in Study's *The geometry of dynames* *[565]* showed that the manifolds of oriented straight lines in Euclidean, elliptic and Lobačevskian spaces can be represented by means of the manifolds of the unit *bivectors*, that is, unit vectors in three-dimensional space whose coordinates are, respectively, dual, split complex, and complex numbers, or, equivalently, spheres of unit radius in dual, split complex and complex three-dimensional Euclidean spaces. In the first of these cases, if the dual angle is $\varphi_0 + \varepsilon\varphi_1$, then the number φ_0 is equal to the angle between the corresponding oriented lines and the number φ_1 is equal to the shortest distance between them. In the second case, if the split complex angle is $\varphi_0 + e\varphi_1$, then the numbers φ_0 and φ_1 are, respectively, equal to the largest and smallest distances between the corresponding lines which, in turn, are, respectively, equal to the appropriate one of the lengths of the two common perpendiculars to the two lines in question. (In the case of Clifford parallels $\varphi_0 = \varphi_1$ and two lines have infinitely many common perpendiculars.) In the third case, if the complex angle is $\varphi_0 + i\varphi_1$, then the numbers φ_0 and φ_1 are respectively, equal to the shortest distance between the two lines—the length of their common perpendicular—and the angle between the planes passing through this perpendicular and the lines. The groups of motions of Euclidean, elliptic, and Lobačevskian spaces R_3, S_3 and $^1S_3$ can be represented by the groups of motions of spheres in the dual, split complex and complex spaces $^0CR_3$, $^1CR_3$ and CR_3, respectively, and every fact of the geometry of these spheres can be interpreted as a fact of the geometry of the spaces R_3, S_3 and $^1S_3$. To each pair of lines in these spaces one can associate the twist motion—whose axis is a common perpendicular of these lines—that makes the first of the two lines coincide with the second. To this motion one can associate the biquaternion which when multiplied by the bivector representing the first line yields the bivector representing the second line; that is why the biquaternions associated with the three spaces R_3, S_3 and $^1S_3$ determine their twist motions (*twists* or *screws*). What motivated the investigations of Clifford, Kotel'nikov, and Study was the desire to answer the question whether or not the geometry of non-Euclidean space contradicted the principles of mechanics. Kotel'nikov and Study developed a theory of *sliding vectors* in these spaces and showed that, just as in the space R_3 a system of forces and a system of instantaneous angular velocities (a force and an instantaneous angular velocity are both sliding vectors) are equivalent, respectively, to a force screw and a kinematic twist, the first of which consists of a force and a couple of forces in a plane perpendicular to it and the second of the angular velocity of rotation about some axis and the translational velocity along this axis (which can be regarded as a couple of angular velocities), so too in elliptic and Lobačevskian spaces S_3 and $^1S_3$ every system of sliding vectors is equivalent to two sliding vectors whose lines of action are two reciprocal polars.

Associative Algebras

All the above generalizations of numbers, due to de Morgan, Ch. Graves, J. T. Graves, Hamilton, Cayley, Grassmann, and Clifford, are instances of the single notion of an *algebra* defined by the American algebraist Benjamin Peirce (1809–1880) in the paper *Linear associative algebras* (Harvard, 1881) *[413]*. Peirce defined an algebra as an n-dimensional linear space on which there is defined an associative multiplication of vectors that is distributive with respect to addition and commutative with respect to multiplication of a vector by a number. The requirement of associativity of multiplication excludes from the class of algebras the system of octaves and the system of vectors in three-dimensional space under ordinary cross product multiplication.

If e_1, e_2, ..., e_n are a basis of an algebra then in order to define its vector multiplication it suffices to prescribe the multiplication of the basis vectors

$$e_i e_j = \sum_k C_{ij}^k e_k. \qquad (10.26)$$

The formulas (10.26) are called the *structure formulas* of the algebra, and the constants C_{ij}^k are called its *structure constants*.

B. Peirce introduced the notion of so-called *nilpotent elements*, that is, elements **e** for which there is a natural number r such that $\mathbf{e}^r = 0$ (such elements are, for example, the dual number ε and all basis elements of the Grassmann numbers), and the notion of *idempotent elements*, that is, elements **e** with $\mathbf{e}^2 = \mathbf{e}$ (such elements are, for example, the double numbers $(1 \pm \mathbf{e})/2$, the Clifford numbers $(1 \pm \mathbf{e}_{123})/2$ and the numbers obtained from them by replacing the element $\mathbf{e}_{123}$ by any basis element whose number of indices is of either of the forms $4m - 1$ and $4m$. Peirce used these concepts to classify complex algebras of small dimensions.

K. Weierstrass developed a general theory of algebras in lectures given as early as 1861 but his investigations were published only in 1884 in the paper *On the theory of magnitudes formed out of n principal units* (Zur Theorie der aus n Haupteinheiten gebildeten Grössen. Leipzig, 1884) *[622, vol. 2, pp. 311–332]*. Weierstrass introduced the notion of a *direct sum* $\mathbf{A}_1 \oplus \mathbf{A}_2 \oplus \cdots \oplus \mathbf{A}_r$ of several algebras $\mathbf{A}_i$: given algebras $\mathbf{A}_i$ with bases $\mathbf{e}_1, \ldots, \mathbf{e}_{n_1}; \mathbf{e}_{n_1+1}, \ldots, \mathbf{e}_{n_2}; \ldots; \mathbf{e}_{n_{r-1}+1}, \ldots, \mathbf{e}_{n_r}$, their direct sum $\mathbf{A}_1 \oplus \mathbf{A}_2 \oplus \cdots \oplus \mathbf{A}_r$ is the algebra whose basis consists of all these elements subject to the condition that the product of elements from different bases is zero. Weierstrass showed that *every commutative algebra without nilpotent elements is the direct sum of several copies of the field* **R** *of real numbers and the field* **C** *of complex numbers*. For example, if n is odd then the algebra $\mathbf{Cy}_n$ of *cyclic numbers* with basis 1, **e**, $\mathbf{e}^2$, $\ldots$, $\mathbf{e}^{n-1}$, where $\mathbf{e}^n = 1$, is the direct sum $\mathbf{C} \oplus \cdots \oplus \mathbf{C} \oplus \mathbf{R}$ of $n - 1$ copies of the field **C** and one copy of the field **R**, and, if n is even, then it is the direct sum $\mathbf{C} \oplus \cdots \oplus \mathbf{C} \oplus \mathbf{R} \oplus \mathbf{R}$ of $(n/2) - 1$ copies of **C** and two copies of **R**. In particular, if $n = 2$, then we see that the algebra $^1\mathbf{C}$ of split complex numbers (where the role of the elements $\mathbf{e}_1$ and $\mathbf{e}_2$ is played by $(1 \pm \mathbf{e})/2$) is the direct sum $^1\mathbf{C} = \mathbf{R} \oplus \mathbf{R}$ of two copies of the field **R**, and if $n = 3$ then we see that the algebra $\mathbf{T}_3$ of *triplets* of de Morgan and Graves is the direct sum $\mathbf{T}_3 = \mathbf{C} \oplus \mathbf{R}$.

In the paper *Generalization of the foundations of ordinary complex functions* (Verallgemeinerung der Grundlagen der gewöhnlichen complexen Functionen. Dresden, 1893) *[504]*, Georg Scheffers (1865–1945), a student of Lie, tried to define analytic functions whose domain and range belong to an algebra. He managed to do this only for commutative algebras ($C_{ij}^k = C_{ji}^k$). In this case he found an analyticity condition for a function $\mathbf{y} = f(\mathbf{x})$, where $\mathbf{x} = \sum_i x_i \mathbf{e}_i$, $\mathbf{y} = \sum_i y_i \mathbf{e}_i$, in the form

$$\sum_h \frac{\partial y_i}{\partial x_h} C_{jk}^h = \sum_h \frac{\partial y_h}{\partial x_i} C_{hk}^j.$$

For complex numbers this criterion reduces to the Cauchy-Riemann conditions. For the algebras of split complex and dual numbers this criterion reduces to the respective conditions

$$\frac{\partial y_1}{\partial x_1} = \frac{\partial y_2}{\partial x_2}, \quad \frac{\partial y_2}{\partial x_1} = \frac{\partial y_1}{\partial x_2} \quad \text{and} \quad \frac{\partial y_1}{\partial x_1} = \frac{\partial y_2}{\partial x_2}, \quad \frac{\partial y_2}{\partial x_1} = 0.$$

Ring Theory

Like fields and skew fields, algebras are special cases of the notion of a *ring*. We mentioned previously various number fields introduced by Galois in the *Memoir on the conditions for solvability of equations in radicals*. In the paper *On number theory. Part of the investigations on the theory of permutations and algebraic equations*. (Sur la théorie des nombres. La partie des études sur la théorie des permutations et équations algébriques. Paris, 1830) *[194]*. Galois introduced another variety of fields. The source of Galois's latter paper was Gauss's classical book *Disquisitiones arithmeticae [196, vol. 1]* in which Gauss introduced the notion of congruence of two numbers modulo a third: numbers a and b are said to be congruent modulo p,

$$a \equiv b \pmod{p}$$

if the remainders of the division of a and p are the same. The notation $\equiv$ emphasizes the analogy between the relations of congruence and equality, both of which are reflexive ($a \equiv a$), symmetric ($a \equiv b$ implies $b \equiv a$), and transitive ($a \equiv b$ and $b \equiv c$ imply $a \equiv c$). If p is a prime, then operations on congruences modulo p are entirely analogous to operations on equalities in the sense that one can add the same number to both sides of the congruence, multiply both of its sides by the same number, and, finally, divide both of its sides by the same number provided it is not a multiple of p, that is, if it is not congruent to zero modulo p. A congruence relation breaks up the integers into disjoint classes of congruent elements called *residue classes* modulo p. There are p residue classes. One set of representatives of the residue classes is $0, 1, 2, \ldots, p - 1$. One can define addition and multiplication of residue classes: the sum of two residue classes A and B with representatives a and b is the residue class with the representative $a + b$, and their product is the residue class with representative ab. It is easy to see that all residue classes under addition, and all residue classes other than the *principal residue class* consisting of multiples of p form commutative groups, and that multiplication of residue classes is distributive over addition. The principal residue class plays the role of zero, the residue class containing 1 plays the role of a multiplicative identity, and the role of the multiplicative inverse A^{-1} of a residue class A is played by the class containing an element b such that $ab \equiv 1$. Thus the set of residue classes modulo a prime has all properties of a number field. This field is called a *residue class field* and is denoted by $\mathbf{F}_p$.

In the paper *On number theory*, Galois considers congruences of the form $F(x) \equiv 0 \pmod{p}$, where $F(x)$ is a polynomial of degree v. If $F(x)$ is irreducible modulo p, that is, if it is not possible to write it as a sum of a product $\varphi(x)\psi(x)$ and a term $p\chi(x)$ that is congruent to zero modulo p, then Galois introduces a root i of the congruence which stands in the same relation to the congruence as the imaginary unit i in relation to the equation $x^2 + 1 = 0$; this is emphasized by the use of the same symbol in both cases. Such roots are now called *Galois imaginaries*. Then Galois considers expressions of the form

$$\alpha = a_0 + a_1 i + a_2 i^2 + \cdots + a_{\nu-1} i^{\nu-1}, \qquad (10.27)$$

where $a_0, a_1, \ldots, a_{\nu-1}$ are integers. The sum and product modulo p of such expressions are again such expressions. If multiplication yields a power i^ν then such a power is expressed in terms of lower powers using the congruence $F(i) \equiv 0 \pmod{p}$. The difference and quotient modulo p of two such expressions are again expressions of this type. Thus the totality of such expressions has the same properties as a number field and is called a *Galois field*. It is denoted by $\mathbf{F}_q$. A Galois field $\mathbf{F}_q$ of the type considered above contains $q = p^\nu$ elements. The field $\mathbf{F}_p$ of residue classes modulo p can be viewed as a Galois field with $\nu = 1$.

   Another line of investigations that led to ring theory had its origin in Gauss's paper *The theory of biquadratic residues II* (1832) *[196, vol. 2, pp. 95–178]*. Here Gauss introduced what we now call *Gaussian integers* $a + bi$, $i^2 = -1$, a, b integers, Gauss showed that addition and multiplication of such numbers again leads to such numbers. For these numbers the concepts of divisibility and primeness make sense. Also, it turns out that prime rational integers, viewed as Gaussian integers, need not be prime. For example, Fermat knew that primes of the form $4n + 1$ can be written as sums of squares $p^2 + q^2$ and thus are products $(p + qi)(p - qi)$.

   Gaussian integers are a special case of *algebraic integers*, that is, numbers of the form (10.27) where $a_0, a_i, \ldots, a_\nu$ are rational integers and i is a root of a polynomial $F(x)$ of degree ν that cannot be written as a product $\varphi(x)\psi(x)$ of polynomials of lower degree. Thus these numbers bear the same relation to rational integers as elements of a Galois field to residue classes modulo a prime. General algebraic integers were first considered by the German mathematician Peter Lejeune Dirichlet (1805–1859) in the paper *On the theory of complex units* (Zur Theorie der complexen Einheiten. Berlin, 1846) *[154, vol. 1, pp. 103–107]*. Dirichlet's investigations on algebraic integers were collected in his *Lectures on number theory* (Vorlesungen über Zahlentheorie. 1863) *[155]*. In 1847 the French mathematician and mechanist Gabriel Lamé (1795–1870) published a paper entitled *General proof of Fermat's theorem on the impossibility of the equality $x^n + y^n = z^n$ in whole numbers* (Démonstration générale du théorème de Fermat sur l'impossibilité en nombres entièrs de l'équation $x^n + y^n = z^n$. Paris, 1847) *[301]* which contained a false proof of *Fermat's last theorem*. In this paper Lamé considered numbers of the form (10.27) where i is a primitive ν-th root of 1. The connection between Fermat's theorem and such numbers is that the $z^n \neq x^n + y^n$ is equivalent to

$$z^n \neq (x + y)(x + \omega y) \ldots (x + \omega^{n-1} y),$$

where ω is a primitive n-th root of unity. Lamé's proof was based on the assumption that numbers of the form $x_0 + x_1 \omega + \cdots + x_{n-1} \omega^{n-1}$, $x_0, \ldots, x_{n-1}$ integers, are uniquely decomposable into prime factors, that is, factors whose only divisors are the units of the integral domain of such numbers.

Lamé's mistake was that he failed to realize that for numbers of this form it is not always true that if a product ab is divisible by a prime p then either a or b must be divisible by p. In the same year Ernst Kummer (1810–1893) noticed Lamé's error. In the paper *On the theory of complex numbers* (Zur Theorie der complexen Zahlen. Berlin, 1847) *[295]*, Kummer suggested that algebraic integers that cannot be written as products of rational integers but have the property that a product ab is divisible by p without a or b being divisible by p be regarded as composite rather than prime numbers but that the prime factors into which they can be decomposed be added as new objects to the field of algebraic numbers under consideration. Following the example of Poncelet, who called the imaginary points which he added to real curves *ideal points*, Kummer called the new objects he introduced *ideal prime factors* (ideale Primfactoren). Actually, Kummer's numbers are real or complex numbers that do not enter into the system considered by Kummer: for example, in the case of numbers of the form $a + b\sqrt{-3}$ the number 4 can be written as $2 \cdot 2$ and as $(1 + \sqrt{-3})(1 - \sqrt{-3})$; in this case the role of ideal prime factors of the number 2 is played by $1 + i$ and $1 - i$. Kummer hoped that ideal factors would make it possible to "save" Lamé's proof and actually used them to prove a large number of special cases of Fermat's last theorem without, however, obtaining a general proof of this theorem.

The theory of algebraic integers and, above all, Kummer's papers gave rise to a general theory of rings and fields in papers by Richard Dedekind (1831–1916). After the death of his teacher Dirichlet, Dedekind prepared for publication the third edition of Dirichlet's *Lectures on number theory* (1879) *[155]* and included in it an appendix (labeled XI) in which he introduced the concept of an *order* (Ordnung) as a set of elements on which there are defined operations of addition and multiplication such that all elements form a commutative group under addition, and multiplication is associative and distributive over addition. Dedekind called an *order* whose nonzero elements form a group a *corpus* (*Körper*). The French term for the German *Körper* was *corps*. The initial English term *corpus* was later displaced by *field*. In view of the multiplicity of meanings of the word *order*, David Hilbert replaced it in his *Theory of algebraic number fields* (Die Theorie der algebraischen Zahlenkörper. Göttingen, 1897) *[226, vol. 1, pp. 63–363]* with the term *ring*.

The rational numbers **Q**, the real numbers **R**, the complex numbers **C**, and the algebraic numbers form infinite fields. The residue classes modulo a prime p form a finite field $\mathbf{F}_p$. The Galois fields $\mathbf{F}_q$ are likewise finite. The quaternions form a skew field **H**. The rational integers **Z**, the split complex numbers $^1\mathbf{C}$, the dual numbers $^\circ\mathbf{C}$ and the algebraic integers form commutative rings. The Grassmann numbers $\mathbf{G}_n$, the Clifford numbers $\mathbf{K}_n$, the Clifford-Study numbers $^1\mathbf{K}_n$, the real matrices $\mathbf{R}_n$, the complex matrices $\mathbf{C}_n$, and the quaternion matrices $\mathbf{H}_n$ form noncommutative rings. We saw that the split complex numbers, the dual numbers, the Clifford numbers for $n \geq 4$, the Grassmann numbers and matrices all have zero divisors. There are finite rings that are not fields, for example, the rings of residue classes modulo a composite number pq: In fact, the product of the residue classes containing p and q, respectively, is the zero class; that is, these residue classes are divisors of zero. Dedekind replaced Kummer's notion of an *ideal factor* with that of an *ideal*, which he defined as a subring of a ring such that the products of its elements by ring elements also belong to this subring. The ideals of the ring of integers are the multiples of fixed integers. The algebraic integers that are multiples of Kummer's *ideal factors* are also ideals, and this is the origin of the term *ideal*.

If the associativity of multiplication in a ring is replaced by the property that every two elements of the ring generate an associative algebra (this property is equivalent to the weak associative laws $x(y^2) = (xy)y$ and $(x^2)y = x(xy)$) then the ring is called an *alternative* ring. The octaves form an alternative skew field **O**. If we replace the unit l in **O** by e, $e^2 = 1$, then we obtain the alternative algebra ^{1}**O** of *split octaves*.

Just as in the case of groups, so too in the case of rings one can speak of *ring homomorphisms* and *ring isomorphisms*. If a ring is a homomorphic image of another ring then the elements of the latter that are mapped by the homomorphism onto the zero element of the former form an ideal. Two algebras are said to be isomorphic if they are isomorphic as rings and the image of the product of each element of one of the algebras by a numerical factor is the product of the image of that element by the same numerical factor.

An extremely important type of ring is a *simple* ring, defined as a ring without nontrivial two-sided ideals (the trivial ideals of a ring are itself and the ideal consisting of zero alone).

Simple Associative Algebras

Examples of simple rings are simple associative algebras, many of which we have encountered previously.

The deep connections and analogies between associative algebras and Lie algebras have been the reason that, toward the end of the 19th century, associative algebras were studied by Lie himself and by his student Scheffers; at the same time these algebras were studied by Weierstrass's student Ferdinand Georg Frobenius (1849–1917).

We have denoted the direct sum of algebras **A** and **B** by $A \oplus B$. Given two algebras **A** and **B** with respective bases $e_1, e_2, \ldots, e_n$ and $f_1, f_2, \ldots, f_m$, the algebra of rank $n \cdot m$ with basis elements $e_i \cdot f_\alpha = f_\alpha \cdot e_i$ is called the *tensor product* of these algebras and is denoted by $A \otimes B$. It is not difficult to see that $\mathbf{R}_m \otimes \mathbf{R}_n = \mathbf{R}_{mn}$, $\mathbf{H} \otimes \mathbf{H} = \mathbf{R}_4$, $\mathbf{C}_n = \mathbf{R}_n \otimes \mathbf{C}$ and $\mathbf{H}_n = \mathbf{R}_n \otimes \mathbf{H}$. We shall call the tensor products $\mathbf{C} \otimes \mathbf{C}$, $\mathbf{C} \otimes {}^\circ\mathbf{C}$, $\mathbf{H} \otimes \mathbf{C}$ and $\mathbf{H} \otimes {}^\circ\mathbf{C}$ the algebras of *bicomplex* and *bidual numbers*, of *biquaternions* and *duoquaternions*, respectively. It is easy to check that $\mathbf{C} \otimes \mathbf{C} = \mathbf{C} \oplus \mathbf{C}$. In the nineties of the 19th century simple associative algebras were studied by the Lie school, by Frobenius and by Fedor Eduardovič Molin [Molien] (1861–1941), who worked in Dorpat and in Tomsk, at first independently of the Lie school and of Frobenius and then in contact with them. In the paper *Reduction of complex number systems to standard forms* (Zurückführung complexer Zahlensysteme auf typische Formen. Leipzig, 1891) *[505]*, Scheffers called algebras *complex number systems*. In the paper *On systems of higher complex numbers* (Über Systeme höherer complexer Zahlen. Dorpat, 1892) *[371]*, Molin called algebras *systems of higher complex numbers*. In *Theory of hypercomplex magnitudes* (Theorie der hypercomplexen Grössen. Leipzig, 1903) *[189]*, Frobenius called algebras *systems of hypercomplex magnitudes*. All three scholars extended the concepts of a *simple* and *semisimple algebra* and óf a *radical*, which first arose in the theory of Lie algebras, to associative algebras. A semisimple algebra is an algebra without nilpotent elements; if there are nilpotent elements in an algebra then they form an ideal, and it is this ideal that is called the radical of the algebra.

Weierstrass's theorem on algebras can be stated as follows: every semisimple commutative algebra is isomorphic to a direct sum of several replicas of the fields R and C. In the previously mentioned paper, Molin gave a criterion for semisimplicity of an algebra analogous to Cartan's criterion for semisimplicity of a Lie algebra and showed that the factor algebra of every algebra by its radical is semisimple, every semisimple algebra is isomorphic to a direct sum of simple algebras, and every simple algebra over the field C is isomorphic to an algebra C_n. In the paper *Bilinear groups and systems of complex numbers* (Les groupes bilinéaires et les systèmes de nombres complexes. Toulouse, 1898) *[96, part 2, vol. 1, pp. 7–105]*, É. Cartan restated the results of Molin and proved analogous theorems for real simple algebras and showed, in particular, that every real simple noncommutative algebra is isomorphic to one of the algebras R_n, C_n, and H_n.

We note that formula (10.12) establishes an isomorphism between the skew field H and a subalgebra of the algebra C_2, and formula (10.13) establishes an isomorphism of the algebras ${}^1H$ and R_2. The latter implies the isomorphism of the algebras $H \otimes C$ and C_2.

Spaces over Algebras

In the previous chapter we encountered projective, elliptic, hyperbolic and sympletic spaces over the field C and the skew field H. Similar spaces can be constructed over other simple and semisimple associative algebras (we will see below that similar spaces can also be constructed over other classes of algebras). The projective space over an algebra A, denoted by AP_n, can be defined by means of projective coordinates x^0, x^1, ..., x^n that are elements of A. If A is a field then the coordinates x^i must not all be equal to zero and if A is not a field then the coordinates x^i must not belong to a twosided ideal of A; the coordinates x^i and $x^i k$, where k is a nonzero element of A, define the same point of AP_n. The collineations of AP_n have the form $'x^i = \sum_j a_j^i f(x^j)$, where $x \to f(x)$ is an automorphism of A. The correlations of AP_n have the form $u_i = \sum_j \varphi(x^j)a_{ij}$, where $\varphi \to \varphi(x)$ is an antiautomorphism of A. The general theory of one-dimensional spaces over algebras was constructed by Walter Benz in *Lectures on the geometry of algebras* (Vorlesungen über die Geometrie der Algebren. Berlin, 1973). Geometries of spaces over algebras of matrices were introduced by Nicoló Spampinato (1892–1971) in the paper *On the geometry of a line space considered as a hypercomplex S_1* (Sulla geometria dello spazio rigato considerato come un S_1 ipercomplesso. Naples, 1935) *[552]* and by Hua Loo Keng (1910–1985) in *Geometries of matrices* (New York, 1945–1947) *[235a]* and in *The geometry of symmetric matrices over the field of real numbers* (Geometriya simmetričeskih matric nad polem deĭstvitel'nyh čisel. Moscow, 1946) *[235b]* for one-dimensional spaces over symmetric matrices. The geometry of two-dimensional spaces over matrices was introduced by Carmela Carbonaro in the paper *The line S_5 considered as hypercomplex S_2 connected with the regular complex algebra of order 4 ($L'S_5$ rigato considerato come un S_2 ipercomplesso legato all'algebra complesso regolare d'ordine 4. Catania, 1936) *[87]*. The general theory of spaces over algebras of matrices was constructed by Maqsud Ali Simran oglu Javadov (Džavadov) (1902–1972) in the papers *Projective spaces over algebras* (Proektivnye prostranstva nad algebrami. Baku, 1957) *[162]* and *Non-Euclidean spaces over algebras* (Neevklidovy geometrii nad algebrami. Baku, 1957) *[163]*.

The geometry of rectangular matrices was constructed by Hua Loo Keng and the author in *The geometry of rectangular matrices and its application to real projective and non-Euclidean geometry* (Geometriya pryamougol'nyh matric i ee priloženiya k vesčestvennoi proektivnoi i neevklidovoi geometrii. Peking, Kazan, 1957) *[235c]*.

Geometries over the algebras $C \otimes C$ and $H \otimes C$ were constructed by Nazim Tanriverdi oglu Abbasov in the papers *Bicomplex elliptic spaces* (Bikompleksnye elliptičeskie prostranstva. Baku, 1962) *[1]* and *Biquaternion elliptic spaces* (Bikvaternionnye elliptičeskie prostranstva. Baku, 1963) *[2]*. For algebras with zero divisors the axioms of projective space are satisfied only in the "basic case" and there exist pairs of *contiguous points* through which pass more than one straight line and pairs of *contiguous straight lines* that intersect in more than one point. In affine spaces over algebras with zero divisors there are, in addition to continuous straight lines, *diverging straight lines* that can be mapped by translation to contiguous ones. Over such algebras one can also construct spaces of *fractional dimension*. Such spaces were introduced by Il'ya Adamovič Čahtauri (b. 1946), in the paper *Projective and elliptic spaces of integral and fractional dimension over algebras of matrices* (Proektivnye i elliptičeskie prostranstva celoĭ i drobnoĭ razmernosti nad algebrami matric. Tbilisi, 1971) *[105]*.

The general theory of spaces over rings with zero divisors was founded by the Rumanian mathematician and poet Dan Barbilian (Ion Barbu, 1895–1961) in the paper *On the axiomatics of plane projective ring geometry* (Zur Axiomatik der projektiven ebenen Ringgeometrie. 1940–1941) *[36]*. Barbilian called non-contiguous points *points in the clear state* and contiguous points *points in the spectral state*. The term "contiguous" was introduced by Wilhelm Klingenberg (see *[286]*).

Spaces with contiguous points and lines are instances of *Hjelmslev spaces*— spaces with more than one line through two points and more than one point of intersection of two lines. These spaces were introduced by the Danish geometer Juhannes Hjelmslev (1873–1950) in the paper *Introduction to the general congruence theory* (Einleitung in die allgemeine Kongruenzlehre. Copenhagen, 1929–1949) *[232]*.

Spaces over algebras have interesting interpretations in real spaces. Thus the projective space ^1CP_n over the algebra of split complex numbers can be interpreted in a pair of real spaces P_n. Write a split complex number $z = x + ey$ as $z = z_+e_+ + z_-e_-$, where z_+ and z_- are reals and $e_+ = (1 + e)/2$ and $e_- = (1 - e)/2$. Note that e_+ and e_- are idempotents ($e_+^2 = e_+, e_-^2 = e_-, e_+e_- = 0$). Now associate to every point x of ^1CP_n the points $x_+(x_+^i)$ and $x_-(x_-^i)$ in two P_n. Two points x and y are contiguous if $x_+^i = x_-^i$ or $y_+^i = y_-^i$. Similarly, two lines $u_0x^0 + u_1x^1 + u_2x^2 = 0$ and $v_0x^0 + v_1x^1 + v_2x^2 = 0$ in $^1CP_2$ are contiguous if $u_i^+ = v_i^+$ or $u_i^- = v_i^-$. There are similar interpretations of the spaces ^1CS_n and $^1C^lS_n$ in pairs of spaces S_n and ^lS_n, respectively.

The space $^1C\bar{S}_n$ is interpreted in the space P_n. Specifically, every point $x(x^i)$ of $^1C\bar{S}_n$ is interpreted by an 0-*couple* (point + hyperplane) of P_n: if $x^i = X^ie_+ + U_ie_-$, then to every point $x(x^i)$ of $^1C\bar{S}_n$ we associate the 0-couple consisting of the point $X(X^i)$ and the hyperplane $U(U_i)$ (i.e., the hyperplane $\sum_i U_iX^i = 0$), and the right part of the formula (9.6) is equal to the cross ratio (XY, UV) of the points X, Y and the points of intersection of the line XY with the hyperplanes U, V. This interpretation is presented in the author's book *[465, p. 655]*.

The space $\mathbf{R}_{m+1}P_n$ is interpreted in the space P_{nm+n+m}: every point $X(X^i)$ of $\mathbf{R}_{m+1}P_n$ is interpreted by an $(m - 1)$-dimensional plane $X_0X_1 \ldots X_m$ of P_{nm+n+m} such that the coordinates of the α-th point X_α form the α-th columns of the matrices X^i. Two points X and Y are contiguous if the corresponding planes lie in an M-plane, $M < 2m - 1$, and two lines K and L are represented by $(2m - 1)$-dimensional planes if these lines lie in a two-dimensional plane; they are contiguous if the corresponding planes intersect. If $nm + n + m < N < (n + 1)m + (n + 1) + m$, we obtain, analogously, the real interpretation in P_n of a space of fractional dimension. If $m = 2$ we obtain the interpretation of the space ^1HP_n by lines of P_{2n+1}. If we replace the group of collineations of P_{2n+1} by the group of motions of S_{2n+1} or the group of symplectic transformations of Sp_{2n+1}, then we obtain, respectively, the interpretations of the manifolds of lines of S_{2n+1} and Sp_{2n+1} in the spaces $^1H\bar{S}p_n$ and $^1H\bar{S}_n$. In the latter case, the right part of the formula (9.6) is equal to the symplectic invariant of two lines of Sp_{2n+1}. This interpretation is presented in the author's book *[465, p. 663]*. Similarly, we find (see *[1, 2]*) that the Hermitian elliptic space $(\mathbf{C} \otimes \mathbf{C}) \widetilde{\bar{S}}_n$ over the algebra $\mathbf{C} \otimes \mathbf{C}$ (the conjugation $\alpha \to \widetilde{\bar{\alpha}}$ in an algebra $\mathbf{A} \otimes \mathbf{B}$ is the product of the conjugations $\alpha \to \bar{\alpha}$ and $\alpha \to \tilde{\alpha}$ in the algebras $\mathbf{A}$ and $\mathbf{B}$ respectively) can be interpreted as the pairs of points of the complex space $\mathbf{C}\bar{S}_n$, and that the analogous space $(\mathbf{H} \otimes \mathbf{C}) \widetilde{\bar{S}}_n$ over the algebra $\mathbf{H} \otimes \mathbf{C}$ can be interpreted as the manifold of lines of the complex space $\mathbf{C}\bar{S}_{2n+1}$.

Linear Representations of Groups

If an isomorphic or homomorphic image of a group G is a group of real or complex matrices then we speak of a *linear representation* of the group G. Linear representations of Lie groups play an important role in the theory of Lie groups as well as in its applications.

The theory of linear representations of finite groups is closely linked to the theory of linear representations of algebras. The latter are defined very much like linear representations of groups. Every algebra has a linear representation: to obtain it we need only associate to an element $a = \sum_i a^i e_i$ of the algebra $\mathbf{A}$ the matrix (A_j^i) of the linear transformation $y = ax$ on the algebra. Since

$$ax = \left(\sum_i a^i e_i\right)\left(\sum_j x^j e_j\right) = \sum_i \sum_j a^i x^j (e_i e_j) = \sum_j \sum_k \left(\sum_i C_{ij}^k a^i\right) x^j e_k,$$

the elements of the matrix (A_j^i) and the coordinates a^i of the element a are connected by the relation $A_j^k = \sum_i C_{ij}^k a^i$. This isomorphism is called the *regular representation* of the algebra $\mathbf{A}$.

To obtain a linear representation of a group it suffices to find a group of elements of an algebra that is an isomorphic or homomorphic image of that group.

A complete theory of finite commutative groups was developed by Frobenius in the paper *On group characters* (Über Gruppencharaktere. Leipzig, 1896) *[190]*. If a group with typical element g is represented by complex matrices $G = (G_j^i)$, then Frobenius uses the function $G(g)$ from the group into the group of matrices to define a *character* $\chi(g)$ as the numerical function of the group element whose value at g is the trace $\sum_i G_i^i$ of the matrix $G(g)$. In the case of a commutative group the representing matrices are of order 1, that is, are numbers, so that, in this case, the characters can be defined as homomorphic maps of the group to the group of complex numbers of absolute value 1. Frobenius showed that the number of different characters of a commutative group G, including the *principal character* $\chi_0(g) = 1$, is equal to the number of elements of the group; that the product $\chi_1(g)\chi_2(g)$ of two characters is again a character; and that under this multiplication the characters form a group with identity element $\chi_0(g)$. Hence, in this case, the group of characters is isomorphic to the group G. In particular, the characters of the cyclic group $1, \theta, \theta^2, \ldots, \theta^{n-1}$ with $\theta^n = 1$ are given by $\chi_h(\theta^k) = e^{2\pi hki/n}$. Frobenius's proof was based on the investigation of an algebra whose basis elements formed a group isomorphic to the given group. When he constructed the regular representation of this algebra, Frobenius noticed that, for a suitable choice of a basis in the algebra, the group of matrices representing the basis elements of the algebra consisted of the matrices of the linear transformations $x'_h = e^{2\pi hki/n} x_h$.

In the paper *On the condition of reducibility for any group of linear substitutions* (London, 1905) *[84]*, the English algebraist William Burnside (1852–1927) extended Frobenius's results to finite noncommutative groups. Like Frobenius before him, Burnside considered an algebra whose basis elements formed a group isomorphic to the given group. Burnside also constructed the regular representation of the algebra and noticed that in this case too the linear transformations forming the regular representation decompose into linear representations in the invariant subspaces of the algebra, but now these subspaces are no longer all one-dimensional as in the case of commutative groups. The linear representations in the invariant subspaces of an algebra, determined by its regular representation, exhaust all possible linear representations of this algebra. Also, among these representations the representation by matrices of order k is repeated k times. It follows that the orders $k_0, k_1, \ldots, k_r$ of the matrices forming all possible linear representations of a finite group and the order n of the group are connected by the relation

$$k_0^2 + k_1^2 + \cdots + k_r^2 = n.$$

One of these linear representations is the representation that associates to each group element the number 1. To this "principal representation" there corresponds the number $k_0 = 1$.

Linear Representations of Lie Groups

The theory of linear representations of Lie groups was founded by Élie Cartan in the paper *Projective groups that leave no plane manifold invariant* (Les groupes projectifs qui ne laissent invariante aucune multiplicité plane. Paris, 1913) *[96, part 1, vol. 1, pp. 355–398]*. In the paper *Theory of representations of continuous semisimple groups by means of linear transformations [625, pp. 262–366]* Weyl found all linear representations of simple and semisimple Lie groups. The most important of the linear representations of the groups O_n, and ^lO_n are the transformations of the projective coordinates x^i of points, of the Plücker coordinates (9.2) of straight lines, and of the analogously defined *Grassmann coordinates* of planes of all dimensions resulting from the motions of elliptic and hyperbolic spaces S_{n-1} and $^lS_{n-1}$ and the so-called spinor representations of the groups of motions of these spaces. Earlier we defined the algebras $\mathbf{K}_n$ of Clifford numbers. The algebras $\mathbf{K}_1, \mathbf{K}_2$, and $\mathbf{K}_3$ are isomorphic to the fields $\mathbf{R}$ and $\mathbf{C}$ and the skew field $\mathbf{H}$, respectively, and the algebras $\mathbf{K}_n$ for $n > 3$ are further generalizations of the complex numbers and of the quaternions. In his *Investigations of sums of squares* (1886) *[328]*, R. Lipschitz showed that the groups O_n have two-valued representations given by the transformations (10.6) where ξ and ξ' are linear combinations of the elements $\mathbf{e}_1, \mathbf{e}_2, \ldots, \mathbf{e}_n$ of the algebra $\mathbf{K}_{n+1}$ and α is a linear combination of basis elements of these algebras with even numbers of indices.

There is a similar representation for matrices of the group CO_n and for elements of the algebra $\mathbf{K}_n \otimes \mathbf{C}$. In the paper *Spinors in n-dimensions* (Baltimore, 1935) *[78]* Weyl and his student Richard Brauer (b. 1901) showed that the algebras $\mathbf{K}_{2n+1} \otimes \mathbf{C}$ are isomorphic to the algebras of complex matrices $\mathbf{C}_{2^n}$ and that algebras $\mathbf{K}_{2n} \otimes \mathbf{C}$ are isomorphic to the direct sums of algebras $\mathbf{C}_{2^{n-1}} \oplus \mathbf{C}_{2^{n-1}}$. The two-valued representation of the groups CO_{2n+1} by means of elements of the algebras $\mathbf{K}_{2n+1} \otimes \mathbf{C}$, that is, by means of matrices of the algebra $\mathbf{C}_{2^n}$, is a spinor representation; the two-valued representation of the groups CO_{2n} by means of elements of the algebras $\mathbf{K}_{2n} \otimes \mathbf{C}$, that is, by means of pairs of matrices of the algebra $\mathbf{C}_{2^{n-1}}$, determines two two-valued representations of the groups CO_{2n} by means of matrices of the algebra $\mathbf{C}_{2^{n-1}}$ that are spinor representations.

We note that there is a close connection between spinor representations of the groups O_n and CO_n and the local isomorphisms, mentioned in the previous chapter, between groups in the classes B_1, D_2, B_2 and D_3 and groups in the classes $A_1, B_1 \otimes B_1, C_2$ and A_3.

Linear representations of complex simple Lie groups make possible the determination of linear representations of compact and noncompact real groups of which they are the complex forms.

The Clifford algebras $\mathbf{K}$ and the Clifford-Study algebras ${}^l\mathbf{K}_n$ determining two-valued representations of the groups O_n and ${}^lO_n$ are also simple or semisimple algebras. Their structure is somewhat more complicated then the structure of the algebras $\mathbf{K}_n \otimes \mathbf{C}$. In the paper *Complex numbers* (Nombres complexes. Paris, 1908) (a fourfold revision of E. Study's paper with the same title) *[96, part 2, pp. 107–246]*, É. Cartan formulated the structure of the algebras $\mathbf{K}_n$ and ${}^l\mathbf{K}_n$ as follows:

if $h = 1 - i_1^2 - i_2^2 - \cdots - i_n^2$ then the systems under consideration have the forms

$$S_m \text{ if } h = \pm 1 \pmod 8, \qquad IS_m \text{ if } h \equiv \pm 2 \pmod 8,$$

$$QS_m \text{ if } h = \pm 3 \pmod 8, \qquad 2S_m \text{ if } h \equiv 0 \pmod 8,$$

$$\text{and } 2QS_m \text{ if } h \equiv 4 \pmod 8$$

[96, part 2, p. 242].

Cartan's $i_1, i_2, \ldots, i_n$ are our $\mathbf{e}_1, \mathbf{e}_2, \ldots, \mathbf{e}_{n-1}$, so that $h = n - 2l$; Cartan's $S_m, IS_m, QS_m, 2S_m$, and $2QS_m$ are, respectively, our $\mathbf{R}_m, \mathbf{C}_m, \mathbf{H}_m, \mathbf{R}_m \oplus \mathbf{R}_m$ and $\mathbf{H}_m \oplus \mathbf{H}_m$.

Cartan's result, communicated without proof, was proved by the author in the book *Non-Euclidean geometries* (Moscow, 1955) *[465, pp. 452–458]* and can be formulated as follows. Let $h = n - 2l$. Then each of the algebras K_{2n+1} and $^lK_{2n+1}$ is isomorphic to one of the algebras R_{2^n} and $H_{2^{n-1}}$ according as $h \equiv \pm 1$ or $\pm 3 \pmod 8$ and each of the algebras K_{2n} and $^lK_{2n}$ is isomorphic to one of the algebras $C_{2^{n-1}}$, $R_{2^{n-1}} \oplus R_{2^{n-1}}$ and $H_{2^{n-2}} \oplus H_{2^{n-2}}$ according as $h \equiv \pm 2$, 0 or 4 (mod 8). In *Lectures on the theory of spinors* (Leçons sur la théorie des spineurs. Paris, 1938) *[95]* Cartan showed that spinor representations may be regarded as transformations of suitably defined coordinates of the *plane generators of maximal dimension* of the absolutes of the elliptic and hyperbolic spaces S_n and ^lS_n.

In *[160, 161]*, M. Javadov suggested another type of geometric interpretation of spinor representations. In Javadov's interpretation the spinor representations are regarded as transformations of suitably defined coordinates of the *points* of the absolutes of elliptic and hyperbolic spaces (see also *[461]* and *[473]*).

Quasisimple and k-quasisimple Associative Algebras

If we apply to simple and semisimple associative algebras the same passage to the limit that enabled us to go from simple Lie algebras to quasisimple ones then we obtain *quasisimple associative algebras* and, if we repeat it k times, *k-quasisimple associative algebras*. The classification of quasisimple associative algebras obtained from simple ones by passage to the limit was carried out by Mihail Petrovič Zamahovskiĭ (b. 1942) in the paper *Quasisimple algebras, quasimatrices and spinor representations of quasi-non-Euclidean motions* (Kvaziprostye algebry, kvazimatricy i spinornye predstavleniya kvasineevklidovyh dviženiĭ. Kazan, 1969) *[477, pp. 64–65]*. Such algebras are the algebra ^0C of dual numbers, the algebra ^0H of semiquaternions $a + bi + c\varepsilon + d\eta$, $i^2 = -1$, $\varepsilon^2 = 0$, $i\varepsilon = -\varepsilon i = \eta$, the algebra ^{10}H of split semiquaternions $a + be + c\varepsilon + d\eta$, $e^2 = 1$, $\varepsilon^2 = 0$, $e\varepsilon = -\varepsilon e = \eta$, a, b, c, d real, the tensor product $H \otimes {}^0C$, the algebras ^0C_n, ^0H_n, $^{10}H_n$ and $H_n \otimes {}^0C$ of matrices over these algebras, and the algebras of *quasimatrices*, isomorphic to the subalgebras of matrices of the form $\begin{bmatrix} A & B\varepsilon \\ C\varepsilon & D \end{bmatrix}$ in the algebras $^0C_{2n}$, $C_{2n} \otimes {}^0C$, and $H_{2n} \otimes {}^0C$, where A, B, C, D are, respectively, elements of matrices in the algebras R_n, C_n and H_n. We shall denote the algebras of quasimatrices by R_n^m, C_n^m and H_n^m, respectively. The algebras K_n^m and $^{kl}K_n^m$, obtained by an analogous process from the algebras K_n and ^lK_n, are also quasisimple; these algebras were defined by the author in his *Non-Euclidean geometries* *[465, pp. 540–541]* for $m = 0$ and, in the general case, by Tat'yana Glebovna Orlovskaya (b. 1944) *[477, pp. 68–70]*. By means of these algebras one can define spinor representations of the groups of motions of the quasielliptic and quasihyperbolic spaces that admit geometric representations similar to the interpretations of Cartan and Javadov. The classification of two-quasisimple associative algebras obtained from simple algebras was carried out by Inna Ivanovna Kolokol'ceva (b. 1942) in the paper *Biquasisimple algebras* (Bikvaziprostye algebry. Tomsk, 1973) *[480]* (see also the paper *Biquasisimple algebras, biquasimatrices and spinor representations of biquasinoneuclidean motions* (Bikvaziprostye algebry, bikvazimatricy i spinornye predstavleniya bikvazineevklidovyh dvizeniĭ. Kazan, 1975) by N. T. Abbasov and I. I. Kolokol'ceva *[3]*). Such an algebra is the algebra $^{00}\mathbf{H}$ of 1/4-quaternions $a + b\varepsilon + c\eta + d\omega$, $\varepsilon^2 = \eta^2 = 0$, $\varepsilon\eta = -\eta\varepsilon = \omega$, a, b, c, d real, studied by Albina Borisovna Rudenko *[480]*.

One can also define projective, elliptic, hyperbolic and symplectic spaces over quasisimple and k-quasisimple algebras analogous to the spaces over simple algebras defined in the previous chapter. The aim here is to provide geometric interpretations of simple Lie groups. The fundamental groups of these spaces are quasisimple and k-quasisimple Lie groups, and they form geometric interpretations of such groups. These interpretations were introduced for quasisimple groups by L. M. Karpova *[479]* and for two-quasisimple groups by Natalya Serafimovna Denisova (b. 1946) in the paper *Biquasisimple Lie groups* (Bikvaziprostye gruppy Li. Moscow, 1973) *[141]*.

Alternative Algebras

The alternative algebra $^1\mathbf{O}$ of split octaves was introduced by the American mathematician Leonard Eugene Dickson (1874–1954) in the book *Algebras and their number systems* (Algebren und ihre Zahlensysteme. Zürich-Leipzig, 1827) *[147]*. The elements of $^1\mathbf{O}$, as elements of $\mathbf{O}$, can be represented as pairs of quaternions (α, β) multiplied in accordance with a rule similar to the rule (10.8), namely

$$(\alpha, \beta)(\gamma, \delta) = (\alpha\gamma + \bar{\delta}\beta, \delta\alpha + \beta\bar{\gamma}). \tag{10.28}$$

Split octaves are also called *Cayley-Dickson numbers, antioctaves* and *pseudooctaves*. In the paper *Theory of alternative rings* (Theorie der alternativen Ringe. Hamburg, 1930) *[650]* the German mathematician Max Zorn showed that the algebra $^1\mathbf{O}$ can be represented by *vector-matrices* $\begin{pmatrix} \alpha & \mathbf{a} \\ \mathbf{b} & \beta \end{pmatrix}$, where α and β are real numbers and $\mathbf{a}$ and $\mathbf{b}$ are vectors in three-dimensional Euclidean space, with the multiplication rule

$$\begin{pmatrix} \alpha & \mathbf{a} \\ \mathbf{b} & \beta \end{pmatrix}\begin{pmatrix} \gamma & \mathbf{c} \\ \mathbf{d} & \delta \end{pmatrix} = \begin{pmatrix} \alpha\gamma - \mathbf{a}\mathbf{c} & \alpha\mathbf{c} + \delta\mathbf{a} + \mathbf{b} \times \mathbf{d} \\ \mathbf{b}\gamma + \beta\mathbf{d} + \mathbf{a} \times \mathbf{c} & \beta\delta - \mathbf{b}\mathbf{d} \end{pmatrix}.$$

In this paper Zorn showed that the algebras $\mathbf{O}$ and $^1\mathbf{O}$ and the algebra $\mathbf{O} \otimes \mathbf{C}$ of *bioctaves* are the only *simple alternative algebras* and that every *semisimple alternative algebra* is isomorphic to a direct sum of these algebras and a simple associative algebra.

There are also *quasisimple alternative algebras*. Such algebras, obtained by passage to the limit from simple ones, are the algebra $^0\mathbf{O}$ of *semioctaves*, that can be defined either as the algebra of pairs (α, β) of quaternions with the multiplication rule

$$(\alpha, \beta)(\gamma, \delta) = (\alpha\gamma, \delta\alpha + \beta\bar{\gamma}) \tag{10.28'}$$

or as the algebra of pairs of semiquaternions with the multiplication rule (10.8), and the algebra $^{10}\mathbf{O}$ of *split semioctaves*, that can be defined either as the algebra of pairs of split quaternions with the multiplication rule (10.28') or as the algebra of pairs of semiquaternions with the multipication rule (10.28). The elements of the algebras $^0\mathbf{O}$ and $^{10}\mathbf{O}$ can be written in the form $\alpha + \beta e$, where α and β are pairs of quaternions and split quaternions, respectively. There are similar definitions of the two-quasisimple alternative algebras $^{00}\mathbf{O}$ and $^{100}\mathbf{O}$ of $\frac{1}{4}$-octaves and split $\frac{1}{4}$-octaves, respectively, and of the three-quasisimple alternative algebra $^{000}\mathbf{O}$ of $\frac{1}{8}$-octaves.

Just as there is a close connection between the simple Lie groups of the infinite series and simple associative algebras, so too there is a close connection between the simple Lie groups of the exceptional classes and simple alternative algebras. Just as every automorphism of the algebras $\mathbf{H}$ and $^1\mathbf{H}$ is of the form (10.6), that is, just as the groups of these automorphisms are locally isomorphic to the groups O_3 and $^1O_3$, respectively, so too—as Cartan showed in the paper *Real simple finite continuous groups*—a compact group in the class G_2 is a group of automorphisms of the algebra $\mathbf{O}$. It is easy to show that a noncompact group in G_2 is a group of automorphisms of the algebra $^1\mathbf{O}$. Similar proofs show that the quasisimple groups of the class G_2 are groups of so-called metric automorphisms of the algebras $^0\mathbf{O}$ and $^{10}\mathbf{O}$.

We mentioned earlier that the octaves with modulus 1 ($\alpha\bar{\alpha} = 1$) form a nonassociative analogue of a group called a *loop*. The analogous elements of the algebras $^1\mathbf{O}$, $^0\mathbf{O}$ and $^{10}\mathbf{O}$ also form loops. In the paper *Analytic loops* (Analitičeskie lupy. Moscow, 1955) *[352]*, the Soviet mathematician A. I. Mal'cev developed a theory of loops that are analogs of Lie groups; the corresponding analogs of Lie algebras are now called *Mal'cev algebras* (see, for example, *[496]*). The concepts of simplicity, semisimplicity, and so on, apply to analytic loops and Mal'cev algebras. In the paper *Simple Mal'cev algebras over fields of characteristic zero* (San Francisco, 1962) *[497]*, the American mathematician A. A. Sagle showed that loops of elements of modulus 1 of the algebras $\mathbf{O}$, $^1\mathbf{O}$ and $\mathbf{O} \otimes \mathbf{C}$ are the only real analytic loops. It follows that the only semisimple real analytic loops are the loops of elements of modulus 1 of the algebras $^0\mathbf{O}$ and $^{10}\mathbf{O}$ and of the tensor products $\mathbf{O} \otimes {}^0\mathbf{C}$, $^1\mathbf{O} \otimes {}^0\mathbf{C}$ and $^0\mathbf{O} \otimes \mathbf{C}$.

Jordan Algebras

Jordan algebras, introduced by the German physicist and mathematician Paskual Jordan (1902–1980) in the paper *On a class of nonassociative hyper-*

complex algebras (Über eine Klasse nichtassoziativer hypercomplexer Algebren. Göttingen, 1933) *[252]*, have important geometric applications. Jordan
algebras are commutative and nonassociative algebras with multiplication
$A \circ B = B \circ A$ satisfying the condition $(A \circ (B \circ B)) \circ B = (A \circ B) \circ (B \circ B)$. For
each associative and certain nonassociative algebras $\mathbf{A}$ there is the Jordan
algebra $\mathbf{A}^+$ with multiplication $A \circ B = 1/2(AB + BA)$. For Jordan algebras
one defines simple and quasisimple algebras in the same way as for associative,
alternative and Lie algebras. The classification of complex simple Jordan
algebras was carried out by Abraham Adrian Albert (1905–1972) in the paper
A structure theory for Jordan algebras (Baltimore, 1947) *[12]* (see also the
papers *[247]* and *[248]* of Nathan Jacobson (b. 1910) and F. D. Jacobson);
the classification of real Jordan algebras was carried out by Isaĭ L'vovič
Kantor (b. 1936) in the paper *Transitive-differential groups of transformations
and invariant connections on homogeneous spaces* (Tranzitivno-differencial'nye
gruppy preobrazovaniĭ i invariantnye svyaznosti na odnorodnyh prostranstvah. Moscow, 1966) *[262]*; and the classification of quasisimple Jordan
algebras was carried out by M. P. Zamahovskiĭ in the paper *Bireductive
spaces, simple and quasisimple Jordan algebras* (Bireduktyvnye prostranstva,
prostye i kvaziprostye iordanovy algebry. Moscow, 1972) *[476]*. If in the
algebra $\mathbf{A}$ there is the involution $A \to A^T$, $A \to \bar{A}^T$ or $A \to \tilde{A}^T$, then the
elements $A = A^T$, $A = \bar{A}^T$ and $A = \tilde{A}^T$ also form Jordan algebras designated
by SA^+, $\bar{S}A^+$ and $\tilde{S}A^+$, respectively; if $\mathbf{A}$ is a matrix algebra and these involutions are replaced by $A \to E_l A^T E_l$, $A \to E_l \bar{A}^T E_l$, $A \to E_l \tilde{A}^T E_l$ $(E_l = (\varepsilon_i \delta_{ij}))$ or
by $A \to J A^T J$ $(A \in \mathbf{R}_n$ or $\mathbf{C}_n$, J is a quasidiagonal matrix with submatrices
$\begin{bmatrix} 0 & 1 \\ -1 & 0 \end{bmatrix}$ along the main diagonal) and $A \to i\bar{A}^T i$ $(A \in \mathbf{H}_n)$, then we denote
these Jordan algebras as $^l SA^+$, $^l \bar{S}A^+$, $^l \tilde{S}A^+$, $Sp A^+$, $\bar{S}p A^+$, respectively. In our
notations all simple complex Jordan algebras are: (A) $\mathbf{C}_n^+$, (B) SC_n^+, (C) $Sp C_n^+$,
(D) $(\mathbf{K}_n \otimes \mathbf{C})^+$, (E) $(\mathbf{O}_3 \otimes \mathbf{C})^+$ and all simple real Jordan algebras are:
(A) SC_n^+, $^l SC_n^+$, $\mathbf{R}_n^+$, $\mathbf{H}_{n/2}^+$, (B) SR_n^+, $^l SR_n^+$, $Sp H_{n/2}$, (C) SH_n, $^l SH_n$, $Sp R_{2n}$, (D)
$\mathbf{K}_n^+$, $^l \mathbf{K}_n^+$, (E) $\bar{S}\mathbf{O}_3$, $^l \bar{S}\mathbf{O}_3$, $\bar{S}^1 \mathbf{O}_3$. In the algebras $\mathbf{K}_n^+$ and $^l \mathbf{K}_n^+$ one considers
only elements of the form $a + \sum a_i e_i$.
   Jordan algebras are closely connected with projective, non-Euclidean and
symplectic geometries. In Table V we give in the first line the simple real
Jordan algebra, in the second line—the spaces whose fundamental groups are
isomorphic to groups of automorphisms of these algebras, in the third line—
the spaces whose fundamental groups are isomorphic to groups of linear
transformations of these algebras, and in the fourth line—the spaces whose
fundamental groups are isomorphic to groups to fractional linear transformations of these algebras.
   The theory of transformations of simple and quasisimple Jordan algebras
was constructed by Raisa Porfir'evna Vyplavina (b. 1938), I. I. Kolokol'ceva
and Viktor Viktorovič Malyutin (b. 1945) in the paper *Fractional linear
transformations of Jordan algebras* (Drobno-lineĭnye preobrazovaniya ior-
danovyh algebr. Kazan 1974) *[473]*. The second and third lines of this table
have no meaning for $\mathbf{K}_1^+ = \mathbf{K}_1 = \mathbf{C}$ and $^1\mathbf{K}_1^+ = {}^1\mathbf{K}_1 = {}^1\mathbf{C}$ but the fourth line

Table V

A				B		
SC_n^+	$^lSC_n^+$	R_n^+	H_n^+	SR_n^+	$^lSR_n^+$	SpH_n^+
$C\bar{S}_{n-1}$	$C^l S_{n-1}$	P_{n-1}	HP_{n-1}	S_{n-1}	$^l S_{n-1}$	$H\bar{S}p_{n-1}$
CP_{n-1}		$^1CP_{n-1}$	$(H \otimes {}^1C)P_{n-1}$	P_{n-1}		HP_{n-1}
$C^n\bar{S}_{2n-1}$		P_{2n-1}	HP_{2n-1}	Sp_{2n-1}		$H^n\bar{S}_{2n-1}$

C			D		E		
SH_n^+	$^lSH_n^+$	SpR_{2n}^+	K_n^+	$^lK_n^+$	SO_3^+	$^1SO_3^+$	$S^1O_3^+$
$H\bar{S}_{n-1}$	$H^l S_{n-1}$	Sp_{2n-1}			$O\bar{S}_2$	$O^1\bar{S}_2$	$^1O\bar{S}_2$
HP_{n-1}		P_{n-1}	S_{n-1}	$^l S_{n-1}$	OP_2		$^1OP_2$
$H\bar{S}p_{2n-1}$		$^n S_{2n-1}$	$^1 S_{n+2}$	$^{l+1} S_{n+2}$	$O\bar{S}p_5$		$^1O\bar{S}p_5$

preserves its meaning for these algebras: the fractional linear transformations of the planes of complex and split complex variables

$$z' = \frac{az + b}{cz + d} \qquad (10.29)$$

represent the motions of the spaces $^1S_3$ and $^2S_3$, respectively. These representations follow from the connection between the absolute of $^1S_3$ and *inversive geometry* in the plane R_2, mentioned in Klein's *Erlangen program* (see p. 343), and the analogous connection between the absolute of $^2S_3$ and the geometry in the plane $^1R_2$, because the transformations (10.29) of complex and split complex variables represent the *circular transformations* of the planes R_2 and $^1R_2$, respectively. We note that inversion in a circle:

$$Az\bar{z} + B\bar{z} + \bar{B}z + C = 0 \quad (A = \bar{A}, C = \bar{C}) \qquad (10.30)$$

(A, C real, B complex or split complex) in these planes has the form

$$z' = \frac{\bar{B}z + C}{-Az - B} \qquad (10.31)$$

and inversion in a circle

$$Az\bar{z} + B\bar{z} + \bar{B}z + C = 0 \quad (A = -\bar{A}, C = -\bar{C}) \qquad (10.32)$$

—the form

$$z' = \frac{\bar{B}z - C}{Az + \bar{B}}. \qquad (10.33)$$

There is an analogous interpretation for the motions of the space $^1R_3$ and the transformations (10.29) in the plane of a dual variable, but equations

(10.30) and (10.32) are not equivalent—(10.30) represents a pair of parallel *isotropic lines* of the *flag plane* F_2 and (10.32) represents a *cycle* (having the form of a parabola). The transformations (10.31) and (10.33) represent inversions in a pair of lines and a cycle, respectively. Therefore every *quadratic birational transformation* (see p. 347) *is a combination of inversions* (10.31) *and* (10.33) *in the planes of complex, split complex and dual variables and collineations* (see the paper *Quadratic Cremona transformations and complex numbers* (Kvadratičnye kremonovy preobrazovaniya i kompleksnye čisla. Moscow, 1952) by the author and Zalman Alterovič Skopec (1917–1984) *[484]*. In this paper we do not consider the quasisimple Jordan algebras other than the algebra ^0C but all these algebras and their geometric interpretations are considered in the paper *[473]*. In M. A. S. Javadov's (Džavadov) papers *[160, 161]* the fractional linear transformations of the algebras $\mathbf{K}_n$ and $^l\mathbf{K}_n$ for $n \geq 1$ form the geometric interpretation of spinor representations of groups of motions of the spaces S_n and ^lS_n.

Note that if in the decomposition (9.7) of the Lie algebra of a simple Lie group $\lambda = 1$, then the subspace g_{-1} has the structure of a Jordan algebra. In particular, matrices of the Jordan algebras $\mathbf{R}_n^+$, $S\mathbf{R}_n^+$ and $Sp\mathbf{R}_n^+$ are matrix coordinates of $(m - 1)$-dimensional planes of P_{2n-1}, isotropic planes of Sp_{2n-1} and planes on the absolute of $^nS_{2n-1}$. The fractional linear transformations of symmetric matrices forming the Jordan algebra $S\mathbf{R}_n$ were studied by Hua Loo Keng *[235b]*. Hua Loo Keng has shown that these transformations represent the symplectic transformations of the space Sp_{2n-1}; in this case, the matrices of $S\mathbf{R}_n^+$ are matrix coordinates of isotropic $(m - 1)$-dimensional planes of Sp_{2n-1} called *Lagrangian submanifolds* of this space (for the origin of this term and for important connections between symplectic geometry and mechanics see the paper of A. Weinstein *[623]*).

Geometry of the Exceptional Lie Groups

The alternative and Jordan algebras are closely connected with the exceptional Lie groups.

The compact group G_2 is the group of automorphisms of the algebra $\mathbf{O}$ of octaves and the noncompact group G_2 is the group of automorphisms of the algebra $^1\mathbf{O}$ of split octaves. These groups leave invariant the real axes of those algebras and the seven-dimensional planes of elements $\alpha = -\bar{\alpha}$ which are, respectively, the spaces R_7 and $^3R_7$; these groups are transitive on the unit spheres in those spaces. If we identify the antipodal points of these spheres, then we obtain the spaces S_6 and $^3S_6$. These spaces, whose fundamental groups are thought of as the groups of automorphisms of $\mathbf{O}$ and $^1\mathbf{O}$ respectively, are called *G-elliptic* and *G-hyperbolic spaces* and are denoted by Sg_6 and $^3Sg_6$.

The geometry of the spaces Sg_6 and $^3Sg_6$ was studied by Nadežda Nikolaevna Adamuško (b. 1940) in the paper *Geometry of the simple and quasi-*

simple groups G_2 (Geometriya prostyh i kvaziprostyh grupp G_2. Moscow, 1969) *[6]* and by Rimma Gumarovna Tlupova (b. 1956) in the paper *Curves in G-elliptic 6-space* (Krivye v G-elliptičeskom 6-prostranstve. Tbilisi, 1980) *[486]*. The structures of the algebras **O** and ^{1}**O** induce nonintegrable almost-complex structures at all points of Sg_6 and almost-complex or almost-product structures at all points of $^3Sg_6$. The only symmetry figures of these spaces are the *holomorphic two-dimensional planes* of the almost-complex and almost-product structures. The α_1-figures of the space $^3Sg_6$ are the *points* of the absolute, the α_2-figures are the *special lines* on the absolute; the geometry of these figures was studied by Gerhard Johan Schellekens in the paper *On a hexagonic structure* (Amsterdam, 1962) *[506]*.

Compact and noncompact groups in the class F_4 can be viewed as groups of motions of the octave elliptic plane $\mathbf{O}\bar{S}_2$, *the octave hyperbolic plane* $\mathbf{O}^1\bar{S}_2$ *and the split octave elliptic plane* $^1\mathbf{O}\bar{S}_2$. These planes were introduced by, respectively, Hans Freudenthal in the paper *Octaves, exceptional groups and octave geometry* (Oktaven, Ausnahmegruppen und Oktavengeometrie. Utrecht, 1951) *[185]*, Jacques Tits in the papers *The octave projective plane and the exceptional Lie groups* (Le plan projectif des octaves et les groupes de Lie exceptionnels. Brussels, 1953) *[582]*, *On the classification of semisimple algebraic groups* (Sur la classification des groupes algébriques semisimples. Paris, 1959) *[585]*, and by David Borisovič Persic (b. 1940) in the paper *Geometries over antioctaves* (Geometrii nad antioktavami. Moscow, 1967) *[416]*. Freudenthal coordinatizes the points of the plane $\mathbf{O}S_2$ by elements of the Jordan algebra $\bar{S}O_3^+$. The latter are Hermitian symmetric matrices (x_{ij}), $x_{ij} = \bar{x}_{ji}$, with $x_{ij}x_{ji} = x_{ii}x_{jj}$. These conditions ensure that all x_{ij} belong to a single associative subalgebra. Therefore this coordinatization is equivalent to a coordinatization by three coordinates x_0, x_1, x_2 from one associative subalgebra defined up to a multiplication $x_i \to x_i l$, where l belongs to the same subalgebra. Under these conditions $x_{ij} = x_i \bar{x}_j$. In this notation the theory of the planes $\mathbf{O}\bar{S}_2$, $\mathbf{O}^1\bar{S}_2$ and $^1\mathbf{O}\bar{S}_2$ is analogous to the theory of the spaces $\mathbf{H}\bar{S}_n$, $\mathbf{H}^l\bar{S}_n$ and $^1\mathbf{H}\bar{S}_n$, respectively.

The two noncompact groups in the class E_6 can be viewed as the groups of collineations of the octave and split octave projective planes $\mathbf{O}P_2$ and $^1\mathbf{O}P_2$ introduced in the papers *[185]* and *[416]* by Freudenthal and Persic, respectively. The coordinatizations of the planes $\mathbf{O}P_2$ and $^1\mathbf{O}P_2$ are the same as those of the planes $\mathbf{O}\bar{S}_2$, $\mathbf{O}^1\bar{S}_2$ and $^1\mathbf{O}S_2$.

In the paper *The compact simple Lie group E_6 as the group of motions of the complex octave non-Euclidean plane* (Kompaktnaya prostaya gruppa Li E_6 kak gruppa dviženiĭ kompleksnoĭ oktavnoĭ neevklidovoĭ ploskosti. Baku, 1954) *[471]*, the author showed that the compact group in the class E_6 can be viewed as the group of motions of the bioctave elliptic plane $(\mathbf{O} \otimes \mathbf{C})\tilde{\bar{S}}_2$ defined in a manner analogous to that of the plane $\mathbf{O}\bar{S}_2$. The non-Euclidean planes $(\mathbf{O} \otimes \mathbf{C})^1\tilde{\bar{S}}_2$, $(^1\mathbf{O} \otimes \mathbf{C})\tilde{\bar{S}}_2$, $(\mathbf{O} \otimes {}^1\mathbf{C})\tilde{\bar{S}}_2$ and $(^1\mathbf{O} \otimes {}^1\mathbf{C})\tilde{\bar{S}}_2$ are defined in a similar manner.

What explains the absence of multidimensional analogs of these planes is that, as noted by Hilbert in his *Foundations of geometry [230]*, Desargues's theorem on homologous triangles holds only in geometries in which the co-efficients form an associative system. Since in spaces of dimension greater than two Desargues's theorem is a consequence of the incidence axioms of projective geometry (in Hilbert's work, the axioms in groups I and II and the strong parallel axiom), the spaces on nonassociative skew fields in which the incidence axioms of projective geometry hold must be of dimension two.

The geometric interpretation of the exceptional simple groups E_7 and E_8 was realized by H. Freudenthal in a series of papers under the common title *The connections between E_7 and E_8 and the octave plane* (Beziehungen der E_7 and E_8 zur Oktavenebene. Amsterdam, 1954–1959) *[186]*, by the author in the paper *Geometric interpretation of the compact simple groups in the class E* (Geometričeskaya interpretaciya kompaktnyh prostyh grupp Li klassa E. Moscow, 1956) *[472]*, and by I. L. Kantor in the paper *Models of exceptional Lie algebras* (Modeli osobyh algebr Li. Moscow, 1973) *[264]*. In the paper *[186]* Freudenthal introduced the so-called *Freudenthal magic square*

$$\begin{array}{|cccc|}\hline B_1 & A_2 & C_3 & F_4 \\ A_2 & A_2 \times A_2 & A_5 & E_6 \\ C_3 & A_5 & D_6 & E_7 \\ F_4 & E_6 & E_7 & E_8 \\ \hline \end{array}$$

The first line of this square is the groups of motions of the elliptic planes S_2, $C\bar{S}_2$, $H\bar{S}_2$ and $O\bar{S}_2$; the second one the groups of collineations of the projective planes P_2, CP_2, HP_2 and OP_2; the third one the groups of symplectic transformations of the spaces Sp_5, $C\bar{S}p_5$, $H\bar{S}p_5$ and $O\bar{S}p_5$. (The latter space needs to be defined only to the extent that the elements of the Jordan algebra $\bar{S}O_3$ are interpreted as isotropic two-dimensional planes of this space.) Freudenthal calls the geometries of the fourth line *metasymplectic geometries* and considers four geometric figures of these geometries: *symplecta*, namely sets of isotropic two-dimensional planes of the spaces Sp_5, $C\bar{S}p_5$, $H\bar{S}p_5$ and $O\bar{S}p_5$, respectively, *planes*, namely isotropic planes in these spaces, *lines* and *points*, namely lines and points of these projective planes. In the paper *Metasymplectic geometries as geometries on absolutes of Hermitian planes* (Metasimplektičeskie geometrii kak geometrii na absolutah érmitovyh ploskosteĭ. Moscow, 1983) *[463]* (see also *[485]*). Tamara Andreevna Stepaško (b. 1949) has shown that these four figures are elements of parabolic spaces; specifically, *for F_4 points are α_4-figures, lines are α_3-figures, planes are α_2-figures, symplecta are α_1-figures; for E_6 points are (α_1, α_6)-figures, lines are (α_3, α_5)-figures, planes are α_4-figures, symplecta are α_2-figures; for E_7 points are α_6-figures, lines are α_4-figures, planes are α_3-figures, symplecta are α_1-figures;*

for E_8 points are α_1-figures, lines are α_6-figures; planes are α_7-figures, symplecta are α_8-figures. For split groups the sets of points can be considered as absolutes of the planes ${}^1O\bar{\tilde{S}}_2$, $({}^1O \otimes {}^1C)\bar{\tilde{S}}_2$, $({}^1O \otimes {}^1H)\bar{\tilde{S}}_2$ and $({}^1O \otimes {}^1O)\bar{\tilde{S}}_2$, for the groups considered by Freudenthal—as absolutes of the planes ${}^1O\bar{\tilde{S}}_2$, $({}^1O \otimes {}^1C)\bar{\tilde{S}}_2$, $({}^1O \otimes H)\bar{\tilde{S}}_2$ and $({}^1O \otimes O)\bar{\tilde{S}}_2$ (see Figures 110 FI, EI, II, V, VI, VIII, IX; all the figures considered by Freudenthal in connection with the groups he investigated were real). The decompositions (9.7) defined a fibration of these absolutes, the tangent spaces to lines and planes lie completely in the tangent space to the base of this fibration, the tangent space to the symplecta cuts the tangent space to this base in 4-, 8-, 16- and 32-dimensional planes and the tangent space to the fiber-in a real line. There are similar interpretations of the quasisimple Lie groups of the exceptional classes. The Euclidean planes $O\bar{R}_2$, $(O \otimes C)\bar{\tilde{R}}_2$, $(O \otimes H)\bar{\tilde{R}}_2$ and $(O \otimes O)\bar{\tilde{R}}_2$ were constructed by the author in the paper *A geometric interpretation of the quasisimple exceptional Lie groups in the classes E_7 and E_8* (Geometričeskaya interpretaciya kvaziprostyh osobyh grupp Li klassov E_7 i E_8. Moscow, 1973) *[472]*. In *[417, 418]* D. B. Persic constructed geometric interpretations of the quasisimple groups in the class F_4 as non-Euclidean planes over the algebras ${}^0O$ and ${}^{01}O$ and in *[297]* Tat'yana Anatol'evna Kuznecova (b. 1942) constructed analogous interpretations of the quasisimple groups in the class E_6.

Applications of Linear Representations of Lie Groups to Physics

Linear representations of Lie groups have important applications to quantum mechanics. To this problem is devoted the famous book of H. Weyl *Theory of groups and quantum mechanics* (Gruppentheorie und Quantummechanik. Leipzig, 1931) *[630]*.

Here we shall consider two cases of these applications—applications of spinor representations of the group of Lorentz transformations and of the group of conformal transformations of Minkowski's space. The first of these groups is isomorphic to the group of rotations of Minkowski's space-time ${}^1R_4$ and to the group of motions of the Lobačevskian space ${}^1S_3$, and the second to the group of motions of the hyperbolic space ${}^2S_5$. The spinor representation of the first group is realized by the matrices of the algebra C_2, the spinor group of the group of motions of ${}^1S_3$ is the group CSL_2 and spinors of this group are the vectors of complex two-dimensional space. If we replace in the formula (10.29) the complex numbers z and z' by the ratios z^0/z^1 and ${}'z^0/{}'z^1$, then we obtain the formulas of the spinor representation of the group of motions of ${}^1S_3$ in the form

$$'z^0 = az^0 + bz^1, \qquad 'z^1 = cz^0 + dz^1. \tag{10.34}$$

The point z of the complex plane and the spinor $\{z^0, z^1\}$ correspond to a point of the absolute of ${}^1S_3$ and to an isotropic vector of the space ${}^1R_4$:

$$s^0 = \frac{1}{2}(z\bar{z} + 1), \qquad s^1 = \frac{1}{2}(z + \bar{z}),$$

$$s^2 = \frac{1}{2i}(z - \bar{z}), \qquad s^3 = \frac{1}{2}(z\bar{z} - 1)$$

or

$$s^0 = z^0\bar{z}^0 + z^1\bar{z}^1, \qquad s^1 = z^0\bar{z}^1 + z^1\bar{z}^0,$$

$$s^2 = \frac{1}{i}(z^0\bar{z}^1 + z^1\bar{z}^0), \qquad s^3 = (z^0\bar{z}^0 + z^1\bar{z}^1). \tag{10.35}$$

P.A.M. Dirac in his *The principles of quantum mechanics* [152] has shown that the existence of the *spin* of an electron finds expression in the division of the wave functions ψ^α of the electron into two groups such that the integral of the sum of the moduli of the values of the functions in either group is equal to the probability that the particle is in the given domain and its spin is, respectively, positive or negative. Therefore the wave field of an electron is given at every point of the space-time domain by four complex-valued functions ψ^0, ψ^1, ψ^2, ψ^3 such that under a Lorentz transformation the functions ψ^0, ψ^1 transform like the coordinates of a spinor of $^1R_4$, i.e., by the formulas (10.34), and ψ^2, ψ^3 like $\bar{\psi}^0$, $\bar{\psi}^1$. The formulas (10.35) for the spinor $\{\psi^0, \psi^1\}$ have the form

$$\left.\begin{array}{ll} s^1 = \psi^0\bar{\psi}^1 + \psi^1\bar{\psi}^0, & s^2 = i(\psi^1\bar{\psi}^0 - \psi^0\bar{\psi}^1) \\ s^3 = \psi^0\bar{\psi}^0 - \psi^1\bar{\psi}^1, & s^4 = \psi^0\bar{\psi}^0 + \psi^1\bar{\psi}^1. \end{array}\right\} \tag{10.36}$$

Thus the functions ψ^0, ψ^1 define the isotropic vector **s** with coordinates (10.36) and the functions ψ^2, ψ^3 define the isotropic vector **t** with coordinates

$$\left.\begin{array}{ll} t^1 = \psi^2\bar{\psi}^3 + \psi^3\bar{\psi}^2, & t^2 = i(\psi^2\bar{\psi}^3 - \psi^3\bar{\psi}^2), \\ t^3 = \psi^2\bar{\psi}^2 - \psi^3\bar{\psi}^3, & t^4 = \psi^2\bar{\psi}^2 + \psi^3\bar{\psi}^3. \end{array}\right\} \tag{10.37}$$

At every point of the wave field of an electron, there is defined the vector **j** of current density, whose coordinates j^1, j^2, j^3 are the coordinates of the spatial vector of current density and j^4 is the charge density *[447, pp. 332–334]*. These coordinates have the form

$$\left.\begin{array}{l} j^1 = \psi^0\bar{\psi}^1 + \psi^1\bar{\psi}^0 + \psi^2\bar{\psi}^3 + \psi^3\bar{\psi}^2, \\ j^2 = i(\psi^1\bar{\psi}^0 - \psi^0\bar{\psi}^1 + \psi^2\bar{\psi}^3 - \psi^3\bar{\psi}^2), \\ j^3 = \psi^0\bar{\psi}^0 - \psi^1\bar{\psi}^1 + \psi^2\bar{\psi}^2 - \psi^3\bar{\psi}^3, \\ j^4 = \psi^0\bar{\psi}^0 + \psi^1\bar{\psi}^1 + \psi^2\bar{\psi}^2 + \psi^3\bar{\psi}^3. \end{array}\right\} \tag{10.38}$$

The comparison of the formula (10.38) with (10.36) and (10.37) shows that

$$\mathbf{j} = \mathbf{s} + \mathbf{t}$$

i.e. the vector **j** of current density is the sum of isotropic vectors **s** and **t**. Thus at every point of the wave field of an electron there are defined two two-dimensional planes—the plane of the vectors **s** and **t** and a plane orthogonal to it. Also, at every point of the electromagnetic field in space-time there are defined two orthogonal planes: at every point of this field there is defined the tensor $F^{ij} - F^{ji}$ connected with the vectors **E** and **H** of the electric and magnetic fields by the relations $F^{0i} = cE^i$, $F^{32} = H^1$, $F^{13} = H^2$, $F^{21} = H^3$. But the skew symmetric tensor F^{ij} is equivalent to a force screw, or a kinematic twist, in the space $^1S_3$ in the hyperplane at infinity of $^1R_4$, and, as was shown by A. P. Kotel'nikov and E. Study (see p. 398), every such twist is equivalent to two sliding vectors whose lines of action are two reciprocal polars. Therefore, at every point of the electromagnetic field in the space-time there are defined two orthogonal two-dimensional planes whose lines at infinity are these reciprocal polars. Four lines—two real and two imaginary—of the intersection of these planes with the light cone at this point form the *light tetrad*. Therefore, the lines of the isotropic vectors **s** and **t** and two imaginary lines of intersection of the completely orthogonal plane with the same cone can be called the *electronic tetrad*. The lengths of the two sliding vectors at infinity defined by the electromagnetic field are two angles in $^1R_4$. As such, they are invariant under conformal transformations of this space, and *conformal transformations of space-time preserve the electromagnetic field.*

The spinor representation of the group of motions of the space $^2S_5$, which is isomorphic to the group of conformal transformations of $^1R_4$, is realized by the matrices of the algebra C_4. The spinor group of motions of $^2S_5$ is the pseudounitary group C^2SU_4 and the spinors of this group are the vectors of complex four-dimensional space called *twistors*. The isomorphism of the group of motions of $^2S_5$ and the group of motions of $C^2\bar{S}_3$ represented by these matrices is one of the geometric interpretations of the isomorphism of the simple Lie groups D_3 and A_3 (see Figure 111d). The theory of twistors was constructed by R. Penrose in *Relativistic symmetry groups* (Boston, 1974) *[415].*

Finite Geometry

Finite geometry, that is, geometry of spaces containing finite numbers of points, lines and planes, was founded by Gino Fano (1871–1952). In the paper *On the fundamental postulates of projective geometry* (Sui postulati fondamentali della geometria proiettiva. Roma, 1892) *[177a]* Fano gave an example of a geometry independent of axioms of continuity, namely the projective space over the field F_p of residues classes modulo p; in modern notation the space F_pP_k. O. Veblen and W. Bussey in the paper *Finite projective geometries* (New York, 1906) *[612a]* defined the projective space F_qP_n over an arbitrary Galois field F_q ($q = p^v$) and the affine space F_qE_n over the same field. The number of points of F_qE_n is q^n. Since the cellular decomposition of the space over an arbitrary field is

$$P_n = E_n + E_{n-1} + \cdots + E_1 + E_0, \qquad (10.39)$$

the number of points of the space $F_q P_n$ is

$$\text{card } F_q P_n = q^n + q^{n-1} + \cdots + q + 1 = \frac{q^{n+1} - 1}{q - 1}. \qquad (10.40)$$

This formula is a special case of the general formula for the number of m-dimensional planes of the space $F_q P_n$

$$\text{card } F_q P_{n,m} = \frac{(q^{n+1} - 1)(q^n - 1)\ldots(q^{n-m+1} - 1)}{(q^{m+1} - 1)(q^m - 1)\ldots(q - 1)}. \qquad (10.41)$$

In particular, for lines of $F_q P_3$ the formula (10.41) gives

$$\text{card } F_q P_{3,1} = \frac{(q^4 - 1)(q^3 - 1)}{(q^2 - 1)(q - 1)} = (q^2 + 1)(q^2 + q + 1). \qquad (10.42)$$

Note that the famous problem proposed by Thomas Pennington Kirkman (1806–1895) in his paper *Triads made with* 15 *things* (London, 1850) *[279]* admits a geometric interpretation in $F_2 P_3$: in the problem it is required to make a week's schedule of walks for 15 schoolgirls such that every day, every girl be in a "triad" with two other girls. The schoolgirls must be considered as points of $F_2 P_3$ and the "triads" as lines of this space. Formula (10.42) for $q = 2$ shows that in this space card $F_2 P_{3,1} = 35$. The problem reduces to division of this space into 7 modes on 5 lines. The Freudenthal diagram for octave units (Figure 113) is an example of the plane $F_2 P_2$ in which the lines are the 3 sides and 3 altitudes of a triangle and the circle inscribed in it.

There are also finite affine and projective spaces different from $F_q E_n$ and $F_q P_n$. In the case of $F_q E_n$ the number q is called the *order* of the finite affine space with q points on every line and in the case of $F_q P_n$—the order of the projective space with $q + 1$ points on every line. If the order q is equal to $4r + 1$ or $4r + 2$ (r a natural number) and is not a sum of two squares of natural numbers, then no two-dimensional projective plane of order q exists; specifically, no finite two-dimensional projective planes of order 6, 14, 21, 22 exist. Planes of order 2, 3, 4, 5, 7, 8, 11, 13, 16, 17, 19, 23, 25 exist. The question of existence of planes of order 10 and 12 is open. Two-dimensional projective planes can be constructed using Euler's Latin squares *[491]*.

In $F_q P_n$, $q \neq 2^\nu$, there are involutive correlations $u_i = \sum a_{ij} x^i$, where $a_{ij} = a_{ji}$ or $a_{ij} = -a_{ji}$. In the first case the correlation is a polarity of the quadric $\sum_i \sum_j a_{ij} x^i x^j = 0$. Since every nondegenerate quadric in $F_q P_{2n}$ reduces to the form

$$\sum_i x^i x^{2n-i-1} + (x^{2n})^2 = 0, \qquad (10.43)$$

and every nondegenerate quadric in $F_q P_{2n-1}$ reduces to one of the forms

$$\sum_i x^i x^{2n-i-1} = 0 \qquad (10.44)$$

and

$$\sum_i x^i x^{2n-i-3} + (x^{2n-2})^2 + \alpha(x^{2n-1})^2 = 0, \qquad (10.45)$$

where α is a nonsquare in F_q, it follows that in $F_q P_{2n}$ there exists one kind of non-Euclidean space—the space $F_q S_{2n}$ with absolute reduced to the form (10.43), and in $F_q P_{2n-1}$ there exist two kinds of *non-Euclidean spaces*—the spaces $F_q{}^1 S_{2n-1}$ and $F_q S_{2n-1}$ whose absolutes reduce to the forms (10.44) and (10.45), respectively. Just as in the real case, these absolutes can be considered as *conformal spaces* $F_q C_{2n-1}$, $F_q{}^1 C_{2n-2}$ and $F_q C_{2n-2}$. Beniamino Segre (1903–1973), in the paper *Galois geometries* (Le geometrie di Galois. Milan, 1960) *[518a]*, found the cardinalities of all m-dimensional planes on these quadrics. In particular, for $m = 0$

$$\text{card } F_q C_{2n-1} = \text{card } F_q P_{2n-1} = \frac{q^{2n} - 1}{q - 1}, \qquad (10.46)$$

$$\text{card } F_q{}^1 C_{2n} = \frac{(q^{n+1} - 1)(q^n + 1)}{q - 1}, \qquad (10.47)$$

$$\text{card } F_q C_{2n} = \frac{(q^{n+1} + 1)(q^n - 1)}{q - 1}. \qquad (10.48)$$

In particular, $\text{card } F_q C_1 = q + 1$, $\text{card } F_q C_2 = q^2 + 1$, $\text{card } F_q{}^1 C_2 = (q + 1)^2$, $\text{card } F_q C_3 = q^3 + q^2 + q + 1$, $\text{card } F_q C_4 = (q^3 + 1)(q + 1)$, $\text{card } F_q{}^1 C_4 = (q^2 + 1)(q^2 + q + 1) = \text{card } F_q P_{3,1}$. The latter equality is a consequence of the finite analogue of the Plücker transfer.

The *circles* of $F_q C_{2n-1}$, $F_q C_{2n}$ and $F_q{}^1 C_{2n}$ are intersections of the corresponding quadrics in $F_q P_{2n}$ and $F_q P_{2n+1}$ with two-dimensional planes. The *m-dimensional spheres* are intersections of the same quadrics with $(m + 1)$-dimensional planes. The planes of $F_q C_2$ are *extensions* of the planes $F_q E_2$ by one point, and the circles through this point are affine lines extended by this point. Therefore, on every circle of $F_q C_2$ there are $q + 1$ points.

If in $F_q P_{2n-1}$ there is given an involutive correlation $u_i = \sum_j a_{ijk}{}^j x^j$, $a_{ij} = -a_{ji}$, then we obtain the *symplectic space* $F_q Sp_{2n-1}$. In $F_{q^2} P_n$ there is an involutive correlation $u_i = \sum a_{ij}(x^i)^q$, $a_{ij} = (a_{ji})^q$. This correlation defines the *Hermitian non-Euclidean space* $F_{q^2} \bar{S}_n$ which is the finite analogue of $C\bar{S}_n$.

The groups of collineations of $F_q P_n$, of motions of $F_q S_{2n}$, $F_q S_{2n+1}$, $F_q{}^1 S_{2n-1}$ and $F_{q^2} \bar{S}_n$ and of the symplectic transformations of $F_q Sp_{2n-1}$ *are simple finite groups* and together with the *cyclic groups* Z_p exhaust all infinite series of such groups. All these groups are generated by reflections and can be characterized by *Coxeter diagrams* shown in Figure 107 ((a) for $F_q P_n$ and $F_{q^2} \bar{S}_n$, (b) for $F_q S_{2n}$ and $F_q Sp_{2n-1}$, (c) for $F_q S_{2n-1}$ and $F_q{}^1 S_{2n-1}$). These groups are also the groups of automorphisms of Lie algebras over F_q and can be considered as finite analogues of Lie groups with *Satake diagrams* shown in

Figure 110a ($A\,I$ for $F_q P_n$, lower $A\,III$ for $F_{q^2}\bar{S}_n$, lower $B\,I$ for $F_q S_{2n}$, $C\,I$ for $F_q Sp_{2n-1}$, middle and lower $D\,I$ for $F_q S_{2n-1}$ and $F_q^1 S_{2n-1}$). As for real Lie groups in these spaces one can define α_i-figures and $(\alpha_{i_1},\ldots,\alpha_{i_n})$-figures and the Satake diagrams can be interpreted as images of these figures. Such definitions were given by Valentina Georgievna Alyab'eva (b. 1946) in the paper *Geometric interpretation of simple finite groups and their parabolic subgroups* (Geometričeskaya interpretaciya prostyh konečnyh grupp i ih paraboličeskih podgrupp. Har'kov, 1983) *[20]*. There is a large literature devoted to finite geometries, particularly to the geometrics of two-dimensional projective and conformal planes (see the book of Peter Dembowski *Finite geometries [140]*).

The simple finite groups not in the infinite series—the so-called *sporadic groups*—also admit geometric interpretations. Over the field F_q one can define analogues of the exceptional Lie groups G_2, F_4, E_6, E_7, E_8. These groups are also groups generated by reflections and can be characterized by *Coxeter diagrams* (Figure 107, (d) ($r = 6$), (f), (h), (i), (j)) and by *Satake diagrams* (Figure 110b; G for G_2, $F\,I$ for F_4, $E\,I$ and $E\,II$ for two kinds of E_6, $E\,V$ for E_7 and $E\,VIII$ for E_8).

The orders or the simple finite groups which are fundamental groups of the spaces $F_q P_n$, $F_q S_{2n}$, $F_q Sp_{2n-1}$, and $F_q S_{2n-1}$ and of the sporadic groups (which are analogues of the exceptional Lie groups) are equal to

$$\frac{1}{u} q^N \Pi(q^{a_i} - 1),$$

where N is the number of positive roots, u is the order of the centers of the groups of matrices over the field F_q representing these groups, and the numbers a_i are, as Claude Chevalley noted in the paper *On certain simple groups* (Sur certains groupes simples. Sendai, 1954) *[113]*, the same numbers that appear in the expression (9.7) for the Poincaré polynomial of the corresponding compact simple group. The planes $F_q E_2$, $F_q P_2$ and $F_q C_2$ are the cases of *Steiner triple systems* introduced by the German geometer Jacob Steiner (1796–1863) in *A combinatorial problem* (Kombinatorische Aufgabe. Berlin, 1853) *[555]*. The Steiner triple system $S(v, k, t)$ is a set of v elements in which there are *blocks* consisting of k elements and every t elements belong to one and only one block; the number of blocks is $b = \binom{v}{t} \Big/ \binom{k}{t}$. The plane $F_q E_2$ is the Steiner triple system $S(q^2, q, 2)$, the plane $F_q P_2$ is $S(q^2 + q + 1, q + 1, 2)$ and the plane $F_q C_2$ is $S(q^2 + 1, q + 1, 3)$ (the blocks of $F_q E_2$ and $F_q P_2$ are lines, of $F_q C_2$—circles). $F_q C_2$ is the one-point extension of $F_q E_2$. The one-point extension of the system $S(v, k, t)$ is the system $S(v + 1, k + 1, t + 1)$.

The *Mathieu groups* M_{11}, M_{12}, M_{22}, M_{23}, M_{24} are simple finite groups introduced by the French mathematician Émile Mathieu (1835–1890) in the papers *Memoir on the study of functions of many variables* [Mémoire sur l'étude des fonctions de plusieurs quantités, Paris, 1861] *[356]* and *On a five-times-transitive function of 24 variables* (Sur la fonction cinq fois transitive de 24 quantités. *Paris, 1873) [357]*. They are transitive groups of transformations of the Steiner triple systems $S(11,5,4)$, $S(12,6,5)$, $S(22,6,3)$, $S(23,7,4)$ and $S(24,8,5)$, respectively. They are all one-point extensions: the system $S(11,5,4)$ of $F_3 C_2 = S(10,4,3)$; the system $S(12,6,5)$ of $S(11,5,4)$; the system $S(22,6,3)$ of $F_4 P_2 = S(21,5,2)$; the system $S(23,7,4)$ of $S(22,6,5)$; and the system $S(24,8,5)$ of $S(23,7,4)$. These interpretations were studied by Vladimir Vasil'evič Afanas'ev (b. 1950) in $[7,8,9]$ (Yaroslavl', 1979–1981). The blocks of these systems for $t = 3$ are called $circles$, for $t = 4$ $spheres$, and for $t = 5$ $hyperspheres$. The theorems of the geometries of these systems are analogous to the theorems of conformal geometry.

The Steiner triple systems are the cases for more general $block\ designs$, very important for mathematical statistics, and $incidence\ structures$ (see $[491,\ 140,\ 619]$).

Finite spaces can also be constructed over finite rings with divisors of zero. Nadežda Ivanovna Haritonova (b. 1951), in the paper $Projective\ and\ elliptic\ spaces\ over\ a\ Galois\ square$ (Proektivnye i elliptičeskie prostranstva nad kvadratom Galua. Čeboksary, 1976) $[216a]$, constructed projective and non-Euclidean spaces over a $Galois\ square$ isomorphic to the direct sum $F_q \oplus F_q$; these spaces are interpreted as products of two copies of the corresponding space over F_q. Jacques Thas, in the paper $The\ m\text{-}dimensional$ $projective\ space\ P_m(M_n(GF(q)))\ over\ the\ total\ matrix\ algebra\ M_n(GF(q))\ of$ $n \times n\text{-}matrices\ with\ elements\ in\ the\ Galois\ field\ GF(q)$, (Rome, 1971) $[577a]$, constructed the projective space over the matrix algebra $(F_q)_n$ and proved (independently of $[87]$ and $[162]$) that the space $(F_q)_n P_m$ admits an interpretation as a manifold of $(n - 1)$-dimensional planes of $F_q P_{nm+m-1}$. Sof'ya Borisovna Kapralova (b. 1948), in the paper $Dual\ Galois\ spaces$ (Dual'nye prostranstva Galua. Kazan, 1979) $[264a]$, constructed the projective space $F_q(\varepsilon) P_n$ over the dual extension of F_q, i.e. the ring of elements $a + b\varepsilon$, $\varepsilon^2 = 0$, a, b elements of F_q. The extension $F_q(\varepsilon)$ is intermediate between F_{q^2}, an analogue of a complex extension, and $F_q \oplus F_q$, an analogue of a split complex extension. Whereas, in view of (10.41),

$$\text{card}\, F_{q^2} P_{n,m} = \frac{(q^{n+1} - 1)(q^{n+1} + 1)(q^n - i)(q^n + 1)\ldots(q^{n-m+1} - 1)(q^{n-m+1} + 1)}{(q^{m+1} - 1)(q^{m+1} + 1)(q^m - 1)(q^m + 1)\ldots(q - 1)(q + 1)}$$

and

$$\text{card}\, (F_q \oplus F_q) P_{n,m} = \frac{(q^{n+1} - 1)^2 (q^n - 1)^2 \ldots (q^{n-m+1} - 1)^2}{(q^{m+1} - 1)^2 (q^m - 1)^2 \ldots (q - 1)^2},$$

in the space $F_q(\varepsilon) P_n$

$$\text{card}\, F_q(\varepsilon) P_{n,m} = \frac{(q^{n+1} - 1)q^{n+1}(q^n - 1)q^n \ldots (q^{n-m+1} - 1)q^{n-m+1}}{(q^{m+1} - 1)q^{m+1}(q^m - 1)q^m \ldots (q - 1)q}.$$

Analogues of simple Lie groups and of projective, non-Euclidean and symplectic spaces can also be constructed over other fields, such as the field Q_p of p-adic numbers and fields of algebraic numbers. These groups were

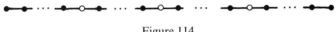

Figure 114

studied by J. Tits in *Classification of semisimple algebraic groups* (Providence, 1966) *[584]*. Clearly, this classification is restricted to the classification of simple groups for which there are analogues of the Satake diagrams for simple real Lie groups. In contradistinction to the field **R**, over these fields there are division algebras that are skew fields of arbitrary dimension m^2; just as the complex extension of the skew field **H** yields C_2, extension of these fields to algebraically closed fields results in all these division algebras becoming matrix algebras. The analogue of the Satake diagram for the projective space over this algebra has the form shown in Figure 114: the white points of this diagram correspond to points, lines and m-dimensional planes of this space, the black points—to the imaginary α_i-figures. The geometry of the projective line over the field $\mathbf{Q}_p$ for $p = 2$ was studied by Jean Pièrre Serre (b. 1926) in the paper *Trees, amalgams and SL_2 [523]*.

Finite Geometries and Betti Numbers

We have mentioned the applications of finite geometries to mathematical statistics. Finite geometries have also been applied to coding theory (see *[327a]*). We have mentioned the remarkable connection between the Poincaré polynomials of compact simple Lie groups and the orders of the finite groups of the same classes found by C. Chevalley *[113]*. We shall now discuss an analogous connection in simpler cases.

If the cellular decomposition of an N-dimensional compact topological manifold M_N is of the form

$$M_N = a_N E_N + a_{N-1} E_{N-1} + \cdots + a_1 E_1 + a_0 E_0, \qquad (10.49)$$

where the E_k are the k-dimensional cells and the coefficients a_k are integers, we shall mean by the *cellular polynomial of M_N* the polynomial

$$p_N(M_N) = a_N t^N + a_{N-1} t^{N-1} + \cdots + a_1 t + a_0.$$

Since the cellular decomposition of the n-dimensional sphere $\sum_n$ is of the form $E_n + E_0$, and that of the projective space P_n is of the form $E_n + E_{n-1} + \cdots + E_1 + E_0$, the cellular polynomials of $\sum_n$ and of P_n are, respectively, of the form

$$p_n(\textstyle\sum_n) = t^n + 1;$$
$$p_n(P_n) = t^n + t^{n-1} + \cdots + t + 1. \qquad (10.50)$$

The cellular decomposition of the manifold $P_{n,m}$ of m-dimensional planes of the space P_n was found by C. Ehresmann in the previously mentioned paper

On the topology of certain homogeneous spaces [164]. As Ehresmann has shown the coefficient a_k in the cellular decomposition (10.49) of this manifold is equal to the number of sets of $m + 1$ integers $(n_0, n_1, \ldots, n_m)$ such that $n_0 + n_1 + \cdots + n_m = k$ and $n - m \geq n_m \geq n_{m-1} \geq \cdots \geq n_1 \geq n_0 \geq 0$. In this case $N = (m + 1)(n - m)$, and the cellular polynomial can be written in the form

$$p_{(m+1)(n-m)}(P_{n,m}) = \frac{(t^{n+1} - 1)(t^n - 1)\ldots(t^{n-m+1} - 1)}{(t^{m+1} - 1)(t^m - 1)\ldots(t - 1)}. \qquad (10.51)$$

Comparison of formula (10.51) with (10.41) shows that the expression (10.51) coincides with card $F_t P_{n,m}$ (see [462]).

Ehresmann has also noted that the nonzero Betti numbers b_{2k} of the manifold $CP_{n,m}$ are equal to the numbers a_k (the Betti numbers b_{2k-1} of this manifold are equal to zero). (It is clear that for CP_n, $b_{2k} = 1$, $b_{2k-1} = 0$ are analogously connected with the coefficients a_k of the polynomial $p_n(P_n)$, and that for $\sum_n$, $b_k = a_k$. This connection between the Betti numbers of these simply-connected manifolds and the cardinalities of the corresponding finite algebraic manifolds points to the cause of the remarkable connection noted by C. Chevalley.

Quantum Physics and Geometry

Finite and discrete geometries have not only theoretical and narrow practical interest. The development of modern physics shows that the theory of physical space also leads to finite and discrete ideas. We saw in chapter 7 that the study of elementary particles of quantum physics is carried out by means of linear operators in infinite-dimensional Hilbert space. This implies that it is advantageous to apply to quantum physics the theory of representations of the Lorentz group and other Lie groups by means of these operators.

A number of difficulties of modern quantum physics leads to the idea of revision of the *Cantor continuum* that lies at the foundations of Euclidean, non-Euclidean, Riemannian, and pseudo-Riemannian geometries. In this connection some physicists have come back to Mach's idea that "space and time represent just an imaginary surface and, in all likelihood, consist of discontinuous but not sharply distinguishable elements" *[347, p. 446].* The question of the need to replace the notion of a continuous space by a concrete one in order to avoid difficulties due to "divergences" was posed by Victor Amazaspovič Ambarcumian (b. 1908) (who later became a famous astronomer) and Dmitriĭ Dmitrievič Ivanenko (b. 1904) in the paper *On the question of avoiding infinite feedback of an electron* (Zur Frage nach Vermeidung der unendlichen Selbstwirking des Elektrons. Leipzig, 1930) *[21].* Attempts to construct a "quantum space and time" have been made by Soviet as well as other physicists (see the book of A. A. Sokolov and D. D. Ivanenko

[546, pp. 486 and 601], the book of Silberstein *[539]*, the papers of Snyder *[544]*, and V. L. Averbah and B. V. Medvedev *[32]* and the monograph of A. N. Vyal'cev *[616]*.

The failure of these attempts is explained by the fact that the notion that space and time "consist of discontinuous elements" is just as onesided as the notion of a purely continuous space. Our future notions of space and time will, undoubtedly, reflect to a much greater degree than contemporary models the unity of continuity and discreteness. The deep connections between discrete and continuous geometry discovered by modern mathematicians offer hope of further progress in this direction (see *[13]*).

References

1. Abbasov, N. T. Bicomplex elliptic spaces—Uč. zap. Azerb. gos. un-ta, seriya fiz.-mat. nauk, 1962, **2**, 3–9 (Russian).
2. Abbasov, N. T. Biquaternion elliptic spaces.—Uč. zap. Azerb. gos. un-ta, ser. fiz.-mat nauk, 1963, **2**, 3–9 (Russian).
3. Abbasov, N. T., Kolokol'ceva, I. I. Biquasisimple algebras, biquasimatrices and spinor representations of biquasinon-Euclidean motions.—Izv. vuzov, Matematika, 1975, **11**, 1–8 (Russian).
4. Abel, N. H. Oeuvres complètes, vol. 1–2. Publ. par L. Sylow et S. Lie. Christiania, 1881.
5. Abū'l-Wafa, al-Būzjānī. The book about what a tradesman must know of geometric construction.—Fiziko-mat. nauki v stranah Vostoka, 1966, **1**, 56–140. Translated by S. A. Krasnova (Russian).
6. Adamuško, N. N. Geometry of simple and quasisimple groups G_2.—Uč. zap. MOPI im. N. K. Krupskoĭ, 1969, **253**, 23–42 (Russian).
7. Afanas'ev, V. V. Construction of extended finite projective planes.—Konstr. alg. geom. Yaroslavl', 1979, **180**, 3–10 (Russian).
8. Afanas'ev, V. V. Extension of extended finite projective planes.—Konstr. alg. geom. Yaroslavl', 1980, **190**, 10–21 (Russian).
9. Afanas'ev, V. V. The triply extended finite projective plane $E^3 P(2,4)$.—Konstr. alg. geom. Yaroslavl', 1981, **194**, 16–24 (Russian).
10. Ahmedov, A. On the author of the Roman edition of "Euclid's Elements."—Izv. AN UzSSR, seriya fiz.-mat. nauk, 1971, **15** (No. 5), 9–12 (Russian).
10a. Ahmedov, A. A., ad-Dabbāgh, J., Rosenfeld, B. A. Istanbul manuscripts of al-Khwārizmī's treatises.—Erdem, J. of the Atatürk Culture Center, 1987, **3**, N7, 163–186.
11. Ahmedov, A., Rozenfel'd, B. A. "Cartography." One of the earliest preserved works of Bīrūnī.—in: Mathematics in the medieval East. Ed. by S. H. Siraždinov. Taškent, Fan, 1978, 127–153 (Russian).
12. Albert, A. A. A structure theory for Jordan algebras.—Ann. Math., 1947, **48**, 546–567.
13. Aleksandrov, A. D. Lenin's dialectics and mathematics.—Priroda, 1951, **N1**, 5–15 (Russian).
14. Alexandroff, P. S. Zur Begründung der n-dimensionalen Topologie.—Math. Ann., 1925, **94**, 296–308.

15. Alexandroff, P. S., Hopf, H. Topologie. Berlin, 1935.
16. Alexandroff, P., Urysohn, P. Mémoire sur les espaces topologiques compacts.—Verhandelingen der Konink. Akad. van Wettensch, Amsterdam, 1929, **14**, 1–96.
17. Alfonso. Rectifier of the curved.—Ed. and transl. into Russian by G. M. Gluskina, comm. by G. M. Gluskina, S. Ya. Lur'e, and B. A. Rozenfel'd. Moscow, Nauka, 1983.
18. [Alfonso X]. Libros del saber de astronomia del rey Alfonso de Castilla.—Capil., anot., y coment. par D. Manuel Rico y Sinobas, vol. 3, Madrid, 1864.
19. Alhazenis Arabis. Opticae thesaurus. F. Risner (Ed.). Basileae, 1572.
20. Alyab'eva, V. G. Geometric interpretation of finite simple groups and their parabolic subgroups.—Ukr. geom. sbornik., 1983, **26**, 3–6 (Russian).
21. Ambarzumian, V., Iwanenko, D. Zur. Frage nach Vermeidung der unendlichen Selbstrückwirkung des Elektrons.—Z. Phys., 1930, **64**, 563–567.
22. Andreeva, L. V., Šestyreva L. V. Limit symplectic spaces.—Uč. zap. Kolomen. ped. inst., 1964, **8**, 23–44 (Russian).
23. Apollonium de Perge. Les coniques. Trad. P. Ver Eecke. Paris, 1959.
24. Apollonius of Perga. Treatise on conic sections.—Ed. T. L. Heath. N.Y., Bernes–Nolle, n.d.
25. Archimedes. Works.—Ed. and transl. by T. L. Heath. N.Y.: Dover, 1953.
26. Archimedis opera omnia cum commentariis Eutocii.—Ed. et lat. vert. J. L. Heiberg. Vol. 1–3. Lipsiae: Teubner, 1880–1881.
27. Argand, R. Essai sur une manière de représenter les quantités imaginaires dans les constructions géométriques. Paris, 1874.
28. Aristotelis opera cum commentariis Averrois, t. 1–4. Venetiis, 1560.
29. Aristotle. Works.—Translated under the editorship of W. D. Ross, vol. 1–12, Oxford Univ. Press, 1928–1952.
30. Astrahanceva, L. N. History of the algebra of triplets.—Trudy XIV nauč. konf. aspirantov i mladših naučnyh sotrudnikov IIEiT AN SSSR, sekciya mat.-meh., 1971, 70–77 (Russian).
31. Autolikos von Pitane. Der rotierende Kugel und Aufgang und Untergang der Gestirne. Übers. A. Czwalina. Leipzig, 1931.
32. Averbah, V. L., Medvedev, B. V. On the theory of quantized spacetime.—Dokl. AN SSSR, 1949, **64**, **N1**, 41–44 (Russian).
33. Avicenne. Le livre de science, vol. 1–2.—Trad. par M. Achena et H. Massé. Paris, 1955–1958.
34. [Azarchel]. Saphaeae recentis res doctrinae patris Abrysakh Azarchelis summi astronomi a Joanne Schonero. Norimbergae, 1534.
35. Baltzer, R. Die Elemente der Mathematik, Bd. 1–2. Dresden, 1867.
36. Barbilian, D. Zur Axiomatik der projektiven ebenen Ringgeometrie.—Jahresber. Deut. Math. Verein. 1940, **50**, 179–229; 1941, **51**, 34–76.
37. Barlow, W. Über die geometrischen Eigenschaften starrer Strukturen und ihre Anwendung auf Krystalle.—Z.f. Krystallographie und Mineralogie, 1894, **23**, 1–63.
38. Bašmakova, I. G., Slavutin, E. I. Vieta's calculus of triangles and the investigation of diophantine equations.—Istoriko-mat. issl., 1976, **21**, 78–101 (Russian).
39. Bašmakova, I. G., Slavutin, E. I. A history of diophantine analysis from Diophantus to Fermat.—Moscow, Nauka, 1984 (Russian).
40. Battaglini, G. Sulla geometria immaginaria di Lobatschewsky.—Giorn. mat., 1867, **5**, 217–231.

41. Bellavitis, G. Calcolo delle equipollenze.—Ann. delle Scienze del Regno Lombardo-Veneto. Padova, 1835, **5**, 244–259.

42. Beltrami, E. Opere matematiche, vol. 1.—Milano: Napoli, 1902.

42a. Benz, W. Vorlesungen über die Geometrie der Algebren. Berlin: Springer, 1933.

43. Berezin, F. A. Laplace operators on semisimple Lie groups and on certain symmetric spaces.—Usp. mat. nauk, 1957, **12**, N1, 152–156 (Russian).

44. Berezin, F. A. and Gel'fand, I. M. Remarks on the theory of spherical functions on symmetric Riemannian manifolds.—Trudy Mosk. mat. ob-va, 1956, **5**, 311–351 (Russian).

45. Berger, M. Classification des espaces homogènes symétriques irreductibles.—C. r. Acad. sci. Paris, 1955, **240**, 2370–2372.

46. Berger, M. Structure et classification des espaces symétriques à groupe d'isométrie semisimple.—C. r. Acad. sci. Paris, 1955, **241**, 1966–1968.

47. Berger, M. Les espaces symétriques non compacts.—Ann. scient. Ecole norm. supér., 1957, (3), **74**, N 2, 85–117.

48. Berggren, J. L. Al-Bīrūnī on a plane map of the sphere.—J. for the History of Arabic Science, 1982, **6**, 48–80.

48a. Berggren, J. L. Episodes in the Mathematics of Medieval Islam, Springer-Verlag, 1986.

49. Bergson, H. Essai sur les données immédiates de la conscience.—Paris: Alcan, 1911.

50. Bernoulli Joh. Der Briefwechsel von Joh. Bernoulli, Bd. 1, Basel, 1955.

51. Bertrand, L. Eléments de géométrie. Paris, 1812.

52. [Bêrûnî]. Epître de Bêrûnî contenant le répertoire des ouvrages de Muhammad b. Zakariya ar-Râzî.—Publ. par. P. Kraus. Paris, 1936.

53. Bērūnī Abū Rayḥān. Selected works v. 4–7.—Translations by U. I. Karimov, P. G. Bulgakov, B. A. Rozenfel'd and others. Taškent, Fan, 1973–1987 (Russian).

54. Bērūnī and Ibn Sīnā. Correspondence.—Translated by Yu. N. Zavadovskiĭ. Taškent, Fan, 1973 (Russian).

55. Betti, E. Opere matematiche, v. 1–2. Milano, 1903–1913.

56. Bianchi, L. Sui simboli a quatro indici e sulla curvatura de Riemann.—Rend. Accad. Lincei, 1902, (5), **11^{II}**, 3–7.

57. Al-Bīrūnī. Al-Qānūnu'l-Mas'ūdī (Canon Masudicus). Hyderabad: Osmania Univ. 1954–1956.

58. Al-Bīrūnī. The chronology of ancient nations.—Translated by E. Sachau, London, 1879.

59. Al-Bīrūnī Abū'l Rayḥān. The Book of Instruction in the Elements of the art of astrology.—Translated and edited by R. R. Wright, London: Luzac, 1934.

60. Al-Bērūnī Abū Rayḥān. Selected works v. 1–3.—Translations by M. A. Sal'e, A. B. Halidov, Yu. N. Zavadovskiĭ and P. G. Bulgakov. Taškent, Izd-vo AN UzSSR, Fan, 1957–1966.

61. Blaschke, W. Euklidische Kinematik und nichteuklidische Geometrie.—Z. Math. Phys., 1911, **60**, 61–91.

62. Blaschke, W., Müller, R. Ebene Kinematik. München, 1956.

63. Van der Blij, F. History of the octaves.—Simon Stevin, 1961, **34**, N 3, 106–125.

64. Bol, P. G. Selected papers.—Translated by I. M. Rabinovič. Riga, Izd-vo AN Latv SSR, 1961 (Russian).

65. Bolyai János. Appendix a tér tudománya.—Szerkesetette, beveretéssel, magyarázatakkal Kárteszi F. Budapest: Akademiai Kiado, 1937.

66. Bolyai, John. The science of absolute space.—Translated by G. B. Halsted. Supplement I to [71].

67. Bolyai, W. Theoria parallelarum. Maros-Vasarhelyini, 1804.

68. Bolyai, W. Tentamen juventutem studiosam in elementa matheseos purae, elementaris ac sublimioris, methodo intuitiva, evidentiaque huic propria, introducendi, cum appendice triplici. Maros-Vasarhelyini, 1832.

69. Bolyai, W. Kurzes Grundriss eines Versuches 1) die Arithmetik logischstreng darzustellen, 2) in der Geometrie die Begriffe scharf zu bestimmen. Maros-Vasarhely, 1851.

70. Bombelli, R. L'Algebra. Bologna, 1572.

71. Bonola, R. Non-Euclidean geometry. A critical and historical study of its development.—Translated by N. S. Carslaw. N.Y.: Dover, 1955.

72. Borel, A. Le plan projectif des octaves et les sphères comme espaces homogènes.—C. r. Acad. sci. Paris, 1950, 230, 1378–1380.

73. Borelli, G. A. Euclides restitutus. Pisis, 1658.

74. Bortolotti, E. Sulle forme differenziale spezializzate.—Rend. Accad. Lincei, 1930, (6), 12, 541–547.

75. Boscovich, R. J. Trigonometriae sphaericae constructio. Romae, 1737.

76. Boullough, R. K., Caudry, P. J. and others. Solitons.—Berlin: Springer, 1980.

77. Bounyakovsky, V. Considérations sur quelques singularités qui se présentent dans la construction de la géométrie non euclidienne.—Mém. sci. math. phys. Acad. S.-Pétersb., 1872, (7) 18, N 7, 1–16.

77a. Brahmagupta. The Khaṇḍakhādyaka (the astronomical treatise) with commentary of Bhattotpala, vol. 1.—Edited and translated by B. Chatterjee, Calcutta, 1970.

77b. Braunmühl A. v. Vorlesungen über die Geschichte der Trigonometrie.—Bd 1. Leipzig, 1900.

78. Brauer, R., Weyl, H. Spinors in n dimensions.—Amer. J. Math., 1935, 57, 425–449.

79. Brioschi, F. Opere, v. 1–5. Milano, 1901–1905.

80. Brouwer, L. E. Beweis der Invarianz der Dimensionszahl.—Math. Ann., 1911, 70, 161–165.

81. Brouwer, L. E. Die Theorie der endlichen kontinuierlichen Gruppen unabhängig von den Axiomen von Lie.—Math. Ann., 1909, 67, 246–267; 1910, 69, 181–207.

82. Bunyakovskiĭ, V. Ya. Parallel lines.—SPb., Izd-vo AN, 1853 (Russian).

83. Burali-Forti, C. Fondamenti per la geometria differenziale di una superficie col metodo vettoriale generale.—Rend. Circolo mat. Palermo, 1912, 33, 1–40.

84. Burnside, W. On the condition of reducibility for any group of linear substitutions.—Proc. London Math. Soc., 1905, 3, 430–434.

85. Busurina, A. E. Evolution of the problem of parallel lines up to the time of Euclid.—Istoriya i metodologiya estestv. nauk, 1971, 11, 161–171 (Russian).

86. Cantor, G. Ueber die Ausdehnung eines Satzes aus der Theorie der trigonometrischen Reihen.—Math. Ann., 1872, 5, 123–132.

87. Carbonaro, C. $L'S_5$ rigato considerato come un S_2 ipercomplesso legato all'algebra complessa regolare d'ordine 4.—Atti Accad. Gioenia di sci. nat. Catania (6), 1936, 1, N15, 1–27.

88. Cardanus, H. Ars magna sive de regulis algebraicis. Norimbergae, 1545.

89. Cardano, G. The Great Art of the Rules of Algebra (Ars Magna).—Translated and edited by T. R. Witmer, Cambridge, Mass.: MIT Press, 1968.

90. Carnot, L. De la corrélation des figures en géométrie. Paris, 1801.
91. Carnot, L. Géométrie de position. Paris, 1803.
92. Carnot, L. Mémoire sur la relation qui existe entre les distances respectives de cinq points quelconques pris dans l'espace; suivi d'un essai sur la théorie des transversales. Paris, 1806.
93. Carra de Vaux, B. L'Almagèste d'Abû-l Wafâ Albûzjânî.—J. Asiatique (8), 1892, **19**, 408–471.
94. Cartan, E. Leçons sur la géométrie projective complexe. Paris, 1931.
95. Cartan, É. Leçons sur la théorie des spineurs, vol. 1–2.—Paris: Hermann, 1938.
96. Cartan, É. Oeuvres complètes, part 1–3. Paris, 1952–1955.
97. Cartan, É. Les systèmes différentiels extérieurs et leurs applications géométriques. Paris, 1945.
98. Cataldi, P. A. Operetta delle linee rette equidistanti et non equidistanti. Bologna, 1603.
99. Cataldi, P. A. Aggiunta all'operetta delle linee equidistanti et non equidistanti. Bologna, 1604.
100. Cauchy, A. Oeuvres complètes, 1st ser., 12 vols, 2nd ser., 15 vols. Paris, 1882–1970.
101. Cavalieri, B. Directorium generale uranometricum. Bononiae, 1632.
102. [Cavalerius]. Geometria indivisibilibus continuorum nova quadam ratione promota.—Auctore F. Bonaventura Cavalerio. Mediolan. Bononiae: Ferroni, 1635.
103. Cayley, A. Collected mathematical papers, v. 1–13. Cambridge, 1889–1898.
104. Čahlenkova, T. G. Geometry of m-Euclidean spaces.—Izv. vuzov, Matematika, 1958, **N1**, 174–178 (Russian).
105. Čahtauri, I. A. Projective and elliptic spaces of integral and fractional dimension over matrix algebras.—Soobščeniya AN GSSR, 1971, **63**, **N1**, 541–544 (Russian).
106. Chasles, M. Les trois livres de Porismes d'Euclide établis pour la première fois d'après de notice et les lemmes de Pappus. Paris, 1860.
107. Chasles, M. Traité de géométrie supérieure. Paris, 1852.
108. Chasles, M. Aperçu historique sur l'origine et le développement des méthodes en Géometrie. Paris: Gauthier–Villars, 1889.
109. Čebyšev, P. L. Selected mathematical papers. M.-L. Gostehizdat, 1946 (Russian).
110. Černikov, N. A. Lectures on Lobačevskian geometry and the theory of relativity. Novosibirsk: Nauka, 1965 (Russian).
111. Chevalley, C. Two theorems on solvable topological groups. Michigan lectures in topology.—Ann. Arbor, 1941, p. 291–292.
112. Chevalley, C. The theory of Lie groups. v. 1, Princeton: Princeton Univ. Press, 1946.
113. Chevalley, C. Sur certains groupes simples.—Tôhoku Math. J., 1955, **7**, 14–66.
114. Christoffel, E. B. Gesammelte mathematische Abhandlungen. Bd. 1–2. Leipzig—Berlin, 1910.
115. Clairaut, A. C. Éléments de géométrie. Paris, 1741.
116. Clairaut, A. C. Sur les courbes que l'on forme en coupant une surface courbe quelconque par un plan donné position.—Hist. Mém. Acad. Sci. Paris, 1731 (1733), 483–493.
117. Clavius, Ch. Euclidis Elementorum libri XV. Romae, 1574.
118. Clebsch, A. Über diejenigen ebenen Kurven, deren Koordinaten rationale Funktionen eines Parameters sind.—J. reine und angew. Math., 1865, **64**, 43–65.

119. Clebsch, A. Über die Singularitäten algebraischer Kurven.—J. reine und angew. Math., 1865, **64**, 98–100.
120. Clebsch, A., Lindemann, F. Vorlesungen über Geometrie, Bd. 1–2. Leipzig, 1875–1876.
121. Clifford, W. K. The common sense of the exact sciences.—N.Y.: Knopf, 1946.
122. Clifford, W. K. Mathematical Papers. N.Y., 1968.
123. Clifford, W. K. Lectures and Essays, v. 1–2. London, 1901.
124. Comte, A. Cours de philosophie positive. Vol. 1–6. Paris: Ballière, 1877.
125. Coolidge, J. L. Geometry of the complex domain. Oxford: The Clarendon Press, 1924.
126. Copernicus, N. On the revolutions/Transl. by E. Rosen.—Complete works in 2 vols. Warsaw-Cracow: Pol. Sci. Publ., 1978.
127. Coxeter, H. S. M. Integral Cayley numbers.—Duke Math. Journal, 1946, **13**, N 4, 561–578.
128. Coxeter, H. S. M. Discrete groups generated by reflections.—Ann. of Math. (2), 1934, **35**, 588–621.
129. Coxeter, H. S. M. The polytope 2_{21} whose twenty seven vertices correspond to the lines on the general cubic surface.—Amer. Journal of Math. 1940, **76**, 457–486.
129a. Coxeter, H. S. M. The space–time continuum.—Hist. math., 1975, **2**, 289–298.
130. Coxeter, H. S. M. Regular Polytopes. N.Y.: Dover, 1973.
131. Crelle, A. L. Théorie des parallèles.—J. reine und angew. Math., 1835, **11**, 198.
132. Cremona, L. Le transformazioni geometriche delle figure piane. Opere. Bologna, 1862–1865.
132a. Crowe, M. J. A history of vector analysis. 1st ed. U. of Notre Dame Press, 1967; revised 2nd ed. Dover, 1985.
133. D'Aguillon, F. Opticorum libri VI. Antwerpiae, 1613.
134. D'Alembert, J. le R. Essai d'une nouvelle théorie sur la resistance des fluides. Paris, 1752.
135. D'Alembert, J. le R. Dimension. In: Encyclopédie ou dictionnaire raisonné des sciences, des arts et des métiers, v. 4. Paris, 1764, 1009–1010.
136. Darboux, G. Sur une classe remarquable de courbes et des surfaces algébriques et sur la théorie des imaginaires. Bordeaux, 1873.
137. David, L. von. Die beiden Bolyai. Basel: Birkhäuser, 1951.
138. Debarnot, M.-Th. Les clefs de l'astronomie d'Abū al-Rayḥān Muḥammad b. Aḥmad al-Bīrūnī. La trigonométrie sphérique chez les Arabes de l'Est à la fin du Xe siècle. Thèse. Paris, 1980.
139. Defrise, P. Analyse géométrique de la cinematique des milieux continues.—Publ. Inst. roy. météorol. Belg. sèr. B, 1953, N 6.
140. Dembowski, P. Finite geometries. Berlin: Springer, 1968.
141. Denisova, N. S. Biquasisimple Lie groups. In: Geometry of homogeneous spaces (sb. trudov MGPI im. V. I. Lenina). M., 1973, 27–41 (Russian).
142. Desargues, G. Oeuvres, v. 1. M. Poudra (Ed.). Paris, 1864.
143. Descartes, R. Oeuvres, v. 1–12, Publ. Ch. Adam et P. Tannery. Paris, 1897–1910.
144. Descartes, R. Geometry.—Translated by D. E. Smith and M. L. Latham. N.Y.: Dover 1954.
145. Descartes, R. The philosophical writings of Descartes. Translated by J. Cottingham, R. Stoothof and D. Murdoch, vols. 1–2, Cambridge Univ. Press 1985.

146. Des Cartes, R. Geometria. F. Schooten (Transl.). Lugduno—Batavii, J. Maire, 1649.

147. Dickson, L. E. Algebren und ihre Zahlensysteme. Zürich—Leipzig, 1927.

148. Diels, H. Ancilla to the pre-Socratic philosophers.—Translated by K. Freeman. Cambridge: Harvard Univ. Press, 1948.

149. Dilgan, H. Böyük matematikci Ömer Hayyâm. Istanbul, 1959.

150. Dilgan, H. Démostration du Ve Postulat d'Euclide par Schams ed-Din Samarkandi. Introduction de l'ouvrage Aschkal-üt-teessis.—Rev. histoire sci. et leur appl., 1960, **13**, N 3, 191–196.

151. Diophantus of Alexandria. A study in the history of Greek algebra. T. L. Heath, Cambridge, 1885.

151a. Dioyhantus of Alexandria. Arithmetics and Book on Polynomial Numbers. Transl. by I. N. Veselovskiĭ; introd. and comm. I. G. Bašmakova. M., Nauka, 1974 (Russian).

152. Dirac, P. A. M. The principles of quantum mechanics. Oxford: The Clarendon Press, 1958.

153. Dirac, P. A. M. The physical interpretation of quantum mechanics.—Proc. Roy. Soc., 1942, **180**, 1–40.

154. Dirichlet, P. G. L. Werke, 1–2. Berlin, 1889–1897.

155. Dirichlet, P. G. L. Vorlesungen über Zahlentheorie. Herausg. R. Dedekind. Braunschweig: Vieweg 1863.

156. Dreÿer, J. L. E. Das Planisphaerium des Claudius Ptolemaeus.—Isis, 1927, **9**, 255–278.

157. Dupin, Ch. Développements de géométrie. Paris, 1813.

158. Du Val, P. Geometrical note on de Sitter's world.—Philos. Mag., 1924, **6** (47), 930–938.

159. Dynkin, E. B. The structure of semisimple Lie algebras.—American Math. Soc. Translations. (1). 1962, **9**, 328–469.

160. Džavadov, M. A. Representation of conformal transformations in Euclidean and pseudo-Euclidean spaces of an arbitrary number of variables as linear fractional transformations.—Dokl. AN SSSR, 1952, 86, **4**, 653–656 (Russian).

161. Džavadov, M. A. Geometric interpretation of spinor representations of groups of motions of non-Euclidean spaces.—Uč. zap. Azerb. gos. un-ta, 1957, **N11**, 3–18 (Russian).

162. Džavadov, M. A. Projective spaces over algebras.—Uč. zap. Azerb, gos. un-ta, 1957, **N2**, 3–18 (Russian).

163. Džavadov, M. A. Non-Euclidean geometries over algebras.—Uč. zap. Azerb. gos. un-ta, 1957, **N4**, 3–16 (Russian).

164. Ehresmann, Ch. Sur la topologie de certains espaces homogènes.—Ann. of Math. 1934, **35**, 396–443.

165. Einstein, A. Geometrie und Erfahrung. Berlin: Springer, 1921.

166. Einstein, A. Über den Aether.—Verhandl. Schweiz. naturforsch. Gesellschaft, 1924, **105**, 85–93.

167. Einstein, A. Ernst Mach.—Phys. Z. 1916, **17**, 101–105.

168. Engel, F., Stäckel, P. Die Theorie der Parallellinien von Euklid bis auf Gauss. Leipzig, 1895.

169. Engels, F. Dialectic of nature.—Translated by C. C. Dutt. Moscow: Progress, 1982.

170. Engels, F. Anti-Dühring. Herr Eugen Dühring's revolution in science. Moscow: Progress, 1977.

171. Enriques, F. Introduzione alla geometria sopra le superficie algebraiche. Roma, 1896.

172. Euclidis. Opera omnia, v. 1–8. J. L. Heiberg (Ed.). Lipsiae, 1883–1916.

173. [Euclid]. The thirteen books of Euclid's *Elements*, vol. 1–3.—Translated with introduction and commentary by T. L. Heath. Cambridge Univ. Press, 1926.

174. Euclidis. Elementa ex interp. F. Commandini. Pesarii, 1572.

175. Euclidis. Elementorum geometricorum libri tredecim ex traditione doctissimi Nasiridini Tusini. Romae, 1594.

176. Euler, L. Opera omnia (1). Opera mathematica, t. 1–29. Leipzig–Berlin–Zürich, 1911–1956.

177. Euler, L. Opera omnia (2). Opera mechanica et astronomica, t. 1–29. Leipzig–Berlin–Zürich, 1922–1969.

177a. Fano, G. Sur postulati fondamentali della geometria proiettiva.—Giornale di matem, 1892, **30**, 106–132.

178. Al-Fārābī. Mathematical treatises.—Translations by M. F. Bokštein, S. A. Krasnova, A. Kubesov and others. Alma-Ata: Nauka, 1972 (Russian).

179. Fedenko, A. S. Symmetric spaces with simple noncompact fundamental group.—Dokl. AN SSR, 1956, **108**, 1026–1028 (Russian).

180. Fedenko, A. S. Symmetric spaces with simple fundamental groups.—Uč. zap. Belorus. gos. un-ta, seriya mat., 1959, N3 (52), 3–25 (Russian).

181. Fedorov, E. S. Symmetry and structure of crystals.—M., Izd-vo AN SSSR, 1949 (Russian).

182. Fermat, P. Oeuvres, v. 1–4. P. Tannery (Ed.). Paris, 1891–1922.

182a. Field J. V., Gray J. J. The geometrical work of Girard Desargues. N.Y.: Springer-Verlag, 1987.

183. Fomin, V. E. Differential geometry of Banach manifolds.—Kazan: Izd-vo KGU, 1983 (Russian).

184. Frechet, M. Sur quelques points du calcul fonctionnel.—Rend. Circolo mat. Palermo, 1906, **22**, 1–74.

185. Freudenthal, H. Oktaven, Ausnahmegruppen und Oktavengeometrie.—Geom. Dedic. 1985, **19**, 1–63.

186. Freudenthal, H. Beziehungen der E_7 und E_8 zur Oktavenebene.—Proc. Koninkl. nederl. Akad. wet., 1964, **A57**, 218–230, 363–368; 1955, **A58** 151–157, 277–285; 1959, **A62**, 165–201, 447–474; 1963, **A66**, 457–487.

187. Friedmann, A. A. Über die Krümmung des Raumes.—Z. Phys., 1922, **10**, 377–386.

188. Friedmann, A. A. Über die Möglichkeit einer Welt mit konstanter negativer Krümmung des Raumes.—Z. Phys., 1924, **21**, 326–330.

189. Frobenius, F. G. Theorie der hyperkomplexen Grössen.—Sitzungsber. Preuss. Akad. Wiss., 1903, **N24**, 504–537, 634–645.

190. Frobenius, F. G. Über Gruppencharaktere.—Monatsber. Preuss. Akad. Wiss., 1896, **N40**, 985–1021.

191. Fubini, G. G. Sulle metriche definite da una forma Hermitiana.—Atti Ist. Veneto Sci., 1903, **63**, 502–513.

192. Fubini, G. G. Il parallelismo di Clifford negli spazî ellittici.—Ann. Scuola norm. super. Pisa, 1900, **9**, 1–74.

193. Fuss, N. De proprietatibus quibusdam ellipseos in superficie sphaericae descriptae.—Nova acta Acad. Sci. Petropol., (1785), 1788, **3**, 90–99.

194. Galois, E. Écrit ét mémoires mathématiques.—Ed. R. Bourgne et J.-P. Azra. Paris: Gauthier–Villard, 1962.

195. Garig, G. E. [Harig, G. E.] The Tartaglia-Cardano controversy about cubic equations and its social background.—Arhiv istoriĭ nauki i tehniki, 1935, **7**, 67–104 (Russian).
196. Gauss, C. F. Werke, Bd. 1–12. Göttingen, 1870–1927.
198. Gel'fand, I. M. Lectures on linear algebra.—Translated by A. Shenitzer. N.Y.–London, Interscience Publishers, 1961.
199. Gel'fand, I. M. and Naĭmark M. A. Unitary representations of the classical groups. M., 1950 (Russian).
200. Gemma Frisius, R. De astrolabio catholico et usu eiusdem. Antwerpiae, 1548.
201. Gersonid, Lev [Gersonides Leo, Levi ben Gerson]. Commentaries on the introductions in Euclid's *Elements*.—Translated by I. G. Polskiĭ. Istoriko-mat. issl., 1958, **11**, 763–782 (Russian).
202. Gibbs, J. Vector analysis. New Haven, 1925.
203. Giordano, Vitale. Euclide restituto, overo gli antichi elementi geometrici ristaurati et facilitati, libri XV. Romae, 1680.
204. Girard, A. Invention nouvelle en l'algèbre. Amsterdam, 1629.
205. Glagolev, N. A. Riemannian manifolds with projective structure.—Mat. Sb. 1925, **32**, 177–191 (Russian).
206. Gleason, A. Groups without small subgroups.—Ann. Math., 1952, **56**, N 2, 193–212.
207. Gluškov, V. M. Structure of locally bicompact groups and Hilbert's Fifth Problem.—Usp. mat. nauk, 1957, **12**, N2 (Russian).
208. Goldstein, B. R. The Astronomy of Levi ben Gerson (1288–1344). N.Y.e.a.: Springer, 1985.
209. Gosset, T. On the regular and semi-regular figures in space of *n*-dimensions.—Messenger of Math. 1900, **29**, 43–48.
210. Grant, E. A source book in medieval science. Cambridge, 1974.
211. Grassmann, H. Gesammelte mathematische und physikalische Werke, Bd. 1–3. Leipzig, 1894–1911.
212. Graves, Ch. On algebraic triplets.—Proc. Royal Irish Acad., 1847, **3**, 51–54, 57–64, 80–84, 105–108.
212a. Grigoryan, É. S., Abu-l-Fazl. Tabrizi's treatise *On a proof of the well-known postulate of Euclid*. In: *Some questions involved in research dealing with the history of mathematics and mechanics in Azerbayjan*. Baku, 1971 (Russian).
213. Gur'ev, S. E. An attempt to prefect the elements of geometry.—SPb, Izd-vo AN, 1798 (Russian).
214. Hamilton, W. R. The mathematical papers, v. 1–3. Cambridge, 1931–1967.
215. Hamilton, W. R. Lectures on quaternions. Dublin, 1853.
216. Hamilton, W. R. Elements of quaternions. N.Y., 1969.
216à. Haritonova N. I. Projective and elliptic spaces over a Galois square. Aktual'nye problemy geometrii i ee priloženiĭ. Vol. 2.—Čeboksary, Čuvaš. Gos. Univ., 1976, 44–49 (Russian).
217. Hausdorff, F. Mengenlehre. Berlin–Leipzig: De Gruyter, 1935.
218. Hausdorff, F. Set theory.—Translated by J. R. Aumann e.a. N.Y.: Chelsea, 1957.
219. Heath, T. L. Mathematics in Aristotle. Oxford Univ. Press, 1939.
220. Heaviside, O. Electromagnetic theory, v. 1–2. London, 1951.
221. Hegel, G. W. F. Philosophy of nature.—Translated by A. V. Miller, Oxford U. Press, 1970.

221a. Heisenberg, W. Remarks on Einstein's sketch of a unified field theory (Russian). In: Collected works, series C, group II.—R. Piper Verlag, München, Zürich, 1984, 120–125.

222. Helgason, S. Differential geometry, Lie groups and symmetric spaces. N.Y. e.a.: Academic Press, 1978.

223. Helmholtz, H. Wissenschaftliche Abhandlungen. Bd. 2, Leipzig: Barth, 1882.

224. Heronis Alexandrini. Opera quae supersunt omnia, t. 1–5. J. L. Heiberg, L. Nix, W. Schmidt, H. Schöne (Eds.). Lipsiae, 1899–1914.

225. Hesse, O. Über ein Übertragungsprinzip.—J. für reine u. angew. Math. 1866, Bd. 66, 15–21.

226. Hilbert, D. Gesammelte Abhandlungen. Bd. 1–3. Berlin, 1932–1935.

227. Hilbert, D. Grundzüge einer allgemeinen Theorie der linearen Integralgleichungen.—Göttinger Nachr., 1904, 213–259, 49–91; 1905, 307–338; 1906, 157–227, 439–480; 1910, 335–417.

228. Hilbert, D. Grundzüge einer allgemeinen Theorie der linearen Integralgleichungen. Leipzig–Berlin, 1924.

229. Hilbert, D. The Foundations of geometry.—Translated by L. Unger, Open Court, La Salle, Ill., 1971.

230. Hilbert, D. Grundlagen der Geometrie. Stuttgart: Teubner, 1962.

231. Die Hilbertsche Probleme.—Unter der Red. von P. S. Alexandrov. Leipzig: Akad. Verlag, 1971.

231a. [Hilbert] Mathematical developments arising from Hilbert problems, edited by F. E. Browder, Proc. of Symposia in Pure Math., vol. XXVIII, AMS, Providence, R.I., 1976.

232. Hjelmslev, J. Einleitung in die allgemeine Kongruenzlehre.—Mitt. Danske Vid. Selsk., mat.-fys. medd. 1929, **8, 10**; 1942, **19**; 1945, **22**; 1949, **25**.

233. Hoppe, R. Die regelmässigen linear begrenzten Figuren jeder Anzahl von Dimensionen.—Arch. Math. u Phys., 1882, **68**, 151–165.

234. Hoüel, J. Essai critique sur les principes fondamentaux de la géométrie. Paris, 1867.

235. Hudson, H. Cremona transformations in the plane and in space. Cambridge Univ. Press, 1927.

235a. Hua, L. K. Geometries of matrices. Sci. Record, 1945, **1**, 262–267, Trans. AMS, 1947, **61**, 193–255.

235b. Hua, L. K. The geometry of symmetric matrices over the field of real numbers. Dokl. AN SSSR, 1946, **53**, 95–97, 195–196 (Russian).

235c. Hua, L. K. and Rosenfeld, B. A. Geometry of rectangular matrices and its applications to real projective and non-Euclidean geometry Sci. Sinica, 1957, **6**, N 6, 995–1011 (Russian).

236. Ibn Abi Useibi'a. 'Ujûn el-anbâ fî tabaqât el-atibba, Bd. 1–2. Hrsg. A. Müller, Königsberg, 1884.

237. Ibn al-Haytham. Resolution of doubts in Euclid's *Elements* and interpretation of its special meanings. Kitābfī Ḥall shukūk Kitāb Uqlīdis fi 'l-Uṣūl wa-sharḥ ma'ānīh. Frankfurt-am-Main. Inst. for the history of Arabic-Islamic Science, 1985.

238. Ibn al-Haytham al-Ḥasan. Optics. Kitab al-Manāẓir. Vol. 1–4.—Ed. and translated by A. I. Sabra. Kuwait, 1983.

239. [Ibn 'Irāq]. Rasā'il Abī Naṣr ilā-l-Bīrūnī. Hyderabad: Osmania Univ. 1365h [1946].

240. [Ibn el-Nadîm]. Kitâb al-Fihrist von Abû'l-Farâg Muh. b. Isḥaq bekannt unter dem Namen Ibn Abî Ja'qûb el-Nadîm, Bd. 1. G. Flügel, J. Rödiger und A. Müller (Hrsg.). Leipzig, 1871.

241. Ibn Qurra. Ein Werk über ebene Sonnenuhren. Hrsg., übers. und erläutert von K. Garbers. Quellen und Studien zur Geschichte der Mathematik, Astronomie und Physik, Abt. A, Bd. 4. Berlin, 1937.

242. [Ibn Sinan]. The Works of Ibrahim Ibn Sinan.-Ed. A. S. Saidan. Kuwait, 1983.

243. Ikhwān al-Ṣafā'wa Khullān al Wafā'. Rasā'il, vol. 1–4. Beyruth, 1376h [1957].

244. Initius Algebras.—Abhandl. Geschichte math. Wiss., 1902, 13, 435–609.

245. Inönü, E., Wigner, E. P. On the contraction of groups and their representations.—Proc. Nat. Acad. Sci. USA, 1953, 39, N 6, 510–524.

246. Jacobi, C. G. J. Gesammelte Werke, Bd. 1–7. Berlin, 1881–1891.

247. Jacobson, N. Structure and representations of Jordan algebras.—Amer. Math. Soc. Colloquium Publs, 39. Providence, 1968.

248. Jacobson, F. D., Jacobson, N. Classification and representations of semi-simple Jordan algebras.—Trans. Amer. Math. Soc., 1949, 65, N 2, 141–169.

248a. Jaouiche, K. La théorie des parallèles en pays d'Islam.—Paris: Vrin, 1986.

249. Jordan, C. Traité sur les substitutions et des équations algébriques. Paris, 1870.

250. Jain, L. C., Patni, G. C. Basic mathematics.—Exact sciences from Jaina sources, vol. 1, Jaipur, 1982.

251. Jordan, C. Oeuvres. Paris, 1961–1964.

252. Jordan, P. Über eine Klasse nichtassoziativer hyperkomplexer Algebren.—Nachr. Ges. Wiss. Göttingen, 1933, 569–575.

253. Jung. F. Ableitungsbildung im räumlichen Größenfelde. Berlin, 1908.

254. Kagan, V. F. Lobačevskiĭ.—M.-L., Izd-vo AN SSSR, 1948 (Russian).

255. Kagan, V. F. Foundations of geometry, v. 1–2. Odessa, 1905–1907 (Russian).

256. Kagan, V. F. Foundations of geometry. A history of the evolution of the study of foundations of geometry, v. 1–2. M., 1949–1956 (Russian).

257. Kähler, E. Über eine bemerkenswerte Hermitesche Metrik.—Abhandl. Math. Seminars Hamburg. Univ., 1933, 9, 173–186.

258. Kallenberg, G. W. M. Differential geometry of a particular group of projective transformations.—Proc. Koninkl. Nederl. Akad. wet., 1957, A60, N 2, 147–156.

259. Kaluza, T. Zum Unitätsproblem der Physik.—Sitzungsber. Preuss. Akad. Wiss., 1921, 966–972.

260. Kant, I. Werke. Bd. 1. Berlin: Cassirer, 1912.

261. Kant, I. Critique of pure reason.—Transl. by J. M. D. Meiklejohn. London: Bell, 1930.

262. Kantor, I. L. Transitive-differential groups of transformations on homogeneous spaces.—Trudy sem. po vektornomu i tenzornomu analizu pri MGU, 1966, 13, 301–398 (Russian).

263. Kantor, I. L. Some generalizations of Jordan algebras.—Trudy seminara po vektornomu i tenzornomu analizu pri MGU, 1972, 16, 407–499 (Russian).

264. Kantor, I. L. Models of exceptional Lie algebras.—Dokl. AN SSSR, 1973, 208, 1276–1279 (Russian).

264a. Kapralova, S. B. Dual Galois spaces.—Trudy geometričeskogo seminara KGU, 1979, 12, 38–44 (Russian).

265. Karimullin, A. G. and Laptev, B. L. What did Lobačevskiĭ read? Kazan: Izd-vo KGU, 1979 (Russian).

266. Karpova, L. M., Tagi-Zade, A. K. Al-Ṣaghānī's treatise *The book of projection of the sphere to the plane of the astrolabe.*—Trudy XV nauč. konf. aspirantov i

mladših naučnyh sotrudnikov IIEiT AN SSSR, sekciya mat.-meh., 1972, 77–81 (Russian).

267. Karpova, L., Rosenfeld, B. A. The treatise of Thâbit bin Qurra on sections of a cylinder and on its surface.—Arch. internat. histoire sci., 1974, **24**, N 94, 66–92.

268. Kartan (Cartan), E. Geometry of Lie groups and symmetric spaces.—Transl. by B. A. Rozenfel'd. M., IL., 1949 (Russian).

269. al-Kāshī, Ghiyāth al-Dīn Jamshīd Mas'ūd. The key to arithmetic. Treatise on the circumference.—Transl. by B. A. Rozenfel'd. M., Gostehizdat, 1956 (Russian).

270. Kennedy, E. S. Late medieval planetary theory.—Isis, 1966, **57**, N 3, 365–378.

271. Kepler, J. Gesammelte Werke, Bd. 1–18. München, 1937–1969.

272. Khayyam, Omar. Treatises.—Translated by B. A. Rozenfel'd. M., Izd-vo vost. lit-ry, 1962 (Russian).

273. Khairetdinova, N. G. The trigonometric treatise of an anonymous Isfahan scholar.—Istoriko-mat. issl., 1966, **17**, 444–464 (Russian).

274. al-Khaisam ibn [Ibn al-Haytham] Book of commentaries on the introductions to Euclid's *Elements.*—Istoriko-mat. issl., 1958, **11**, 743–762, 777–780 (Russian).

275. [al-Khāzin]. Muḥammad ibn al-Ḥasan. Epistle devoted to the proof that the sides of two square numbers whose sum is a square cannot both be odd, and that either both are even or one is even and the other odd. This is followed by an epistle devoted to the construction of triangles with rational sides.—Translated by B. A. Rozenfel'd. In: Iz istorii fiziko-mat, nauk na srednevekovom Vostoke. Naučnoe nasledstvo, v. 6. M.: Nauka, 1983, 161–174 (Russian).

276. Khayyam, Omar. Explanation of the difficulties in Euclid's postulates.—Edited by A. I. Sabra. Alexandria, 1961.

277. Al-Khayyāmī ʿOmar. Discussion of difficulties in Euclid.—Translated by A. R. Amir-Moéz. Scripta mathematica, 1959, **24**, N.Y., 275–303.

278. Humāī, Jalāt al-Din. Khayyāmī-nāmah, vol. 1. Tehran, 1346 s.h. [1967].

278a. Killing W. Zusammensetzung der stetigen endlichen Transformationsgruppen.—Math. Ann., 1888, **31**, 252–290; 1889, **33**, 1–48; 1889, **34**, 57–122; 1890, **36**, 161–189.

279. Kirkman, T. R. Triads made with 15 things.—Cambridge and Dublin math. J. 1850, **5**, 260.

280. Klamroth, M. Über die Auszüge aus griechischen Schriftstellern bei al-Ja'qubi.—Z. Dtsch. Morgenländ. Ges., 1888, **42**, 1–44.

281. Klein, F. Vorlesungen über das Ikosaeder und die Auflösung der Gleichungen vom fünften Grade. Leipzig, 1884.

282. Klein, F. Gesammelte mathematische Abhandlungen. Bd 1–3, Berlin: Springer, 1921–1923.

283. Klein, F. Vorlesungen über die Entwicklung der Mathematik im 19 Jahrhundert. Bd 1. Berlin: Springer, 1926.

284. Klein, F. Vorlesungen über nicht-euklidische Geometrie. Berlin: Springer, 1928.

285. Klimanova, T. M. Unitary semielliptic spaces.—Izd. AN AzSSR, seriya fiz.-mat. i tehn. nauk, 1963, **3**, 21–29 (Russian).

286. Klingenberg, W. Euklidische Ebenen mit Nachbarelementen.—Math. Z. 1954/55, **61**, 374–385.

287. Koecher, M. Imbedding of Jordan algebras into Lie algebras.—Amer. J. Math., 1967, **39**, N 3, 787–816; 1968, **40**, N 2, 476–510.

288. Kotel'nikov, A. P. The twist calculus and some of its applications to geometry and mechanics.—Kazan, Izd-vo Kazan'sk. un-ta, 1895 (Russian).

289. Kotel'nikov, A. P. Projective theory of vectors.—Kazan Izd-vo Kazan'sk. unta, 1899 (Russian).

290. Kotel'nikov, A. P. The principle of relativity and Lobačevskian geometry. A chapter in the book: In memoriam N. I. Lobačevskiĭ, v. 2. Kazan, 1927, 37–66 (Russian).

291. Koyré, A. From the closed world to the infinite universe. Baltimore, 1957.

292. Koyré, A. Études d'histoire de la pensée philosophique. Paris, 1961.

293. Kramar, F. D. Elements of vector algebra in Wallis' mechanics.—Vopr. ist. estestv. i teh., 1967, 21, 103–108 (Russian).

294. Krein, M. G. Helices in infinite-dimensional Lobačevskian space and Lorentz transformations.—Usp. mat. nauk, 1948, 3, N3, 158–160 (Russian).

295. Kummer, E. Zur Theorie der complexen Zahlen.—J. reine und angew. Math., 1847, 35, 319–326.

296. Kuratowski, K. L'opération $\bar{A}$ de l'analysis situs.—Fundamenta math., 1922, 3, 182–199.

297. Kuznecova, T. A. Bioctave geometries and their analogues.—Ukr. geometr. Sb., 1975, 17, 92–107 (Russian).

298. Lagrange, J. L. Oeuvres, v. 1–14. Paris, 1867–1892.

299. Lagrange, J. L. Mécanique analytique, Vol. 1–2. Paris, 1853.

300. Laguerre, E. Oeuvres, v. 1–2. Paris, 1898–1905.

301. Lamé, G. Démonstration générale du théorème de Fermat sur l'impossibilité en nombres entières de l'équation $x^n + y^n = z^n$.—J. math. pure et appl., 1847, 24, 310–316.

302. Lapkovskiĭ, A. K. Relativistic kinematics, non-Euclidean spaces and exponential mapping. Minsk: Nauka i tehnika, 1985.

303. Laplace, P. S. Oeuvres, v. 1–14. Paris, 1878–1912.

304. Laptev, B. L. The theory of parallels in Lobačevskiĭ's early papers.—Istoriko-mat. issl., 1951, 4, 201–229 (Russian).

305. Laptev, B. L. Lambert as a geometer.—Istoriko-mat. issl. 1980, 25, 248–260 (Russian).

306. Lasker, E. Zur Theorie der Moduln und Ideale.—Math. Ann., 1905, 60, 20–116.

307. Laugwitz, D. Differentialgeometrie ohne Dimensionsaxiom. Math. Z. 1954, 61, N1, 100–118; N2, 134–144.

308. Lefschetz, S. Algebraic topology. N.Y.: Amer. Math. Soc., 1942.

309. Legendre, A. M. Éléments de géométrie. 1ère ed. Paris, 1794.

310. Legendre, A. M. Éléments de géométrie. 3éme ed. Paris, 1800.

311. Legendre, A. M. Éléments de géométrie. 12éme ed. Paris, 1823.

312. Leibnizens mathematische Schriften, Bd. 1–7. London–Berlin–Halle, 1849–1863.

312a. Leibnitz G. W. New essays concerning human understanding.—Translated by A. G. Langley, Chicago–London, 1916.

313. Lenin, V. I. Philosophical notebooks.—Coll. works, vol. 38, M.: Foreign Languages Publishing House, 1963.

314. Leibniz, G. W. Ein Dialog zur Einführung in die Arithmetik und Algebra. Nach der Originalschrift herausg., übers. und komm. von E. Knobloch. Stuttgart: Fromann-Holzboog, 1926.

315. Leibniz, G. W. Philosophical papers and letters.—Translated by L. E. Loemker. Dordrecht–Boston: Reidel, 1976.

316. Lenin, V. I. Materialism and empiriocriticism. Moscow: Progress, 1977.

317. Léonard de Pisa. Le livre de nombres carrés. Trad. P. Ver Eecke. Bruges, 1952.
318. Leonardo Pisano. Scritti, v. 1–2. Publ. de B. Boncompagni. Roma, 1857–1862.
319. Levi, E. Sulla struttura dei gruppi finiti e continui.—Atti Accad. Torino, 1905, **40**, 3–17.
320. Levi-Civita, T. Nozione di parallelismo in uno varieta qualunque e consequente spezificazione geometrica della curvatura Riemanniana.—Rend. Circolo mat. Palermo, 1917, **42**, 173–205.
321. Lexell, A. J. Solutio problematis geometrici ex doctrina sphericorum.—Acta Acad. sci. Petropol., (1781), 1784, **5**:**1**, 112–126.
322. Lexell, A. J. De proprietatibus circulorum in superficie sphaerica descriptorum.—Acta Acad. sci. Petropol., (1782), 1786, **6**:**1**, 58–103.
323. L'Huillier, S. Mémoire sur la polyédrométrie; conténant une démonstration directe du théorème d'Euler sur les polyèdres et une examen des divers exceptions auxquelles ce théorème est assujetti.—Ann. math. pures et appl., 1812–1813, **3**, 169–191.
324. Libri, G. Histoire des sciences mathématiques en Italie, v. 1–2. Paris, 1833–1838.
325. Lie, S. Gesammelte Abhandlungen, Samlede avhandlingen. Bd. 1–7. Hrsg. F. Engel und P. Heegaard. Leipzig–Oslo, 1934–1960.
326. Lie, S. Bemerkungen zu v. Helmholtz Arbeit "Über die Tatsachen die der Geometrie Zugrunde liegen." Sitzungsberichte Sächs. Ges. Wiss. Math.-Phys. Kl. 1886, **38**, 337–342.
327. Lie, S., Engel, F. Theorie der Transformatonsgruppen, Bd. 1–3. Leipzig, 1888–1893.
327a. Lint, J. H. van. Coding theory.—N.Y.: Springer-Verlag, 1973.
328. Lipschitz, R. Untersuchungen über die Summen von Quadraten. Bonn, 1886.
329. Listing, L. B. Vorstudien zur Topologie.—Göttinger Studien. Göttingen, 1847, 811–875.
330. Listing, J. B. Der Census räumlicher Komplexe, oder Verallgemeinerung des Euler'schen Satzes von den Polyedern.—Abhandl. Kgl. Ges. Wiss. Göttingen, 1862, **10**, 97–180.
331. Lobatschefsky, N. I. Études géométriques sur la théorie des parallèles. Trad. J. Hoüel. Paris, 1866.
332. Lobachevski, N. Geometrical researches on the theory of parallels.—Translated by G. D. Halsted. Supplement II to [71].
333. Lobačevskiĭ, N. I. The Complete Works. v. 1–5. M., Gostehizdat, 1946–1951 (Russian).
334. Lobačevskiĭ, N. I. New investigations in the theory of parallel lines.—Translated by A. V. Letnikov. Mat. sb., 1868, **3**, 78–120 (Russian).
335. Lobatschewsky, N. I. Pangeometria. Trad. G. Battaglini—Giorn. mat., 1867, **5**, 273–330.
336. Lomonosov, M. V. Selected philosophical works. M., Socekgiz, 1950 (Russian).
337. Loos, O. Symmetric Spaces. Vol. 1–2. N.Y.-Amsterdam: Benjamin, 1969.
338. Lopšic, A. M. Some problems of tensor algebra in linear dimensionless spaces. Trudy seminara po vektornomu i tenzornomu analizu pri MGU, 1948, **6**, 365–419 (Russian).
339. Lopšic, A. M. Problems of projective, affine and descriptive geometry in dimensionless space.—Trudy III vsesovuzn. mat. s″ezda. 1956, **2**, 140 (Russian).
340. Lorentz, H. A., Einstein, A., Minkowski, H., Weyl, H. The principle of relativity. N.Y.: Dover, 1923.

341. Luckey, P. Beiträge zur Erforschung der islamischen Mathematik, I.—
      Orientalia (N.S.), 1948, **17**, 490–510.
341a. Luckey, P. Zur Entstehung der Kugeldreieckrechnung. Dtsch. Math., 1941, **5**,
      405–446.
342. Lucretius, Carus Titus. De rerum natura.—Edited and translated by C. Bailey.
      Vol. 1. Oxford: Clarendon Press, 1947.
343. Lur'e, S. Ya. The theory of infinitesimals in the works of ancient atomists.—
      M.-L., Izd-vo AN SSSR, 1935 (Russian).
344. Lur'e, S. Ya. Democritus. Texts, translations and studies. L., Nauka, 1970
      (Russian).
345. Mach, E. The science of mechanics: a critical and historical account of its
      development. Translated by Th. J. McCormack. La Salle, Ill., 1960.
346. Mach, E. Die Geschichte und die Wurzel des Satzes von der Erhaltung der
      Arbeit. Leipzig, 1909.
347. Mach, E. Knowledge and error.—Translated by Th. J. McCormack, P.
      Foulkes. Dordrecht–Boston, 1976.
348. [al-Māhānī] Abhandlungen von Abū 'Abdallāh al-Māhānī über die Bestim-
      mung des Azimuts-in: Luckey, P. Die Schrift des Ibrāhīm b. Sinān b. Iābit über
      die Schatteninstrumente. Inaugural-Dissertation. Tübingen, 1944, 200.
349. Maimonides. Guide of the Perplexed.—Translated by S. Pines. Chicago Univ.
      Press, 1963.
350. Mal'cev, A. I. Foundations of linear algebra.—Translated by T. C. Brown.
      Freeman, 1963.
351. Mal'cev, A. I. Solvable topological groups.—Mat. sb., 1946, **19**, N2, 165–174
      (Russian).
352. Mal'cev, A. I. Analytic loops.—Mat. sb., 1955, **36**, N3, 569–575 (Russian).
353. Mamedbeïli, G. D. Muḥammad Naṣīr al-Dīn al-Ṭūsī on the theory of
      parallel lines and the theory of ratios: Baku, Izd-vo AN AzSSR, 1959 (Russian).
354. Markina, L. M. Dual Hermitian non-Euclidean spaces.—Uč. zap. MOPI im.
      N.K. Krupskoĭ. 1969, **262**, 147–165 (Russian).
355. Marx, K. Mathematical manuscripts.—Translated by S. A. Yanovskaya and
      others. M., Nauka, 1968 (Russian).
356. Mathieu, É. Mémoire sur l'étude des fonctions de plusieurs quantités.—J.
      Math. pur. appl. 1861, **6**, 241–323.
357. Mathieu, É. Sur la fonction cinq fois transitive de 24 quantités.—J. Math. pur.
      appl. 1873, **18**, 25–46.
358. Matvievskaya, G. P. Number theory in the medieval Near and Middle East.
      Taškent, Fan, 1967 (Russian).
359. Maxwell, C. A treatise on electricity and magnetism. Cambridge, 1873.
360. Medvedev, F. A. The first monograph on functional analysis.—Istoriko-mat.
      issl. 1973, **18**, 71–93 (Russian).
361. [Menelaos]. Die Sphärik von Menelaos aus Alexandrien in der Verbesserung
      von Abū Naṣr Manṣūr b. 'Alī b. 'Iraq mit Untersuchungen zur Geschichte des
      Textes bei den islamischen Mathematiker. Übers. M. Krause. Berlin, 1936.
362. Menger, K. Dimensionstheorie. Leipzig–Berlin, 1928.
363. Meschkowski, H. Probleme des Unendlichen. Werk und Leben Georg Cantors.
      Braunschweig, 1967.
364. Meyerson, E. Le déduction relativiste. Paris, 1925.
365. Michel, H. Traité de l'astrolabe. Paris, 1947.

366. Minding, F. Bemerkung über die Abwickelung krummer Linien von Flächen.—
Journal für die reine und angew. Math. 1830, **6**, 159–161.

367. Minding, F. Wie sich entscheiden lässt ob zwei gegebene krumme Flächen auf
einander abwickelbar sind oder nicht; nebst Bemerkungen übe die Flächen von
unverenderlichem Krümungsmasse.—J. reine u. angew. Math. 1839, **19**, 370–
387.

368. Minding, F. Beiträge zur Theorie der kürzesten Linien auf krummen Flächen.
—J. reine u. angew. Math. 1840, **20**, 323–327.

369. Möbius, A. F. Gesammelte Werke, Bd. 1–4. Leipzig, 1885–1887.

370. Moisil, G. Sur les géodésiques des espaces de Riemann singuliers.—Bull. Math.
Soc. Roum. Sci., 1940, **42**, 33–52.

371. Molien, T. Ueber Systeme höherer complexer Zahlen. Dorpat, 1892.

372. Monge, G. Géométrie descriptive. Paris: Coursier, 1820.

373. Monge, G. Application de l'analyse à la géométrie. Paris: Bachelier, 1850.

374. Monge, G. Mémoire sur les développées des rayons de courbure et les différents
genres d'inflection des courbes à double courbure.—Mém. divers savants, 1785,
10.

375. Montgomery, D., Zippin, L. Small subgroups in finite dimensional groups.—
Ann. Math., 1952, **56**, N 2, 213–241.

376. Montgomery, D., Zippin, L. Topological transformation groups. N. Y., 1955.

377. Morgan, A. de. On the foundation of algebra.—Trans. Cambridge Philos. Soc.,
1847, **8**, N 3, 241–254.

378. Morozov, V. V. On nonsemisimple maximal subgroups of simple groups.
(doc. diss.) Kazan, 1943 (Russian).

379. Murdoch, J. Euclid. Dictionary of scientific biography, v. 4. N.Y., 1971, 414–
459.

380. [al-Nayrīzī]. The treatise of al-Faḍl ibn Ḥātim al-Nayrīzī on the proof of the
well-known postulate of Euclid.—Translated by A. A. Abdurahmanov and
B. A. Rozenfel'd.—Istoriko-mat. issl. 1982, **26**, 325–329 (Russian).

381. Nallino, C. Al-Battānī sive Albatenii opus astromicum, t. 1–3. Mediolani,
1899–1907.

382. Nau, F. Le traité de l'astrolabe plan de Sévère Sebokt.—J. Asiatique, 1899 (9),
13, 56–101, 238–303.

383. Neugebauer, O. The early history of the astrolabe.—Isis, 1949, **40**, 240–256.

384. Neugebauer, O. The exact sciences in antiquity. Brown Univ. Press, Providence,
R.I., 1957.

385. Neumann, J. von. Collected Works, v. 1–6. Oxford–London–N.Y.–Paris,
1961–1963.

386. Neumann, J. von. Mathematical foundations of quantum mechanics.—Translated by R. T. Beyer. Princeton Univ. Press, 1955.

387. [Newton]. Sir Isaac Newton's mathematical principles of natural philosophy
and his system of the world.—Translated by A. Motte in 1729. Translation
revised by F. Cajori. Cambridge Univ. Press, 1934.

388. Newton, I. Enumeratio linearum tertii ordinis.—Ed. J. Stirling, Paris, 1797.

389. Nikitina, L. S. Semiantiquaternion spaces.—Uč. zap. MOPI im. N.K. Krupskoĭ,
1969, **262**, 166–190 (Russian).

390. Noether, M. Zur Theorie des eindeutigen Entsprechens algebraischer Gebilde
von beliebig vielen Dimensionen.—Math. Ann., 1870, **2**, 293–316; 1875, **8**,
495–533.

391. Noether, M. Extension du théorème de Riemann—Roch aux surfaces algébriques.—C. r. Acad. sci. Paris, 1886, **103**.

392. Nomizu, K. Invariant affine connections on homogenous spaces.—Amer. J. Math., 1954, **76**, N 1, 33–65.

393. Norden, A. P. Spaces with affine connection. M.-L. Gostehizdat, 1950 (Russian).

394. Norden, A. P. On the interpretation of spaces with degenerating metric.—Dokl. AN SSSR, 1945, **50**, 57–60 (Russian).

395. Norden, A. P. Generalized geometry of two-dimensional ruled space.—Mat. sb., 1946, **18**(60), 139–152 (Russian).

396. Norden, A. P. Gauss and Lobačevskiĭ.—Istoriko-mat. issl., 1956, **9**, 145–146 (Russian).

397. Oresme, N. A treatise on the uniformity and difformity of intensities known as Tractatus de configurationibus qualitatum et motum.—Edited and translated by M. Clagett. Madison: Univ. of Wisconsin Press, 1968.

398. Osipovskiĭ, T. F. On space and time.—Istoriko-mat. issl., 1952, **5**, 9–17 (Russian).

399. Ostrogradskiĭ, M. V. Selected works.—Translated by V. I. Antropova, T. N. Klado and others. L., Izd-vo AN SSSR, 1958.

400. Ostrogradsky, M. V. Note sur la théorie de la chaleur.—Mém. Acad. Sci. St.-Petersb., VI sér., Sc. math., phys. et nat. 1831, **1**, 129–138.

401. Ostrogradsky, M. V. Mémoire sur le calcul des variations des intégrales multiples.—Mém. Acad. Sci. St.-Petersb., VI sér., Sc. math. et phys., 1835, t. 1, 35–58.

402. Ovsyannikov. L. V. Group analysis of differential equations. M.: Nauka, 1978 (Russian).

403. Pacioli, L. Summa de arithmetica, geometria, proportioni et proportionalita. Venetiae, 1494.

404. Pappi Alexandrini collectiones quae supersunt, t. 1–3. F. Hultsch. Ed. Berolini, 1876–1879.

405. Parnasskiĭ, I. V. Axiomatic development of three-dimensional parabolic geometry.—Uč. zap. Orlov. gos. ped. in-ta, 1956, **2**, N2 (Russian).

406. Parnasskiĭ, I. V. On degenerate Riemannian geometries.—Trudy II nauč. konf. mat. kafedr ped. in-tov Povolž'ya. Kuĭbyšev, 1962, 176–181 (Russian).

407. Pascal, B. Oeuvres complètes. Paris: Éd. du Seuil, 1983.

408. Pasch, M. Vorlesungen über neuere Geometrie. Leipzig, 1882.

409. Pauli, W. On Dirac's new method of field quantization.—Rev. Modern Phys., 1943, **15**, N 3, 175–207.

410. Peano, G. Calcolo geometrico secondo l'Ausdehnungslehre di Grassmann preceduto dalle operazioni della logica deduttiva. Torino, 1888.

411. Peano, G. I principii di geometria logicamente esposti. Torino, 1889.

412. Pearson, K. The grammar of science. London, 1900.

413. Peirce, B. Linear associative algebras.—Amer. J. Math., 1881, **4**, 97–221.

414. Peletarius, J. De occulto parte numerorum quam algebram vocant. Parisiis, 1560.

415. Penrose, R. Relativistic symmetry groups. In: Group theory of nonlinear problems. Dordrecht–Boston: Reidel, 1974, 1–58.

416. Persic, D. B. Geometries over antioctaves.—Izv. AN SSSR, seriya mat., 1967, **31**, 1263–1270 (Russian).

417. Persic, D. B. Geometry over degenerate octaves.—Dokl. AN SSSR, 1967, **173**, 1010–1013 (Russian).

418. Persic, D. B. Degenerate octaves and the projective plane. In: Voprosy diff. i neevklid. geom. (Uč. zap. MGPI im. V. I. Lenina), 1967, 299–328 (Russian).
419. Petrosyan, G. B., Rozenfel'd B. A. Aghānis' proof of Euclid's Fifth Postulate.— Izv. AN Arm. SSR, seriya fiz.-mat. nauk, 1960, 13, N1, 153–164 (Russian).
420. Pieri, M. Della geometria elementare come sistema ipotetico deduttivo. Torino, 1899.
421. Pimenov, R. I. Application of semi-Riemannian geometry to unified field theory. —Dokl. AN SSSR, 1964, 157, N4, 795–798 (Russian).
422. Pimenov. R. I. On the definition of semi-Riemannian spaces.—Vestn. Leningr, un-ta, 1965, 14, N1, 137–140 (Russian).
423. Pincherle, S. Opere scelte, t. 1–3. Roma, 1954–1962.
424. Pines, S. Beiträge zur islamischen Atomenlehre. Inaugural-Dissertation. Berlin, 1936.
425. Plato's Cosmology. The Timaeus of Plato.—Transl. and comm. by F. M. Cornford. London–New York, 1937.
426. [Plato]. The Dialogues of Plato. vol. 1.—Translated by B. Jovett. Oxford Univ. Press 1953.
427. Platonis quae supersunt opera. T. 1–8, Lipsiae: Weigel, 1821–1825.
428. Plücker, J. Gesammelte wissenschaftliche Abhandlungen, Bd. 1–2. Leipzig, 1895–1896.
429. Plücker, J. Neue Geometrie des Raumes gegründet auf die Betrachtung der geraden Linien als Raumelement. Leipzig, 1868.
430. Plücker, J. Theorie der algebraischen Curven. Bonn, 1839.
431. Poincaré, H. Oeuvres, v. 1–11. Paris, 1916–1956.
432. Poincaré, H. Les géométries non-euclidiennes.—Revue générale des sciences. 1891, 2, N: 23, 769–774.
433. Poincaré, H. The foundations of science/Transl. by G. B. Halsted. Lancaster: Science Press, 1946.
434. Poncelet, J. V. Traité des propriétés projectives des figures, v. 1–2. Paris, 1865–1866.
435. Pont, J.-C. La topologie algébrique des origines à Poincaré. Paris, 1974.
435a. Pont, J.-C. L'Aventure des parallèles. Histoire de la geométrie non-euclidienne: precurseurs et attardés. Berne: Lang, 1986.
436. Pontrjagin, L. Topological groups/Transl. F. Lehmer. Princeton: Univ. Press, 1946.
437. Pontrjagin, L. The theory of topological commutative groups.—Ann. of Math. 1934, 35, 361–388.
438. Pontryagin, L. S. Hermitian operators in a space with indefinite metric.— Izv. AN SSSR, seriya mat., 1944, 8, N6, 243–280 (Russian).
439. Pontryagin, L. S. On Betti numbers of compact Lie groups.—Dokl. AN SSSR, 1935, 433–435 (Russian).
440. Proclus. A commentary on the first Book of Euclid's Elements. Translated, with introduction and Notes, by Glenn R. Morrow, Princeton: Princeton Univ. Press, 1970.
441. Ptolemy's Almagest. Translated and annot. by G. J. Toomer, N.Y. e.a.: Springer, 1984.
442. [Ptolémée]. Composition Mathématique de Claude Ptolémée.—Ed. et trad. par N. Halma. Vol. 1–2. Paris. 1813–1816.
443. Ptolemaeus, C. Opera quae extant omnia, v. 1–2. J. L. Heiberg (Ed.). Leipzig, 1905–1907.

444. Putyata, T. V., Laptev, B. L., Rozenfel'd, B. A., Fradlin, B. L. Aleksandr Petrovič
      Kotel'nikov. 1865–1944. M., Nauka, 1968 (Russian).
445. Qurbānī Abū'l-Qāsim. Riyāḍidānān-i Irānī az Khwārizmī tā Ibn Sīnā. Teheran,
      1350 s.h. [1971].
446. Rached, R. Entre arithmétique et algèbre. Recherches sur l'histoire des mathé-
      matiques Arabes. Paris: Belles-lettres, 1984.
447. Raševskiĭ, P. K. Riemannian geometry and tensor analysis. Nauka, 1967
      (Russian).
448. Raševskiĭ, P. K. Scalar field in a fiber space.—Trudy seminava po vektornomu i
      tenzornomu analizu pri MGU, 1948, **6**, 225–248 (Russian).
449. Raševskiĭ, P. K. Symmetric spaces with affine connection and with torsion.—
      Trudy seminara po vektornomn i tenzornomu analizu pri MGU, 1950, **8**,
      182–192 (Russian).
450. Regiomontanus, J. On triangles.—Ed. and transl. B. Hughes. Madison–
      Milwaukee–London, 1950.
451. Reye, T. Die Geometrie der Lage, Bd. 1–3. Leipzig, 1909.
452. Rham, G. de. Sur l'analysis situs des variétés à n dimensions.—J. math. pures
      et appl. 1931, **10**, 115–200.
453. Ricci, G. Opera, t. 1–2. Roma, 1956–1957.
454. Riemann, B. Gesammelte mathematische Werke und wissenschaftlicher Nach-
      lass/Herausg. von H. Weber unter Mitwirkung von R. Dedekind. N.Y.: Dover,
      1953.
455. Riese, A. Die Coss.—In: B. Berlet. Adam Riese, sein Leben, seine Rechenbücher
      und seine Art zu rechnen. Leipzig–Frankfurt a. M., 1892, 33–62.
456. Riess, F. Oeuvres complètes.—Összegyüjtött munkái, v. 1–2. Paris–Budapest,
      1960.
457. Richter-Bernburg, L. Al-Bīrūnī's Maqāla fī tasṭīḥ al-ṣuwar wa talṭīkh al-
      kuwar.—J. for the History of Arabic science. 1982, **6**, 113–122.
458. Roch, G. Über die Anzahl der willkürlichen Konstanten in algebraischen
      Funktionen.—J. reine und angew. Math., 1865, **64**, 372–376.
459. Rožanskiĭ. Anaxagoras. At the sources of ancient science. M., Nauka, 1972
      (Russian).
460. Rozenfel'd, B. A. Geometric interpretation of quasisimple exceptional Lie
      groups of classes E_7 and E_8.—Soviet Math. Dokl. 1973, **14**, N:4, 1016–1020.
461. Rozenfel'd, B. A., Kolokol'ceva, I. I., Malyutin, V. V. Geometric interpretations
      of linear fractional transformations of Jordan algebras.—Soviet Math. Dokl.
      1974, **15**, N:5, 1252–1256.
462. Rozenfel'd, B. A., Magasumov, G. S. Cellular decompositions of manifolds of
      forms of simplicity.—Soviet Math. Dokl. 1983, **27**, N:1, 37–39.
463. Rozenfel'd, B. A., Stepaško, T. A. Metasymplectic geometries as geometries
      on absolutes of Hermitian planes.—Soviet Math. Dokl. 1983, **27**, N:1, 132–
      135.
464. Rozenfel'd, B. A. Multidimensional spaces. M., Nauka, 1966 (Russian).
465. Rozenfel'd, B. A. Non-Euclidean geometries. M., Gostehizdat, 1955 (Russian).
466. Rozenfel'd, B. A. Non-Euclidean spaces. M., Nauka, 1969 (Russian).
467. Rozenfel'd, B. A. New data on the author of the Roman edition of Naṣīr al Dīn
      al Ṭūsī's Euclid's Elements.—Vopr. ist. yestestv. i tehniki, 1973, **N1, 36**.
468. Rosenfeld, B. A. Analytical principle of continuity. Archives Internationales
      d'Histoire des Sciences, 1965, **N70–71**, 3–22.

469. Rozenfel'd, B. A. The algebraic treatise of al-Samaw'al.—Istoriko-mat. issl., 1975, **20**, 125–149 (Russian).

470. Rozenfel'd, B. A. Figures of simplicity and semisimplicity.—Trudy seminara po vektornomu i tenzornomu analizu pri MGU, 1963, **12**, 269–285 (Russian).

471. Rozenfel'd, B. A. The compact simple group E_6 as a group of motions of the complex octave noneuclidean plane.—Dokl. AN AzSSR, 1954, **10**, N12, 829–833 (Russian).

472. Rozenfel'd, B. A. Geometric interpretation of compact simple Lie groups in the class E.—Dokl. AN SSSR, 1956, **106**, 600–603 (Russian).

472a. Rozenfel'd, B. A. Geometric interpretation of quasisimple exceptional Lie groups in the classes E_7 and E_8.—Dokl. AN SSSR, 1973, **211**, 2, 289–292 (Russian).

473. Rozenfel'd, B. A., Vyplavina, R. P., Kolokol'ceva, I. I., Malyutin, V. V. Linear fractional transformations of Jordan algebras.—Izv. vuzov, Mat., 1974, **5**, 169–184 (Russian).

474. Rozenfel'd, B. A., Dobrovolskiĭ, I. G., Sergeeva, N. D. On the astronomical treatises of al-Farghānī.—Istoriko-astronom. issl., 1972, **11**, 191–210 (Russian).

475. Rozenfel'd, B. A., Zamahovskiĭ, M. P. Simple and semisimple Jordan algebras.—Izv. vuzov, Matematika, 1971, **8**, 111–121 (Russian).

476. Rozenfel'd, B. A., Zamahovskiĭ, M. P. Bireductive spaces, simple and semisimple Jordan algebras.—Trudy seminara po vektornomu i tenzornomu analizu pri MGU, 1972, **16**, 251–266 (Russian).

477. Rozenfel'd, B. A., Zamahovskiĭ, M. P., Orlovskaya, T. G., Semenova, I. N. Semisimple algebras, quasimatrices and spinor representations of quasinon-Euclidean motions. Izv. vuzov, Matematika, 1968, **4**, 62–73.

478. Rozenfel'd, B. A., Karpova, L. M. Symmetric semi-Riemannian spaces.—Izv. vuzov, Matematika, 1964, **1**, 100–116.

479. Rozenfel'd, B. A., Karpova, L. M. Flag group and contraction of Lie groups.—Trudy seminara po vektornomu i tenzornomu analizu pri MGU, 1966, **13**, 168–202 (Russian).

480. Rozenfel'd, B. A., Kolokol'ceva, I. I., Rudenko A. B. Biquasisimple algebras. Geometr. sb., **17**.—Trudy Tomskovo gos. un-ta, 1973, **246**, 20–31 (Russian).

481. Rozenfel'd, B. A., Kubesov, A., Sobirov, G. Who was the author of the Roman edition of Naṣīr al-Dīn al-Ṭūsī's *Euclid's Elements?*—Vopr. istorii estestv. i tehniki, 1966, **20**, 52–53 (Russian).

482. Rozenfel'd, B. A., Rožanskaya, M. M. Geometric transformations and variables in the work of Ibrāhīm ibn Sinān.—Istorīya i metodol. estestv. nauk, 1970, **9**, 178–181 (Russian).

483. Rozenfel'd, B. A., Rožanskaya, M. M., Sokolovskaya, Z. K. Abū'l-Rayḥān al-Bīrūnī. M., Nauka, 1973 (Russian).

484. Rozenfel'd, B. A., Skopec, Z. A. Quadratic Cremona transformations and complex numbers.—Dokl. AN SSSR, 1952, **83**, N6, 801–804 (Russian).

485. Rozenfel'd, B. A., Stepaško, T. A. Topological structure of manifolds of figures of simplicity.—Trudy geom. seminara KGU, 1982, **14**, 61–69 (Russian).

486. Rozenfel'd, B. A., Tlupova, R. G. Curves in G-elliptic space.—Trudy mat. inst. im. A. Razmadze AN GSSR, 1980, **64**, 94–104 (Russian).

487. Rozenfel'd, B. A., Yuškevič, A. P. Omar Khayyam. M. Nauka, 1965 (Russian).

488. Rozenfel'd, B. A., Yuškevič, A. P. The theory of parallels in the medieval East in 9–14th centuries. M., Nauka, 1983 (Russian).

489. Ruffini, P. Opere mathematiche, t. 1, Roma, 1953.
490. Rumyanceva, L. V. Quaternion symplectic geometry.—Trudy seminara po vektornomu i tenzornomu analizu pri MGU, 1963, **12**, 287–314 (Russian).
491. Ryser, H. J. Combinatorial mathematics. Math. Assoc. of America, 1963.
492. Sabit ibn Korra [Thābit ibn Qurra]. Mathematical treatises. Translated by J. al-Dabbāgh, L. M. Karpova, G. P. Matvievskaya, B. A. Rozenfel'd, A. Y. Sansour, T. D. Stolyarova, N. G. Khairetdinova.—Naučnoe nasledstvo, v. 8, M.: Nauka, 1984 (Russian).
493. Sabra, A. I. Thābit ibn Qurra on Euclid's parallels postulate.—J. Warburg and Courtauld Inst., 1968, **31**, 12–32.
494. Sabra, A. I. Simplicius' proof of Euclid's parallels postulate.—J. Warburg and Courtauld Inst., 1969, **32**, 1–24.
495. Saccheri, G. Euclides ab omni naevo vindicatus.—Ed. and transl. G. B. Halsted. Chicago–London, 1920.
496. Sagle, A. A. Malcev algebras.—Trans. Amer. Math. Soc., 1961, **101**, 426–458.
497. Sagle, A. A. Simple Malcev algebras over fields of characteristic zero.—Pacific J. Math., 1962, **12**, 1057–1078.
498. Saint-Venant, J. C. B. Sur les sommes et différences géométriques et leur application pour la simplification de la exposition de la mécanique.—C. r. Acad. sci. Paris, 1845, **21**, 20–116.
499. al Salar Husam al-Din. Premises for the proof of the postulate on parallel lines stated by Euclid in the beginning of the first book. Translated by B. A. Rozenfel'd and N. G. Khairetdinova.—Istoriko-mat. issl., 1974, **19**, 285–293 (Russian).
500. al-Samarqandī, Shams al-Dīn. Basic assumptions. Translated by B. A. Rozenfel'd.—Istoriko-mat. issl., 1961, **14**, 598–602 (Russian).
501. al-Samarqandī, Shams al-Dīn. Ashkāl al-ta'sīs. Istanbul, 1273h [1857].
502. As-Samaw'al. Al-Bahir en algèbre. S. Ahmad, R. Rashed (Eds.). Damas., 1972.
503. Satake, I. On representations and compactifications of symmetric Riemannian spaces. Princeton, 1960, Ann. of Math., **71**, 77–110.
504. Scheffers, G. Verallgemeinerung der Grundlagen der gewöhnlichen komplexen Funktionen.—Sitzungsber. Sächs. Ges. Wiss. Math.-phys. Kl., 1893, **45**, 828–842.
505. Scheffers, G. Zurückführung complexer Zahlensysteme auf typische Formen.—Math. Ann., 1891, **39**, 293–390.
506. Schellekens, G. J. On a hexagonic structure.—Proc. Koninkl. nederl. Acad. wet., 1962, **A65**, 201–234.
507. Schläfli, L. Gesammelte mathematische Abhandlungen, Bd. 1–2. Basel, 1950–1953.
508. Schönfliess, A. Krystallsysteme und Krystallstruktur. Leipzig, 1891.
509. Schoute, P. H. Mehrdimensionale Geometrie, Bd. 1–2. Leipzig, 1902–1905.
510. Schouten, J. A. Ricci-Kalkül. Berlin, 1924.
511. Schouten, J. A. Über die verschiedenen Arten der Übertragung die einer Differentialgeometrie zugrunde gelegt werden können.—Math. Z., 1922, **23**, 56–81.
512. Schouten, J. A. Über unitäre Geometrie.—Proc. Koninkl. nederl. Acad. wet., 1929, **32**, 457–465.
513. Schubert, F. T. De projectione sphaeroidis ellipticae geographica.—Nova Acta Acad. Sci. Petropol., (1787) 1789, **5**, 130–146.
514. Schubert, H. Kalkül der abzählenden Geometrie. Leipzig, 1877.
515. Schubert, H. Die *n*-dimensionalen Verallgemeinerungen der fundamentalen Anzahlen unseren Raumes.—Math. Ann., 1886, **26**, 26–51.

516. Schur, F. Grundlagen der Geometrie. Leipzig, 1909.

517. Schur, F. Über den Zusammenhang der Räume konstanten Krümmungsmasses mit den projektiven Räumen.—Math. Ann., 1886, **29**, 537–567.

518. Schweikart, F. K. Die Theorie der Parallellinien nebst dem Vorschlage ihrer Verbannung aus der Geometrie. Leipzig, 1908.

518a. Segre, B. Le geometrie di Galois. Annali di matem., 1959, **48**, 1–96.

519. Segre, C. Un nuovo campo di ricerche geometriche.—Atti Accad. Sci. Torino, 1889, **25**, 276–301, 430–457, 592–612; 1890, **26**, 40–56.

520. Segre, C. Sulla teoria e sulle classificazioni delle omografie in uno spazio lineare ad un numero qualunque di dimensioni.—Mem. Accad. Lincei, fis.-mat. ser., 1884, **19**, 127–148.

521. Semenova, J. N. Limit projective spaces.—Uč. zap. Kolomen. ped. in-ta, 1964, **8**, 165–174 (Russian).

522. Sergeeva, N. D., Karpova, L. M. Al-Farghānī's proofs of the fundamental theorem of stereographic projection.—Vopr. istorii yestestv. i tehn., 1973, **N3**, 50–53 (Russian).

523. Serre, J. P. Arbres, amalgames, SL_2.—Astèrisque, Soc. Mat. France **46** (1977).

524. Sesiano, J. Books IV–VII of Diophantus' *Arithmetica* in the Arabic translation attributed to Quṣṭā ibn Lūqā. N.Y. e.a. Springer, 1982.

525. Severi, F. Trattato di geometria algebraica. Bologna, 1926.

526. Severi, F. Sul teorema de Riemann—Roch e sulle serie continue appartenenti ad una superficie algebraica.—Atti Accad. Sci. Torino, 1905, **40**, 766–776.

527. Šaripov, A. D. Little-known pages of the correspondence between Bīrūnī and ibn-Sīnā. Obšěestv. nauki v Uzbekistane, 1965, **11**, 35–48.

528. al-Shīrāzī, Quṭb al-Dīn. Durra al-tāj li ghurra al-Dubāj. Vol. 1–4, Tehran, 1317–1320 s.h. [1939–1941].

529. al-Shīrāzī, Quṭb al Din. Commentaries on the treatise on the motion of rolling and the relation between the straight and the curved. Translated by J. al-Dabbāgh.—In: Iz istorii fiziko-matematičeskih nauk na srednevekovom Vostoke. Naučnoe nasledstvo, v.6., M.: Nauka, 1983, 175–228 (Russian).

530. Širokov, A. P. On symmetric spaces determined by algebras.—Izv. vuzov, Matematika, 1963, **5**, 159–171.

531. Širokov, A. P. Spaces over associative unitary algebras.—Uč. zap. Kazansk. gos. un-ta, 1963, 123, **1**, 222–247 (Russian).

532. Širokov, P. A. Tensor calculus. Kazan: Izd. Kazansk. un-ta, 1961 (Russian).

533. Širokov, P. A. Selected papers on geometry. Kazan. Izd. Kazansk. un-ta, 1966 (Russian).

534. Ščerbatskoĭ, F. I. The philosophy of Buddhism. Petrograd, 1919 (Russian).

535. Shkolenok, G. A. Geometrical constructions equivalent to non-linear algebraic transformations of the plane in Newton's early papers.—Arch. History Exact Sci., 1972, **9**, N 1, 22–44.

536. Schmidt, O. Yu. Selected papers, v. 1–3. M.: Izd-vo AN SSSR, 1959–1960 (Russian).

537. al-Sijzī, Abu Saʿīd Aḥmad. Book on the measurement of spheres by means of spheres. Translated by B. A. Rozenfel'd and R. S. Safarov.—Istoriko-mat. issl. 1985, **29**, 326–333 (Russian).

538. Silberstein, L. Projective geometry of Galilean space-time.—Philos. Mag., 1925, **10**, 681–696.

539. Silberstein, L. Discrete space-time. Toronto, 1936.

540. Sīnā, Abū ʿAlī ibn. The mathematical chapters of the *Book of knowledge*.—

Translated by B. A. Rozenfel'd and N. A. Sadovskiĭ. Dušanbe, Irfon, 1967 (Russian).

541. Sitter, W. de. On Einstein's theory of gravitation and its astronomical consequences.—Monthly Notices of Roy. Astron. Soc., 1917, **78**, 3–30.

542. Smirnov, S. V. The astrolabe of the Moscow museum of Eastern cultures.— Istoriko-astronom. issl., 1969, **10**, 311–330 (Russian).

543. Smorodinskiĭ, Ya. A. Geometric representation of the kinematics of collisions. In: Voprosy fiziki elementarnyh častic. Erevan: Izd-vo AN Arm. SSR, 1963, 242–271 (Russian).

544. Snyder, H. Quantized space-time.—Phys. Rev., 1947, **71**, 38–41.

545. (Sobranie pravil nauki astronomiĭ) Collection of the rules of the science of astronomy, book 3. Translated by N. G. Khairetdinova.—Fiziko-mat. nauki v stranah Vostoka, 1969, **2**, 147–190 (Russian).

546. Sokolov, A. A., Ivanenko, D. D. Quantum field theory. M.-L., Gostehizdat, 1952 (Russian).

548. Sommerfeld, A. Über die Zusammensetzung der Geschwindigkeiten in der Relativitätstheorie.—Phys. Z., 1909, **10**, N 22, 826–829.

549. Sommerville, D. M. Y. Bibliography of non-Euclidean geometry, including the theory of parallels, the foundation of geometry and space of n dimensions. N.Y., 1970.

550. Sommerville, D. M.Y. An introduction to the geometry of n dimensions. London, 1929.

551. Sommerville, D. M. Y. Classification of geometries with projective metrices.— Proc. Edinburgh Math. Soc., 1910–1911, **28**, 25–41.

552. Spampinato, N. Sulla geometria dello spazio rigato considerato come un S_1 ipercomplesso.—Atti Accad. Sci. fis.-mat. Napoli (2), 1935, **20**, N:11, 1–25.

552a. Spivak, M. A. A comprehensive introduction to differential geometry, v. 2. Berkeley: Publish or Perish, 1979.

553. Staudt Ch. von. Geometrie der Lage. Nürnberg, 1847.

554. Stcherbatsky, Th. [Ščerbatskoĭ, F. I.]. The central conception of Buddhism and the meaning of the the word "Dharma." London: Royal As. Soc., 1923.

555. Steiner, J. Combinatorische Aufgabe.—J. reine und angew. Math., 1853, **45**, 181–182.

556. Stekloff, W. Sur la théorie de fermeture des systèmes de fonctions orthogonales dépendent d'un nombres quelconque de variables.—Zap. Imp. AN (8), fiz.-mat. otd., 1911, **30**, **4**, 1–86.

557. Stevin, S. The principal works, v. 1–3. Amsterdam, 1955–1961.

558. Stifel, M. Die Coss Christoff Rudolffs mit schönem Exempeln der Coss, durch Michael Stifel gebessert und sehr gemehrt. Königsberg, 1553.

559. Stifel, M. Arithmetica integra. Norimbergae, 1544.

560. Steenrod, N. The topology of fiber bundles. Princeton, 1951.

561. Stipanić, E. Problem paralela kod Federika Grisogona. Mat. vestnik, Nova serija, 1973, **10** (25), **4**, 369–377.

562. Stolyarova, T. D. Statics in the Near and Middle East (kand. dis.). M., 1973 (Russian).

563. Stringham, W. I. Regular figures in n-dimensional space.—Amer. J. of math., 1880, **3**, 1–14.

564. Strubecker, K. Differentialgeometrie des isotropen Raumes.—Sitzungsber. Akad. Wiss. Wien, Abt. II^a, 1941, **150**, 1–53.

565. Study, E. Geometrie der Dynamen. Leipzig, 1903.

566. Study, E. Kürzeste Wege im komplexen Gebiete.—Math. Ann., 1905, **60**, 321–377.

567. Study, E. Ein Seitenstück zur Theorie der linearen Transformationen einer komplexen Veränderlichen.—Math. Z., 1923, **18**, 55–86, 201–229; 1924, **21**, 45–71, 174–194.

568. Study, E. Über nicht-Euklidische und Linien-Geometrie.—Jahresber. Dtsch. Math. Verein., 1902, **11**, 313–340.

569. Sude, B. H. Ibn al-Haytham's commentary on the premises of Euclid's *Elements*, Books I–VI. Princeton, 1974.

570. Suidae Lexicon, Bd. 1–5. A. Adler (Ed.). Leipzig, 1928–1938.

571. Suter, H. Über die Projektion der Sternbilder und der Länder von al-Bīrūnī.—Abhandl. Geschichte Naturwiss. und Med., 1922, **4**, 79–93.

572. Suter H. Abhandlung über die Ausmessung der Parabel von Ibrāhīm b. Sinān b. Thābit.—Vierteljarhschr. Naturforsch. Ges. Zürich, 1918, **63**, N 1–2, 214–228.

573. Tagi-zade, A. K. The astrolabes of al-Ṣaghānī, al-Bīrūnī, al-Sijzī and al-Zarqālī. Trudy XIII Meždunar. kongresa po istorii nauki, 1974, 3–4, 143–146 (Russian).

574. Tannery, P. Mémoires scientifiques, v. 1–17. Toulouse–Paris, 1912–1950.

575. Tapero, T. B. On the history of infinite-dimensional spaces.—Trudy XVII nauč. konf. aspirantov i mladšyh naučnyh sotr. IIEiT AN SSR, sekciya mat.-meh., 1975, 130–138 (Russian).

576. Tapero, T. B. Metric invariants of pairs of planes in infinite-dimensional space. Sovrem. geom., L., 1978, 103–113 (Russian).

577. Tapero, T. B. Common perpendiculars of pairs of planes in infinite-dimensional Euclidean space.—Funkcionalnyĭ analiz, Ul'yanovsk, 1979, **13**, 131–143 (Russian).

577a. Thas, J. The m-dimensional projective space $P_m(M_n(GF(q)))$ over the total matrix algebra $M_n(GF(q))$ of $n \times n$ matrices with elements in the Galois field $GF(q)$. Rome, 1971.

578. Théodose de Tripoli. Les sphériques. Trad. P. Ver Eecke. Paris, 1959.

579. Tits, J. Sur certaines classes d'espaces homogènes de groupes de Lie.—Mem. Acad. Roy. Belg. Sci., 1955, **29**, **N3**.

580. Tits, J. Les groupes de Lie exceptionnels et leur interprétation géométrique.—Bull. Soc. Math. Belge, 1956–1957, **8**, 48–81.

581. Tits, J. Sur la géométrie des R-espaces.—J. Math. pur. appl., 1957, **36**, N:1, 17–38.

582. Tits, J. Le plan projectif des octaves et les groupes de Lie exceptionnels.—Bull. Acad. roy. Belg, Sci., 1953, **39**, 309–329.

583. Tits, J. Le plan projectif des octaves et les groupes exceptionnels E_6 et E_7—Bull. Acad. roy. Belg. Sci., 1954, **40**, 29–40.

584. Tits, J. Classification of algebraic semisimple groups. Proc. of Symposia in Pure Math., vol. IX, AMS, Providence, R.I., 1966, 33–62.

585. Tits, J. Sur la classification des groupes algébriques semi-simples.—C. R. Acad. Sci. Paris, 1959, **249**, 1438–1440.

586. Tleuberdiev, S. K. Mathematical atomism in the medieval East.—Vopr. istorii estestv. i teh. 1984, **N1**, 88–90 (Russian).

587. Tllašev, H. New data on the history of mathematics in Central Asia in 13–15th centuries (avtoref. kand. dis.). Taškent, 1973 (Russian).

588. Tóth, I. Das Parallelenproblem in Corpus Aristotelicum.—Arch. History Exact. Sci., 1967, **3**, N 4/5, 249–422.

589. Toupin, R. A. World invariant kinematics.—Arch. Rational Mech. and Analysis, 1958, **1**, N 3, 181–211.

590. el-Toussy Nassiruddin. Traité du quadrilatère.—Éd. et trad. par Alexandre-Pacha Carathéodory. Constantinople, 1891.

591. Tropfke, J. Geschichte der Elementar-Mathematik, Bd. 1–7, 3. Aufl. Berlin–Leipzig, 1930–1937.

592. al-Ṭūsī, Naṣīr al-Dīn. Jawāmiʾ al-ḥisāb biʾl-takht waʾl-turāb. Ed. A. S. Saidan.—al-Abhāth. 1968, **20**, 91-64, 213–292.

593. al-Ṭūsī, Naṣīr al-Dīn. A treatise that heals the doubt raised by parallel lines.—Translated by B. A. Rozenfelʾd. Istoriko-mat. issl., 1960, **13**, 483–532 (Russian).

594. al-Ṭūsī, Naṣīr al-Dīn. Collection of arithmetical computations with board and dust.—Translated by S. A. Ahmedov and B. A. Rozenfelʾd. Istoriko-mat. issl., 1963, **15**, 431–444 (Russian).

595. al-Ṭūsī, Naṣīr al-Dīn. Majmuʿ al-rasāʾil, v. 1–2. Hyderabad, 1308–1309h (1939–1940).

596. al Ṭūsī Naṣīr al-Dīn. Taḥrir Uqlīdis fī-l-handasa. Teheran, 1298h (1881).

597. Tychonoff, A. N. Ein Fixpunktsatz.—Math. Ann., 1935, **111**, 767–776.

598. Unguru, S. A thirteenth century "proof" of the parallel postulate.—Historia mathematica, 1978, **5**, 205–210.

599. Urysohn, P. Mémoire sur les multiplicités Cantoriennes.—Fund. math., 1926, **8**, 225–351, Varh. Kon. Acad. wet. Amsterdam, 1928, **13**, N4, 1–172.

599a. Varāhamihira. The Pañcasiddhāntika. Translated and comm. by O. Neugebauer and D. Pingree. Vol. 1–2. København, 1970–1971.

600. Varičak, V. Darstellung der Relativitätstheorie im dreidimensionalen Lobatschefskijschen Raume. Zagreb: Narodni Novini, 1924.

601. Varičak, V. Über die nichteuklidische Interpretation der Relativtheorie.—Jahresber. Deut. Math. Verein, 1912, **21**, 103–122.

602. Veblen, O. Projektive Relativitätstheorie. Berlin, 1933.

603. Vieta, F. Opera mathematica. Lugduno Batavorum, 1646.

604. [Vitello]. Vitellonis filii Thuringo-Poloni Opticae libri decem. Basileae, 1572.

605. Vitruvius … On architecture. Vol. 1–2. London–N.Y.: Heideman, Putnam's sons, 1934.

606. Vitruvii Pollionis de architectura libri decem cum commentariis D. Barbari. Venetiis: Franciscius et Crugher, 1577.

607. Wachter, F. L. Demonstratio axiomatis geometrici in Euclidis undecimi. Danzig, 1817.

608. van der Waerden, B. L. Science awakening.—Transl. A. Dresden, Noordhoff, Groningen, 1954.

609. van der Waerden, B. L. Topologische Begründung der abzählenden Geometrie.—Math. Ann., 1929, **102**, 337–362.

610. van der Waerden, B. L. Die Klassifikation der einfachen Lieschen Gruppen.—Math. Zeitschr. 1933, Bd. **37**, 446–462.

612. Veblen, O., Whitehead, J. H. C. The foundations of differential geometry. Cambridge, 1932.

612a. Veblen, O., Bussey, W. Finite projective geometries. Trans. AMS, 1906, **7**, 241–259.

613. Veselovskiĭ, I. N. Non-Euclidean geometry in antiquity. Istoriya i metodol. estestsv. nauk, 1971, **11**, 152–160 (Russian).

614. Vitruvius. Ten books on architecture. With comments by D. Barbaro.—Trans-

lated by A. I. Venediktov, V. P. Zubov and F. A. Petrovskiĭ. M., Izd-vo Akad. arhitek., 1938 (Russian).

615. Vizgin, V. P. Unified field theories in the first third of the 20th century. M.: Nauka, 1985 (Russian).

616. Vyalcev, A. N. Discrete space-time. M.: Nauka, 1965 (Russian).

617. Wallis, J. Opera mathematica, t. 1–3. Oxoniae, 1693–1695.

618. Wallis, J. A treatise of Algebra, both historical and practical. London, 1685.

619. Wan Zhe-xian, Yang Ben-fu, Dai Zong-dao, Feng Xu-ning. Notes on finite geometries and the construction of PBIB-designs, I–IV.—Scientia Sinica, 1964, **13**, N 3, 515–517, N:6, 1006–1007, N:12, 2001–2004.

620. Waring, E. Miscellanea analytica de aequationibus algebraicis et curvarum proprietatibus. Cantabrigae, 1762.

621. Weber, H. Algebra. Leipzig, 1898.

622. Weierstrass, K. Mathematische Werke, Bd. 1–7. Berlin, 1894–1927.

623. Weinstein, A. Symplectic geometry.—Bull. Amer. Math. Soc. (N.S.), 1981, **5, 1**, 1–13.

624. Wessel, C. Essai sur la représentation de la direction. Copenhague, 1897.

625. Weyl, H. Theorie der Darstellung kontinuierlicher halbeinfacher Gruppen durch lineare Transformationen—Math. Zeitsehr, 1924. Bd. **23**, 271–304, 1925, Bd. **24**, 328–395.

626. Weyl, H. Space-Time-Matter/Transl. H. L. Brose. N.Y.: Dover, 1950.

627. Weyl, H. Selecta. Basel. Stuttgart: Birkhäuser, 1956.

628. Weyl, H. The classical groups. Their invariants and representations. Princeton Univ. Press, 1939.

629. Weyl, H. Symmetry. Princeton Univ. Press, 1952.

630. Weyl, H. Gruppentheorie und Quantenmechanik. Leipzig: Hirzel, 1931.

631. Whiteside, D. T. The mathematical papers of Isaac Newton, v. 1–6. Cambridge, 1967–1974.

632. Whitney, H. Sphere-spaces.—Proc. Nat. Acad. Sci. USA, 1935, **21**, 462–468.

633. Whitney, H. On the theory of sphere bundles.—Proc. Nat. Acad. Sci. USA, 1940, **26**, 148–153.

634. Whitney, H. On the topology of differentiable manifolds. Michigan Lectures in Topology. Ann. Arbor, 1941.

635. Wolf, K. B., Boyer, C. B. The algebra and group deformations $I^m[SO(n) \otimes SO(m)] \Rightarrow SO(n, m)$, $I^m[U(n) \otimes U(m)] \Rightarrow U(n, m)$ and $I^m[Sp(n) \otimes Sp(m)] \Rightarrow Sp(n, m)$ for $1 < m < n$.—J. Math. Phys., 1974, **15**, N 12, 2096–2101.

636. Yaglom, I. M. Geometric transformations, v. 8, 21 (transl. A. Shields), 24 (transl. A. Shenitzer) in the New Math. Library series, Random House, N.Y., 1962, 1968, 1973.

637. Yaglom, I. M. A simple non-Euclidean geometry and its physical basis.—Translated by A. Shenitzer. Springer Verlag, New York, 1979.

638. Yaglom, I. M. Projective metrics in the plane and complex numbers. Trudy seminara po vektornomu i tenzornomu analizu pri MGU, 1949, **7**, 276–318 (Russian).

639. Yaglom, I. M. Complex numbers in geometry.—Translated by E. J. F. Primrose. Academic Press, New York, 1968.

640. Yaglom, I. M., Rozenfel'd, B. A., Yasinskaya, E. U. Projective metrics.—Translated by E. J. F. Primrose, Sov. Math. Surveys, 1966, **19**, n5, 49–107.

641.   Yamabe, H. On the conjecture of Iwasawa and Gleason—Ann. of Math., 1953, **58**, N:1, 48–54.

642.   Yamabe, H. A generalization of a theorem of Gleason.—Ann. of Math., 1953, **58**, N:2, 351–365.

643.   Yaniševskii, É. P. A historical note on the life and work of N. I. Lobačevskiĭ.— Izd-vo Kazan'sk. un-ta, 1868 (Russian).

644.   Yankevič, Č. [Jankiewicz, Cz.]. On degenerate Riemannian geometries.—Byul. Polsk. akad. nauk, otd. 3, 1954, **2**, **7**, 305–308 (Russian).

645.   Yano, K., Bochner, S. Curvature and Betti numbers. Princeton Univ. Press, 1953.

646.   Yasinskaya, E. U. Semi-Euclidean and seminon-Euclidean spaces.—Dokl. AN SSSR, 1961, **137**, **N6**, 1327–1330 (Russian).

647.   Zeuthen, H. G. Lehrbuch der abzählenden Methoden der Geometrie. Leipzig, 1914.

647a. Železina, I. I. Line-geometry in degenerate non-Euclidean spaces.—Dokl. AN SSSR, 1956, **106**, **N6**, 959–962 (Russian).

648.   Zaharov, B. E., Manakov, S. E., Novikov, S. P., Pitaevskiĭ, L. P. The theory of solitons. The inverse problem method. M.: Nauka, 1980 (Russian).

649.   Zeldovič, Ya. B. A. A. Friedmann's theory of the expanding universe.—Usp. fiz. nauk, 1963, **80**, **N3**, 357–390 (Russian).

650.   Zorn, M. Theorie der alternativen Ringe.—Abhandl. Math. Seminar Hamburg. Univ., 1931, **8**, 123–147.

651.   Zubov, V. P. Bradwardine's treatise *On the continuum*.—Istoriko-mat. issl., 1960, **13**, 385–440 (Russian).

652.   Zubov, V. P. Evolution of atomic conceptions up to the beginning of the 19th century. M., Nauka, 1965 (Russian).

Index

DATE DUE
